VOLUME ONE HUNDRED AND TWENTY THREE

ADVANCES IN CANCER RESEARCH

Hyaluronan Signaling and Turnover

VOLUME ONE HUNDRED AND TWENTY THREE

ADVANCES IN CANCER RESEARCH

Hyaluronan Signaling and Turnover

Edited by

MELANIE A. SIMPSON
Fred and Pamela Buffett Cancer Center
Department of Biochemistry
University of Nebraska
Lincoln, Nebraska, USA

PARASKEVI HELDIN
Ludwig Institute for Cancer Research Ltd,
Uppsala Biomedicinska Centrum BMC, Sweden

ELSEVIER

AMSTERDAM • BOSTON • HEIDELBERG • LONDON
NEW YORK • OXFORD • PARIS • SAN DIEGO
SAN FRANCISCO • SINGAPORE • SYDNEY • TOKYO
Academic Press is an imprint of Elsevier

Academic Press is an imprint of Elsevier
525 B Street, Suite 1800, San Diego, CA 92101-4495, USA
225 Wyman Street, Waltham, MA 02451, USA
32 Jamestown Road, London, NW1 7BY, UK
The Boulevard, Langford Lane, Kidlington, Oxford, OX5 1GB, UK
Radarweg 29, PO Box 211, 1000 AE Amsterdam, The Netherlands

First edition 2014

ISBN: 978-0-12-800092-2
ISSN: 0065-230X

For information on all Academic Press publications
visit our website at store.elsevier.com

CONTENTS

Contributors *xi*
Preface *xv*

1. Emerging Roles for Hyaluronidase in Cancer Metastasis and Therapy **1**

Caitlin O. McAtee, Joseph J. Barycki, and Melanie A. Simpson

1. Introduction 2
2. Of Mole Rats and Men: Insights About HA and Cancer 3
3. Hyaluronidase Expression in Cancer 6
4. Hyaluronidase Function and the Metastatic Process 11
5. Hyaluronidase Targeting in Cancer Therapy and Imaging 18
6. Conclusions and Future Perspective 26

Acknowledgments 27
References 27

2. Targeting Hyaluronic Acid Family for Cancer Chemoprevention and Therapy **35**

Vinata B. Lokeshwar, Summan Mirza, and Andre Jordan

1. Introduction 36
2. Targeting HA Production 38
3. Targeting HA Signaling 43
4. HA as a Carrier for Drug Delivery 45
5. Targeting HA Receptors 48
6. Targeting HAase 53
7. Conclusion 54

Acknowledgments 55
References 55

3. Aberrant Posttranscriptional Processing of Hyaluronan Synthase 1 in Malignant Transformation and Tumor Progression **67**

Sophia Adamia, Jitra Kriangkum, Andrew R. Belch, and Linda M. Pilarski

1. Splicing and Cancer 68
2. Control of Pre-mRNA Splicing 69
3. Impact on Cancer of Alterations in Splicing Machinery 71
4. Aberrant Splicing of Hyaluronan Synthase 1 72

5. Clinical Impact of Aberrant HAS1 Splicing 75
6. Genetic Variations in HAS1 76
7. Functional Impact of HAS1Vs 76
8. Functional Outcomes of HAS1Vs in Transfectants 80
9. HAS1Vs and Mitotic Catastrophe 83
10. Conclusion 88
Acknowledgments 88
References 88

4. Hyaluronan Synthases Posttranslational Regulation in Cancer 95

Davide Vigetti and Alberto Passi

1. Introduction 96
2. HA Synthesis 98
3. HA Catabolism 101
4. HAS2 Phosphorylation, AMPK, and Its Dual Effects on Tumors 103
5. O-GlcNAcylation and Cancer 108
6. Conclusions 113
Acknowledgment 113
References 113

5. Hyaluronan-Coated Extracellular Vesicles—A Novel Link Between Hyaluronan and Cancer 121

Kirsi Rilla, Hanna Siiskonen, Markku Tammi, and Raija Tammi

1. Introduction 122
2. Extracellular Vesicles as Novel Communicators Between Cells 124
3. EVs Promote Tumor Progression 126
4. HA Synthesis Enhances Shedding of Extracellular Vesicles 129
5. HA–EVs as Predictors, Targets, and Carriers of Therapy 137
6. Conclusions 139
Acknowledgments 139
References 140

6. Hyaluronan in the Healthy and Malignant Hematopoietic Microenvironment 149

Sophia K. Khaldoyanidi, Valentina Goncharova, Barbara Mueller, and Ingrid U. Schraufstatter

1. Introduction 151
2. BM MSC, Their Derivatives, and HA 153
3. Macrophages and HA in the BM 163

4. Endothelial Cells and HA in BM 166
5. Nerve Cells and HA in BM 173
6. Role of HA in the BM Microenvironment in Hematological Malignancies 174
7. Conclusion 177
References 178

7. Hyaluronan Regulation of Endothelial Barrier Function in Cancer 191

Patrick A. Singleton

1. Introduction 192
2. HA Regulation of Vascular Integrity 194
3. HA Regulation of Endothelial Barrier Function During Tumor Angiogenesis 197
4. HA Regulation of Endothelial Barrier Function During Cancer Metastasis 198
5. Potential Therapeutic Effects of HMW-HA in Inhibiting Endothelial Barrier Disruption During Cancer Progression 200
6. Conclusions 201
Acknowledgments 203
References 203

8. HAS2 and CD44 in Breast Tumorigenesis 211

Paraskevi Heldin, Kaustuv Basu, Inna Kozlova, and Helena Porsch

1. Introduction 211
2. Molecular Classification of Breast Cancer 213
3. Role of Stromal HA in Tumor Progression 215
4. Expression of HAS Genes and Breast Cancer Malignancy 218
5. HA–CD44 Interactions: A Regulatory Network During TGFβ-Mediated EMT 220
6. Conclusions 223
Acknowledgments 223
References 224

9. CD44 is a Multidomain Signaling Platform that Integrates Extracellular Matrix Cues with Growth Factor and Cytokine Signals 231

Véronique Orian-Rousseau and Jonathan Sleeman

1. Introduction 232
2. Ligation of ECM Components by CD44 234
3. CD44: Coreceptor for Cell-Surface Receptors 238
4. CD44 as a Multidomain Signal Integration Platform 244
5. Concluding Remarks 248
References 249

10. Hyaluronan–CD44 Interaction Promotes Oncogenic Signaling, microRNA Functions, Chemoresistance, and Radiation Resistance in Cancer Stem Cells Leading to Tumor Progression **255**

Lilly Y.W. Bourguignon, Marisa Shiina, and Jian-Jian Li

1. Introduction 256
2. Regulation of Tumor Progression by HA/CD44 258
3. Activation of CSCs by HA/CD44 261
4. Conclusion 269

Acknowledgments 269
References 270

11. Advances and Advantages of Nanomedicine in the Pharmacological Targeting of Hyaluronan-CD44 Interactions and Signaling in Cancer **277**

Spyros S. Skandalis, Chrisostomi Gialeli, Achilleas D. Theocharis, and Nikos K. Karamanos

1. Introduction 278
2. Importance of Targeting Hyaluronan–CD44 in Tumors 279
3. Therapeutic Interventions/Strategies That Target Hyaluronan and/or CD44 to Perturb Hyaluronan-CD44 Interactions in Tumors 284
4. Advances in Nanomedicine Related with Hyaluronan-CD44 Targeting 295
5. Concluding Remarks/Conclusions 304

Acknowledgments 304
References 305

12. Hyaluronan/RHAMM Interactions in Mesenchymal Tumor Pathogenesis: Role of Growth Factors **319**

Katerina Kouvidi, Dragana Nikitovic, Aikaterini Berdiaki, and George N. Tzanakakis

1. Introduction 320
2. The Role of Hyaluronan and Its Receptors in Fibrosarcoma 323
3. GF Signaling and ECM Organization in Fibrosarcoma Pathogenesis 332
4. Concluding Remarks 337

References 338

13. CD147: Regulator of Hyaluronan Signaling in Invasiveness and Chemoresistance **351**

G. Daniel Grass, Lu Dai, Zhiqiang Qin, Chris Parsons, and Bryan P. Toole

1. Introduction 352
2. Structure and Pleiotropic Functions of CD147 353

3. CD147-Induced HA Synthesis and Signaling 356
4. CD147–HA Axis in Cellular Invasion 356
5. CD147–HA Axis in Chemoresistance 361
6. Induction of the CD147–HA Axis by Kaposi's Sarcoma-Associated Herpesvirus 363
7. Conclusions 364
References 366

Index *375*

CONTRIBUTORS

Sophia Adamia
Medical Oncology, Dana-Farber Cancer Institute, Harvard Medical School, Boston, Massachusetts, USA

Joseph J. Barycki
Department of Biochemistry, University of Nebraska, Lincoln, Nebraska, USA

Kaustuv Basu
Ludwig Institute for Cancer Research, Science for Life Laboratory, Uppsala University, Uppsala, Sweden

Andrew R. Belch
Department of Oncology, University of Alberta and Cross Cancer Institute, Edmonton, Alberta, Canada

Aikaterini Berdiaki
Laboratory of Anatomy-Histology-Embryology, School of Medicine, University of Crete, Heraklion, Greece

Lilly Y.W. Bourguignon
Department of Medicine, University of California at San Francisco & Endocrine Unit (111N), VA Medical Center, San Francisco, California, USA

Lu Dai
Department of Medicine, Stanley S. Scott Cancer Center, Louisiana State University Health Sciences Center, New Orleans, Louisiana, USA

G. Daniel Grass
Department of Regenerative Medicine & Cell Biology, Medical University of South Carolina, Charleston, South Carolina, USA

Chrisostomi Gialeli
Biochemistry, Biochemical Analysis and Matrix Pathobiology Research Group, Laboratory of Biochemistry, Department of Chemistry, University of Patras, and Foundation for Research and Technology, Institute of Chemical Engineering Sciences (FORTH/ICE-HT), Patras, Greece

Valentina Goncharova
Torrey Pines Institute for Molecular Studies, San Diego, California, USA

Paraskevi Heldin
Ludwig Institute for Cancer Research, Science for Life Laboratory, and Department of Medical Biochemistry and Microbiology, Uppsala University, Uppsala, Sweden

Andre Jordan
Sheila and David Funte Program in Cancer Biology, University of Miami-Miller School of Medicine, Miami, Florida, USA

Nikos K. Karamanos
Biochemistry, Biochemical Analysis and Matrix Pathobiology Research Group, Laboratory of Biochemistry, Department of Chemistry, University of Patras, and Foundation for Research and Technology, Institute of Chemical Engineering Sciences (FORTH/ICE-HT), Patras, Greece

Sophia K. Khaldoyanidi
Torrey Pines Institute for Molecular Studies, San Diego, California, USA

Katerina Kouvidi
Laboratory of Anatomy-Histology-Embryology, School of Medicine, University of Crete, Heraklion, Greece

Inna Kozlova
Ludwig Institute for Cancer Research, Science for Life Laboratory, Uppsala University, Uppsala, Sweden

Jitra Kriangkum
Department of Oncology, University of Alberta and Cross Cancer Institute, Edmonton, Alberta, Canada

Jian-Jian Li
Department of Radiation Oncology, University of California Davis, Sacramento, California, USA

Vinata B. Lokeshwar
Department of Urology; Department of Cell Biology, and Miami Clinical Translational Institute, University of Miami-Miller School of Medicine, Miami, Florida, USA

Caitlin O. McAtee
Department of Biochemistry, University of Nebraska, Lincoln, Nebraska, USA

Summan Mirza
Department of Urology, University of Miami-Miller School of Medicine, Miami, Florida, USA

Barbara Mueller
Torrey Pines Institute for Molecular Studies, San Diego, California, USA

Dragana Nikitovic
Laboratory of Anatomy-Histology-Embryology, School of Medicine, University of Crete, Heraklion, Greece

Véronique Orian-Rousseau
Karlsruhe Institute of Technology, Institute of Toxicology and Genetics, Karlsruhe, Germany

Chris Parsons
Department of Medicine, and Department of Microbiology, Immunology & Parasitology, Stanley S. Scott Cancer Center, Louisiana State University Health Sciences Center, New Orleans, Louisiana, USA

Alberto Passi
Dipartimento di Scienze Chirurgiche e Morfologiche, Università degli Studi dell'Insubria, Varese, Italy

Linda M. Pilarski
Department of Oncology, University of Alberta and Cross Cancer Institute, Edmonton, Alberta, Canada

Helena Porsch
Ludwig Institute for Cancer Research, Science for Life Laboratory, Uppsala University, Uppsala, Sweden

Zhiqiang Qin
Department of Microbiology, Immunology & Parasitology, Stanley S. Scott Cancer Center, Louisiana State University Health Sciences Center, New Orleans, Louisiana, USA

Kirsi Rilla
Institute of Biomedicine, University of Eastern Finland, Kuopio, Finland

Ingrid U. Schraufstatter
Torrey Pines Institute for Molecular Studies, San Diego, California, USA

Marisa Shiina
Department of Medicine, University of California at San Francisco & Endocrine Unit (111N), VA Medical Center, San Francisco, California, USA

Hanna Siiskonen
Institute of Biomedicine, and Department of Dermatology, Kuopio University Hospital and University of Eastern Finland, Kuopio, Finland

Melanie A. Simpson
Department of Biochemistry, University of Nebraska, Lincoln, Nebraska, USA

Patrick A. Singleton
Department of Medicine, Section of Pulmonary and Critical Care, and Department of Anesthesia and Critical Care, The University of Chicago, Chicago, Illinois, USA

Spyros S. Skandalis
Biochemistry, Biochemical Analysis and Matrix Pathobiology Research Group, Laboratory of Biochemistry, Department of Chemistry, University of Patras, Patras, Greece

Jonathan Sleeman
Karlsruhe Institute of Technology, Institute of Toxicology and Genetics, Karlsruhe, and Centre for Biomedicine and Medical Technology Mannheim (CBTM), Universitätsmedizin Mannheim, University of Heidelberg, Mannheim, Germany

Markku Tammi
Institute of Biomedicine, University of Eastern Finland, Kuopio, Finland

Raija Tammi
Institute of Biomedicine, University of Eastern Finland, Kuopio, Finland

Achilleas D. Theocharis
Biochemistry, Biochemical Analysis and Matrix Pathobiology Research Group, Laboratory of Biochemistry, Department of Chemistry, University of Patras, Patras, Greece

Bryan P. Toole
Department of Regenerative Medicine & Cell Biology, and Hollings Cancer Center, Medical University of South Carolina, Charleston, South Carolina, USA

George N. Tzanakakis
Laboratory of Anatomy-Histology-Embryology, School of Medicine, University of Crete, Heraklion, Greece

Davide Vigetti
Dipartimento di Scienze Chirurgiche e Morfologiche, Università degli Studi dell'Insubria, Varese, Italy

PREFACE

Hyaluronan (HA) is a normal component of human extracellular matrices that modulates tissue hydration, elasticity, and cellular fate. Excessive accumulation of HA is observed in a number of pathological conditions, including inflammation, wound healing, and cancer. The underlying mechanisms of HA accumulation in solid tumors, and the increased level of circulating HA in some blood borne malignancies, are still not fully understood. Steady-state HA levels are not static but are impacted by the dynamic interplay of the HA synthase (HAS) enzymes that synthesize and secrete HA at the plasma membrane, the hyaluronidase (Hyal) enzymes that hydrolyze HA, and the cell surface receptors that mediate internalization and processing of HA. In this volume of *Advances in Cancer Research*, we focused on the latest research in regulation of HA synthesis and turnover, and the consequences of excess HA as a differential trigger of intracellular signaling pathways.

Hyaluronidase activity is associated with enhanced permeability of tissue, interstitium, or vasculature. Several significant new directions pursued in recent characterizations of these enzymes are discussed in Chapters 1 and 2. In particular, the activity of hyaluronidases, Hyal1 and Hyal2, was found to be essential for the aggressive progression of tumors and could be targeted by systemically applied small-molecule inhibitors to reduce tumor growth. In addition, it has been demonstrated that HA accumulation does not initiate or promote tumor progression, but rather protects against tumor initiation and blocks progression, unless hyaluronidase activity is also upregulated. These studies were especially conclusive when examined in the collective picture of clinical association studies, functional manipulation of human tumor cells tested by preclinical outcomes, and characterization of tumorigenesis in the naked mole rat model, which is cancer impervious until its HA production is inhibited or its hyaluronidase-mediated removal of HA is elevated. Recently, successful clinical trials using controlled application of hyaluronidase to sensitize cancer to chemotherapy are also discussed.

HAS enzymes were first cloned, sequenced, and characterized in the 1990s. Numerous studies of HAS gene expression subsequently showed that HA production is significantly elevated by growth factor and cytokine-induced increases in HAS transcript levels. Chapters 3–5 discuss recent advances that reveal additional complexity in the control of HAS activity. Novel posttranscriptional and posttranslational mechanisms controlling

HA production include alternative splicing, ubiquitination, and reversible phosphorylation or O-linked N-acetylglucosaminylation. These studies define a central connection between HA production and cellular metabolic status. The newly identified presence of HA and HAS3 in exocytosed vesicles called exosomes is also described and summarized with respect to its possible implications for cellular transformation, tumor or stromal cell–cell communication, and metastasis.

The information content of HA as a signal in tumorigenesis is extensively discussed in Chapters 6–13. It is well accepted that HA polymers versus HA oligomers or fragments elicit different responses downstream of ligation by cell surface receptors such as CD44. Chapters 6 and 7 describe the impact of HA signaling on the microenvironment of bone, differentiation of mesenchymal stem cells, and endothelial integrity. The cooperativity of between HA, CD44, and growth factor receptors is detailed in Chapters 8–11, with a particular focus on tumor cell invasion, tumorigenesis, cancer stem cell proliferation, and TGFβ-mediated epithelial-to-mesenchymal transition as a prerequisite for the metastatic spread of cancer cells. Further attention is given to the targeting of these mechanisms for pharmacologic interventions in tumor progression. Chapters 12 and 13 examine additional therapeutic implications of HA cell surface interactions with the receptor for HA-mediated motility (RHAMM) and the extracellular matrix metalloproteinase inducer (EMMPRIN, or CD147).

Melanie A. Simpson
Paraskevi Heldin

CHAPTER ONE

Emerging Roles for Hyaluronidase in Cancer Metastasis and Therapy

Caitlin O. McAtee, Joseph J. Barycki, Melanie A. Simpson[1]
Department of Biochemistry, University of Nebraska, Lincoln, Nebraska, USA
[1]Corresponding author: e-mail address: msimpson2@unl.edu

Contents

1. Introduction 2
2. Of Mole Rats and Men: Insights About HA and Cancer 3
   2.1 HA and hyaluronidase accelerate human cancers 3
   2.2 Naked mole rats resist cancer 5
3. Hyaluronidase Expression in Cancer 6
   3.1 Hyal1 6
   3.2 Hyal2 8
   3.3 Hyal3, Hyal4, and PH-20 10
4. Hyaluronidase Function and the Metastatic Process 11
   4.1 Vesicle trafficking and cell motility 11
   4.2 Vesicle shedding 14
   4.3 Products of hyaluronidase: Fragments versus oligos 16
   4.4 Products of hyaluronidase: Beyond HA 17
5. Hyaluronidase Targeting in Cancer Therapy and Imaging 18
   5.1 Structural and functional features of human Hyals 18
   5.2 Targeting of hyaluronidase for cancer therapy 22
   5.3 Hyaluronidase-targeting agents for tumor imaging 25
6. Conclusions and Future Perspective 26
Acknowledgments 27
References 27

Abstract

Hyaluronidases are a family of five human enzymes that have been differentially implicated in the progression of many solid tumor types, both clinically and in functional studies. Advances in the past 5 years have clarified many apparent contradictions: (1) by demonstrating that specific hyaluronidases have alternative substrates to hyaluronan (HA) or do not exhibit any enzymatic activity, (2) that high-molecular weight HA polymers elicit signaling effects that are opposite those of the hyaluronidase-digested HA oligomers, and (3) that it is actually the combined overexpression of HA synthesizing enzymes with hyaluronidases that confers tumorigenic potential. This review examines the literature supporting these conclusions and discusses novel mechanisms by which

Advances in Cancer Research, Volume 123
ISSN 0065-230X
http://dx.doi.org/10.1016/B978-0-12-800092-2.00001-0

hyaluronidases impact invasive tumor cell processes. In addition, a detailed structural and functional comparison of the hyaluronidases is presented with insights into substrate selectivity and potential for therapeutic targeting. Finally, technological advances in targeting hyaluronidase for tumor imaging and cancer therapy are summarized.

1. INTRODUCTION

Hyaluronan (HA) is abundant as a polymer in joint and tissue matrices (Fraser, Laurent, & Laurent, 1997; Laurent, Laurent, & Fraser, 1996), where its roles in hydration, cushioning, and shock absorption have been well studied. Seemingly contrary to this architectural role, HA has also been well defined as a specific biological stimulus, critical for facilitating cellular proliferation and motility (Toole, 1997). A dramatic illustration of this role was elegantly demonstrated in embryonic heart development, in which HA is both a migration substrate and a signal for EMT, promoting the timed transformation and movement of pericardial cells to form the atrioventricular septum (Camenisch et al., 2000). HA is a simple linear polymer of alternating glucuronic acid and N-acetylglucosamine that can be repeated several hundred times. This relative chemical homogeneity may seem difficult to reconcile with the array of functional outcomes in which HA is implicated. However, advances in the past 20 years have revealed that the information content of the HA stimulus is highly context dependent and may contribute radically different phenotypic outcomes based on processing or degradation by hyaluronidase enzymes and reactive oxygen species, and differential engagement of cell surface receptors and intracellular signaling cascades.

Newly synthesized HA polymers are generated by HA synthases HAS1, HAS2, and HAS3, which are integral plasma membrane enzymes with an intracellular active site that catalyzes alternating monosaccharide addition to an average mass of 100–2000 kDa (Itano et al., 1999), concurrent with extrusion of the polymer to the exterior of the cell. Normal HA synthesis is activated transiently for cell division or motility, after which HA is rapidly cleared from the site by endocytic uptake and/or hyaluronidase-catalyzed hydrolysis. There are five human protein-coding sequences assigned to the hyaluronidase family on the basis of overall homology and active site conservation (*hyal1–hyal4* and PH-20) (Csoka, Frost, & Stern, 2001). Three of these have measurable activity for endolytic HA hydrolysis, one acts primarily on chondroitin sulfate, and activity of the fifth has not yet been determined but it is inactive toward HA. We discuss these enzymes in detail below.

Cellular responses to HA polymers and HA oligosaccharides are executed through multiple mechanisms, involving both cytoskeletal reorganization upon direct binding of HA to surface receptors such as CD44 (Aruffo, Stamenkovic, Melnick, Underhill, & Seed, 1990; Bourguignon, Zhu, Shao, & Chen, 2000; Legg, Lewis, Parsons, Ng, & Isacke, 2002) and receptor-mediated internalization of HA-bound complexes through endosomal pathways (Harada & Takahashi, 2007; Tammi et al., 2001). These complex receptor-mediated events and their aberrant behavior in cancer are the subject of several reviews within this volume. Through these receptors, specific sizes and quantities of HA have opposing impact on cell growth and tissue remodeling. For example, HAS overexpression leads to HA polymer accumulation that can promote tumor growth and/or metastasis (Enegd et al., 2002; Itano et al., 2004; Jacobson, Rahmanian, Rubin, & Heldin, 2002; Kosaki, Watanabe, & Yamaguchi, 1999). These effects of HA are dependent on its steady-state levels, and excess HA polymer suppresses tumor growth (Bharadwaj, Rector, & Simpson, 2007; Itano et al., 2004). Depending on actual chain length, HA oligomers may promote proliferation and angiogenesis or induce apoptosis (Zeng, Toole, Kinney, Kuo, & Stamenkovic, 1998). In this review, we will discuss hyaluronidase-catalyzed processing of HA polymers to shorter fragments and oligomers and their effects on functional outcome.

2. OF MOLE RATS AND MEN: INSIGHTS ABOUT HA AND CANCER

2.1. HA and hyaluronidase accelerate human cancers

Respective functions of HA polymers and HA oligomers, resulting from altered gene expression of HAS or Hyal, respectively, have been carefully dissected in functional studies of cancer progression. The clinical significance of concurrent excess HA and Hyal overexpression in resected or biopsied human tissue specimens confirms the relevance of such mechanisms for human cancer. HA accumulation is a clinical feature of prostate cancers of Gleason sum >4, regardless of the patient's hormone status. However, HA detection is confined to the stromal compartment until later stages of cancer, when HA can be observed in association with abnormal glandular epithelial cells as well (Aaltomaa et al., 2002; Lokeshwar et al., 2001). Overexpression of Hyal1, combined with excess HA detection, clinically predicts prostate cancer biochemical recurrence and reduced 5-year survival (Ekici et al., 2004; Gomez et al., 2009; Posey et al., 2003). Combination of

Hyal1 and HAS2 expression is prognostic for bladder cancer recurrence and expression of Hyal1 is an independent marker for disease-specific mortality in this study (Kramer et al., 2011). Levels of Hyal1 expression measured in clinically invasive resected prostate tumors are tumor promoting in mice (Lokeshwar, Cerwinka, Isoyama, & Lokeshwar, 2005). Similarly, the combined overexpression of HAS and either Hyal1 (Tan, Wang, Li, et al., 2011) or Hyal2 (Udabage, Brownlee, Nilsson, & Brown, 2005) is specifically observed at the invasive front in human breast cancer and associated functionally with breast cancer progression in mice.

Excess HA suppresses tumor growth in the absence of hyaluronidase. Stable HAS overexpression in prostate carcinoma cells that normally make negligible HA and HAS significantly reduces tumor take and tumor growth kinetics in either the subcutaneous (Bharadwaj et al., 2007; Simpson, 2006) or the orthotopic primary injection site (Bharadwaj et al., 2009). In orthotopically implanted animals that bore tumors, tumor vascularity was not appreciably different in HAS-overexpressing tumors and lymph node metastasis was not observed. Accumulation of HA in the tumors suggested poor clearance of HA produced by HAS-overexpressing tumor cells is antiproliferative. In fact, tumor cell proliferation was found to be inversely and temporally coincident with HA production, and these effects could be reversed either by coexpression with Hyal1 or exogenous hyaluronidase addition. Moreover, exogenously added HA did not affect cell growth. This implies that HA production within a tumor must be altered at the level of the tumor cell to impact growth, while the sources of Hyal1 could be numerous.

In contrast to HAS overexpression, HAS + Hyal1 co-overexpression potentiates both tumorigenesis and metastasis (Bharadwaj et al., 2009; Kovar, Johnson, Volcheck, Chen, & Simpson, 2006). Prostate tumor cells that produce endogenous large quantities of HA polymer are more metastatic to lymph nodes when injected intraprostatically in mice, but only if they express Hyal1 (Patel et al., 2002). Low HA-producing prostate tumor cells are normally not metastatic, but when transfected with Hyal1 alone, cells disseminated rapidly to lymph nodes following orthotopic implantation (Kovar et al., 2006). When these cells co-overexpressed HAS and Hyal1, there was a sixfold increase in tumor size and all tumors exhibited lymph node metastasis. Knockdown of HA synthesis abrogated $\approx$90% of spontaneous lymph node metastasis of highly metastatic prostate tumor cells (McCarthy et al., 2005). Thus, HA production by tumor cells in prostate cancer may enhance the aggressive potential of the tumor by providing substrates for Hyal1-dependent autocrine proliferation. However, stromal HA production, activated by cytokines, may also serve to recruit macrophages to tumor sites and thereby enhance angiogenesis indirectly. Macrophages can induce degradation of HA polymers to oligomers through a combination of hyaluronidase-mediated cleavage and reactive oxygen species-induced HA chain scission (Ohnuma et al., 2009).

Mechanistic studies *in vitro* have provided insights to the correlate processes of tumor progression impacted by HAS and Hyal. Coexpression of Hyal1 with HAS2 or HAS3 diminished HA retention, but restored rapid proliferation in culture that was suppressed by HAS, which supports a combined role for excess HA synthesis and processing in maximizing unrestricted growth of prostate cancer cells. Stable HAS transfectants retain HA at the cell surface, grow significantly more slowly in culture, and exhibit 50–90% reduced adhesion and motility on extracellular matrix protein substrates. Adhesion is dependent on differential engagement and cell surface presentation of β1 integrin receptors, which also alter motility. Integrin binding to ECM ligands is linked both to motility and to cell viability, and in fact, HAS-overexpressing cells have higher levels of p21/p27 cyclin-dependent kinase inhibitors and commensurate increased time to S phase in cell cycle analysis (Bharadwaj et al., 2011). These effects were opposite in Hyal1 overexpressing tumor cells, which did not have altered adhesion receptor expression, were more motile and exhibited more rapid cell cycling. It is apparent that uncleared excess HA accumulation as a result of HAS overexpression can lead to alterations in cell surface receptor function.

2.2. Naked mole rats resist cancer

Validation of the respective functions of HA in tumor suppression and hyaluronidase in eliminating protection against tumorigenic insults was provided recently from an unexpected source: the naked mole rat, a mouse-sized hairless organism with a 20–30-year lifespan. Naked mole rats are resistant to cancers of any kind, whether exposed to chemical carcinogens, ultraviolet irradiation, injected tumor cells, or other type of cancer-inducing insult. One protective element in mole rats was found to be an unusually high level of stromal, interstitial, and subcutaneous HA of very high-molecular mass, nearly double the average size of HA polymers found in humans and other rodents (Tian et al., 2013). The quantity of HA production was attributable to elevated expression of a single isozyme orthologue of the human HA synthase HAS2, which also produces HA of high-molecular mass in humans and other mammals. Two substitutions at the HAS2 putative active site, both serine residues that are conserved glutamine residues among other mammalian HAS2 orthologues, may promote increased processivity in the naked mole rat HAS2, but this has not yet been experimentally demonstrated. Also contributing to the steady-state levels of polymeric HA is that the hyaluronidase activity of all tissues tested in the mole rat is significantly lower than in human cells or other rodent tissues. Therefore, the mole rats do not degrade HA appreciably.

Importantly, naked mole rat skin fibroblasts can be transformed by overexpression of constitutively active ras and SV40 large T antigen and will grow in culture, but not in soft agar or in mice. These cells became tumorigenic and grew anchorage independently when their high level of HA polymer production was reduced by knocking down HAS2 or antagonized by increasing HA turnover with overexpression of hyaluronidase (Tian et al., 2013). This is consistent with results discussed above in human breast and prostate cancers overexpressing HAS, which suppresses tumor growth, while concurrent HA synthesis and turnover with both HAS and hyaluronidase present, significantly accelerates tumor growth and metastasis. The authors further confirmed *in vitro* that the growth suppressive response to HA, manifest as early contact inhibition, was dependent upon signaling through the known axis of the CD44 HA receptor, Nrf2, and ultimately $p16^{INK4a}$. These results are important because the respective roles for HAS2 and Hyal2 in naked mole rats were identified through unbiased approaches that validate mechanisms defined in systems that are not truly cancer resistant. Thus, HA turnover as a cause of cancer progression, and the absolute cancer-protective role of intact HA polymers, are concepts that could be firmly established in the intact naked mole rat, which has not been possible in cancer-susceptible organisms.

3. HYALURONIDASE EXPRESSION IN CANCER

3.1. Hyal1

The expression and functional importance of hyaluronidases in cancer has been most widely characterized with respect to the Hyal1 enzyme. Hyal1 is normally expressed in many cell types and is found within cells, partially compartmentalized to vesicles that traffic in a manner distinct from clathrin or caveolin endosomal routes, and partially to lysosomes (McAtee et al., submitted; Puissant et al., 2014). As such, its role in housekeeping levels of HA and glycosaminoglycan (GAG) turnover is well accepted. However,

Hyal1 is also a secreted protein that can be found in tumor interstitial fluid (Lokeshwar et al., 2005; Tan, Wang, Li, et al., 2011) and in conditioned media of tumor cells in culture (Lokeshwar et al., 2005; Simpson, 2006; Tan, Wang, Su, et al., 2011), so its utility as a diagnostic and prognostic biomarker has been extensively validated and exploited.

Genitourinary cancer. Hyal1 is an accurate urinary diagnostic marker for bladder cancer, because it is significantly overexpressed in tumors and shed to the urine of cancer patients (Eissa, Shehata, Mansour, Esmat, & El-Ahmady, 2012; Eissa et al., 2010; Eissa, Zohny, et al., 2012). Hyal1 mRNA was increased 4- to 16-fold in bladder cancer specimens and elevated Hyal1 expression predicted metastasis and mortality (Kramer et al., 2011). In addition, higher expression of Hyal1 correlated with bladder cancer progression to muscle invasion (Kramer et al., 2010). In prostate cancer, Hyal1 expression in biopsy samples predicted recurrence (Gomez et al., 2009).

Colorectal cancer. Hyal1 was detected in serum samples from colorectal carcinoma patients, where serum Hyal1 was reduced in cancer patients compared to levels in normal samples (Kolliopoulos et al., 2013). This phenomenon could be attributable to localization of the free Hyal1 at the primary tumor site rather than in the circulation. A separate study showed that multiple isoforms of Hyal (Hyal1, 2, 3, and PH-20) had increased activity in colorectal cancer patient samples and the expression of Hyal1 and Hyal2 was especially associated with more aggressive stages of cancer (Bouga et al., 2010).

Breast cancer. Clinically, Hyal1 expression can be used to predict invasive breast cancer progression in patients with benign ductal hyperplasias (Poola et al., 2008). Hyal1 overexpression in breast cancer enhances motility and anchorage-independent growth *in vitro*, and angiogenesis *in vivo* (Tan, Wang, Su, et al., 2011). Tumors in mice contained excess HA, resulting from increased HA fragment production by overexpression of Hyal1. Hyal1 levels are high in breast cancer cell lines MDA-MB-231 and MCF-7, and also in metastatic lymph nodes of breast cancer patients (Tan, Wang, Li, et al., 2011). Knockdown of Hyal1 in MDA-MB-231 and MCF-7 lines decreased invasion, adhesion, proliferation, and xenograft tumorigenesis, which confirms its role as a tumor promoter in this cancer type.

Ovarian cancer. Hyaluronidase activity and Hyal1 transcript levels are elevated in mucinous and clear cell epithelial ovarian cancers relative to benign or normal ovarian tissue, concurrent with decreased expression of estrogen receptor ERα (Yoffou et al., 2011). When ERα was overexpressed, Hyal1 expression decreased by approximately half, suggesting a loss of ERα function may be a mechanism by which Hyal1 expression becomes elevated in these tumors (Yoffou et al., 2011). In grade 3 serous ovarian cancer, levels of Hyal1 transcript are lower and HA staining consequently higher in comparison to normal ovary (Nykopp et al., 2009). Thus, the expression of Hyal1 is associated with a range of malignant phenotypes, even within the same tissue.

Pancreatic and lung cancers. A study of pancreatic ductal adenocarcinoma showed a correlation between modest Hyal1 expression and lower survival rates (Cheng, Sato, Kohi, & Yamaguchi, 2013). In the same study, accumulation of HA was found enhanced in cancerous tissues compared to normal tissues. Hyal1 and Hyal3 were also confirmed to have high expression in lung cancer cells when compared to normal control and stromal cells (de Sa et al., 2012). These Hyals, as well as the HAS proteins, are predicted to have a mechanistic role in the invasiveness of lung cancer (de Sa et al., 2013). Defining the correlation between cancer stage, HA production, and Hyal expression could contribute to the development of more individualized treatments for cancer patients.

Hyal1 splice variants. Hyal1 can undergo alternative splicing, and its splice variants have also been studied in cancer. Hyal1 splice variants are enzymatically inactive because they are all missing a 30 amino acid region that is required for wild-type activity (Lokeshwar, Schroeder, Carey, Soloway, & Iida, 2002). The Hyal1-v1 splice variant has been studied in a bladder cancer model, where it is proposed that it forms a complex with wild-type Hyal1 and thus lowers its activity, causing decreased growth and increased apoptosis (Lokeshwar et al., 2006). Differential expression of many of the Hyal splice variants is also associated with cancer outcomes. In a study of lung carcinomas, a better prognosis was associated with higher expression of the Hyal3-v1 splice variant, whereas poor prognosis correlated with expression of Hyal1 wild type (de Sa et al., 2012). Hyal1-v3, Hyal3-v1, and Hyal3-v2 splice variants were also shown to be associated with low tumor recurrence and low Gleason score in prostate cancer (de Sa, Canavez, Silva, Srougi, & Leite, 2009).

3.2. Hyal2

Hyal2 expression has been more recently correlated with progression of multiple cancers with the availability of isozyme-specific antibodies. In some cases, loss of expression has been reported. For example, in a small scale study of human lung cancer, loss of Hyal2 expression correlated with the presence of tumors (Li et al., 2007). Similarly, mRNA levels of Hyal2 were reduced in endometrial cancer relative to normal tissue (Nykopp et al.,

2010). In contrast, expression of Hyal2 was significantly increased in premalignant and malignant melanomas (Siiskonen, Poukka, Tyynela-Korhonen, Sironen, & Pasonen-Seppanen, 2013), and in breast cancer specimens (Tan, Wang, Li, et al., 2011). Hyal2 was specifically expressed at the expanding margins of invasive breast cancer (Udabage et al., 2005).

Epigenetic modifications controlling Hyal2 expression have also been examined in cancer biomarker discovery. The methylation profile of numerous genes including Hyal2 can be used to identify normal versus head and neck squamous cell carcinoma (Langevin et al., 2012). Hyal2 splice variants also show varying expression in gastric cancer samples (Ohnuma et al., 2009). Presence of the splice variant Hyal2ex2-3 was associated with gastric tumor cell lines (Ohnuma et al., 2009). This splice variant is found in the 5′UTR of the Hyal2 gene, which may contribute to its differential expression in cancerous and normal tissues (Ohnuma et al., 2009).

Hyal2 is GPI-anchored at the plasma membrane and associated with cholesterol-rich lipid rafts (Andre et al., 2011), where it acts in complex with CD44 and Hyal1 to promote uptake and endocytic internalization of HA (Harada & Takahashi, 2007). Overexpression of Hyal2 in fibroblasts has been shown to reduce pericellular retention of HA and proteoglycan-rich matrix, partially by increased turnover and partly through loss of CD44 from the cell surface (Duterme, Mertens-Strijthagen, Tammi, & Flamion, 2009). In astrocytoma cells, the overexpression of Hyal2 significantly accelerated tumor growth through enhanced angiogenesis upon implantation in the HA-rich intracerebral microenvironment (Novak, Stylli, Kaye, & Lepperdinger, 1999). As in breast cancer, Hyal2 was associated with increased invasive protrusions in the Hyal2-overexpressing astrocytoma tumors. Results of Hyal2 manipulation in these models are reproduced in the naked mole rat model, in which normally non-tumorigenic skin fibroblasts became highly tumorigenic upon Hyal2 overexpression when implanted in HA-rich tissue (Tian et al., 2013).

Hyal2 has been shown to associate with RON tyrosine kinase in epithelial cells, sequestering it at the plasma membrane and suppressing its activity (Danilkovitch-Miagkova et al., 2003). Upon ligation of Hyal2 by infection with a transforming sheep retrovirus, RON kinase was released and activated EMT through the Akt/ERK pathway. Since the region of Hyal2 required for viral recognition is proximal to the active site (Duh, Dirks, Lerman, & Miller, 2005), it is probable that ligation of Hyal2 by HA or other GAGs triggers signaling through RON kinase. HA-ligated Hyal2 may also signal through the WOX1 (a.k.a. WWOX) apoptosis inducer and Smad4

(Chang et al., 2010; Hsu et al., 2009). In these studies, it was suggested that binding of TGF-β1 to Hyal2 at the plasma membrane caused WOX1 association and complex formation, leading to nuclear translocation and induction of apoptosis through WOX1 association with nuclear transcription factors including p53, ErbB2, and ErbB4 (Hsu et al., 2009). However, the interaction between WOX1 and Hyal2 was found to occur in the Hyal2 catalytic domain, which would not be expected to be available for binding to the intracellular WOX1 if it was engaged on the cell surface by TGF-β1, so the physiological significance of this mechanism needs further examination. It is still possible that the interaction of Hyal2 with WWOX protein at the plasma membrane could prevent it from functioning as a tumor suppressor (Hsu et al., 2009). Overexpression of Hyal2 in fibroblasts also reduces CD44–ERM interaction, which was actually found to result in lower ERM activation and decreased motility of these cells (Duterme et al., 2009). Further study of these complex putative mechanisms will be needed to better define the impact of Hyal2 on cancer progression.

3.3. Hyal3, Hyal4, and PH-20

The other three human hyaluronidase family members, Hyal3, Hyal4, and PH-20, have been studied to a lesser extent in relation to cancer. Hyal3 does not have a GPI attachment site, but Hyal4 and PH-20 are GPI anchored at the plasma membrane. Of these three, only PH-20 has hyaluronidase activity but its expression is testis specific and does not change in multiple tumor cell lines (Patel et al., 2002). Nonetheless, there is evidence for their involvement in the progression of certain types of cancers.

Hyal3 lacks any detectable enzymatic activity to date, but its overexpression was shown to impact hyaluronidase activity in mice by increasing Hyal1 levels (Hemming et al., 2008). Expression of Hyal3 transcripts and splice variants has been examined for potential diagnostic or prognostic value. In lung cancer, expression of the Hyal3-v1 splice variant correlated with lower risk of death from disease (de Sa et al., 2012). In a small study of lung squamous cell carcinoma patients, $\approx$10% of patients carried one of two heterozygous mutations in exon 1 or 2 of the *HYAL3* gene, found only within tumor tissue and not present in normal surrounding tissue (Zhang, Zhu, Fan, & Ji, 2013). Presence of these mutations was correlated with lymph node metastasis, but not with other clinical parameters such as tumor size, grade, or remote metastasis. A modest elevation in Hyal3 transcript was observed in conjunction with those for HAS3 and RHAMM, in nodular basal cell carcinomas (Tzellos et al., 2011). In a panel of breast tumor cell lines, mRNA expression of Hyal3 was associated with relatively low invasive potential (Udabage et al., 2005).

Hyal4 catalyzes degradation of other GAGs than HA, specifically chondroitin sulfate, which could impact tumor cell surface proteoglycan turnover and contribute to cancer progression. Hyal4 is endogenously overexpressed in clear cell renal carcinoma and papillary tumors of the kidney compared to oncocytomas (Chi et al., 2012). Moreover, expression of Hyal4 independently distinguished between benign oncocytoma and renal cell carcinoma, suggesting a possible functional link to progression. An unbiased comparison of chromosome aberrations in low- versus high-grade glioma patients revealed a significant association between high-grade glioma and the overrepresentation of specific portions of chromosome 7, including the *HYAL4* gene (Li et al., 2013).

PH-20 expression has been examined in the context of breast cancer (Madan, Yu, et al., 1999), where it was found in normal, carcinoma *in situ*, invasive, and metastatic breast tissue, but elevation was significantly associated with invasive and metastatic cancer in African American women (Beech, Madan, & Deng, 2002). This was a small study but suggests potential for PH-20 in early detection and prognosis. Similar association between high PH-20 expression and prostate (Madan, Pang, Wilkiemeyer, Yu, & Beech, 1999) or laryngeal cancer (Godin et al., 2000) has also been reported. In breast tumor cells, expression of PH-20 was functionally associated with upregulation of p53 and WOX proteins, which allowed cells to be more susceptible to tumor necrosis factor (Chang, 2002). Like Hyal2, PH-20 enhances the expression of WOX1, which is proapoptotic and therefore can function as a tumor suppressor (Chen et al., 2004).

Overall, the differential expression profile of the hyaluronidase family members and their splice variants could be one of the determining factors in cancer prognosis but more systematic and larger scale studies are needed to understand the significance of these profiles.

4. HYALURONIDASE FUNCTION AND THE METASTATIC PROCESS

4.1. Vesicle trafficking and cell motility

Many studies have investigated how the internalization of HA and its processing enzymes contribute to endocytic trafficking patterns and thus manipulate signaling pathway cascades controlling proliferation and migration, among other processes. HA is internalized from the keratinocyte plasma membrane through receptor mediated or bulk endocytosis by a route not involving clathrin or caveolin (Tammi et al., 2001). Using chemicals that disrupt lysosomal function, there was a shift to a higher molecular weight average fragment size of internalized HA, meaning that a portion of the internalized HA was targeted to lysosomes for degradation after uptake. Excessive or aberrant internalization and trafficking of HA has been demonstrated in tumor cells, which is proposed to result in incomplete HA breakdown products being recycled back to the cell surface (Fig. 1.1), where they act as proangiogenic signaling molecules, or complete breakdown into precursor molecules to be fed into multiple pathways.

Murine macrophages secrete Hyal1 to the extracellular matrix, from whence its reuptake involves the mannose receptor, but is mannose-6-phosphate independent (Puissant et al., 2014). A portion of the internalized Hyal1 undergoes a single cleavage event in endocytic vesicles that are targeted to lysosomes, where cleaved Hyal1 retains enzymatic activity and likely completes degradation of GAGs. In HT1080 human fibrosarcoma cells, treatment with basic fibroblast growth factor increased HA production and decreased Hyal2 expression, which caused build up of high-molecular weight HA in the extracellular matrix (Berdiaki et al., 2009). Treatment of the cells with high-molecular weight HA ($3–4 \times 10^6$ Da) impaired migration and treatment with low-molecular weight HA (31 kDa) enhanced migration in a scratch wound healing assay (Berdiaki et al., 2009). Thus, the net effect of hyaluronidase-mediated HA turnover is increased motility in a variety of cell and tumor types.

HA turnover has been shown to affect levels of plasma membrane proteins, which translated to significant impact on cell adhesion and motility signaling pathways (Bharadwaj et al., 2011, 2009, 2007). In prostate tumor cells, overexpression of HAS3 resulted in lower expression of N-cadherin on the plasma membrane, reduced motility, and delayed cell cycle reentry, while overexpression of Hyal1 produced opposite effects irrespective of HAS3 (Bharadwaj et al., 2009). Prostate tumor cells overexpressing Hyal1 have enhanced endocytic activity as measured by the rate of fluorescently labeled transferrin internalization (McAtee et al., submitted). It is probable that overall endocytic rate affects steady-state levels of receptors at the plasma membrane. This observation would also explain many of the differences in receptor expression and internalization that have been reported to occur with differential expression of Hyals.

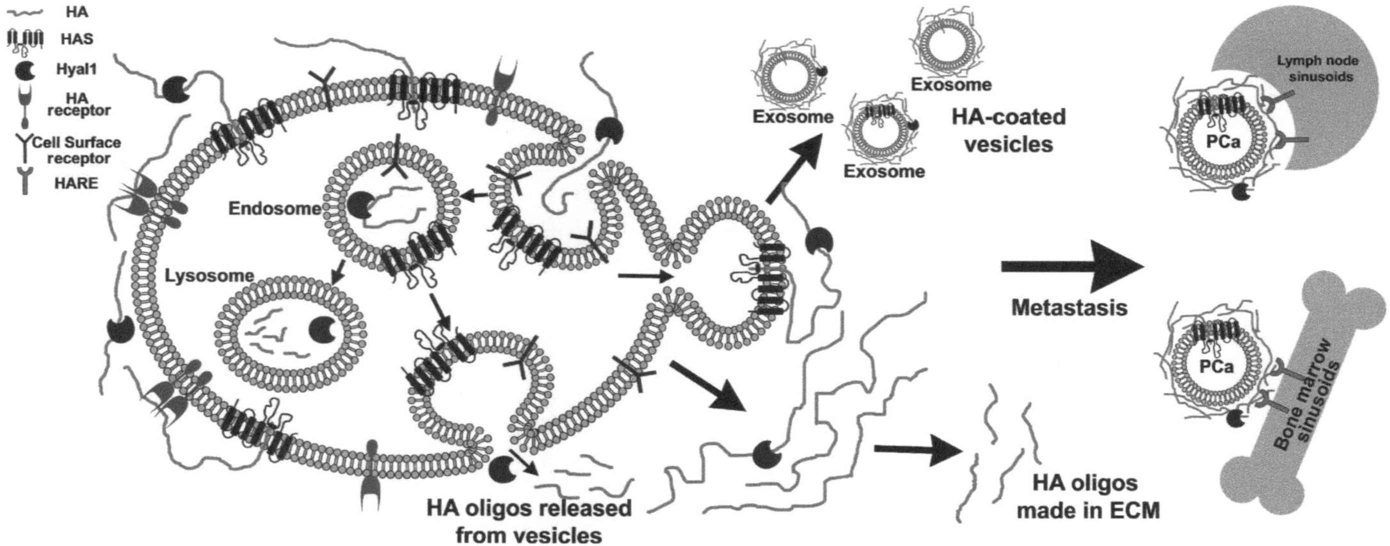

Figure 1.1 Model for HA impact on tumor progression. HA polymers are synthesized at the cell surface by membrane-embedded HAS. Reuptake and/or degradation of HA may require secreted hyaluronidases to generate low MW oligomers of HA. HA is retained by ligation to specific cell surface receptors or residual association with HAS and can act on both tumor cells and associated stromal cells. Tumor cells may signal in HA and/or Hyal1-dependent fashion to endothelial cells of lymphatic vessels, lymph node, or bone marrow sinusoids via other HA receptors. These signals may be released at the primary site to prepare metastatic target tissues and render them hospitable for tumor invasion, or tumor cells bearing HA may generate signals locally upon arrest in metastatic target tissues. The context of the HA signal is assumed to be free extracellular HA polymer or HA oligomer, but HA and/or Hyal1 delivery via exosomes or microvesicles is an emerging possibility. HA internalized by tumor epithelial cells may contribute to cellular transformation, proliferation, motility, and ultimately may be required for sustained tumor growth and metastasis. (See the color plate.)

One way HA homeostasis could affect overall motility of the cell is through the formation and disruption of focal adhesions. In esophageal cancer cells, inhibiting HA synthesis by HAS3 knockdown or by diminishing the HA precursor pool with 4-methylumbelliferone disrupted filopodia and focal adhesions, which subsequently decreased proliferation and migration (Twarock, Tammi, Savani, & Fischer, 2010). These outcomes occurred downstream of HA synthesis inhibition following cleavage of focal adhesion kinase (FAK), which led to decreased activation of ERK. Since FAK plays a pivotal signaling role in multiple functions such as proliferation, motility, and invasion, a better understanding of the complex relationship between FAK, ERK, HA synthesis, and HA turnover will provide insights to the role of HA in cancer progression.

4.2. Vesicle shedding

Increased vesicle shedding is an emerging hallmark of cancerous cells. Tumor cells secrete vesicles that contain proteins, microRNA, and other nucleic acids. Shedding vesicles or microvesicles are produced by budding from the plasma membrane, while the smaller ($\leq$50 nm diameter) exosomes are generated by inward budding within large intracellular organelles called multivesicular bodies that fuse with the plasma membrane to release their exosomal contents. Vesicle shedding was originally proposed to be a mechanism for cellular waste disposal, but strong evidence shows that exosome production and secretion in particular is a tightly regulated process with specific functional relevance. Exosomes are released by tumor cells, have been correlated with cancer progression, and carry relevant biologically active epigenetic regulators capable of transforming target cells. Vesicle shedding suggests a mechanism for local stromal–epithelial crosstalk at the tumor primary site and for communication with cells in remote tissues to promote metastatic susceptibility. HA and Hyals have been implicated in these processes.

Both levels of cellular HA production and hyaluronidase Hyal1 overexpression have been found to correlate with the rate of vesicle secretion from the cell. Cells with high endogenous HAS levels had a higher rate of shedding vesicle release than cells with low HAS expression, and inducing overexpression of HAS3 in a low HAS background increased release of vesicles (Rilla et al., 2013). The origin of these vesicles was a combination of the tips of microvilli, the pinching off of the plasma membrane, and secreted exosomes. The shedding vesicles contained HAS3 protein and retained a perivesicular HA coat. Hyaluronidases were not examined in this study, but independently it has been shown that Hyal1 was contained in both exosomal and microvesicle fractions isolated from conditioned media of prostate tumor cells overexpressing Hyal1 (McAtee et al., submitted). The ratio of exosome- to microvesicle-associated Hyal1 was higher in cells overexpressing wild-type Hyal1 versus a catalytically inactive mutant (E131Q). Thus, the enzymatic activity of Hyal1 is necessary for trafficking of Hyal1 into exosomal vesicles released by the cell.

The appearance of HAS3 and Hyal1 in vesicles is a novel mechanism by which HA homeostasis in tumor cells may influence cancer progression. Tumor cells that overexpress Hyal1 combined with HAS are more tumorigenic and metastatic, through autocrine-enhanced proliferation and motility. However, the well-accepted angiogenic potential of HA oligomers may be partially or largely mediated by vesicle-associated HA delivery. The presence of HA and Hyal1 in exosomes indicates that it could be carried from the primary tumor site through the circulation and arrive at target cells in distant tissues. In this way, a tumor overexpressing HAS or Hyal1 could initiate events to prepare distant tissues for metastasis (Fig. 1.1). HA or Hyal1 at the vesicle surface may facilitate docking and uptake of the vesicle and its contents by other cells. HA and Hyal1 could also produce active HA fragments upon arrest in a remote site, irrespective of cellular uptake, and thereby initiate proliferation or migration. These mechanisms remain to be examined.

Exosomes produced by tumor cells are known to bear unique contents that can affect the proliferation and motility of non-tumorigenic cells. In a study of rat pancreatic adenocarcinoma cell lines, wild-type cells and CD44 knockdown cells secreted exosomes with altered expression of ECM proteases, including hyaluronidase (Mu, Rana, & Zoller, 2013). Tumor-derived exosomes enhanced the migration of rat endothelial, fibroblast, and stromal cell lines. Exosomes are also involved in epithelial-to-mesenchymal transition in cancer progression. Application of exosomes with higher expression of TGF-β to primary fibroblasts can upregulate expression of α-smooth muscle actin and induce formation of a thick HA coat, both hallmarks for myofibroblast differentiation and cancer-associated stromal activation (Webber et al., 2014; Webber, Steadman, Mason, Tabi, & Clayton, 2010). Thus, exosome-associated signals can also affect HA homeostasis of target cells and drive the epithelial to mesenchymal transition.

4.3. Products of hyaluronidase: Fragments versus oligos

The complex information content of HA as a molecular signal in cancer is largely attributable to polymer length and HA quantity, both of which are influenced by hyaluronidase processing. Only Hyal1 and Hyal2 are thought to contribute to processing of HA in most tumors. Both isozymes are capable of generating a range of fragment sizes, and tetrasaccharides are the complete digestion product. Since Hyal2 is GPI anchored, its activity *in vivo* is thought to be limited to short polymer generation. HA oligomers, specifically of 4–25 disaccharides, have been shown to stimulate angiogenesis (West, Hampson, Arnold, & Kumar, 1985), despite the antiproliferative effect of larger HA polymers on endothelial cells that suppresses angiogenesis (Rooney, Kumar, Ponting, & Wang, 1995; West & Kumar, 1989). The antiangiogenic effect of HA polymers on endothelial cells is irreversible once engaged except in the sustained presence of antagonistic HA oligomers (Deed et al., 1997). This normal function of HA may allow tumor cells to directly signal their own vascular development and has been exploited in therapeutic targeting. HA decasaccharides reduced proliferation, motility, and invasion of breast tumor cells and prevented osteolytic lesions in mice (Urakawa et al., 2012).

Several reports have implicated Hyal1 processing of HA in prostate tumor angiogenesis. For example, inhibition of HA polymer synthesis suppressed growth (McCarthy et al., 2005; Simpson, Wilson, & McCarthy, 2002) and reduced vascular density of prostate tumors by $\approx$80%. Seemingly contrary to this finding, excess deposition of HA can suppress angiogenesis of prostate tumors (Bharadwaj et al., 2007). This supports a requirement for HA in angiogenesis, but clearly shows further metabolism of the polymeric form is critical for the angiogenic response. HA fragments (20–30-mers) have been detected in high-grade prostate cancer tissues (Lokeshwar et al., 2001) and knockdown of Hyal1 also impairs angiogenesis (Lokeshwar et al., 2005). HA production in relatively low quantities can promote angiogenesis in prostate tumors (Simpson, 2006), consistent with motility experiments in which low HA concentrations stimulate, while high levels inhibit, migration. This effect suggests there is a threshold HA polymer level that saturates or antagonizes hyaluronidase activity.

HA fragments stimulate cellular chemokinesis. For example, the cervical cancer cell line Hela-S3 exhibited spontaneous chemokinesis that was reduced by knocking down HAS2, Hyal2, or CD44 (Saito et al., 2011). Addition of exogenous high-molecular weight HA (230 and 920 kDa) did not affect this motile phenotype, whereas short HA polymers (23 kDa) were able to restore the chemokinetic process in the Hyal2 knockdown background. Low-molecular weight HA also enhances the migration and proliferation of human papillary thyroid carcinoma cells, acting through the toll-like receptor 4 (Dang et al., 2013). Low-molecular weight HA (3–5 kDa) was shown to associate with CD44 and toll-like receptors to induce an inflammatory response in breast cancer cells (Bourguignon, Wong, Earle, & Xia, 2011). HA size impacts CD44 clustering, which is stimulated by HA polymers and dispersed in the presence of oligomeric HA, leading to altered cell adhesion (Yang et al., 2012). HMW-HA treatment of embryonic fibroblasts stimulates Snail2 expression, and epithelial-to-mesenchymal transition (Craig, Parker, & Camenisch, 2009).

4.4. Products of hyaluronidase: Beyond HA

Chondroitin sulfate is a GAG with similar structure to HA that is covalently attached to proteoglycans at the cell surface and abundantly accumulated in the extracellular matrix of malignant tissue. Sulfated polysaccharides can have differential effects on HA homeostasis. Dextran sulfate has been shown to inhibit degradation of HA by PH-20 *in vitro*, but had complex net effects in cultured cells because the treatment also resulted in lower expression of CD44 and HAS, concurrently with enhanced expression of hyaluronidases (Udabage, Brownlee, Stern, & Brown, 2004). Hyal1, Hyal2, Hyal4, and PH-20 have all been shown to have significant activity toward specific chondroitin sulfates. For example, fragments of chondroitin sulfate E have the ability to activate CD44 signaling, thus promoting tumor cell motility through cytoskeletal rearrangement and increased formation of filopodia (Sugahara et al., 2008).

Recently, quantitative measurement of degraded GAG products by 2-aminobenzidine derivatization and fluorescence monitoring of HPLC anion exchange fractionation has facilitated the direct comparative assessment of substrates for Hyal1, Hyal4, and PH-20 (Honda, Kaneiwa, Mizumoto, Sugahara, & Yamada, 2012; Kaneiwa, Mizumoto, Sugahara, & Yamada, 2010). Because this assay method eliminates some of the pitfalls of other previously used methods, new insights about substrate specificity and enzyme activity parameters have been possible. In particular, there were modest but potentially significant differences in substrate preference depending upon the pH at which activity was assayed. Comparison of enzyme efficiency (V_{max}/K_m) revealed approximately threefold more efficient degradation of chondroitin 4-sulfate (CS-A) relative to HA at the frequently reported optimal pH of 4, and also at 4.5, but catalytic efficiency was comparable for both substrates at pH 3.5 (Honda et al., 2012). Hyal1 was also able to degrade chondroitin 6-sulfate (CS-C) and chondroitin, but this activity was lower by approximately an order of magnitude. At pH 4–4.5, recombinant, bead-immobilized PH-20 was significantly less efficient than Hyal1 in catalyzing depolymerization of any of these substrates, but showed the greatest activity using HA and CS-A, followed by chondroitin. Neither enzyme showed particularly strong size dependence for HA polymer cleavage by this method, which notably quantifies cleavage events stoichiometrically, in contrast to most other methods.

Thus, the existing view in the field that the hyaluronidase activities of this family of enzymes are the most physiologically relevant bears strong consideration. In the acidic context of a tumor, in which excess HA deposition, elevated proteoglycan expression, and aberrant GAG modification of cell surface proteins have all been well documented, the role of the hyaluronidase enzymes in remodeling the extracellular matrix may be equally or more significant because of their ability to cleave other GAGs. This method was also used to demonstrate that Hyal4 has chondroitinase activity, exhibiting a preference for chondroitin 2,6-bissulfate (CS-D) followed by CS-C and CS-A, with no detectable activity toward chondroitin or HA (Kaneiwa et al., 2010). To date, Hyal4 activity has not been associated with cancer, but has not been widely examined.

5. HYALURONIDASE TARGETING IN CANCER THERAPY AND IMAGING

5.1. Structural and functional features of human Hyals

Structure determination and enzymological characterizations of Hyal1 have helped define the key features of human hyaluronidases. In 2007, the crystal structure of human Hyal1 was described, providing the molecular details of the enzyme active site and the orientation of the catalytic domain relative to the C-terminal EGF-like domain (Chao, Muthukumar, & Herzberg, 2007). Given the close pairwise sequence identity between Hyal1 and the other four human Hyals (41–43%), we were able to generate highly credible homology models of each of the enzymes using Phyre 2 (Kelley & Sternberg, 2009). A comparison of the five human Hyals indicates that these enzymes have similar overall folds and conserved active site features.

A pronounced substrate-binding cleft bisects the core of the protein, with several highly conserved residues located at the site of HA cleavage. Using site-directed mutagenesis and steady-state kinetic analysis of recombinant purified protein, the contributions of several residues of Hyal1 to enzymatic activity were previously demonstrated (Zhang et al., 2009). In the proposed substrate-assisted mechanism, Tyr247 and Asp129 polarize the N-acetyl moiety of the N-acetylglucosamine residue to be cleaved, such that it forms an oxyanionic nucleophile to attack and hydrolyze its own glycoside bond. Glu131 protonates the hydroxyl-leaving group and activates an incoming water to release the intramolecular HA intermediate. Arg265 also contributes to Hyal1 activity, as mutation to a leucine severely compromises its ability to cleave HA. However, its precise role in catalysis may be indirect through structural perturbations. Additional mutants at Tyr202 and Ser245 were shown to have full catalytic activity but impaired HA binding. Inspection of the models predicts that Trp321 and Tyr75 will also contribute to HA binding and not catalysis as both residues are remote from the site of cleavage. Recently, the 13 amino acids in the loop between Cys207 and Cys221 in Hyal1 were replaced with four residues of alternating glycine and serine residues in an effort to extend the substrate-binding cleft (Reitinger, Mullegger, Greiderer, Nielsen, & Lepperdinger, 2009). Although comprehensive kinetic analysis was not reported, the engineered mutant had greater enzymatic activity at higher pH values relative to wild-type Hyal1. This observation and the high degree of sequence conservation at the enzyme active site suggest that the observed pH profiles of enzymatic activity for each of the hyaluronidases are unlikely to reflect only the protonation state of catalytic residues, but rather the pH dependence of substrate association with the enzyme.

Comprehensive structural and mechanistic characterizations of the other four human Hyals are lacking. In comparison to Hyal1, human Hyal2 had limited hyaluronidase activity (Lepperdinger, Strobl, & Kreil, 1998; Liu, Duh, Lerman, & Miller, 2003) and no detectable chondroitinase or heparanase activity (Lepperdinger et al., 1998) under the assay conditions employed. Hyal2 was reported to generate larger fragments of HA with an approximate molecular mass of 20 kDa. From a structural standpoint, it is difficult to rationalize why a 20 kDa fragment would not be degraded by the enzyme, given the similarities of the active site clefts of Hyal1 and Hyal2. This limited digestion may reflect structural domains of the HA substrate (Lepperdinger et al., 1998) rather than limitations of the enzyme active site. Alternatively, a soluble form of Hyal2 has been shown to degrade

HA to considerably smaller fragments with extended incubation, and to a similar extent as with PH-20 (Vigdorovich, Strong, & Miller, 2005). Hyal2 has also been shown to serve as a virus entry receptor (Vigdorovich et al., 2005). Residues that mediate this function are not highly conserved among the Hyals and are located adjacent to but distinct from the HA binding cleft (Duh et al., 2005). To date, hyaluronidase activity has not been demonstrated for mammalian Hyal3 (Atmuri et al., 2008; Hemming et al., 2008), despite conservation of key catalytic residues. A cursory examination of the substrate-binding cleft suggests minor amino acid substitutions may preclude HA binding and/or alter substrate specificity. Interestingly, Hyal3 may still promote HA turnover by promoting Hyal1 activity (Hemming et al., 2008). Additional work in this area is needed.

Human Hyal4 was recently shown to be a chondroitin sulfate hydrolase, with limited to no hyaluronidase activity over the length of the assay (Kaneiwa et al., 2010). Subsequent studies confirmed the importance of Glu147 and Tyr218, which are equivalent to Glu131 and Tyr202 in human Hyal1. There are two significant active site differences between Hyal1 and Hyal4 (Kaneiwa et al., 2012). Both Tyr247 and Gln288 are proximal to the site of HA cleavage in Hyal1 (Fig. 1.2). Human Hyal4 in contrast has a glycine (Gly263) and an arginine (Arg305) at these positions. Unpublished results indicate that mutation of Gly263 to a tyrosine did not alter the substrate specificity of human Hyal4 (Kaneiwa et al., 2012), but details of the characterizations are not yet available. In mouse Hyal4, a similar substitution, a mutation of Ser263 to a tyrosine, resulted in an enzyme with both hyaluronidase and chondroitinase activity. Interestingly, mouse Hyal4 has a glutamine residue at position 305 that is structurally equivalent to Gln288 of human Hyal1. Gln288 is adjacent to the C6 of the N-acetylglucosamine and near the C4 as well. Perhaps residues at this position contribute to substrate specificity.

The biological functions of PH-20 have been studied extensively, particularly with respect to fertilization. However, relatively limited mechanistic studies have been reported. Asp146 and Glu148, equivalent to Asp129 and Glu131 in Hyal1, have been shown to significantly contribute to catalysis. Arg211, Glu284, and Arg287 of human PH-20 were also shown to be critical for optimal activity (Vigdorovich et al., 2005). Examination of the homology model of PH-20 indicates these residues are unlikely to be involved directly in substrate binding or catalysis, but are instead key structural residues involved in extensive hydrogen bond networks.

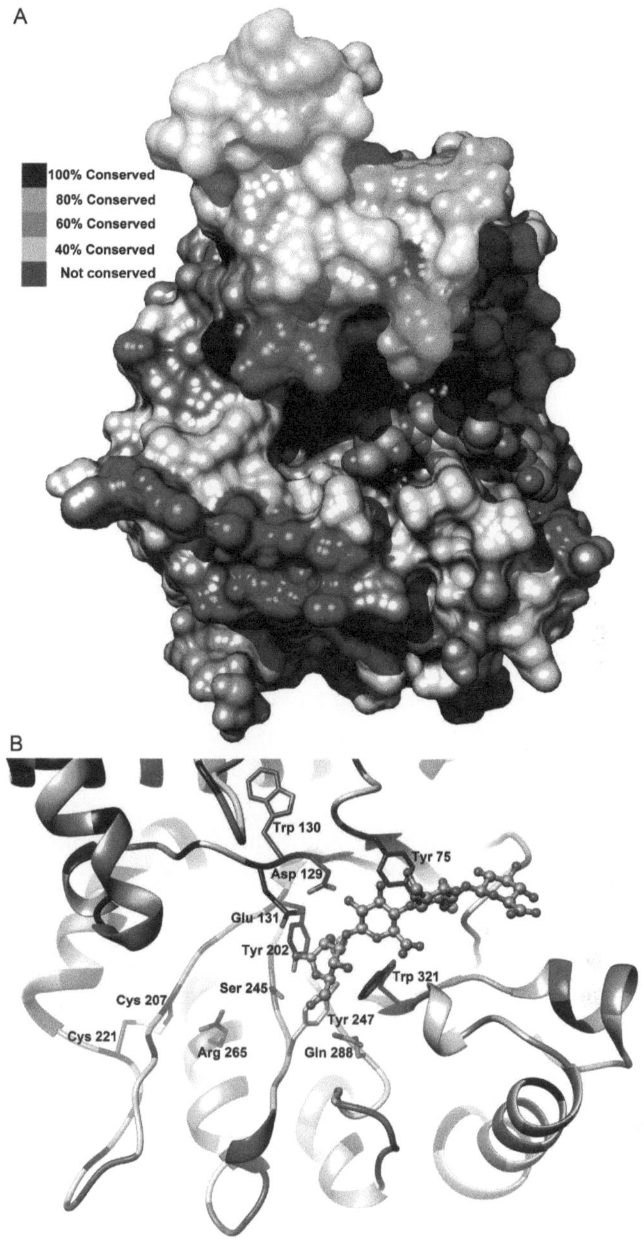

Figure 1.2 The structure of human Hyal1. (A) Surface representation of human Hyal1. A HA tetrasaccharide (solid spheres with carbons colored in green, oxygen in red, and nitrogen in blue) was docked in the enzyme active site in an orientation comparable to that observed in the bee venom hyaluronidase structure (pdb codes: Hyal1 2PE4; bee venom 1FCV). The surface of the protein is colored based on the sequence conservation *(Continued)*

The C-terminal EGF-like domain of Hyal1 packs tightly against the catalytic core of the protein. This domain is proposed to mediate protein–protein interactions that may influence the localization and efficacy of this family of enzymes. An examination of the sequence conservation in this region reveals limited selective pressure and may indicate that different isozymes have different interacting partners or cellular localizations. In Hyal1, deletion of the EGF-like domain eliminated its hyaluronidase activity (Zhang et al., 2009). Of the reported Hyal1 variants, several would largely eliminate the EGF-like domain (Lokeshwar et al., 2006, 2002). As previously discussed (Chao et al., 2007), it is difficult to ascribe functional significance to these variants because the structural data indicate that it is unlikely a folded protein would be produced. This is also the case for the described Hyal3 variants. To date, recombinant or purified versions of Hyal1 or Hyal3 splice variants have not been purified and characterized.

5.2. Targeting of hyaluronidase for cancer therapy

The role of hyaluronidases Hyal1 and Hyal2 in liberating or "activating" protumorigenic and proangiogenic HA fragments is well supported by the functional studies described in the above sections, making it an obvious choice for pharmacological targeting in chemotherapy. Its extracellular or cell surface localization increases its appeal as an accessible target, and general cytotoxicity or off-target effects may be reduced by limiting inhibition to the extracellular space.

Numerous naturally occurring and synthetic compounds have been characterized as inhibitors of PH-20 or Hyal1 *in vitro*. Since PH-20 exhibits relatively high hyaluronidase activity over a broad pH range, inhibition of its activity is frequently screened by loss of absorption by the cationic carbocyanine dye, Stains-All, or by its use to detect product size shifting in gel electrophoresis. Stains-All is sensitive to pH shifts so its signal is significantly diminished in the low pH conditions that are optimal for activity of

Figure 1.2—Cont'd among the five human hyaluronidases, with perfectly conserved residues colored in dark blue and residues with no sequence conservation colored in red. Residues found in four out of five hyaluronidases are colored in light blue, three out of five in gray, and two out of five in pink. The most highly conserved residues are generally located within the enzyme active site or key structural elements. (B) Conserved active site features of human hyaluronidases. A ribbon representation of Hyal1 is color coded as in Panel A, with the HA tetrasaccharide shown in ball and stick representation. Key active site residues are shown in stick representation. Several perfectly conserved residues are clustered at the site of the substrate-assisted cleavage. (See the color plate.)

Hyal1. Consequently, Hyal1 and other acid-active hyaluronidases are typically assayed by chemical derivatization of acetamido groups and measurement of the resulting colorimetric product, known as the Morgan–Elson reaction. This technique has the added advantage of stoichiometrically reporting cleavage events. A third method that has been used directly to compare inhibitors of PH-20 and Hyal1 is a competitive binding assay, or ELISA-like assay, which is a plate-based assay that reports activity by loss of microplate-adsorbed HA polymers detected by a labeled HA-binding protein. The limitation of this and the Stains-All assay is that the HA must be fully degraded before the positive signal is lost, so it is not able to quantify actual cleavage events catalyzed by the enzyme.

Using purified bee venom, bovine testicular, or recombinant human PH-20 hyaluronidases, the most potent inhibition was achieved with vitamin C palmitate, also known as L-ascorbyl 6-hexadecanoate (Botzki et al., 2004; Hofinger et al., 2007), and inhibition was greatest when the length of the alkyl chain was 12–16 carbons. Glycyrrhizic acid was somewhat effective against Hyal1 that was expressed and purified from Drosophila cell culture (Hofinger et al., 2007). PH-20 was also expressed by "autodisplay" on *Escherichia coli* cells and compared directly to bovine testicular hyaluronidase (BTH) for efficacy of known inhibitors (Kaessler, Olgen, & Jose, 2011). BTH is frequently used to screen for new hyaluronidase inhibitors, and this comparison revealed that while BTH was effectively inhibited by vitamin C palmitate and by two indole acetamide/carboxamide derivatives, only vitamin C palmitate significantly inhibited PH-20. Such comparative studies are important in identifying lead compounds for isozyme-selective inhibitors, but it is important to consider that the authors use the Stains-All method, which is non-stoichiometric and loses sensitivity at the low pH optimal for Hyal1 or Hyal2 activity. A series of indole derivatives was characterized with the Morgan–Elson assay, and though the authors only tested BTH (Olgen, Kaessler, Kilic-Kurt, & Jose, 2010; Olgen, Kaessler, Nebioglu, & Jose, 2007), increasing the lipophilicity of the compounds was found to enhance affinity, which is consistent with results of vitamin C ester comparisons above.

Polystyrene sulfonates and sulfated HA showed greater potency than glycyrrhizic acid and comparable inhibition to vitamin C palmitate, using partially purified Hyal1 or PH-20 from cell conditioned media assayed in the competitive plate assay (Isoyama et al., 2006). Sulfated HA partially (Isoyama et al., 2006) or fully (Toida, Ogita, Suzuki, Toyoda, & Imanari, 1999) inhibits Hyal1 and PH-20, but it is not clear from examination of the Hyal1 structure where a sulfate group would be tolerated at the active site, which is relatively locked down and sterically constrained in the model with superimposed HA tetrasaccharide.

Partially sulfated HA polymers (≈300–600 kDa average molecular mass) are the only hyaluronidase inhibitor that has been tested for antitumor efficacy *in vivo*. When given twice weekly (intraperitoneally) beginning at the time of prostate tumor cell injection, this treatment delayed or inhibited subcutaneous tumor growth in mice (Benitez et al., 2011). The authors tested effects of sulfated HA *in vitro* to determine if phenotypic effects that are promoted by enhanced HA turnover were reversed. This was in part the case, since treatment with sulfated HA-induced apoptosis and decreased proliferation, motility, and invasiveness of prostate tumor cell lines. Sulfated HA functioned mainly through inhibiting Akt signaling, and overexpression of Akt or application of non-sulfated HA oligos reversed the sulfated HA effects. However, the study used a subline derived from LNCaP cells, which reportedly lack measurable hyaluronidase activity in conditioned media. Hyal1 expression and activity in the subline were not shown, so it is possible that the response mechanism is not exclusively dependent on hyaluronidase activity, particularly in light of the findings discussed above with respect to other substrates. The authors also reported an effect of the sulfated HA treatment on signaling downstream of HA receptors CD44 and RHAMM, both of which also have the potential to be engaged by the sulfated HA directly. Regardless of targeting specificity, it does appear that sulfated HA has promise in preventing tumor growth, but it will be important for ultimate clinical translation to determine whether sulfated HA can regress established tumors.

Interestingly, PEGPH20 has recently been used to sensitize pancreatic cancer to gemcitabine (currently a first line chemotherapy for this disease), with significantly faster response in mice, though growth of tumors is only delayed by accelerating delivery, and they still reach lethal size. Since HA is highly hydrophilic, its accumulation in tumor stroma increases interstitial fluid pressure so significantly that molecular transfer from the tumor vasculature into the tumor is virtually undetectable. PEGPH20 delivery intravenously prior to gemcitabine treatment was shown to reduce stromal HA and interstitial pressure, thereby allowing the drug to reach the tumor cells and induce cell death (Provenzano et al., 2012). The prospect for use of hyaluronidases in this manner is exciting, but is not without precedent for concern about effects of residual HA degradation products on surviving tumor cell proliferation and motility, as well as tumor angiogenesis.

5.3. Hyaluronidase-targeting agents for tumor imaging

As a cancer biomarker and potential therapeutic target, use of hyaluronidase substrates for noninvasive tumor imaging has received increasing attention. Many investigators have used HA to target nanoparticles, isotopic labels, fluorophores, and other imaging agents to HA receptors that are elevated on the tumor cell surface in many solid epithelial tumors. In addition, the clearance route of HA through lymphatic vessels and accumulation in lymph nodes facilitates its use for imaging tumor-associated lymphatic flux (Proulx & Detmar, 2013; Sharma et al., 2007).

Another novel innovation is the use of HA as a probe for the activity of hyaluronidase. For example, subcutaneous tumors resulting from ovarian tumor cell line injection were effectively imaged by MRI using an intravenously delivered HA-DTPA conjugated to agarose beads, which chelated and partially shielded the contrast agent, gadolinium, subsequently releasing it specifically in the tumor, yielding a change in the imaging parameters (R1 and R2 relaxation rates). Hyal1 and Hyal2 activity in the ovarian tumor cells were both found to contribute *in vitro* to the degradation of the HA carrier, liberating the environmentally sensitive contrast agent (Shiftan et al., 2005). The HA-GdDTPA-bead-targeting agent was further found to be sufficiently sensitive to report apparent kinetics of hyaluronidase activity in the tumors and found the initial activity was localized primarily to the peripheral tumor, possibly concentrated in peritumoral lymphatics (Shiftan & Neeman, 2006). Since hyaluronidases are elevated in the stromal space and/or at the tumor cell surface, and may be involved in the cellular uptake of HA, imaging probes that give no signal until activated by hyaluronidase cleavage offer higher levels of sensitivity.

Besides MRI contrast agents, fluorescent probes have been developed that illustrate the clinical potential for this approach, which may be translatable to an intraoperative setting. The FDA-approved near-infrared fluorescent dye, indocyanin green, is used successfully in the clinic for identifying tumor-involved lymph nodes and metastatic cancer. By encapsulating the dye in a HA nanogel, the dye is more effectively solubilized and stabilized *in vivo*. A proof of concept trial determined that fluorescence intensity could be significantly increased in mice upon hyaluronidase exposure to release the dye, but the agent needs to be validated using tumor xenografts with differential hyaluronidase expression (Kim, Oh, Youn, & Lee, 2014; Mok, Jeong, Kim, & Chung, 2012). An additional intriguing use of HA was to provide the outermost envelope of multilayer liposomes. The liposomes contained the cytotoxic agent paclitaxel and were first coated with arginine/histidine-rich cell-permeating peptides before coating with HA (Jiang et al., 2012). The HA screened the peptides from circulating proteases and targeted the particles to tumor sites enriched in hyaluronidase, whereupon the degradation of HA exposed the peptides, stimulating cellular uptake into vesicles that were dispersed in a pH-dependent manner to release paclitaxel to the cytosol. Specificity of HA targeting in mice was demonstrated by competition of the near-infrared fluorescence signal in the tumor with free HA preinjection and by significantly reduced tumor growth and extended survival time of animals treated with HA-targeted peptide-paclitaxel liposomes. Finally, fluorescent HA-based assays have been tested to improve the sensitivity of hyaluronidase detection in urinary samples, where it is effective in diagnosis of bladder cancer. Fluorescence correlation spectroscopy of autoquenched fluorescein-HA (Rich et al., 2012) and Forster Resonance Energy Transfer (Chib et al., 2013) upon cleavage of fluorescein–rhodamine double-conjugated HA both showed promise for increasing the dynamic range of hyaluronidase detection and improving quantitative cancer staging with urinary samples.

6. CONCLUSIONS AND FUTURE PERSPECTIVE

Hyaluronidase family enzymes have been broadly implicated in a variety of cancers and have demonstrated potential for clinical utility both as biomarkers and therapeutic targets. However, a number of key questions remain to be answered before they can be fully exploited in cancer therapy. For example, systematic studies that examine structure–function relationships among the Hyals will be essential to understand the basis for substrate specificity, binding properties, and high-affinity inhibition. Development of specific strategies to target individual Hyal proteins would be advanced by additional structural data in complexes of each Hyal with a putative substrate or inhibitor, so active site differences can be fully appreciated. Moreover, a more thorough comparison of tissue and tumor-specific effects of each Hyal is needed to clarify when hyaluronidase activity is beneficial (e.g., by removing HA to improve drug delivery) and when this activity is protumorigenic and prometastatic (e.g., Hyal1 and Hyal2). Rigorous comparisons of each Hyal in the context of HA accumulation, as has been done with Hyal1, are needed to determine respective tumorigenic and metastatic proclivity conferred by each Hyal. Finally, more large-scale systematic studies that examine all five Hyals in the same sets of staged cancers would better inform us of the clinical relevance of each Hyal in cancer initiation, progression, and metastasis and give insights about true prognostic power of Hyal expression.

ACKNOWLEDGMENTS

This work was supported by NIH R01 CA165574 and NIH P20 GM103489 to M. A. S.

REFERENCES

Aaltomaa, S., Lipponen, P., Tammi, R., Tammi, M., Viitanen, J., Kankkunen, J. P., et al. (2002). Strong stromal hyaluronan expression is associated with PSA recurrence in local prostate cancer. *Urologia Internationalis, 69*, 266–272.

Andre, B., Duterme, C., Van Moer, K., Mertens-Strijthagen, J., Jadot, M., & Flamion, B. (2011). Hyal2 is a glycosylphosphatidylinositol-anchored, lipid raft-associated hyaluronidase. *Biochemical and Biophysical Research Communications, 411*, 175–179.

Aruffo, A., Stamenkovic, I., Melnick, M., Underhill, C. B., & Seed, B. (1990). CD44 is the principal cell surface receptor for hyaluronate. *Cell, 61*, 1303–1313.

Atmuri, V., Martin, D. C., Hemming, R., Gutsol, A., Byers, S., Sahebjam, S., et al. (2008). Hyaluronidase 3 (HYAL3) knockout mice do not display evidence of hyaluronan accumulation. *Matrix Biology, 27*, 653–660.

Beech, D. J., Madan, A. K., & Deng, N. (2002). Expression of PH-20 in normal and neoplastic breast tissue. *Journal of Surgical Research, 103*, 203–207.

Benitez, A., Yates, T. J., Lopez, L. E., Cerwinka, W. H., Bakkar, A., & Lokeshwar, V. B. (2011). Targeting hyaluronidase for cancer therapy: Antitumor activity of sulfated hyaluronic acid in prostate cancer cells. *Cancer Research, 71*, 4085–4095.

Berdiaki, A., Nikitovic, D., Tsatsakis, A., Katonis, P., Karamanos, N. K., & Tzanakakis, G. N. (2009). bFGF induces changes in hyaluronan synthase and hyaluronidase isoform expression and modulates the migration capacity of fibrosarcoma cells. *Biochimica et Biophysica Acta, 1790*, 1258–1265.

Bharadwaj, A. G., Goodrich, N. P., McAtee, C. O., Haferbier, K., Oakley, G. G., Wahl, J. K., 3rd., et al. (2011). Hyaluronan suppresses prostate tumor cell proliferation through diminished expression of N-cadherin and aberrant growth factor receptor signaling. *Experimental Cell Research, 317*, 1214–1225.

Bharadwaj, A. G., Kovar, J. L., Loughman, E., Elowsky, C., Oakley, G. G., & Simpson, M. A. (2009). Spontaneous metastasis of prostate cancer is promoted by excess hyaluronan synthesis and processing. *American Journal of Pathology, 174*, 1027–1036.

Bharadwaj, A. G., Rector, K., & Simpson, M. A. (2007). Inducible hyaluronan production reveals differential effects on prostate tumor cell growth and tumor angiogenesis. *Journal of Biological Chemistry, 282*, 20561–20572.

Botzki, A., Rigden, D. J., Braun, S., Nukui, M., Salmen, S., Hoechstetter, J., et al. (2004). L-ascorbic acid 6-hexadecanoate, a potent hyaluronidase inhibitor. X-ray structure and molecular modeling of enzyme-inhibitor complexes. *Journal of Biological Chemistry, 279*, 45990–45997.

Bouga, H., Tsouros, I., Bounias, D., Kyriakopoulou, D., Stavropoulos, M. S., Papageorgakopoulou, N., et al. (2010). Involvement of hyaluronidases in colorectal cancer. *BMC Cancer, 10*, 499.

Bourguignon, L. Y., Wong, G., Earle, C. A., & Xia, W. (2011). Interaction of low molecular weight hyaluronan with CD44 and toll-like receptors promotes the actin filament-associated protein 110-actin binding and MyD88-NFkappaB signaling leading to proinflammatory cytokine/chemokine production and breast tumor invasion. *Cytoskeleton (Hoboken), 68*, 671–693.

Bourguignon, L. Y., Zhu, H., Shao, L., & Chen, Y. W. (2000). CD44 interaction with tiam1 promotes Rac1 signaling and hyaluronic acid-mediated breast tumor cell migration. *Journal of Biological Chemistry, 275*, 1829–1838.

Camenisch, T. D., Spicer, A. P., Brehm-Gibson, T., Biesterfeldt, J., Augustine, M. L., Calabro, A., Jr., et al. (2000). Disruption of hyaluronan synthase-2 abrogates normal cardiac morphogenesis and hyaluronan-mediated transformation of epithelium to mesenchyme. *Journal of Clinical Investigation, 106*, 349–360 (see comments).

Chang, N. S. (2002). Transforming growth factor-beta1 blocks the enhancement of tumor necrosis factor cytotoxicity by hyaluronidase Hyal-2 in L929 fibroblasts. *BMC Cell Biology, 3*, 8.

Chang, J. Y., He, R. Y., Lin, H. P., Hsu, L. J., Lai, F. J., Hong, Q., et al. (2010). Signaling from membrane receptors to tumor suppressor WW domain-containing oxidoreductase. *Experimental Biology and Medicine (Maywood, N.J.), 235*, 796–804.

Chao, K. L., Muthukumar, L., & Herzberg, O. (2007). Structure of human hyaluronidase-1, a hyaluronan hydrolyzing enzyme involved in tumor growth and angiogenesis. *Biochemistry, 46*, 6911–6920.

Chen, S. T., Chuang, J. I., Wang, J. P., Tsai, M. S., Li, H., & Chang, N. S. (2004). Expression of WW domain-containing oxidoreductase WOX1 in the developing murine nervous system. *Neuroscience, 124*, 831–839.

Cheng, X. B., Sato, N., Kohi, S., & Yamaguchi, K. (2013). Prognostic impact of hyaluronan and its regulators in pancreatic ductal adenocarcinoma. *PLoS One, 8*, e80765.

Chi, A., Shirodkar, S. P., Escudero, D. O., Ekwenna, O. O., Yates, T. J., Ayyathurai, R., et al. (2012). Molecular characterization of kidney cancer: Association of hyaluronic acid family with histological subtypes and metastasis. *Cancer, 118*, 2394–2402.

Chib, R., Raut, S., Fudala, R., Chang, A., Mummert, M., Rich, R., et al. (2013). FRET based ratio-metric sensing of hyaluronidase in synthetic urine as a biomarker for bladder and prostate cancer. *Current Pharmaceutical Biotechnology, 14*, 470–474.

Craig, E. A., Parker, P., & Camenisch, T. D. (2009). Size-dependent regulation of Snail2 by hyaluronan: Its role in cellular invasion. *Glycobiology, 19*, 890–898.

Csoka, A. B., Frost, G. I., & Stern, R. (2001). The six hyaluronidase-like genes in the human and mouse genomes. *Matrix Biology, 20*, 499–508.

Dang, S., Peng, Y., Ye, L., Wang, Y., Qian, Z., Chen, Y., et al. (2013). Stimulation of TLR4 by LMW-HA induces metastasis in human papillary thyroid carcinoma through CXCR7. *Clinical & Developmental Immunology, 2013*, 712561.

Danilkovitch-Miagkova, A., Duh, F. M., Kuzmin, I., Angeloni, D., Liu, S. L., Miller, A. D., et al. (2003). Hyaluronidase 2 negatively regulates RON receptor tyrosine kinase and mediates transformation of epithelial cells by jaagsiekte sheep retrovirus. *Proceedings of the National Academy of Sciences of the United States of America, 100*, 4580–4585.

de Sa, V. K., Canavez, F. C., Silva, I. A., Srougi, M., & Leite, K. R. (2009). Isoforms of hyaluronidases can be a predictor of a prostate cancer of good prognosis. *Urologic Oncology, 27*, 377–381.

de Sa, V. K., Carvalho, L., Gomes, A., Alarcao, A., Silva, M. R., Couceiro, P., et al. (2013). Role of the extracellular matrix in variations of invasive pathways in lung cancers. *Brazilian Journal of Medical and Biological Research, 46*, 21–31.

de Sa, V. K., Olivieri, E., Parra, E. R., Ab'Saber, A. M., Takagaki, T., Soares, F. A., et al. (2012). Hyaluronidase splice variants are associated with histology and outcome in adenocarcinoma and squamous cell carcinoma of the lung. *Human Pathology, 43*, 675–683.

Deed, R., Rooney, P., Kumar, P., Norton, J. D., Smith, J., Freemont, A. J., et al. (1997). Early-response gene signalling is induced by angiogenic oligosaccharides of hyaluronan in endothelial cells. Inhibition by non-angiogenic, high-molecular-weight hyaluronan. *International Journal of Cancer*, *71*, 251–256.

Duh, F. M., Dirks, C., Lerman, M. I., & Miller, A. D. (2005). Amino acid residues that are important for Hyal2 function as a receptor for jaagsiekte sheep retrovirus. *Retrovirology*, *2*, 59.

Duterme, C., Mertens-Strijthagen, J., Tammi, M., & Flamion, B. (2009). Two novel functions of hyaluronidase-2 (Hyal2) are formation of the glycocalyx and control of CD44-ERM interactions. *Journal of Biological Chemistry*, *284*, 33495–33508.

Eissa, S., Shehata, H., Mansour, A., Esmat, M., & El-Ahmady, O. (2012). Detection of hyaluronidase RNA and activity in urine of schistosomal and non-schistosomal bladder cancer. *Medical Oncology*, *29*, 3345–3351.

Eissa, S., Swellam, M., Shehata, H., El-Khouly, I. M., El-Zayat, T., & El-Ahmady, O. (2010). Expression of HYAL1 and survivin RNA as diagnostic molecular markers for bladder cancer. *Journal of Urology*, *183*, 493–498.

Eissa, S., Zohny, S. F., Shehata, H. H., Hegazy, M. G., Salem, A. M., & Esmat, M. (2012). Urinary retinoic acid receptor-beta2 gene promoter methylation and hyaluronidase activity as noninvasive tests for diagnosis of bladder cancer. *Clinical Biochemistry*, *45*, 402–407.

Ekici, S., Cerwinka, W. H., Duncan, R., Gomez, P., Civantos, F., Soloway, M. S., et al. (2004). Comparison of the prognostic potential of hyaluronic acid, hyaluronidase (HYAL-1), CD44v6 and microvessel density for prostate cancer. *International Journal of Cancer*, *112*, 121–129.

Enegd, B., King, J. A., Stylli, S., Paradiso, L., Kaye, A. H., & Novak, U. (2002). Overexpression of hyaluronan synthase-2 reduces the tumorigenic potential of glioma cells lacking hyaluronidase activity. *Neurosurgery*, *50*, 1311–1318.

Fraser, J. R., Laurent, T. C., & Laurent, U. B. (1997). Hyaluronan: Its nature, distribution, functions and turnover. *Journal of Internal Medicine*, *242*, 27–33.

Godin, D. A., Fitzpatrick, P. C., Scandurro, A. B., Belafsky, P. C., Woodworth, B. A., Amedee, R. G., et al. (2000). PH20: A novel tumor marker for laryngeal cancer. *Archives of Otolaryngology Head & Neck Surgery*, *126*, 402–404.

Gomez, C. S., Gomez, P., Knapp, J., Jorda, M., Soloway, M. S., & Lokeshwar, V. B. (2009). Hyaluronic acid and HYAL-1 in prostate biopsy specimens: Predictors of biochemical recurrence. *Journal of Urology*, *182*, 1350–1356.

Harada, H., & Takahashi, M. (2007). CD44-dependent intracellular and extracellular catabolism of hyaluronic acid by hyaluronidase-1 and -2. *Journal of Biological Chemistry*, *282*, 5597–5607.

Hemming, R., Martin, D. C., Slominski, E., Nagy, J. I., Halayko, A. J., Pind, S., et al. (2008). Mouse Hyal3 encodes a 45- to 56-kDa glycoprotein whose overexpression increases hyaluronidase 1 activity in cultured cells. *Glycobiology*, *18*, 280–289.

Hofinger, E. S., Spickenreither, M., Oschmann, J., Bernhardt, G., Rudolph, R., & Buschauer, A. (2007). Recombinant human hyaluronidase Hyal-1: Insect cells versus Escherichia coli as expression system and identification of low molecular weight inhibitors. *Glycobiology*, *17*, 444–453.

Honda, T., Kaneiwa, T., Mizumoto, S., Sugahara, K., & Yamada, S. (2012). Hyaluronidases have strong hydrolytic activity toward chondroitin 4-sulfate comparable to that for hyaluronan. *Biomolecules*, *2*, 549–563.

Hsu, L. J., Schultz, L., Hong, Q., Van Moer, K., Heath, J., Li, M. Y., et al. (2009). Transforming growth factor beta1 signaling via interaction with cell surface Hyal-2 and recruitment of WWOX/WOX1. *Journal of Biological Chemistry*, *284*, 16049–16059.

Isoyama, T., Thwaites, D., Selzer, M. G., Carey, R. I., Barbucci, R., & Lokeshwar, V. B. (2006). Differential selectivity of hyaluronidase inhibitors toward acidic and basic hyaluronidases. *Glycobiology*, *16*, 11–21.

Itano, N., Sawai, T., Atsumi, F., Miyaishi, O., Taniguchi, S., Kannagi, R., et al. (2004). Selective expression and functional characteristics of three mammalian hyaluronan synthases in oncogenic malignant transformation. *Journal of Biological Chemistry*, *279*, 18679–18687.

Itano, N., Sawai, T., Yoshida, M., Lenas, P., Yamada, Y., Imagawa, M., et al. (1999). Three isoforms of mammalian hyaluronan synthases have distinct enzymatic properties. *Journal of Biological Chemistry*, *274*, 25085–25092.

Jacobson, A., Rahmanian, M., Rubin, K., & Heldin, P. (2002). Expression of hyaluronan synthase 2 or hyaluronidase 1 differentially affect the growth rate of transplantable colon carcinoma cell tumors. *International Journal of Cancer*, *102*, 212–219.

Jiang, T., Zhang, Z., Zhang, Y., Lv, H., Zhou, J., Li, C., et al. (2012). Dual-functional liposomes based on pH-responsive cell-penetrating peptide and hyaluronic acid for tumor-targeted anticancer drug delivery. *Biomaterials*, *33*, 9246–9258.

Kaessler, A., Olgen, S., & Jose, J. (2011). Autodisplay of catalytically active human hyaluronidase hPH-20 and testing of enzyme inhibitors. *European Journal of Pharmaceutical Sciences*, *42*, 138–147.

Kaneiwa, T., Miyazaki, A., Kogawa, R., Mizumoto, S., Sugahara, K., & Yamada, S. (2012). Identification of amino acid residues required for the substrate specificity of human and mouse chondroitin sulfate hydrolase (conventional hyaluronidase-4). *Journal of Biological Chemistry*, *287*, 42119–42128.

Kaneiwa, T., Mizumoto, S., Sugahara, K., & Yamada, S. (2010). Identification of human hyaluronidase-4 as a novel chondroitin sulfate hydrolase that preferentially cleaves the galactosaminidic linkage in the trisulfated tetrasaccharide sequence. *Glycobiology*, *20*, 300–309.

Kelley, L. A., & Sternberg, M. J. (2009). Protein structure prediction on the web: A case study using the Phyre server. *Nature Protocols*, *4*, 363–371.

Kim, S. W., Oh, K. T., Youn, Y. S., & Lee, E. S. (2014). Hyaluronated nanoparticles with pH- and enzyme-responsive drug release properties. *Colloids and Surfaces. B, Biointerfaces*, *116C*, 359–364.

Kolliopoulos, C., Bounias, D., Bouga, H., Kyriakopoulou, D., Stavropoulos, M., & Vynios, D. H. (2013). Hyaluronidases and their inhibitors in the serum of colorectal carcinoma patients. *Journal of Pharmaceutical and Biomedical Analysis*, *83*, 299–304.

Kosaki, R., Watanabe, K., & Yamaguchi, Y. (1999). Overproduction of hyaluronan by expression of the hyaluronan synthase Has2 enhances anchorage-independent growth and tumorigenicity. *Cancer Research*, *59*, 1141–1145.

Kovar, J. L., Johnson, M. A., Volcheck, W. M., Chen, J., & Simpson, M. A. (2006). Hyaluronidase expression induces prostate tumor metastasis in an orthotopic mouse model. *American Journal of Pathology*, *169*, 1415–1426.

Kramer, M. W., Escudero, D. O., Lokeshwar, S. D., Golshani, R., Ekwenna, O. O., Acosta, K., et al. (2011). Association of hyaluronic acid family members (HAS1, HAS2, and HYAL-1) with bladder cancer diagnosis and prognosis. *Cancer*, *117*, 1197–1209.

Kramer, M. W., Golshani, R., Merseburger, A. S., Knapp, J., Garcia, A., Hennenlotter, J., et al. (2010). HYAL-1 hyaluronidase: A potential prognostic indicator for progression to muscle invasion and recurrence in bladder cancer. *European Urology*, *57*, 86–93.

Langevin, S. M., Koestler, D. C., Christensen, B. C., Butler, R. A., Wiencke, J. K., Nelson, H. H., et al. (2012). Peripheral blood DNA methylation profiles are indicative of head and neck squamous cell carcinoma: An epigenome-wide association study. *Epigenetics*, *7*, 291–299.

Laurent, T. C., Laurent, U. B., & Fraser, J. R. (1996). The structure and function of hyaluronan: An overview. *Immunology and Cell Biology*, *74*, A1–A7.

Legg, J. W., Lewis, C. A., Parsons, M., Ng, T., & Isacke, C. M. (2002). A novel PKC-regulated mechanism controls CD44 ezrin association and directional cell motility. *Nature Cell Biology*, *4*, 399–407.

Lepperdinger, G., Strobl, B., & Kreil, G. (1998). HYAL2, a human gene expressed in many cells, encodes a lysosomal hyaluronidase with a novel type of specificity. *Journal of Biological Chemistry*, *273*, 22466–22470.

Li, R., Todd, N. W., Qiu, Q., Fan, T., Zhao, R. Y., Rodgers, W. H., et al. (2007). Genetic deletions in sputum as diagnostic markers for early detection of stage I non-small cell lung cancer. *Clinical Cancer Research*, *13*, 482–487.

Li, Y., Wang, D., Wang, L., Yu, J., Du, D., Chen, Y., et al. (2013). Distinct genomic aberrations between low-grade and high-grade gliomas of Chinese patients. *PLoS One*, *8*, e57168.

Liu, S. L., Duh, F. M., Lerman, M. I., & Miller, A. D. (2003). Role of virus receptor Hyal2 in oncogenic transformation of rodent fibroblasts by sheep betaretrovirus env proteins. *Journal of Virology*, *77*, 2850–2858.

Lokeshwar, V. B., Cerwinka, W. H., Isoyama, T., & Lokeshwar, B. L. (2005). HYAL1 hyaluronidase in prostate cancer: A tumor promoter and suppressor. *Cancer Research*, *65*, 7782–7789.

Lokeshwar, V. B., Estrella, V., Lopez, L., Kramer, M., Gomez, P., Soloway, M. S., et al. (2006). HYAL1-v1, an alternatively spliced variant of HYAL1 hyaluronidase: A negative regulator of bladder cancer. *Cancer Research*, *66*, 11219–11227.

Lokeshwar, V. B., Rubinowicz, D., Schroeder, G. L., Forgacs, E., Minna, J. D., Block, N. L., et al. (2001). Stromal and epithelial expression of tumor markers hyaluronic acid and HYAL1 hyaluronidase in prostate cancer. *Journal of Biological Chemistry*, *276*, 11922–11932.

Lokeshwar, V. B., Schroeder, G. L., Carey, R. I., Soloway, M. S., & Iida, N. (2002). Regulation of hyaluronidase activity by alternative mRNA splicing. *Journal of Biological Chemistry*, *277*, 33654–33663.

Madan, A. K., Pang, Y., Wilkiemeyer, M. B., Yu, D., & Beech, D. J. (1999). Increased hyaluronidase expression in more aggressive prostate adenocarcinoma. *Oncology Reports*, *6*, 1431–1433.

Madan, A. K., Yu, K., Dhurandhar, N., Cullinane, C., Pang, Y., & Beech, D. J. (1999). Association of hyaluronidase and breast adenocarcinoma invasiveness. *Oncology Reports*, *6*, 607–609.

McAtee C. O., Berkebile A., Elowsky C., Fangman T., Wahl J. K., Khalimonchuk O., et al. Hyaluronidase Hyal1 increases tumor cell proliferation and motility through accelerated vesicle trafficking. Submitted.

McCarthy, J. B., Turley, E. A., Wilson, C. M., Price, M., Bullard, K. M., & Beck, M. (2005). Hyaluronan biosynthesis in prostate carcinoma growth and metastasis. In E. A. Balasz, & V. C. Hascall (Eds.), *Hyaluronan and Tumors. Hyaluronan: Structure, metabolism, biological activities, therapeutic applications.* Edgewater, NJ: Matrix Biology Institute. Chapter 4.

Mok, H., Jeong, H., Kim, S. J., & Chung, B. H. (2012). Indocyanine green encapsulated nanogels for hyaluronidase activatable and selective near infrared imaging of tumors and lymph nodes. *Chemical Communications (Cambridge, England)*, *48*, 8628–8630.

Mu, W., Rana, S., & Zoller, M. (2013). Host matrix modulation by tumor exosomes promotes motility and invasiveness. *Neoplasia*, *15*, 875–887.

Novak, U., Stylli, S. S., Kaye, A. H., & Lepperdinger, G. (1999). Hyaluronidase-2 overexpression accelerates intracerebral but not subcutaneous tumor formation of murine astrocytoma cells. *Cancer Research*, *59*, 6246–6250.

Nykopp, T. K., Rilla, K., Sironen, R., Tammi, M. I., Tammi, R. H., Hamalainen, K., et al. (2009). Expression of hyaluronan synthases (HAS1-3) and hyaluronidases (HYAL1-2) in serous ovarian carcinomas: Inverse correlation between HYAL1 and hyaluronan content. *BMC Cancer*, *9*, 143.

Nykopp, T. K., Rilla, K., Tammi, M. I., Tammi, R. H., Sironen, R., Hamalainen, K., et al. (2010). Hyaluronan synthases (HAS1-3) and hyaluronidases (HYAL1-2) in the accumulation of hyaluronan in endometrioid endometrial carcinoma. *BMC Cancer, 10*, 512.

Ohnuma, S., Miura, K., Horii, A., Fujibuchi, W., Kaneko, N., Gotoh, O., et al. (2009). Cancer-associated splicing variants of the CDCA1 and MSMB genes expressed in cancer cell lines and surgically resected gastric cancer tissues. *Surgery, 145*, 57–68.

Olgen, S., Kaessler, A., Kilic-Kurt, Z., & Jose, J. (2010). Investigation of aminomethyl indole derivatives as hyaluronidase inhibitors. *Zeitschrift für Naturforschung C, 65*, 445–450.

Olgen, S., Kaessler, A., Nebioglu, D., & Jose, J. (2007). New potent indole derivatives as hyaluronidase inhibitors. *Chemical Biology & Drug Design, 70*, 547–551.

Patel, S., Turner, P. R., Stubberfield, C., Barry, E., Rohlff, C. R., Stamps, A., et al. (2002). Hyaluronidase gene profiling and role of hyal-1 overexpression in an orthotopic model of prostate cancer. *International Journal of Cancer, 97*, 416–424.

Poola, I., Abraham, J., Marshalleck, J. J., Yue, Q., Lokeshwar, V. B., Bonney, G., et al. (2008). Molecular risk assessment for breast cancer development in patients with ductal hyperplasias. *Clinical Cancer Research, 14*, 1274–1280.

Posey, J. T., Soloway, M. S., Ekici, S., Sofer, M., Civantos, F., Duncan, R. C., et al. (2003). Evaluation of the prognostic potential of hyaluronic acid and hyaluronidase (HYAL1) for prostate cancer. *Cancer Research, 63*, 2638–2644.

Proulx, S. T., & Detmar, M. (2013). Molecular mechanisms and imaging of lymphatic metastasis. *Experimental Cell Research*, http://dx.doi.org/10.1016/j.yexcr.2013.03.009, pii: S0014-4827(13)00117-1.

Provenzano, P. P., Cuevas, C., Chang, A. E., Goel, V. K., Von Hoff, D. D., & Hingorani, S. R. (2012). Enzymatic targeting of the stroma ablates physical barriers to treatment of pancreatic ductal adenocarcinoma. *Cancer Cell, 21*, 418–429.

Puissant, E., Gilis, F., Dogne, S., Flamion, B., Jadot, M., & Boonen, M. (2014). Subcellular trafficking and activity of Hyal-1 and its processed forms in murine macrophages. *Traffic (Copenhagen, Denmark), 15*, 500–515.

Reitinger, S., Mullegger, J., Greiderer, B., Nielsen, J. E., & Lepperdinger, G. (2009). Designed human serum hyaluronidase 1 variant, HYAL1DeltaL, exhibits activity up to pH 5.9. *Journal of Biological Chemistry, 284*, 19173–19177.

Rich, R. M., Mummert, M., Foldes-Papp, Z., Gryczynski, Z., Borejdo, J., Gryczynski, I., et al. (2012). Detection of hyaluronidase activity using fluorescein labeled hyaluronic acid and fluorescence correlation spectroscopy. *Journal of Photochemistry and Photobiology B, 116*, 7–12.

Rilla, K., Pasonen-Seppanen, S., Deen, A. J., Koistinen, V. V., Wojciechowski, S., Oikari, S., et al. (2013). Hyaluronan production enhances shedding of plasma membrane-derived microvesicles. *Experimental Cell Research, 319*, 2006–2018.

Rooney, P., Kumar, S., Ponting, J., & Wang, M. (1995). The role of hyaluronan in tumour neovascularization (review). *International Journal of Cancer, 60*, 632–636.

Saito, T., Kawana, H., Azuma, K., Toyoda, A., Fujita, H., Kitagawa, M., et al. (2011). Fragmented hyaluronan is an autocrine chemokinetic motility factor supported by the HAS2-HYAL2/CD44 system on the plasma membrane. *International Journal of Oncology, 39*, 1311–1320.

Sharma, R., Wang, W., Rasmussen, J. C., Joshi, A., Houston, J. P., Adams, K. E., et al. (2007). Quantitative imaging of lymph function. *American Journal of Physiology, Heart and Circulatory Physiology, 292*, H3109–H3118.

Shiftan, L., Israely, T., Cohen, M., Frydman, V., Dafni, H., Stern, R., et al. (2005). Magnetic resonance imaging visualization of hyaluronidase in ovarian carcinoma. *Cancer Research, 65*, 10316–10323.

Shiftan, L., & Neeman, M. (2006). Kinetic analysis of hyaluronidase activity using a bioactive MRI contrast agent. *Contrast Media & Molecular Imaging, 1*, 106–112.
Siiskonen, H., Poukka, M., Tyynela-Korhonen, K., Sironen, R., & Pasonen-Seppanen, S. (2013). Inverse expression of hyaluronidase 2 and hyaluronan synthases 1-3 is associated with reduced hyaluronan content in malignant cutaneous melanoma. *BMC Cancer, 13*, 181.
Simpson, M. A. (2006). Concurrent expression of hyaluronan biosynthetic and processing enzymes promotes growth and vascularization of prostate tumors in mice. *American Journal of Pathology, 169*, 247–257.
Simpson, M. A., Wilson, C. M., & McCarthy, J. B. (2002). Inhibition of prostate tumor cell hyaluronan synthesis impairs subcutaneous growth and vascularization in immunocompromised mice. *American Journal of Pathology, 161*, 849–857.
Sugahara, K. N., Hirata, T., Tanaka, T., Ogino, S., Takeda, M., Terasawa, H., et al. (2008). Chondroitin sulfate E fragments enhance CD44 cleavage and CD44-dependent motility in tumor cells. *Cancer Research, 68*, 7191–7199.
Tammi, R., Rilla, K., Pienimaki, J. P., MacCallum, D. K., Hogg, M., Luukkonen, M., et al. (2001). Hyaluronan enters keratinocytes by a novel endocytic route for catabolism. *Journal of Biological Chemistry, 276*, 35111–35122.
Tan, J. X., Wang, X. Y., Li, H. Y., Su, X. L., Wang, L., Ran, L., et al. (2011). HYAL1 overexpression is correlated with the malignant behavior of human breast cancer. *International Journal of Cancer, 128*, 1303–1315.
Tan, J. X., Wang, X. Y., Su, X. L., Li, H. Y., Shi, Y., Wang, L., et al. (2011). Upregulation of HYAL1 expression in breast cancer promoted tumor cell proliferation, migration, invasion and angiogenesis. *PLoS One, 6*, e22836.
Tian, X., Azpurua, J., Hine, C., Vaidya, A., Myakishev-Rempel, M., Ablaeva, J., et al. (2013). High-molecular-mass hyaluronan mediates the cancer resistance of the naked mole rat. *Nature, 499*, 346–349.
Toida, T., Ogita, Y., Suzuki, A., Toyoda, H., & Imanari, T. (1999). Inhibition of hyaluronidase by fully O-sulfonated glycosaminoglycans. *Archives of Biochemistry and Biophysics, 370*, 176–182.
Toole, B. P. (1997). Hyaluronan in morphogenesis. *Journal of Internal Medicine, 242*, 35–40.
Twarock, S., Tammi, M. I., Savani, R. C., & Fischer, J. W. (2010). Hyaluronan stabilizes focal adhesions, filopodia, and the proliferative phenotype in esophageal squamous carcinoma cells. *Journal of Biological Chemistry, 285*, 23276–23284.
Tzellos, T. G., Kyrgidis, A., Vahtsevanos, K., Triaridis, S., Printza, A., Klagas, I., et al. (2011). Nodular basal cell carcinoma is associated with increased hyaluronan homeostasis. *Journal of the European Academy of Dermatology and Venereology, 25*, 679–687.
Udabage, L., Brownlee, G. R., Nilsson, S. K., & Brown, T. J. (2005). The over-expression of HAS2, Hyal-2 and CD44 is implicated in the invasiveness of breast cancer. *Experimental Cell Research, 310*, 205–217.
Udabage, L., Brownlee, G. R., Stern, R., & Brown, T. J. (2004). Inhibition of hyaluronan degradation by dextran sulphate facilitates characterisation of hyaluronan synthesis: An in vitro and in vivo study. *Glycoconjugate Journal, 20*, 461–471.
Urakawa, H., Nishida, Y., Knudson, W., Knudson, C. B., Arai, E., Kozawa, E., et al. (2012). Therapeutic potential of hyaluronan oligosaccharides for bone metastasis of breast cancer. *Journal of Orthopaedic Research, 30*, 662–672.
Vigdorovich, V., Strong, R. K., & Miller, A. D. (2005). Expression and characterization of a soluble, active form of the jaagsiekte sheep retrovirus receptor, Hyal2. *Journal of Virology, 79*, 79–86.
Webber, J. P., Spary, L. K., Sanders, A. J., Chowdhury, R., Jiang, W. G., Steadman, R., et al. (2014). Differentiation of tumour-promoting stromal myofibroblasts by cancer exosomes. *Oncogene*, http://dx.doi.org/10.1038/onc.2013.560.

Webber, J., Steadman, R., Mason, M. D., Tabi, Z., & Clayton, A. (2010). Cancer exosomes trigger fibroblast to myofibroblast differentiation. *Cancer Research*, *70*, 9621–9630.
West, D. C., Hampson, I. N., Arnold, F., & Kumar, S. (1985). Angiogenesis induced by degradation products of hyaluronic acid. *Science*, *228*, 1324–1326.
West, D. C., & Kumar, S. (1989). The effect of hyaluronate and its oligosaccharides on endothelial cell proliferation and monolayer integrity. *Experimental Cell Research*, *183*, 179–196.
Yang, C., Cao, M., Liu, H., He, Y., Xu, J., Du, Y., et al. (2012). The high and low molecular weight forms of hyaluronan have distinct effects on CD44 clustering. *Journal of Biological Chemistry*, *287*, 43094–43107.
Yoffou, P. H., Edjekouane, L., Meunier, L., Tremblay, A., Provencher, D. M., Mes-Masson, A. M., et al. (2011). Subtype specific elevated expression of hyaluronidase-1 (HYAL-1) in epithelial ovarian cancer. *PLoS One*, *6*, e20705.
Zeng, C., Toole, B. P., Kinney, S. D., Kuo, J. W., & Stamenkovic, I. (1998). Inhibition of tumor growth in vivo by hyaluronan oligomers. *International Journal of Cancer*, *77*, 396–401.
Zhang, L., Bharadwaj, A. G., Casper, A., Barkley, J., Barycki, J. J., & Simpson, M. A. (2009). Hyaluronidase activity of human Hyal1 requires active site acidic and tyrosine residues. *Journal of Biological Chemistry*, *284*, 9433–9442.
Zhang, R. X., Zhu, J. H., Fan, J., & Ji, X. Q. (2013). Analysis of HYAL3 gene mutations in Chinese lung squamous cell carcinoma patients. *Tumori*, *99*, 108–112.

CHAPTER TWO

Targeting Hyaluronic Acid Family for Cancer Chemoprevention and Therapy

Vinata B. Lokeshwar[*,†,‡,1], **Summan Mirza**[*], **Andre Jordan**[§]
[*]Department of Urology, University of Miami-Miller School of Medicine, Miami, Florida, USA
[†]Department of Cell Biology, University of Miami-Miller School of Medicine, Miami, Florida, USA
[‡]Miami Clinical Translational Institute, University of Miami-Miller School of Medicine, Miami, Florida, USA
[§]Sheila and David Funte Program in Cancer Biology, University of Miami-Miller School of Medicine, Miami, Florida, USA
[1]Corresponding author: e-mail address: vlokeshw@med.miami.edu

Contents

1. Introduction 36
2. Targeting HA Production 38
   2.1 Targeting HA synthases 38
   2.2 Chemical inhibitors of HA synthesis 39
3. Targeting HA Signaling 43
   3.1 HA oligosaccharides 43
4. HA as a Carrier for Drug Delivery 45
5. Targeting HA Receptors 48
   5.1 CD44 48
   5.2 RHAMM 51
6. Targeting HAase 53
7. Conclusion 54
Acknowledgments 55
References 55

Abstract

Hyaluronic acid or hyaluronan (HA) is perhaps one of the most uncomplicated large polymers that regulates several normal physiological processes and, at the same time, contributes to the manifestation of a variety of chronic and acute diseases, including cancer. Members of the HA signaling pathway (HA synthases, HA receptors, and HYAL-1 hyaluronidase) have been experimentally shown to promote tumor growth, metastasis, and angiogenesis, and hence each of them is a potential target for cancer therapy. Furthermore, as these members are also overexpressed in a variety of carcinomas, targeting of the HA family is clinically relevant. A variety of targeted approaches have been developed to target various HA family members, including small-molecule

Advances in Cancer Research, Volume 123
ISSN 0065-230X
http://dx.doi.org/10.1016/B978-0-12-800092-2.00002-2

inhibitors and antibody and vaccine therapies. These treatment approaches inhibit HA-mediated intracellular signaling that promotes tumor cell proliferation, motility, and invasion, as well as induction of endothelial cell functions. Being nontoxic, non-immunogenic, and versatile for modifications, HA has been used in nanoparticle preparations for the targeted delivery of chemotherapy drugs and other anticancer compounds to tumor cells through interaction with cell-surface HA receptors. This review discusses basic and clinical translational aspects of targeting each HA family member and respective treatment approaches that have been described in the literature.

ABBREVIATIONS

4-MU 4-methylumbelliferone
HA hyaluronic acid
HAase hyaluronidase
HAS HA synthase
oHA HA oligosaccharides
sHA sulfated hyaluronic acid
UGA UDP-glucuronic acid

1. INTRODUCTION

Several members of the hyaluronic acid (HA) family of molecules, HA synthases (i.e., HAS1, HAS2, HAS3), HA receptors (i.e., CD44 and RHAMM), and hyaluronidases (mainly HYAL-1), are critical determinants of tumor growth and progression (Adamia, Pilarski, Belch, & Pilarski, 2013; Ghosh, Kuppusamy, & Pilarski, 2009; Golshani et al., 2007; Karbownik & Nowak, 2013; Orian-Rousseau, 2010; Simpson & Lokeshwar, 2008; Sironen et al., 2011). HA family members promote malignant behavior of tumor cells *in vitro*, and tumor growth, metastasis, and angiogenesis in xenograft models (Bharadwaj et al., 2009; Chao, Muthukumar, & Herzberg, 2007; Gurski et al., 2012; Li, Li, Brown, & Heldin, 2007; Lokeshwar, Cerwinka, & Lokeshwar, 2005; Lokeshwar et al., 2006; Siiskonen, Poukka, Tyynela-Korhonen, Sironen, & Pasonen-Seppanen, 2013; Tan et al., 2011). HA family of molecules are also potential diagnostic and prognostic markers for a variety of carcinomas including breast, bladder, endometrial, ovarian, and prostate (Auvinen et al., 2014; Bouga et al., 2010; Chi et al., 2012; Franzmann et al., 2003; Golshani et al., 2007; Gomez et al., 2009; Kramer et al., 2011; Lokeshwar et al., 2000, 2002; Paiva et al., 2005; Yoshida, Matsuda, Naito, & Ishiwata, 2012; Zhang, Chang, & Liu, 2013).

In tumor tissues, HA is contributed by both tumor stroma and tumor cells and induces intracellular signaling by binding to HA receptors. HYAL-1 is the major tumor-derived hyaluronidase (HAase) that is almost exclusively expressed by tumor cells. By degrading HA, HAase/HYAL-1 generates small HA fragments, some of which (~10–25 disaccharide units) are angiogenic (Lokeshwar et al., 2001, 1999; West, Hampson, Arnold, & Kumar, 1985). In experimental model systems such as breast, bladder, prostate, and colon, studies have been mainly focused on modulating the expression of individual HA family molecules and assessing their effects on tumor cell phenotypes both *in vitro* and *in vivo*. Each HA synthase or HYAL-1, either alone or coexpressed, contributes to tumor cell proliferation, motility, and invasion, and to enhanced tumor growth, metastasis, and angiogenesis in xenografts; in contrast, knockdown of these genes inhibits tumor cell functions (Adamia et al., 2013; Bharadwaj et al., 2009; Chao et al., 2007; Ghosh et al., 2009; Golshani et al., 2007; Li et al., 2007). In the case of HYAL-1, promotion of tumor cell function is dose dependent. At levels detected in clinical specimens, HYAL-1 promotes tumor growth, invasion/metastasis, and angiogenesis; however, overexpression of HYAL-1 at levels exceeding those expressed in tumor tissues induces apoptosis and inhibits tumor formation (Lokeshwar, Cerwinka, Isoyama, & Lokeshwar, 2005; Lokeshwar & Selzer, 2008). Therefore, while a limited degradation of the pericellular HA matrix generates angiogenic HA fragments which induce intracellular signaling, complete degradation of the HA matrix, as a result of experimental overexpression of HYAL-1, will inhibit tumor growth and progression. Studies on HA-mediated signaling usually do not distinguish between HA and HA fragments present in the pericellular matrix. However, tumor-associated HA consists of both large HA polymers (mol. wt $\geq 0.5 \times 10^6$ Da) and smaller angiogenic oligosaccharides, the latter correlating with HAase activity in tumor tissues (Franzmann et al., 2003; Lokeshwar et al., 2001). Angiogenic HA fragments have been detected in the urine of patients with bladder cancer and Wilm's tumor, in the saliva of patients with head and neck cancer, and in bladder and prostate tumor tissues (Franzmann et al., 2003; Kumar, West, Ponting, & Gattamaneni, 1989; Lokeshwar, Obek, Soloway, & Block, 1997; Lokeshwar et al., 2001). In contrast to the angiogenic HA fragments, HA oligosaccharides consisting of 2–3 disaccharide units have been shown to have antitumor activity, presumably because they inhibit HA-induced signaling (Ghatak, Misra, & Toole, 2002; Hosono et al., 2007; Toole, Ghatak, & Misra, 2008; Urakawa et al., 2012).

Interaction between pericellular HA/angiogenic fragments and HA receptors induces multiple intracellular pathways. For example, CD44-HA and/or RHAMM-HA signaling promotes cell survival, cancer stemness, motility, and invasion by activating growth factor receptor signaling (e.g., ErbB2, c-Met), PI3/Akt and Erk pathways, small GTPase proteins (i.e., RhoA and Rac1), Ras, NFkB and src signaling, cytoskeleton reorganization, etc., and some of these pathways, in turn, may induce HA synthase and HA receptor expression (Benitez et al., 2011; Benitez, Yates, Shamaldevi, Bowen, & Lokeshwar, 2013; Bernert, Porsch, & Heldin, 2011; Bharadwaj et al., 2011; Bourguignon, Wong, Earle, Krueger, & Spevak, 2010; Dortet, Veiga, Bonazzi, & Cossart, 2010; Hatano et al., 2011; Kim, Park, Lee, & Jeoung, 2008; Lokeshwar et al., 2010; Misra, Toole, & Ghatak, 2006). CD44 and RHAMM also have compensatory roles and therefore, if HA receptors are to be targeted for cancer therapy, silencing of both receptors may be necessary to completely abrogate HA signaling (Benitez et al., 2011; Lokeshwar et al., 2010; Turley & Naor, 2012). HA-mediated signaling events also induce expression of a variety of cytokines and chemokines, COX2, and matrix metalloproteinases, which promote tumor angiogenesis and invasion/metastasis (Chow, Tauler, & Mulshine, 2010; Dunn et al., 2009; Lokeshwar et al., 2010; Misra et al., 2008, 2006; Porsch et al., 2013; Vincent, Jourdan, Sy, Klein, & Mechti, 2001; Voelcker et al., 2008). HA signaling studies reveal that targeting HA and other HA family members by small-molecule inhibitors, genetic manipulation, and vaccination could be exploited for cancer therapy. In this review, we discuss some of these approaches.

2. TARGETING HA PRODUCTION

2.1. Targeting HA synthases

Although targeting of HA synthases has not been exploited for therapeutic purposes, genetic knockdown studies shed light on the crucial roles that HA synthases play in different types of cancers. For example, HAS1 knockdown in bladder cancer cells induces cell cycle arrest in G2-M phase, followed by apoptosis induction via the extrinsic pathway. HAS1 knockdown also cripples the chemotactic ability and invasion of bladder cancer cells and inhibits tumor growth and angiogenesis over fivefold (Golshani et al., 2008). HAS1 downregulation studies also revealed that there is a feedback loop between HA synthesis and the expression of HA receptors. Knockdown of HAS1 causes transcriptional downregulation of CD44 (Golshani et al., 2008).

In the clinical scenario, HAS1 expression is elevated in bladder, prostate, and renal cell carcinomas, correlates with HA presence in tumor tissues, and is an independent predictor of unfavorable outcome (Chi et al., 2012; Golshani et al., 2007). HAS1 mRNA is extensively alternatively spliced, and although the functions of the splice variants are unknown, these splice variants associate with poor survival in multiple myeloma patients. These clinical findings thus corroborate the experimental findings on the functional involvement of HAS1 in tumor growth and progression (Adamia et al., 2005; Kriangkum, Warkentin, Belch, & Pilarski, 2013). Similar to HAS1, HAS2 and HAS3 knockdown inhibits tumor growth and metastasis in breast and osteosarcoma models and recently, HAS2 knockdown has been shown to potentiate radiation-induced DNA damage and apoptosis in cancer cells (Li et al., 2007; Shen et al., 2014; Udabage et al., 2005). HAS2 and HAS3 expression also correlates with clinical outcome, including resistance to chemotherapy and increased risk for anthracycline-related cardiomyopathy (Auvinen et al., 2014; Paiva et al., 2005; Ricciardelli et al., 2013; Wang et al., 2014). However, since cancer cells express more than one HA synthase, targeting HA synthesis using small-molecule inhibitors would be a better approach than inhibiting the expression or function of each individual HA synthase.

2.2. Chemical inhibitors of HA synthesis

2.2.1 4-Methylumbelliferone

4-Methylumbelliferone (4-MU) or 7-hydroxy-4-methylcoumarin is a well-studied inhibitor of HA synthesis (Clarkin, Allen, Wheeler-Jones, Bastow, & Pitsillides, 2011; Kakizaki et al., 2004, 2002; Morohashi et al., 2006; Nakamura et al., 1997; Saito et al., 2013). Mammalian cells synthesize HA, using two building blocks—UDP-glucuronic acid (UGA) and UDP-*N*-acetyl-D-glucosamine. UGA is synthesized by oxidation of UDP-glucose via the enzyme UDP-glucose dehydrogenase. UGA is also a substrate for cellular detoxification enzymes known as UDP-glucuronosyltransferases. In cells treated with 4-MU, UDP-glucuronosyltransferase transfers glucuronic acid onto 4-MU. This depletes the intracellular pool of UGA and HA synthesis is halted. Depending upon the specific isozyme, the K_m values of UDP-glucuronosyltransferase and HAS range from 100 to 900 μM. In several studies using tumor cell lines, the IC_{50} of 4-MU for inhibiting HA synthesis is in the range of 400 μM (Kakizaki et al., 2004, 2002; Lokeshwar et al., 2010; Nakamura et al., 1997). Recently, it has also been shown that 4-MU downregulates HAS2 and HAS3 expression

(Saito et al., 2013). The study showed that 4-MU indeed inhibits both HA and sulfated glycosaminoglycan synthesis in the chick limb bud micromass culture (Clarkin et al., 2011). In the cancer arena, however, 4-MU has been extensively studied as a HA synthesis inhibitor and, in addition to its effect on enzymatic inhibition of HA synthesis, has been shown to downregulate HAS2 and HAS3 expression by 60–80% in some cancer cells (Kultti et al., 2009).

Although 4-MU is known as an HA synthesis inhibitor, it is also widely used in assays to measure enzyme activity or quality control, due to its HA-dependent fluorescent properties. In fact, 4-MU (also known as hymecromone) is a popular term in PubMed, with over 1000 citations. The majority of these articles use 4-MU as a fluorescent indicator of food, water, soil, or liver function (including clinical trials of 4-MU) but also include articles that are completely unrelated to 4-MU (over 300 citations). Fewer than 10% of the articles cited in PubMed on 4-MU are related to its use as an HA synthesis inhibitor, less than 5% are on its use in cancer cells either as a fluorescent indicator or as a HA synthesis inhibitor, and even a smaller number of studies have examined the anticancer potential of 4-MU in preclinical models. Over several decades, 4-MU has been investigated in small clinical trials as a choleretic (bile-inducing) and antispasmodic agent (Abate et al., 2001; Camarri & Marchettini, 1988; Garrett, Venitz, Eberst, & Cerda, 1993; Hoffmann, Schwarz, Pohl, Ziegenhagen, & Kruis, 2005; Quaranta, Rossetti, & Camarri, 1984). In fact, 4-MU is sold in Europe and Asia as a dietary supplement to improve liver health. Although it is a coumarin-derivative, 4-MU lacks the antisperminogenic and antiaromatase activities of coumarin and anticoagulant activity of coumadin or warfarin (Chen, Cho, Karlsberg, Zhou, & Yuan, 2004; Crooke, Fitzpatrick, O'Kennedy, & McCormack, 1997; Keating, 1997; Omarbasha, Fair, & Heston, 1989). According to the NIOSH registry, the LD_{50} for 4-MU ranges from 2.8 to 7.3 g/kg (RTECS #: GN7000000) and at doses where 4-MU shows biological efficacy in controlling tumor growth (200–450 mg/kg/day oral dose), it has no serum or organ toxicity (Lokeshwar et al., 2010).

At concentrations of 0.2–1 mM (i.e., ~35–180 μg/ml), 4-MU inhibits proliferation, motility, and invasion, and causes loss of filopodia and focal adhesions, in a variety of cultured tumor cells (Arai et al., 2011; Hiraga, Ito, & Nakamura, 2013; Lokeshwar et al., 2010; Okuda et al., 2012; Piccioni et al., 2012; Twarock et al., 2011; Uchakina, Ban, & McKallip, 2013; Urakawa et al., 2012). 4-MU inhibits tumor spheroids and osteoclast-like cell formation (Hiraga et al., 2013) and downregulates the expression of both CD44 and RHAMM, suggesting a feedback loop between HA synthesis and HA receptor expression (Lokeshwar et al., 2010). In addition, 4-MU treatment inhibits a variety of HA signaling events, including downregulation of phospho-ErbB2, phospho-Akt, and downstream effectors MMP-2/MMP-9 and IL-8 expression (Fig. 2.1; Lokeshwar et al., 2010; Okuda et al., 2012; Twarock et al., 2011; Urakawa, Nishida, Wasa, et al., 2012). The efficacy of 4-MU as a glucuronate scavenger could contribute to its antitumor effects. For example, it has been shown that 4-MU (1 μM) inhibits glucuronidation and prevents inactivation of androgen in androgen-dependent prostate cancer cells. As a result, 4-MU promotes the androgen-dependent growth of these prostate cancer cells (Wei, Galbenus, Raza, Cerny, & Simpson, 2009). Inhibition

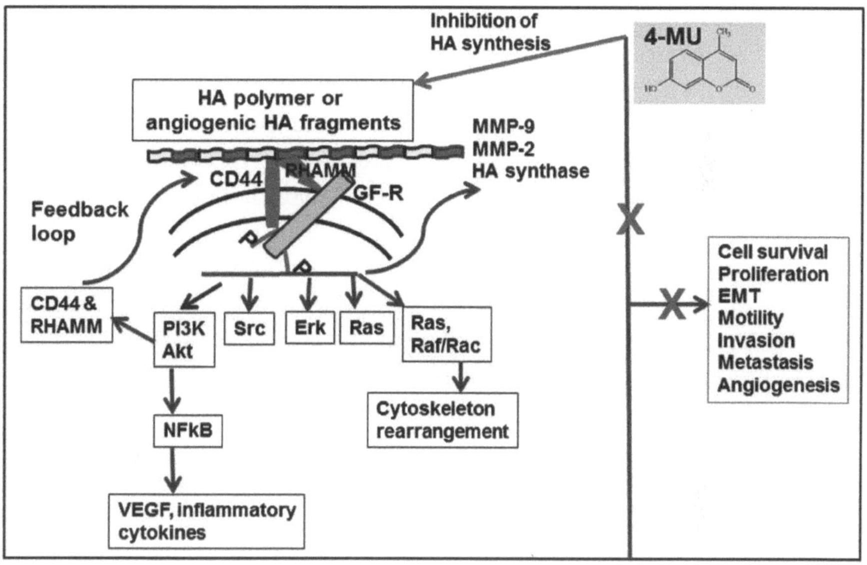

Figure 2.1 Molecular basis for the antitumor activity of 4-MU. Binding of HA receptors to cell surface HA receptors, CD44 and RHAMM, triggers a variety of signaling events, including complex formation between HA receptors and growth factor receptor protein tyrosine kinases, and activation of downstream effectors such as Akt, NFkB, src, Erk, Ras/Raf/Rac-1. These signaling events culminate in the expression of a variety of inflammatory cytokines, VEGF, matrix metalloproteinases (MMP-2, MMP-9), as well as HA synthase and CD44/RHAMM. By inducing these signaling events and downstream effectors, HA drives cell survival, proliferation, epithelial–mesenchymal interaction, invasion, and motility which lead to tumor growth and progression. Since 4-MU inhibits HA synthesis, it blocks the first event in this signaling cascade and hence shows potent antitumor and antimetastatic efficacy. (See the color plate.)

of the glucuronidation of androgen by 4-MU at concentrations as low as 1 μM is intriguing. Based solely on the K_m values of UDP-glucuronyltransferases (as described above), at such a low concentration, 4-MU would not be expected to act as a competitive inhibitor of HA synthesis or of glucuronidation. Therefore, its inhibitory potential may be dependent on cell type and expression level of specific UDP-glucuronosyltransferase isozymes, or 4-MU may have other effects than competitive inhibition of glucuronidation or of HA synthesis. Nevertheless, at the concentrations used in cancer cells and tumor models and also the doses at which it is consumed as a dietary supplement, 4-MU is demonstrated to have antitumor activities. For example, in xenograft studies, oral administration of 4-MU has been shown to inhibit tumor growth and metastasis in prostate, B16 melanoma, skin, liver, osteosarcoma, breast, and esophageal cancer model systems (Bhattacharyya et al., 2009; Kudo et al., 2004; Lokeshwar et al., 2010; Nakazawa et al., 2006; Piccioni et al., 2012; Twarock et al., 2011; Urakawa, Nishida, Wasa, et al., 2012; Yoshihara et al., 2005). Recently, 4-MU was shown to inhibit bone metastasis in a breast cancer model (Hiraga et al., 2013). As in the case of *in vitro* studies, some mouse xenograft studies have used 4-MU orally at doses as high as 1–3 g/kg; however, in other studies, 4-MU has shown remarkable efficacy at 200–400 mg/kg doses (Arai et al., 2011; Bhattacharyya et al., 2009; Hiraga et al., 2013; Kudo et al., 2004; Nakazawa et al., 2006; Okuda et al., 2012; Piccioni et al., 2012; Twarock et al., 2011; Urakawa, Nishida, Wasa, et al., 2012; Yoshihara et al., 2005). Based on the FDA's formula of mouse-to-human dose conversion, 200–400 mg/kg doses in mice equates to 1.1–2.2 g/day doses in humans; these are doses at which 4-MU is consumed for improving liver health (Abate et al., 2001; Camarri & Marchettini, 1988; Garrett et al., 1993; Hoffmann et al., 2005; Quaranta et al., 1984; U.S. Department of Health and Human Services, 2005). Considering 4-MU is consumed as a dietary supplement at similar doses, conducting clinical trials to test the toxicity and efficacy profile of 4-MU as an anticancer agent should be possible.

Although the potential of 4-MU as a single agent has been examined in xenograft studies, only two studies have reported its combination with other agents. 4-MU has been shown to enhance the efficacy of gemcitabine in one pancreatic cancer model at 1 g/kg dose (Nakazawa et al., 2006). More recently, 4-MU has been shown to synergize with Sorafenib, a tyrosine kinase inhibitor, approved by the FDA for the treatment of metastatic renal cell carcinoma (Benitez et al., 2013). In that study, 4-MU synergized with

Sorafenib at concentrations at which 4-MU alone did not inhibit HA synthesis and neither agent alone had any inhibitory effects on renal cell carcinoma cells *in vitro* or *in vivo*. However, the combination inhibited HA synthesis, proliferation, motility, and invasion *in vitro* and completely abrogated tumor growth in a Sorafenib-resistant xenograft model without toxicity (Benitez et al., 2013).

Taken together, 4-MU is an orally bioavailable dietary supplement that inhibits HA synthesis and has shown significant promise as an antitumor and antimetastatic agent. With a favorable toxicity profile and high efficacy, this HA synthesis inhibitor has potential for clinical translation.

2.2.2 Other HA synthesis inhibitors

Although not as effective as 4-MU, D-mannose has been shown to inhibit HA synthesis in a dose-dependent manner. Mannose at ~20 mM concentration inhibits HA synthesis by causing a reduction in the cellular concentration of UDP-*N*-acetylhexosamines (i.e., UDP-*N*-acetyl-D-glucosamine and UDP-*N*-acety-D-galactosamine). Mannose treatment was shown to inhibit dermal fibroblast invasion and to prevent enhanced leukocyte binding to HA (Jokela et al., 2008, 2013). However, antitumor effects of mannose in the context of HA synthesis inhibition have not yet been investigated. A curcumin analogue, 1,5-bis(4-hydroxy-3-methoxyphenyl)-1,4-pentadien-3-one (hylin), has been shown to inhibit multidrug resistance protein 5-mediated export of HA in fibroblasts ($IC_{50} \sim 5\ \mu M$; Prehm, 2013). However, given that curcumin has poor bioavailability *in vivo*, any of its analogues will need to be tested for their bioavailability before examining their efficacy on HA signaling and on cancer models.

3. TARGETING HA SIGNALING

3.1. HA oligosaccharides

In contrast to the effects of large HA polymers and angiogenic HA fragments, HA oligosaccharides (oHA) of <10 disaccharide units have been shown to possess antitumor activity in cancer cell lines of different tissue origins (Alaniz et al., 2006; Cordo Russo et al., 2008; Fuchs et al., 2013; Slomiany, Dai, Bomar, et al., 2009; Slomiany, Dai, Tolliver, et al., 2009; Slomiany, Grass, Robertson, et al., 2009; Ween, Hummitzsch, Rodgers, Oehler, & Ricciardelli, 2011; Yang et al., 2012). These oHA inhibit anchorage-independent growth, motility, and invasion, and induce apoptosis. These phenotypic effects are a direct consequence of the abrogation of HA signaling, especially the PI3/Akt pathway, and also of the association between CD44 and receptor tyrosine kinases (Ghatak et al., 2002; Slomiany, Dai, Tolliver, et al., 2009; Urakawa, Nishida, Knudson, et al., 2012). oHA induce PTEN expression (Ghatak et al., 2002), inhibit CD44 clustering on the plasma membrane, and block the interaction of CD44 with Emmprin, which is elevated in a variety of carcinomas (see Chapter 12). oHA inhibit plasma membrane localization of monocarboxylate transporters (MCT1 and MCT4) with CD44 and Emmprin, resulting in the inhibition of lactate efflux (Slomiany, Grass, Robertson, et al., 2009). In addition, oHA inhibit association of CD44 with drug transporters BCRP (ABCG2) and P-glycoprotein (ABCB1) in the plasma membrane and increase sensitivity of tumor cells to chemotherapeutic agents such as doxorubicin (DOX) (Slomiany, Dai, Bomar, et al., 2009; Slomiany, Grass, Robertson, et al., 2009). In fact, in vincristine-resistant lymphoma cell lines, oHA enhance cytotoxicity of vincristine by downregulating PI3-kinase/Akt signaling and also the activity of Pgp-1 (Cordo Russo et al., 2008). oHA also appear to inhibit versican-induced pericellular matrix formation, which is dependent on HA and CD44 interaction; subsequently, oHA inhibit motility and invasion in ovarian cancer cells (Ween et al., 2011). Recently, large polymeric HA was demonstrated to enhance CXCR4 and CXCL12 signaling and, consequently, cause an increase in Erk phosphorylation and phenotypic readouts, including tumor cell motility, capillary sprouting, and angiogenesis. This HA-increased signaling was due to the association of CXCR4 with CD44, and oHA inhibited this association and related signaling (Fuchs et al., 2013).

Therapeutic efficacy of oHA as an antitumor agent has been examined in a few studies. Initially, systemic administration of oHA was shown to inhibit tumor growth in the subcutaneous B16-F10 murine melanoma model (Ghatak et al., 2002). Similarly, systemic administration of oHA inhibits growth of an experimental malignant peripheral nerve sheath tumor (MPNST) model (Slomiany, Dai, Bomar, et al., 2009). In an intratibial growth model of breast cancer, injection of oHA (i.e., decasaccharide) in the tibia inhibited osteolysis and HA accumulation in the bone metastatic lesions (Urakawa, Nishida, Knudson, et al., 2012). In osteosarcoma models, intratumoral injection of HA octasaccharides reduced HA accumulation in local tumors, which consequently caused a significant inhibition of primary tumor growth and distant lung metastasis (Hosono et al., 2007). One of the advantages of oHA is that they are nonimmunogenic, since they are simply repeating disaccharide units, and inhibit tumor cell functions *in vitro* along with tumor growth and metastasis by abrogating CD44 and HA interaction. As discussed above, oHA can improve response of tumor cells to chemotherapeutic agents. For example, paclitaxel conjugated to oHA is internalized by CD44-overexpressing tumor cells and is 50 times more cytotoxic than when administered alone (Journo-Gershfeld, Kapp, Shamay, Kopecek, & David, 2012). Similarly, at concentrations at which oHA alone are not effective, they synergize with DOX and reduce the growth of malignant peripheral nerve sheath tumors (MPNST; Slomiany, Dai, Bomar, et al., 2009).

Taken together, as a single agent, oHA have shown promise in inhibiting CD44-mediated HA signaling in tumor cells and tumor growth at primary and secondary metastatic sites. Interestingly, little work has been done to demonstrate whether or not oHA have similarly inhibitory activity on abrogating HA/RHAMM interaction. Since in recent literature, RHAMM expression has shown promise as a potential biomarker for cancer diagnosis and prognosis, examination of oHA effects on RHAMM-mediated HA signaling will be helpful for the clinical translation of oHA as a cancer treatment. Based on the published studies discussed above, oHA may have better clinical utility in enhancing the efficacy of chemotherapeutic drugs by decreasing pericellular HA matrix. However, before oHA can advance into clinical trials as an anticancer drug or as an adjuvant to chemotherapeutic drugs, pharmacokinetic and pharmacodynamic studies will need to be conducted in preclinical models to determine the therapeutic window. Since the differences in the length of HA oligosaccharides that have angiogenic/tumor-promoting activity and oHA that have antitumor activity are only a few disaccharide units, reliable methods to synthesize and separate oHA of defined length (6–10 disaccharides) on a large scale will need to be developed for clinical translation.

4. HA AS A CARRIER FOR DRUG DELIVERY

HA has been used for targeted anticancer drug deliveries in the form of HA-coated nanoparticles. Being a hydrophilic large polymer of repeating disaccharide units, HA can either be directly conjugated to drugs or can form self-assembling micelles to generate amphilic nanoparticles. This allows efficient delivery of hydrophobic drugs to target sites, thus improving their bioavailability and half-life. Additional advantages of HA nanosystems include low to no immunogenicity, noninflammatory properties, and good biocompatibility, bioavailability, and biodegradability (Choi, Saravanakumar, Park, & Park, 2012). HA also shows great versatility, as it bears many functional groups that can be used for various modifications and conjugations. All of these properties make HA a central constituent of multifunctional nanoparticles to deliver combination and synergistic therapies. Most importantly, since HA has a strong binding affinity for RHAMM and CD44—two receptors that are commonly overexpressed in various solid tumors and cancer stem cells—the HA nanoparticles can be used for the targeted delivery of chemotherapeutic drugs and other novel treatments (Jaracz, Chen, Kuznetsova, & Ojima, 2005). Many studies have shown successful use of HA in nanoparticle formations, both *in vitro* and *in vivo*.

Direct conjugation of HA to anticancer drugs is a simple yet effective nanoparticle formulation used to improve treatment efficacy. This method involves covalently linking HA to a drug, often using a bond that can readily be cleaved after reaching the target site. The anticancer drug then gains improvement in parameters such as stability, circulation time, solubility, and tumor-targeting specificity (Choi et al., 2012; Jaracz et al., 2005). Small organic or inorganic ions, when conjugated with HA, show increased stability and can be targeted to tumors. For example, the HA-selenium nanoparticle system shows increased stability in circulation and antitumor efficacy (Ren et al., 2013). The common chemotherapy drugs DOX and paclitaxel have also demonstrated increased efficacy and reduced dose-limiting toxicity when incorporated in HA nanoparticles. In the DMBA-induced rat mammary model, HA-paclitaxel significantly increased the antitumor efficacy of paclitaxel than when it was administered alone (Al-Ghananeem et al., 2009).

HA has also been used to generate amphiphilic polymeric nanoparticles that can self-assemble. In these nanoparticles, amphiphilic nanoparticles' hydrophobic ends encapsulate the hydrophobic anticancer drug, and the hydrophilic ends form a protective coating in the form of a spherical micelle. Such coating protects the drug from degradation or absorption until the micelle's contents are released within the target cell (Choi et al., 2010, 2012). To improve the amphilic nature of the HA nanoparticles, HA has been successfully conjugated to various hydrophobic compounds. For example, HA has been conjugated to hydrophobic groups such as 5β-cholanic acid and polyethylene glycol to encapsulate Chlorin e6, a hydrophobic photosensitizer used in photodynamic therapy and imaging. Chlorin e6 was successfully loaded onto HA-based micelles with a high loading capacity and exhibited both improved cancer imaging capabilities and tumor inhibition under laser irradiation (Yoon et al., 2012). Similarly, intravenously delivered DOX conjugated to HA-coated superparamagnetic iron oxide nanoparticles was found to accumulate at much higher levels and distribute more widely in the tumor tissue than DOX delivered alone. This resulted in decreased tumor growth of human SKOV-3 xenografts in mice and increased their lifespan (El-Dakdouki et al., 2014).

HA has also been conjugated to other groups that provide more than the ability to self-assemble. For example, HA has been conjugated with poly(L-histidine) for selective delivery of chemotherapeutic drugs such as DOX. Since poly(L-histidine) is a pH-responsive polymer that destabilizes through acidic protonation, the HA-poly(L-histidine) nanoparticles allow sustained drug release specifically in tumor tissues, which naturally have a lower pH (Qiu et al., 2013). Yet another option for the HA nanosystem modification is the addition of other tumor-targeting moieties. As folate receptors are overexpressed in certain tumor types, one study conjugated folate to the outer shell of a HA-octadecyl(C18)-based nanosystem. While both the folate receptor-conjugated and nonconjugated HA-octadecyl (C18) nanoparticles were internalized by MCF-7 breast cancer cells via CD44-mediated endocytosis, folate receptors further enhanced the endocytosis of folate-HA-C18 nanoparticles. However, in this study the *in vivo* efficacy of both nanoparticles was not evaluated (Liu et al., 2011).

As HA holds many functional groups that can be substituted, there is potential for the conjugation of more than one treatment to HA nanosystems, allowing for synergistic combination therapies. In one example, HA was conjugated with all-*trans* retinoid acid, a hydrophobic anticancer drug, to form micelles around paclitaxel. This HA nanosystem was found to have low nonspecific toxicity and high antitumor efficacy in the B16 melanoma model (Yao, Zhang, Zhou, Liu, & Zhang, 2013). Another example of synergistic therapy via HA nanoparticles involves adsorption of a photodynamic therapy agent, hematoporphyrin monomethyl ether, onto HA-derivatized carbon nanotubes. Carbon nanotubes alone have been used previously for photothermal therapy and drug delivery. When conjugated to HA, the carbon nanotubes offer a surface for aromatic hematoporphyrin monomethyl ether to associate with the core of the nanosystem. This combined nanosystem has increased solubility, neutral pH, and targeted delivery. The HA-CNT-HMME nanosystem exhibited effective tumor control in the B16 melanoma model using both photothermal and photodynamic therapies with synergistic effects and reduced nontumor toxicity (Shi et al., 2013).

Finally, hydrophilic treatments can also benefit from HA-based nanocarriers through the formation of ionic nanocomplexes in place of covalent bonds (Choi et al., 2012). The positively charged protein TRAIL can associate with HA through simple mixing, and the HA-TRAIL complexes display increased stability in circulation and higher bioavailability to improve the anticancer effects of TRAIL (Na et al., 2008). Even RNA interference therapy can benefit from forming nanocomplexes with HA, despite sharing a negative charge. Compared to the naked siRNAs that are easily degraded or excreted, and with viral delivery vectors having potential toxicity and immunogenicity drawbacks, HA nanoparticles appear to be an appealing alternative carrier for siRNAs. One study conjugated HA to various polyamines, mainly polyethyleneimine (PEI). This conjugation, when mixed with siRNAs, formed a self-assembling system with a negatively charged surface and a positively charged core. Through associations with HA-polyethylene glycol nanosystems (HA-PEG), the study found a successful balance of charges in the nanosystem of HA-PEG/HA-PEI/siRNA. This combination allowed for successful tumor-specific uptake via CD44 receptor-mediated endocytosis, along with facilitated endosome escape to release siRNAs into the tumor cell cytoplasm (Ganesh, Iyer, Morrissey, & Amiji, 2013). This system showed the highest target gene knockdown, attesting once more to the versatility, biocompatibility, and importance of HA in nanosystems.

In summary, HA nanoparticles are an attractive delivery system for effective deliveries of antitumor agents, as HA serves both as a tumor-targeting moiety and a drug carrier that combats side effects. For these reasons, HA can potentially be exploited for cancer therapy.

5. TARGETING HA RECEPTORS

5.1. CD44

Since the discovery that CD44 is a stem cell marker in addition to being a HA receptor, targeting CD44 for anticancer therapy has been attempted using DNA vaccines, anti-CD44 monoclonal antibody, and nanoparticle-mediated delivery of CD44siRNA approaches.

5.1.1 CD44 vaccines

CD44 cDNA vaccine has been delivered by implanting virtual lymph nodes in immunocompetent animals to generate anti-CD44 antibodies. These virtual lymph nodes are generated by insertion of CD44-standard or

CD44-variant cDNA into a silicone tube filled with 2.5 cm long segment of hydroxylated-polyvinyl acetate wound dressing sponge followed by subcutaneous injection in mice. Using this model, CD44-standard form vaccination was shown to reduce autoimmune encephalomyelitis by induction of apoptosis (Garin et al., 2007). In a similar approach with virtual lymph nodes, effects of CD44 vaccination on tumor growth and lung metastasis were evaluated in a mouse mammary adenocarcinoma model (DA3 cells). The mouse models were injected with virtual lymph nodes containing human CD44 variants (v3–v10) or CD44-standard cDNAs. Immunization resulted in the expression of antibodies against human CD44-variant and CD44-standard forms. In 75% of the animals, mice expressing CD44-variant (v3–v10) antibodies showed a greater inhibition of tumor growth than mice expressing CD44-standard antibodies. The remaining 25% of animals showed reduced tumor growth, and metastasis was eliminated in all animals. In this study, human CD44 cDNA was used to avoid interference from mouse serum proteins; however, this raises a question of whether the injection of human CD44 cDNA in a clinical setting will generate enough antibody response. It is also intriguing that CD44-standard form did not generate the same immune response against tumors (Wallach-Dayan, Rubinstein, Hand, Breuer, & Naor, 2008). Vaccination of animals with dendritic cells and B16 cells coated with anti-CD44 antibodies inhibited lung metastasis (50% reduction) and delayed tumor growth of B16 melanoma cells; 60% of mice remained tumor-free for 8 months. In these models, vaccination induced CD8 T cell activation (Pilon-Thomas, Verhaegen, Kuhn, Riker, & Mule, 2006).

5.1.2 CD44 siRNA delivery

CD44 siRNAs are usually delivered to tumor cells using nanoparticles. For example, biodegradable poly(D,L-lactide-*co*-glycolide) acid nanoparticles have been used to simultaneously deliver CD44 and FAK siRNAs to ovarian cancer xenografts. Knockdown of both genes reduced tumor growth by inhibiting angiogenesis and proliferation in tumors (Zou et al., 2013). More recently, a nanoscale-based drug delivery system was tested in an ovarian cancer xenograft model. The nanoscale delivery system contained a modified polypropylenimine dendrimer as a carrier, paclitaxel, a synthetic analog of luteinizing hormone-releasing hormone peptide for targeting tumor cells, and siRNA targeted to CD44 mRNA. This dendrimer was able to downregulate CD44 mRNA and protein expression and inhibit tumor growth without toxicity, suggesting that the targeted delivery of CD44 siRNA along with chemotherapeutic agents may be explored for cancer therapy (Shah et al., 2013).

5.1.3 Targeting CD44 for delivering antitumor therapies

Overexpression of CD44 in tumor cells has also been exploited to deliver cytotoxic drugs and siRNAs using HA-coated self-assembling nanoparticles or liposomes. For example, DOX-loaded, HA-coated silica nanoparticles containing a highly fluorescent core to target CD44 have been recently developed for targeting tumor cells. However, their targeted delivery, uptake, and efficacy in experimental tumor models have not yet been evaluated (El-Dakdouki, Pure, & Huang, 2013). Similarly, ONCOFID™-S, a new bioconjugate HA that targets CD44-overexpressing tumor cells, can deliver cytotoxic agents and has shown high antitumor efficacy both *in vitro* and *in vivo* (Serafino et al., 2011). In the ONCOFID™-S bioconjugate, antitumor drugs are covalently bound to HA. These HA-encapsulated siRNA nanoparticles, chemotherapeutic agent-encapsulated liposomes, or covalently bound HA-bioconjugates are selectively taken up by tumors in systemic delivery, and reduce specific gene expression in several xenograft models. The uptake of these nanoparticles or liposomes by tumor cells occurs through HA-receptor-mediated internalization, since incubation of tumor cells with soluble HA inhibits uptake of these materials. This suggests that the delivery of these nanoparticles is occurring through HA-receptor-mediated internalization of nanoparticles and liposomes (Dalla Pozza et al., 2013; Ganesh, Iyer, Gattacceca, Morrissey, & Amiji, 2013).

5.1.4 Targeting of CD44 protein

Anti-CD44 monoclonal antibodies have been used to inhibit tumor growth in experimental model systems. Targeting of CD44 by specific antibodies induces apoptosis in T lymphoma cells by the intrinsic pathway, which, in turn, is dependent on the activation of CD44-associated protein phosphatase 2A (Rajasagi et al., 2010). In chronic lymphocytic leukemia cells, that express high levels of CD44, a humanized monoclonal antibody specific for CD44 (RG7356) was found to be cytotoxic to leukemia B cells without affecting the viability of normal B cells. Systemic administration of this antibody caused complete clearance of leukemia xenografts. Interestingly, the effects of the antibody were not neutralized in the presence of HA, suggesting that CD44 may have functions other than binding to HA that also play a role in leukemia cell growth (Zhang et al., 2013).

As in the case of HA-coated nanoparticles or liposomes, anti-CD44 antibodies have also been used to either image tumors or for delivering chemotherapeutic agents to experimental tumor models. For example, a chimeric monoclonal antibody U36 (cMAb U36) and its F(ab′)2 and Fab′ fragments that recognize the CD44v6 isoform have shown potential for radioimmunotherapy and radioimmunotargeting of experimental tumors (Sandstrom et al., 2012, 2008). In clinical studies, anti-CD44 antibodies have been used to deliver radioisotopes or mertansine for treatment of CD44-expressing tumors. In these studies, disease stabilization was observed in breast or head and neck tumor patients; however, dose-limiting toxicity was observed with distribution of antibody in the skin, where high levels of CD44 are expressed (Platt & Szoka, 2008). In phase I studies, maximum tolerated dose, safety, and efficacy of the immunoconjugate BIWI 1 (bivatuzumab mertansine), which consists of a highly potent antimicrotubule agent coupled to an anti-CD44v6 monoclonal antibody, was evaluated in head and neck cancer patients. In this study, while three patients showed a partial response, binding of BIWI to CD44v6 on skin keratinocytes mediated serious skin toxicity with a fatal outcome, leading to early termination of this trial (Riechelmann et al., 2008). The HA binding domain is highly conserved in all CD44 isoforms, and therefore, attempts have been made to target this domain using specific monothiophosphate-modified aptamers. These aptamers bind specifically to CD44-expressing tumor cells with high efficacy. However, the *in vivo* bioavailability of these aptamers has not been examined in detail (Somasunderam et al., 2010).

Taken together, CD44 has been targeted for cancer therapy and cancer imaging through vaccination, specific antibodies, siRNAs, and aptamers. However, since targeting of CD44 using CD44v6 monoclonal antibodies in clinical trials caused serious adverse reactions, risk assessment of CD44 targeting by any one of the approaches discussed above needs to be carefully evaluated before using that approach for cancer therapy.

5.2. RHAMM

Since RHAMM is overexpressed in a variety of carcinomas, including acute leukemias, RHAMM vaccines have been tested in phase I and II clinical trials for chronic lymphocytic leukemia and acute myeloid leukemia (AML). Initially, a vaccine was constructed by injecting dendritic cells transfected with *in vitro* transcribed RHAMM mRNA into mice implanted with glioma cells. The vaccine improved the survival of tumor-bearing mice by threefold and histological analysis showed activation of CD4+, CD8+, and CD25+ T lymphocytes (Amano et al., 2007). RHAMM is known to elicit both humoral and cellular immune responses. Efficacy of a peptide-based RHAMM vaccine has been examined in phase I/II clinical trials. In a phase I clinical trial, 10 patients with AML, myelodysplastic syndrome, or multiple myeloma, expressing HLA-A2 and RHAMM expression on tumor cells, were vaccinated with a RHAMM peptide R3 (aa 165–173; ILSLELMKL). Subcutaneous administration of this vaccine increased R3-specific CD8+ T cells and three patients showed partial response, including a significant reduction of blasts in the bone marrow (Schmitt et al., 2008). High dose of the same R3-peptide vaccine again showed immunological responses, including an increase in R3-specific CD8+ cells and regulatory T cells. About 35% of patients also showed a partial clinical response. However, both the clinical and immunological responses were lower than when the vaccine was administered at a lower dose (1 mg vs. 0.3 mg; Greiner et al., 2010). The vaccine had similar immunological (i.e., increase in R3-specific CD8+ cells) and clinical response in patients with chronic lymphocytic leukemia (Giannopoulos et al., 2010). Interestingly, although patients with CML are characterized as having immunosuppression, the vaccine caused profound changes (mostly an increase) in different T cell subsets and in cytokines that are involved in immune response (i.e., TGF-β, IL-10, IL-2, and TNF) during the vaccination period (Giannopoulos, Wlasiuk, Dmoszynska, Rolinski, & Schmitt, 2011). However, despite the early success with the R3-peptide vaccine, it was recently shown that only baseline RHAMM expression was present in both leukemic stem cells (CD34+CD38−) and hematopoietic stem cells (CD34+CD38−) from healthy control cells. In addition, activation of T cells also increased RHAMM expression. Therefore, the study concluded that RHAMM is not a suitable target for immunotherapy (Snauwaert et al., 2012). Considering CD44 vaccine trials were terminated due to toxicity and RHAMM is ubiquitously expressed and essential for normal cellular functions, RHAMM vaccination for cancer therapy may need more work before it can be translated into clinic.

In summary, although targeting of CD44 and RHAMM appears to be a conceptually attractive strategy, considering its effects on normal cells, and the immune system, the risk versus benefit must be carefully evaluated before clinical usage.

6. TARGETING HAase

As discussed above and in Chapter 1, HAase has been shown to be a critical determinant of cancer growth, metastasis, and angiogenesis. HAase activity and HYAL-1 expression are potentially independent predictors of clinical outcome in several carcinomas. Therefore, targeting HAase activity and/or HYAL-1 expression could potentially be an effective strategy for cancer therapy. Many synthetic and naturally occurring compounds have been tested as HAase inhibitors. These include high molecular mass poly(styrene-4-sulfonate), gossypol, sodium aurothiomalate, fenoprofen, glycerrhizic acid, fatty acids, plant-derived compounds, heparin, and O-sulfated HA (sHA). The activities of most of these inhibitors have been tested against testicular HAase. For example, more than 60 years ago, O-sHA was reported as an inhibitor of testicular HAase (Balazs, Hogberg, & Laurent, 1951; Mio & Stern, 2002). As a result, HAase inhibitors were tried as contraceptives (Balazs et al., 1951; Joyce, Mack, Anderson, & Zaneveld, 1986; Mio & Stern, 2002). In a study of 21 inhibitors, sHA polymers with varying degrees of sulfation were found to be the most effective inhibitors of HYAL-1 (IC_{50} 0.0083–0.019 μM) (Isoyama et al., 2006). The presence of sulfate, and not the degree of sulfation, was found to be the determinant of HAase inhibitory activity displayed by sHA. sHA derivatives inhibited HYAL-1 by a mixed inhibition mechanism (i.e., competitive + uncompetitive) and were 5- to 17-fold better as uncompetitive inhibitors than as competitive inhibitors (Isoyama et al., 2006). Since sHA is a natural compound and is derived from HA, its antitumor activity has been examined in prostate cancer models. sHA with 2.75 degrees of sulfation was found to inhibit prostate cancer cell growth through induction of apoptosis. sHA inhibited HA signaling, with inhibition of PI3-kinase/Akt pathway as the major HA signaling target. The study also showed that there is a feedback loop between Akt and HA receptors. As a result of downregulated Akt activation following sHA treatment, HA receptors were transcriptionally downregulated. Moreover, sHA inhibited the growth of an androgen-independent prostate cancer model, by inhibition of angiogenesis and increased apoptosis in tumor tissues. The tumor inhibitory effects of sHA were accompanied by lack of any serum or organ toxicity (Benitez et al., 2011). The model of sHA action on tumor cells is shown in Fig. 2.2.

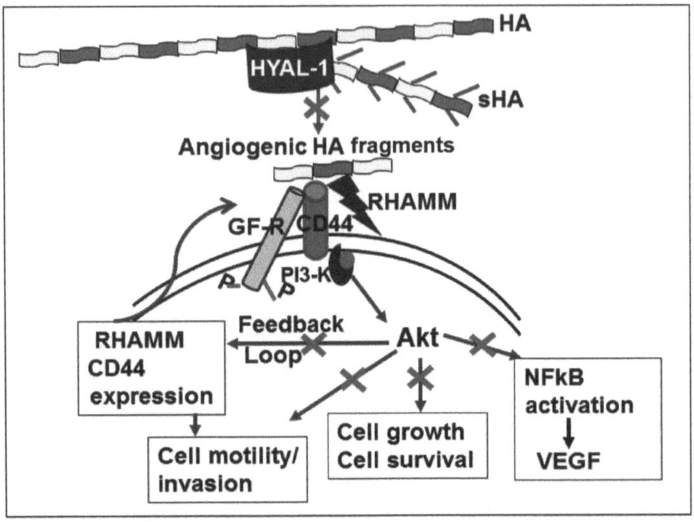

Figure 2.2 Molecular basis for the antitumor activity of sHA. Tumor-derived HAase, HYAL-1, degrades tumor-associated pericellular HA into angiogenic fragments. These fragments induce HA signaling events described in Fig. 2.1. By inhibiting HYAL-1 activity, sHA blocks the formation of angiogenic HA fragments and hence inhibits HA signaling. The antitumor activity of sHA has been shown to be mediated by inhibition of PI3-kinase/Akt activation and downstream effectors, leading to inhibition of cell survival, proliferation, motility/invasion, and angiogenesis. *Adapted from Benitez et al. (2011).* (See the color plate.)

Taken together, targeting of tumor-derived HAase may be an attractive strategy for controlling tumor growth and progression.

7. CONCLUSION

The tumor-associated HA-HAase system is an attractive target for cancer therapy, since it regulates major intracellular signaling pathways such as PI3-kinase/Akt, Erk, ras, src, and processes such as epithelial–mesenchymal transition, that impact tumor cell proliferation, invasive phenotype, and angiogenic potential. Among the targeted therapies described in this review, targeting of HA synthesis by a small molecule, 4-MU, seems most promising. Several studies have demonstrated that 4-MU effectively inhibits tumor growth, metastasis, and angiogenesis in experimental tumor models and inhibits a variety of HA signaling pathways. Since 4-MU is already consumed as a dietary supplement at doses comparable to those at which it is effective as an antitumor agent in experimental models, it can be rapidly translated into clinic for cancer therapy. oHA of

<10 disaccharides have also shown efficacy in controlling tumor growth and metastasis; however, the challenge for their use in cancer therapy will be to carefully control the length of oHA, since slightly longer HA oligosaccharides can promote tumor growth and metastasis by stimulating HA signaling. The use of HA nanosystems also holds promise for the targeted and safe delivery of chemotherapeutic drugs and other anticancer compounds to tumor cells. Since HA nanoparticles can improve the half-life of anticancer agents and focus the delivery to cells overexpressing HA receptors, many anticancer therapies can potentially be more effective at smaller doses and, thus, benefit from reduced drug-related toxicities. Targeting of HA receptors by antibodies or vaccines has been tested in clinical trials. Although this is potentially an effective approach, serious toxicities have been associated with targeting of HA receptors. Tumor-derived HAase, HYAL-1, has shown potential to be an independent prognostic indicator for disease progression, recurrence, metastasis, and survival and, therefore, a natural target for cancer therapy. However, only the HAase inhibitor, sHA, has been tested and found to have antitumor activity without toxicity in one cancer model. The growing body of literature is demonstrating the involvement of HA family of molecules in tumor growth and progression, and their role as biomarkers in virtually all types of cancers. This will be a great impetus for testing the experimental therapeutic strategies discussed in this review, as well as furthering many new HA-targeted therapies to enter clinical trials for specific tumor types.

ACKNOWLEDGMENTS

This work was supported by grants NCI/NIH 1R01CA176691-01 (V. B. L.), NCI/NIH R01 72821-14 (V. B. L.), and 1R21CA184018-01 (V. B. L.).

REFERENCES

Abate, A., Dimartino, V., Spina, P., Costa, P. L., Lombardo, C., Santini, A., et al. (2001). Hymecromone in the treatment of motor disorders of the bile ducts: A multicenter, double-blind, placebo-controlled clinical study. *Drugs Under Experimental and Clinical Research*, *27*(5–6), 223–231.

Adamia, S., Pilarski, P. M., Belch, A. R., & Pilarski, L. M. (2013). Aberrant splicing, hyaluronan synthases and intracellular hyaluronan as drivers of oncogenesis and potential drug targets. *Current Cancer Drug Targets*, *13*(4), 347–361.

Adamia, S., Reiman, T., Crainie, M., Mant, M. J., Belch, A. R., & Pilarski, L. M. (2005). Intronic splicing of hyaluronan synthase 1 (HAS1): A biologically relevant indicator of poor outcome in multiple myeloma. *Blood*, *105*(12), 4836–4844. http://dx.doi.org/10.1182/blood-2004-10-3825.

Alaniz, L., Garcia, M. G., Gallo-Rodriguez, C., Agusti, R., Sterin-Speziale, N., Hajos, S. E., et al. (2006). Hyaluronan oligosaccharides induce cell death through PI3-K/Akt pathway independently of NF-kappaB transcription factor. *Glycobiology*, *16*(5), 359–367. http://dx.doi.org/10.1093/glycob/cwj085.

Al-Ghananeem, A. M., Malkawi, A. H., Muammer, Y. M., Balko, J. M., Black, E. P., Mourad, W., et al. (2009). Intratumoral delivery of Paclitaxel in solid tumor from biodegradable hyaluronan nanoparticle formulations. *AAPS PharmSciTech*, *10*(2), 410–417. http://dx.doi.org/10.1208/s12249-009-9222-5.

Amano, T., Kajiwara, K., Yoshikawa, K., Morioka, J., Nomura, S., Fujisawa, H., et al. (2007). Antitumor effects of vaccination with dendritic cells transfected with modified receptor for hyaluronan-mediated motility mRNA in a mouse glioma model. *Journal of Neurosurgery*, *106*(4), 638–645. http://dx.doi.org/10.3171/jns.2007.106.4.638.

Arai, E., Nishida, Y., Wasa, J., Urakawa, H., Zhuo, L., Kimata, K., et al. (2011). Inhibition of hyaluronan retention by 4-methylumbelliferone suppresses osteosarcoma cells in vitro and lung metastasis in vivo. *British Journal of Cancer*, *105*(12), 1839–1849. http://dx.doi.org/10.1038/bjc.2011.459.

Auvinen, P., Rilla, K., Tumelius, R., Tammi, M., Sironen, R., Soini, Y., et al. (2014). Hyaluronan synthases (HAS1-3) in stromal and malignant cells correlate with breast cancer grade and predict patient survival. *Breast Cancer Research and Treatment*, *143*(2), 277–286. http://dx.doi.org/10.1007/s10549-013-2804-7.

Balazs, E. A., Hogberg, B., & Laurent, T. C. (1951). The biological activity of hyaluron sulfuric acid. *Acta Physiologica Scandinavica*, *23*(2–3), 168–178. http://dx.doi.org/10.1111/j.1748-1716.1951.tb00806.x.

Benitez, A., Yates, T. J., Lopez, L. E., Cerwinka, W. H., Bakkar, A., & Lokeshwar, V. B. (2011). Targeting hyaluronidase for cancer therapy: Antitumor activity of sulfated hyaluronic acid in prostate cancer cells. *Cancer Research*, *71*(12), 4085–4095. http://dx.doi.org/10.1158/0008-5472.CAN-10-4610.

Benitez, A., Yates, T. J., Shamaldevi, N., Bowen, T., & Lokeshwar, V. B. (2013). Dietary supplement hymecromone and sorafenib: A novel combination for the control of renal cell carcinoma. *Journal of Urology*, *190*(1), 285–290. http://dx.doi.org/10.1016/j.juro.2012.12.011.

Bernert, B., Porsch, H., & Heldin, P. (2011). Hyaluronan synthase 2 (HAS2) promotes breast cancer cell invasion by suppression of tissue metalloproteinase inhibitor 1 (TIMP-1). *Journal of Biological Chemistry*, *286*(49), 42349–42359. http://dx.doi.org/10.1074/jbc.M111.278598.

Bharadwaj, A. G., Goodrich, N. P., McAtee, C. O., Haferbier, K., Oakley, G. G., Wahl, J. K., 3rd., et al. (2011). Hyaluronan suppresses prostate tumor cell proliferation through diminished expression of N-cadherin and aberrant growth factor receptor signaling. *Experimental Cell Research*, *317*(8), 1214–1225. http://dx.doi.org/10.1016/j.yexcr.2011.01.026.

Bharadwaj, A. G., Kovar, J. L., Loughman, E., Elowsky, C., Oakley, G. G., & Simpson, M. A. (2009). Spontaneous metastasis of prostate cancer is promoted by excess hyaluronan synthesis and processing. *American Journal of Pathology*, *174*(3), 1027–1036. http://dx.doi.org/10.2353/ajpath.2009.080501.

Bhattacharyya, S. S., Paul, S., Mandal, S. K., Banerjee, A., Boujedaini, N., & Khuda-Bukhsh, A. R. (2009). A synthetic coumarin (4-methyl-7 hydroxy coumarin) has anti-cancer potentials against DMBA-induced skin cancer in mice. *European Journal of Pharmacology*, *614*(1–3), 128–136. http://dx.doi.org/10.1016/j.ejphar.2009.04.015.

Bouga, H., Tsouros, I., Bounias, D., Kyriakopoulou, D., Stavropoulos, M. S., Papageorgakopoulou, N., et al. (2010). Involvement of hyaluronidases in colorectal cancer. *BMC Cancer*, *10*, 499. http://dx.doi.org/10.1186/1471-2407-10-499.

Bourguignon, L. Y., Wong, G., Earle, C., Krueger, K., & Spevak, C. C. (2010). Hyaluronan-CD44 interaction promotes c-Src-mediated twist signaling, microRNA-10b expression, and RhoA/RhoC up-regulation, leading to Rho-kinase-associated cytoskeleton activation and breast tumor cell invasion. *Journal of Biological Chemistry*, *285*(47), 36721–36735. http://dx.doi.org/10.1074/jbc.M110.162305.

Camarri, E., & Marchettini, G. (1988). Hymecromone in the treatment of symptoms following surgery of the bile ducts. *Recenti Progressi in Medicina*, *79*(5), 198–202.

Chao, K. L., Muthukumar, L., & Herzberg, O. (2007). Structure of human hyaluronidase-1, a hyaluronan hydrolyzing enzyme involved in tumor growth and angiogenesis. *Biochemistry*, *46*(23), 6911–6920. http://dx.doi.org/10.1021/bi700382g.

Chen, S., Cho, M., Karlsberg, K., Zhou, D., & Yuan, Y. C. (2004). Biochemical and biological characterization of a novel anti-aromatase coumarin derivative. *Journal of Biological Chemistry*, *279*(46), 48071–48078. http://dx.doi.org/10.1074/jbc.M406847200.

Chi, A., Shirodkar, S. P., Escudero, D. O., Ekwenna, O. O., Yates, T. J., Ayyathurai, R., et al. (2012). Molecular characterization of kidney cancer: Association of hyaluronic acid family with histological subtypes and metastasis. *Cancer*, *118*(9), 2394–2402. http://dx.doi.org/10.1002/cncr.26520.

Choi, K. Y., Chung, H., Min, K. H., Yoon, H. Y., Kim, K., Park, J. H., et al. (2010). Self-assembled hyaluronic acid nanoparticles for active tumor targeting. *Biomaterials*, *31*(1), 106–114. http://dx.doi.org/10.1016/j.biomaterials.2009.09.030.

Choi, K. Y., Saravanakumar, G., Park, J. H., & Park, K. (2012). Hyaluronic acid-based nanocarriers for intracellular targeting: Interfacial interactions with proteins in cancer. *Colloids and Surfaces B: Biointerfaces*, *99*, 82–94. http://dx.doi.org/10.1016/j.colsurfb.2011.10.029.

Chow, G., Tauler, J., & Mulshine, J. L. (2010). Cytokines and growth factors stimulate hyaluronan production: Role of hyaluronan in epithelial to mesenchymal-like transition in non-small cell lung cancer. *Journal of Biomedicine and Biotechnology*, *2010*, 485468. http://dx.doi.org/10.1155/2010/485468.

Clarkin, C. E., Allen, S., Wheeler-Jones, C. P., Bastow, E. R., & Pitsillides, A. A. (2011). Reduced chondrogenic matrix accumulation by 4-methylumbelliferone reveals the potential for selective targeting of UDP-glucose dehydrogenase. *Matrix Biology*, *30*(3), 163–168. http://dx.doi.org/10.1016/j.matbio.2011.01.002.

Cordo Russo, R. I., Garcia, M. G., Alaniz, L., Blanco, G., Alvarez, E., & Hajos, S. E. (2008). Hyaluronan oligosaccharides sensitize lymphoma resistant cell lines to vincristine by modulating P-glycoprotein activity and PI3K/Akt pathway. *International Journal of Cancer*, *122*(5), 1012–1018. http://dx.doi.org/10.1002/ijc.23122.

Crooke, D., Fitzpatrick, B., O'Kennedy, R., & McCormack, T. (1997). Coumarins-multifaceted molecules with many analytical and other applications. In R. O'Kennedy, & R. D. Thome (Eds.), *Coumarins biology, applications and mode of action* (pp. 303–331). Chichester: Wiley.

Dalla Pozza, E., Lerda, C., Costanzo, C., Donadelli, M., Dando, I., Zoratti, E., et al. (2013). Targeting gemcitabine containing liposomes to CD44 expressing pancreatic adenocarcinoma cells causes an increase in the antitumoral activity. *Biochimica et Biophysica Acta*, *1828*(5), 1396–1404. http://dx.doi.org/10.1016/j.bbamem.2013.01.020.

Dortet, L., Veiga, E., Bonazzi, M., & Cossart, P. (2010). CD44-independent activation of the Met signaling pathway by HGF and InlB. *Microbes and Infection*, *12*(12–13), 919–927. http://dx.doi.org/10.1016/j.micinf.2010.07.009.

Dunn, K. M., Lee, P. K., Wilson, C. M., Iida, J., Wasiluk, K. R., Hugger, M., et al. (2009). Inhibition of hyaluronan synthases decreases matrix metalloproteinase-7 (MMP-7) expression and activity. *Surgery*, *145*(3), 322–329. http://dx.doi.org/10.1016/j.surg.2008.11.008.

El-Dakdouki, M. H., Pure, E., & Huang, X. (2013). Development of drug loaded nanoparticles for tumor targeting. Part 1: Synthesis, characterization, and biological evaluation in 2D cell cultures. *Nanoscale*, *5*(9), 3895–3903. http://dx.doi.org/10.1039/c3nr33777d.

El-Dakdouki, M. H., Xia, J., Zhu, D. C., Kavunja, H., Grieshaber, J., O'Reilly, S., et al. (2014). Assessing the in vivo efficacy of doxorubicin loaded hyaluronan nanoparticles. *ACS Applied Materials & Interfaces*, *6*(1), 697–705. http://dx.doi.org/10.1021/am404946v.

Franzmann, E. J., Schroeder, G. L., Goodwin, W. J., Weed, D. T., Fisher, P., & Lokeshwar, V. B. (2003). Expression of tumor markers hyaluronic acid and hyaluronidase (HYAL1) in head and neck tumors. *International Journal of Cancer*, *106*(3), 438–445. http://dx.doi.org/10.1002/ijc.11252.

Fuchs, K., Hippe, A., Schmaus, A., Homey, B., Sleeman, J. P., & Orian-Rousseau, V. (2013). Opposing effects of high- and low-molecular weight hyaluronan on CXCL12-induced CXCR4 signaling depend on CD44. *Cell Death and Disease*, *4*, e819. http://dx.doi.org/10.1038/cddis.2013.364.

Ganesh, S., Iyer, A. K., Gattacceca, F., Morrissey, D. V., & Amiji, M. M. (2013). In vivo biodistribution of siRNA and cisplatin administered using CD44-targeted hyaluronic acid nanoparticles. *Journal of Controlled Release*, *172*(3), 699–706. http://dx.doi.org/10.1016/j.jconrel.2013.10.016.

Ganesh, S., Iyer, A. K., Morrissey, D. V., & Amiji, M. M. (2013). Hyaluronic acid based self-assembling nanosystems for CD44 target mediated siRNA delivery to solid tumors. *Biomaterials*, *34*(13), 3489–3502. http://dx.doi.org/10.1016/j.biomaterials.2013.01.077.

Garin, T., Rubinstein, A., Grigoriadis, N., Nedvetzki, S., Abramsky, O., Mizrachi-Koll, R., et al. (2007). CD44 variant DNA vaccination with virtual lymph node ameliorates experimental autoimmune encephalomyelitis through the induction of apoptosis. *Journal of Neurological Sciences*, *258*(1–2), 17–26. http://dx.doi.org/10.1016/j.jns.2007.01.079.

Garrett, E. R., Venitz, J., Eberst, K., & Cerda, J. J. (1993). Pharmacokinetics and bioavailabilities of hymecromone in human volunteers. *Biopharmaceutics and Drug Disposition*, *14*(1), 13–39.

Ghatak, S., Misra, S., & Toole, B. P. (2002). Hyaluronan oligosaccharides inhibit anchorage-independent growth of tumor cells by suppressing the phosphoinositide 3-kinase/Akt cell survival pathway. *Journal of Biological Chemistry*, *277*(41), 38013–38020. http://dx.doi.org/10.1074/jbc.M202404200.

Ghosh, A., Kuppusamy, H., & Pilarski, L. M. (2009). Aberrant splice variants of HAS1 (Hyaluronan Synthase 1) multimerize with and modulate normally spliced HAS1 protein: A potential mechanism promoting human cancer. *Journal of Biological Chemistry*, *284*(28), 18840–18850. http://dx.doi.org/10.1074/jbc.M109.013813.

Giannopoulos, K., Dmoszynska, A., Kowal, M., Rolinski, J., Gostick, E., Price, D. A., et al. (2010). Peptide vaccination elicits leukemia-associated antigen-specific cytotoxic CD8+ T-cell responses in patients with chronic lymphocytic leukemia. *Leukemia*, *24*(4), 798–805. http://dx.doi.org/10.1038/leu.2010.29.

Giannopoulos, K., Wlasiuk, P., Dmoszynska, A., Rolinski, J., & Schmitt, M. (2011). Peptide vaccination induces profound changes in the immune system in patients with B-cell chronic lymphocytic leukemia. *Folia Histochemica et Cytobiologica*, *49*(1), 161–167.

Golshani, R., Hautmann, S. H., Estrella, V., Cohen, B. L., Kyle, C. C., Manoharan, M., et al. (2007). HAS1 expression in bladder cancer and its relation to urinary HA test. *International Journal of Cancer*, *120*(8), 1712–1720. http://dx.doi.org/10.1002/ijc.22222.

Golshani, R., Lopez, L., Estrella, V., Kramer, M., Iida, N., & Lokeshwar, V. B. (2008). Hyaluronic acid synthase-1 expression regulates bladder cancer growth, invasion, and angiogenesis through CD44. *Cancer Research*, *68*(2), 483–491. http://dx.doi.org/10.1158/0008-5472.CAN-07-2140.

Gomez, C. S., Gomez, P., Knapp, J., Jorda, M., Soloway, M. S., & Lokeshwar, V. B. (2009). Hyaluronic acid and HYAL-1 in prostate biopsy specimens: Predictors of biochemical recurrence. *Journal of Urology*, *182*(4), 1350–1356. http://dx.doi.org/10.1016/j.juro.2009.06.070.

Greiner, J., Schmitt, A., Giannopoulos, K., Rojewski, M. T., Gotz, M., Funk, I., et al. (2010). High-dose RHAMM-R3 peptide vaccination for patients with acute myeloid leukemia, myelodysplastic syndrome and multiple myeloma. *Haematologica*, *95*(7), 1191–1197. http://dx.doi.org/10.3324/haematol.2009.014704.

Gurski, L. A., Xu, X., Labrada, L. N., Nguyen, N. T., Xiao, L., van Golen, K. L., et al. (2012). Hyaluronan (HA) interacting proteins RHAMM and hyaluronidase impact prostate cancer cell behavior and invadopodia formation in 3D HA-based hydrogels. *PLoS One*, 7(11), e50075. http://dx.doi.org/10.1371/journal.pone.0050075.

Hatano, H., Shigeishi, H., Kudo, Y., Higashikawa, K., Tobiume, K., Takata, T., et al. (2011). RHAMM/ERK interaction induces proliferative activities of cementifying fibroma cells through a mechanism based on the CD44-EGFR. *Laboratory Investigation*, *91*(3), 379–391. http://dx.doi.org/10.1038/labinvest.2010.176.

Hiraga, T., Ito, S., & Nakamura, H. (2013). Cancer stem-like cell marker CD44 promotes bone metastases by enhancing tumorigenicity, cell motility, and hyaluronan production. *Cancer Research*, *73*(13), 4112–4122. http://dx.doi.org/10.1158/0008-5472.CAN-12-3801.

Hoffmann, R. M., Schwarz, G., Pohl, C., Ziegenhagen, D. J., & Kruis, W. (2005). Bile acid-independent effect of hymecromone on bile secretion and common bile duct motility. *Deutsche Medizinische Wochenschrift*, *130*(34–35), 1938–1943. http://dx.doi.org/10.1055/s-2005-872606.

Hosono, K., Nishida, Y., Knudson, W., Knudson, C. B., Naruse, T., Suzuki, Y., et al. (2007). Hyaluronan oligosaccharides inhibit tumorigenicity of osteosarcoma cell lines MG-63 and LM-8 in vitro and in vivo via perturbation of hyaluronan-rich pericellular matrix of the cells. *American Journal of Pathology*, *171*(1), 274–286. http://dx.doi.org/10.2353/ajpath.2007.060828.

Isoyama, T., Thwaites, D., Selzer, M. G., Carey, R. I., Barbucci, R., & Lokeshwar, V. B. (2006). Differential selectivity of hyaluronidase inhibitors toward acidic and basic hyaluronidases. *Glycobiology*, *16*(1), 11–21. http://dx.doi.org/10.1093/glycob/cwj036.

Jaracz, S., Chen, J., Kuznetsova, L. V., & Ojima, I. (2005). Recent advances in tumor-targeting anticancer drug conjugates. *Bioorganic and Medicinal Chemistry*, *13*(17), 5043–5054. http://dx.doi.org/10.1016/j.bmc.2005.04.084.

Jokela, T. A., Jauhiainen, M., Auriola, S., Kauhanen, M., Tiihonen, R., Tammi, M. I., et al. (2008). Mannose inhibits hyaluronan synthesis by down-regulation of the cellular pool of UDP-N-acetylhexosamines. *Journal of Biological Chemistry*, *283*(12), 7666–7673. http://dx.doi.org/10.1074/jbc.M706001200.

Jokela, T. A., Kuokkanen, J., Karna, R., Pasonen-Seppanen, S., Rilla, K., Kossi, J., et al. (2013). Mannose reduces hyaluronan and leukocytes in wound granulation tissue and inhibits migration and hyaluronan-dependent monocyte binding. *Wound Repair and Regeneration*, *21*(2), 247–255. http://dx.doi.org/10.1111/wrr.12022.

Journo-Gershfeld, G., Kapp, D., Shamay, Y., Kopecek, J., & David, A. (2012). Hyaluronan oligomers-HPMA copolymer conjugates for targeting paclitaxel to CD44-overexpressing ovarian carcinoma. *Pharmaceutical Research*, *29*(4), 1121–1133. http://dx.doi.org/10.1007/s11095-012-0672-1.

Joyce, C. L., Mack, S. R., Anderson, R. A., & Zaneveld, L. J. (1986). Effect of hyaluronidase, beta-glucuronidase and beta-N-acetylglucosaminidase inhibitors on sperm penetration of the mouse oocyte. *Biology of Reproduction*, *35*(2), 336–346.

Kakizaki, I., Kojima, K., Takagaki, K., Endo, M., Kannagi, R., Ito, M., et al. (2004). A novel mechanism for the inhibition of hyaluronan biosynthesis by 4-methylumbelliferone. *Journal of Biological Chemistry*, *279*(32), 33281–33289. http://dx.doi.org/10.1074/jbc.M405918200.

Kakizaki, I., Takagaki, K., Endo, Y., Kudo, D., Ikeya, H., Miyoshi, T., et al. (2002). Inhibition of hyaluronan synthesis in Streptococcus equi FM100 by 4-methylumbelliferone. *European Journal of Biochemistry*, *269*(20), 5066–5075.

Karbownik, M. S., & Nowak, J. Z. (2013). Hyaluronan: Towards novel anti-cancer therapeutics. *Pharmacological Reports*, *65*(5), 1056–1074.

Keating, G. J. (1997). The chemistry and occurrence of coumarins. In R. O'Kennedy, R. O'Kennedy, & R. D. Thome (Eds.), *Coumarins biology, applications and mode of action* (pp. 23–65). Chichester: Wiley.

Kim, Y., Park, Y. W., Lee, Y. S., & Jeoung, D. (2008). Hyaluronic acid induces transglutaminase II to enhance cell motility; role of Rac1 and FAK in the induction of transglutaminase II. *Biotechnological Letters*, *30*(1), 31–39. http://dx.doi.org/10.1007/s10529-007-9496-1.

Kramer, M. W., Escudero, D. O., Lokeshwar, S. D., Golshani, R., Ekwenna, O. O., Acosta, K., et al. (2011). Association of hyaluronic acid family members (HAS1, HAS2, and HYAL-1) with bladder cancer diagnosis and prognosis. *Cancer*, *117*(6), 1197–1209. http://dx.doi.org/10.1002/cncr.25565.

Kriangkum, J., Warkentin, A., Belch, A. R., & Pilarski, L. M. (2013). Alteration of introns in a hyaluronan synthase 1 (HAS1) minigene convert Pre-mRNA [corrected] splicing to the aberrant pattern in multiple myeloma (MM): MM patients harbor similar changes. *PLoS One*, *8*(1), e53469. http://dx.doi.org/10.1371/journal.pone.0053469.

Kudo, D., Kon, A., Yoshihara, S., Kakizaki, I., Sasaki, M., Endo, M., et al. (2004). Effect of a hyaluronan synthase suppressor, 4-methylumbelliferone, on B16F-10 melanoma cell adhesion and locomotion. *Biochemical and Biophysical Research Communications*, *321*(4), 783–787. http://dx.doi.org/10.1016/j.bbrc.2004.07.041.

Kultti, A., Pasonen-Seppanen, S., Jauhiainen, M., Rilla, K. J., Karna, R., Pyoria, E., et al. (2009). 4-Methylumbelliferone inhibits hyaluronan synthesis by depletion of cellular UDP-glucuronic acid and downregulation of hyaluronan synthase 2 and 3. *Experimental Cell Research*, *315*(11), 1914–1923. http://dx.doi.org/10.1016/j.yexcr.2009.03.002.

Kumar, S., West, D. C., Ponting, J. M., & Gattamaneni, H. R. (1989). Sera of children with renal tumours contain low-molecular-mass hyaluronic acid. *International Journal of Cancer*, *44*(3), 445–448.

Li, Y., Li, L., Brown, T. J., & Heldin, P. (2007). Silencing of hyaluronan synthase 2 suppresses the malignant phenotype of invasive breast cancer cells. *International Journal of Cancer*, *120*(12), 2557–2567. http://dx.doi.org/10.1002/ijc.22550.

Liu, Y., Sun, J., Cao, W., Yang, J., Lian, H., Li, X., et al. (2011). Dual targeting folate-conjugated hyaluronic acid polymeric micelles for paclitaxel delivery. *International Journal of Pharmaceutics*, *421*(1), 160–169. http://dx.doi.org/10.1016/j.ijpharm.2011.09.006.

Lokeshwar, V. B., Cerwinka, W. H., Isoyama, T., & Lokeshwar, B. L. (2005). HYAL1 hyaluronidase in prostate cancer: A tumor promoter and suppressor. *Cancer Research*, *65*(17), 7782–7789. http://dx.doi.org/10.1158/0008-5472.CAN-05-1022.

Lokeshwar, V. B., Cerwinka, W. H., & Lokeshwar, B. L. (2005). HYAL1 hyaluronidase: A molecular determinant of bladder tumor growth and invasion. *Cancer Research*, *65*(6), 2243–2250. http://dx.doi.org/10.1158/0008-5472.CAN-04-2805.

Lokeshwar, V. B., Estrella, V., Lopez, L., Kramer, M., Gomez, P., Soloway, M. S., et al. (2006). HYAL1-v1, an alternatively spliced variant of HYAL1 hyaluronidase: A negative regulator of bladder cancer. *Cancer Research*, *66*(23), 11219–11227. http://dx.doi.org/10.1158/0008-5472.CAN-06-1121.

Lokeshwar, V. B., Lopez, L. E., Munoz, D., Chi, A., Shirodkar, S. P., Lokeshwar, S. D., et al. (2010). Antitumor activity of hyaluronic acid synthesis inhibitor 4-methylumbelliferone in prostate cancer cells. *Cancer Research*, *70*(7), 2613–2623. http://dx.doi.org/10.1158/0008-5472.CAN-09-3185.

Lokeshwar, V. B., Obek, C., Pham, H. T., Wei, D., Young, M. J., Duncan, R. C., et al. (2000). Urinary hyaluronic acid and hyaluronidase: Markers for bladder cancer detection and evaluation of grade. *Journal of Urology*, *163*(1), 348–356.

Lokeshwar, V. B., Obek, C., Soloway, M. S., & Block, N. L. (1997). Tumor-associated hyaluronic acid: A new sensitive and specific urine marker for bladder cancer. *Cancer Research*, *57*(4), 773–777.
Lokeshwar, V. B., Rubinowicz, D., Schroeder, G. L., Forgacs, E., Minna, J. D., Block, N. L., et al. (2001). Stromal and epithelial expression of tumor markers hyaluronic acid and HYAL1 hyaluronidase in prostate cancer. *Journal of Biological Chemistry*, *276*(15), 11922–11932. http://dx.doi.org/10.1074/jbc.M008432200.
Lokeshwar, V. B., Schroeder, G. L., Selzer, M. G., Hautmann, S. H., Posey, J. T., Duncan, R. C., et al. (2002). Bladder tumor markers for monitoring recurrence and screening comparison of hyaluronic acid-hyaluronidase and BTA-Stat tests. *Cancer*, *95*(1), 61–72. http://dx.doi.org/10.1002/cncr.10652.
Lokeshwar, V. B., & Selzer, M. G. (2008). Hyalurondiase: Both a tumor promoter and suppressor. *Seminars in Cancer Biology*, *18*(4), 281–287. http://dx.doi.org/10.1016/j.semcancer.2008.03.008.
Lokeshwar, V. B., Young, M. J., Goudarzi, G., Iida, N., Yudin, A. I., Cherr, G. N., et al. (1999). Identification of bladder tumor-derived hyaluronidase: Its similarity to HYAL1. *Cancer Research*, *59*(17), 4464–4470.
Mio, K., & Stern, R. (2002). Inhibitors of the hyaluronidases. *Matrix Biology*, *21*(1), 31–37.
Misra, S., Obeid, L. M., Hannun, Y. A., Minamisawa, S., Berger, F. G., Markwald, R. R., et al. (2008). Hyaluronan constitutively regulates activation of COX-2-mediated cell survival activity in intestinal epithelial and colon carcinoma cells. *Journal of Biological Chemistry*, *283*(21), 14335–14344. http://dx.doi.org/10.1074/jbc.M703811200.
Misra, S., Toole, B. P., & Ghatak, S. (2006). Hyaluronan constitutively regulates activation of multiple receptor tyrosine kinases in epithelial and carcinoma cells. *Journal of Biological Chemistry*, *281*(46), 34936–34941. http://dx.doi.org/10.1074/jbc.C600138200.
Morohashi, H., Kon, A., Nakai, M., Yamaguchi, M., Kakizaki, I., Yoshihara, S., et al. (2006). Study of hyaluronan synthase inhibitor, 4-methylumbelliferone derivatives on human pancreatic cancer cell (KP1-NL). *Biochemical and Biophysical Research Communications*, *345*(4), 1454–1459. http://dx.doi.org/10.1016/j.bbrc.2006.05.037.
Na, S. J., Chae, S. Y., Lee, S., Park, K., Kim, K., Park, J. H., et al. (2008). Stability and bioactivity of nanocomplex of TNF-related apoptosis-inducing ligand. *International Journal of Pharmaceutics*, *363*(1–2), 149–154. http://dx.doi.org/10.1016/j.ijpharm.2008.07.013.
Nakamura, T., Funahashi, M., Takagaki, K., Munakata, H., Tanaka, K., Saito, Y., et al. (1997). Effect of 4-methylumbelliferone on cell-free synthesis of hyaluronic acid. *Biochemistry and Molecular Biology International*, *43*(2), 263–268.
Nakazawa, H., Yoshihara, S., Kudo, D., Morohashi, H., Kakizaki, I., Kon, A., et al. (2006). 4-Methylumbelliferone, a hyaluronan synthase suppressor, enhances the anticancer activity of gemcitabine in human pancreatic cancer cells. *Cancer Chemotherapy and Pharmacology*, *57*(2), 165–170. http://dx.doi.org/10.1007/s00280-005-0016-5.
Okuda, H., Kobayashi, A., Xia, B., Watabe, M., Pai, S. K., Hirota, S., et al. (2012). Hyaluronan synthase HAS2 promotes tumor progression in bone by stimulating the interaction of breast cancer stem-like cells with macrophages and stromal cells. *Cancer Research*, *72*(2), 537–547. http://dx.doi.org/10.1158/0008-5472.CAN-11-1678.
Omarbasha, B., Fair, W. R., & Heston, W. D. (1989). Effect of coumarin on the normal rat prostate and on the R-3327H prostatic adenocarcinoma. *Cancer Research*, *49*(11), 3045–3049.
Orian-Rousseau, V. (2010). CD44, a therapeutic target for metastasising tumours. *European Journal of Cancer*, *46*(7), 1271–1277. http://dx.doi.org/10.1016/j.ejca.2010.02.024.
Paiva, P., Van Damme, M. P., Tellbach, M., Jones, R. L., Jobling, T., & Salamonsen, L. A. (2005). Expression patterns of hyaluronan, hyaluronan synthases and hyaluronidases indicate a role for hyaluronan in the progression of endometrial cancer. *Gynecologic Oncology*, *98*(2), 193–202. http://dx.doi.org/10.1016/j.ygyno.2005.02.031.

Piccioni, F., Malvicini, M., Garcia, M. G., Rodriguez, A., Atorrasagasti, C., Kippes, N., et al. (2012). Antitumor effects of hyaluronic acid inhibitor 4-methylumbelliferone in an orthotopic hepatocellular carcinoma model in mice. *Glycobiology*, *22*(3), 400–410. http://dx.doi.org/10.1093/glycob/cwr158.

Pilon-Thomas, S., Verhaegen, M., Kuhn, L., Riker, A., & Mule, J. J. (2006). Induction of anti-tumor immunity by vaccination with dendritic cells pulsed with anti-CD44 IgG opsonized tumor cells. *Cancer Immunology, Immunotherapy*, *55*(10), 1238–1246. http://dx.doi.org/10.1007/s00262-005-0104-8.

Platt, V. M., & Szoka, F. C., Jr. (2008). Anticancer therapeutics: Targeting macromolecules and nanocarriers to hyaluronan or CD44, a hyaluronan receptor. *Molecular Pharmaceutics*, *5*(4), 474–486. http://dx.doi.org/10.1021/mp800024g.

Porsch, H., Bernert, B., Mehic, M., Theocharis, A. D., Heldin, C. H., & Heldin, P. (2013). Efficient TGFbeta-induced epithelial-mesenchymal transition depends on hyaluronan synthase HAS2. *Oncogene*, *32*(37), 4355–4365. http://dx.doi.org/10.1038/onc.2012.475.

Prehm, P. (2013). Curcumin analogue identified as hyaluronan export inhibitor by virtual docking to the ABC transporter MRP5. *Food and Chemical Toxicology*, *62*, 76–81. http://dx.doi.org/10.1016/j.fct.2013.08.028.

Qiu, L., Li, Z., Qiao, M., Long, M., Wang, M., Zhang, X., et al. (2013). Self-assembled pH-responsive hyaluronic acid-g-poly(L-histidine) copolymer micelles for targeted intracellular delivery of doxorubicin. *Acta Biomaterialia*, *10*(5), 2024–2035. http://dx.doi.org/10.1016/j.actbio.2013.12.025.

Quaranta, S., Rossetti, S., & Camarri, E. (1984). Double-blind clinical study on hymecromone and placebo in motor disorders of the bile ducts after cholecystectomy. *Clinica Terapeutica*, *108*(6), 513–517.

Rajasagi, M., von Au, A., Singh, R., Hartmann, N., Zoller, M., & Marhaba, R. (2010). Anti-CD44 induces apoptosis in T lymphoma via mitochondrial depolarization. *Journal of Cellular and Molecular Medicine*, *14*(6B), 1453–1467. http://dx.doi.org/10.1111/j.1582-4934.2009.00909.x.

Ren, Y., Zhao, T., Mao, G., Zhang, M., Li, F., Zou, Y., et al. (2013). Antitumor activity of hyaluronic acid-selenium nanoparticles in Heps tumor mice models. *International Journal of Biological Macromolecules*, *57*, 57–62. http://dx.doi.org/10.1016/j.ijbiomac.2013.03.014.

Ricciardelli, C., Ween, M. P., Lokman, N. A., Tan, I. A., Pyragius, C. E., & Oehler, M. K. (2013). Chemotherapy-induced hyaluronan production: A novel chemoresistance mechanism in ovarian cancer. *BMC Cancer*, *13*, 476. http://dx.doi.org/10.1186/1471-2407-13-476.

Riechelmann, H., Sauter, A., Golze, W., Hanft, G., Schroen, C., Hoermann, K., et al. (2008). Phase I trial with the CD44v6-targeting immunoconjugate bivatuzumab mertansine in head and neck squamous cell carcinoma. *Oral Oncology*, *44*(9), 823–829. http://dx.doi.org/10.1016/j.oraloncology.2007.10.009.

Saito, T., Tamura, D., Nakamura, T., Makita, Y., Ariyama, H., Komiyama, K., et al. (2013). 4-Methylumbelliferone leads to growth arrest and apoptosis in canine mammary tumor cells. *Oncology Reports*, *29*(1), 335–342. http://dx.doi.org/10.3892/or.2012.2100.

Sandstrom, K., Haylock, A. K., Spiegelberg, D., Qvarnstrom, F., Wester, K., & Nestor, M. (2012). A novel CD44v6 targeting antibody fragment with improved tumor-to-blood ratio. *International Journal of Oncology*, *40*(5), 1525–1532. http://dx.doi.org/10.3892/ijo.2012.1352.

Sandstrom, K., Nestor, M., Ekberg, T., Engstrom, M., Anniko, M., & Lundqvist, H. (2008). Targeting CD44v6 expressed in head and neck squamous cell carcinoma: Preclinical characterization of an 111In-labeled monoclonal antibody. *Tumour Biology*, *29*(3), 137–144. http://dx.doi.org/10.1159/000143399.

Schmitt, M., Schmitt, A., Rojewski, M. T., Chen, J., Giannopoulos, K., Fei, F., et al. (2008). RHAMM-R3 peptide vaccination in patients with acute myeloid leukemia, myelodysplastic syndrome, and multiple myeloma elicits immunologic and clinical responses. *Blood, 111*(3), 1357–1365. http://dx.doi.org/10.1182/blood-2007-07-099366.

Serafino, A., Zonfrillo, M., Andreola, F., Psaila, R., Mercuri, L., Moroni, N., et al. (2011). CD44-targeting for antitumor drug delivery: A new SN-38-hyaluronan bioconjugate for locoregional treatment of peritoneal carcinomatosis. *Current Cancer Drug Targets, 11*(5), 572–585.

Shah, V., Taratula, O., Garbuzenko, O. B., Taratula, O. R., Rodriguez-Rodriguez, L., & Minko, T. (2013). Targeted nanomedicine for suppression of CD44 and simultaneous cell death induction in ovarian cancer: An optimal delivery of siRNA and anticancer drug. *Clinical Cancer Research, 19*(22), 6193–6204. http://dx.doi.org/10.1158/1078-0432.CCR-13-1536.

Shen, Y. N., Shin, H. J., Joo, H. Y., Park, E. R., Kim, S. H., Hwang, S. G., et al. (2014). Inhibition of HAS2 induction enhances the radiosensitivity of cancer cells via persistent DNA damage. *Biochemical and Biophysical Research Communications, 443*(3), 796–801. http://dx.doi.org/10.1016/j.bbrc.2013.12.026.

Shi, J., Ma, R., Wang, L., Zhang, J., Liu, R., Li, L., et al. (2013). The application of hyaluronic acid-derivatized carbon nanotubes in hematoporphyrin monomethyl ether-based photodynamic therapy for in vivo and in vitro cancer treatment. *International Journal of Nanomedicine, 8*, 2361–2373. http://dx.doi.org/10.2147/IJN.S45407.

Siiskonen, H., Poukka, M., Tyynela-Korhonen, K., Sironen, R., & Pasonen-Seppanen, S. (2013). Inverse expression of hyaluronidase 2 and hyaluronan synthases 1–3 is associated with reduced hyaluronan content in malignant cutaneous melanoma. *BMC Cancer, 13*, 181. http://dx.doi.org/10.1186/1471-2407-13-181.

Simpson, M. A., & Lokeshwar, V. B. (2008). Hyaluronan and hyaluronidase in genitourinary tumors. *Frontiers in Bioscience, 13*, 5664–5680.

Sironen, R. K., Tammi, M., Tammi, R., Auvinen, P. K., Anttila, M., & Kosma, V. M. (2011). Hyaluronan in human malignancies. *Experimental Cell Research, 317*(4), 383–391. http://dx.doi.org/10.1016/j.yexcr.2010.11.017.

Slomiany, M. G., Dai, L., Bomar, P. A., Knackstedt, T. J., Kranc, D. A., Tolliver, L., et al. (2009). Abrogating drug resistance in malignant peripheral nerve sheath tumors by disrupting hyaluronan-CD44 interactions with small hyaluronan oligosaccharides. *Cancer Research, 69*(12), 4992–4998. http://dx.doi.org/10.1158/0008-5472.CAN-09-0143.

Slomiany, M. G., Dai, L., Tolliver, L. B., Grass, G. D., Zeng, Y., & Toole, B. P. (2009). Inhibition of functional hyaluronan-CD44 interactions in CD133-positive primary human ovarian carcinoma cells by small hyaluronan oligosaccharides. *Clinical Cancer Research, 15*(24), 7593–7601. http://dx.doi.org/10.1158/1078-0432.CCR-09-2317.

Slomiany, M. G., Grass, G. D., Robertson, A. D., Yang, X. Y., Maria, B. L., Beeson, C., et al. (2009). Hyaluronan, CD44, and emmprin regulate lactate efflux and membrane localization of monocarboxylate transporters in human breast carcinoma cells. *Cancer Research, 69*(4), 1293–1301. http://dx.doi.org/10.1158/0008-5472.CAN-08-2491.

Snauwaert, S., Vanhee, S., Goetgeluk, G., Verstichel, G., Van Caeneghem, Y., Velghe, I., et al. (2012). RHAMM/HMMR (CD168) is not an ideal target antigen for immunotherapy of acute myeloid leukemia. *Haematologica, 97*(10), 1539–1547. http://dx.doi.org/10.3324/haematol.2012.065581.

Somasunderam, A., Thiviyanathan, V., Tanaka, T., Li, X., Neerathilingam, M., Lokesh, G. L., et al. (2010). Combinatorial selection of DNA thioaptamers targeted to the HA binding domain of human CD44. *Biochemistry, 49*(42), 9106–9112. http://dx.doi.org/10.1021/bi1009503.

Tan, J. X., Wang, X. Y., Li, H. Y., Su, X. L., Wang, L., Ran, L., et al. (2011). HYAL1 overexpression is correlated with the malignant behavior of human breast cancer. *International Journal of Cancer, 128*(6), 1303–1315. http://dx.doi.org/10.1002/ijc.25460.

Toole, B. P., Ghatak, S., & Misra, S. (2008). Hyaluronan oligosaccharides as a potential anticancer therapeutic. *Current Pharmaceutical Biotechnology, 9*(4), 249–252.

Turley, E. A., & Naor, D. (2012). RHAMM and CD44 peptides-analytic tools and potential drugs. *Frontiers in Bioscience, 17*, 1775–1794.

Twarock, S., Freudenberger, T., Poscher, E., Dai, G., Jannasch, K., Dullin, C., et al. (2011). Inhibition of oesophageal squamous cell carcinoma progression by in vivo targeting of hyaluronan synthesis. *Molecular Cancer, 10*, 30. http://dx.doi.org/10.1186/1476-4598-10-30.

U.S. Department of Health and Human Services, Food and Drug Administration, Center for Drug Evaluation and Research. (2005). *Guidance for industry estimating the maximum safe starting dose in initial clinical trials for therapeutics in adult healthy volunteers.* http://www.fda.gov/downloads/Drugs/GuidanceComplianceRegulatoryInformation/Guidances/ucm078932.pdf.

Uchakina, O. N., Ban, H., & McKallip, R. J. (2013). Targeting hyaluronic acid production for the treatment of leukemia: Treatment with 4-methylumbelliferone leads to induction of MAPK-mediated apoptosis in K562 leukemia. *Leukemia Research, 37*(10), 1294–1301. http://dx.doi.org/10.1016/j.leukres.2013.07.009.

Udabage, L., Brownlee, G. R., Waltham, M., Blick, T., Walker, E. C., Heldin, P., et al. (2005). Antisense-mediated suppression of hyaluronan synthase 2 inhibits the tumorigenesis and progression of breast cancer. *Cancer Research, 65*(14), 6139–6150. http://dx.doi.org/10.1158/0008-5472.CAN-04-1622.

Urakawa, H., Nishida, Y., Knudson, W., Knudson, C. B., Arai, E., Kozawa, E., et al. (2012). Therapeutic potential of hyaluronan oligosaccharides for bone metastasis of breast cancer. *Journal of Orthopaedic Research, 30*(4), 662–672. http://dx.doi.org/10.1002/jor.21557.

Urakawa, H., Nishida, Y., Wasa, J., Arai, E., Zhuo, L., Kimata, K., et al. (2012). Inhibition of hyaluronan synthesis in breast cancer cells by 4-methylumbelliferone suppresses tumorigenicity in vitro and metastatic lesions of bone in vivo. *International Journal of Cancer, 130*(2), 454–466. http://dx.doi.org/10.1002/ijc.26014.

Vincent, T., Jourdan, M., Sy, M. S., Klein, B., & Mechti, N. (2001). Hyaluronic acid induces survival and proliferation of human myeloma cells through an interleukin-6-mediated pathway involving the phosphorylation of retinoblastoma protein. *Journal of Biological Chemistry, 276*(18), 14728–14736. http://dx.doi.org/10.1074/jbc.M003965200.

Voelcker, V., Gebhardt, C., Averbeck, M., Saalbach, A., Wolf, V., Weih, F., et al. (2008). Hyaluronan fragments induce cytokine and metalloprotease upregulation in human melanoma cells in part by signalling via TLR4. *Experimental Dermatology, 17*(2), 100–107. http://dx.doi.org/10.1111/j.1600-0625.2007.00638.x.

Wallach-Dayan, S. B., Rubinstein, A. M., Hand, C., Breuer, R., & Naor, D. (2008). DNA vaccination with CD44 variant isoform reduces mammary tumor local growth and lung metastasis. *Molecular Cancer Therapeutics*, 7(6), 1615–1623. http://dx.doi.org/10.1158/1535-7163.MCT-07-2383.

Wang, X., Liu, W., Sun, C. L., Armenian, S. H., Hakonarson, H., Hageman, L., et al. (2014). Hyaluronan synthase 3 variant and anthracycline-related cardiomyopathy: A report from the children's oncology group. *Journal of Clinical Oncology, 32*(7), 647–653. http://dx.doi.org/10.1200/JCO.2013.50.3557.

Ween, M. P., Hummitzsch, K., Rodgers, R. J., Oehler, M. K., & Ricciardelli, C. (2011). Versican induces a pro-metastatic ovarian cancer cell behavior which can be inhibited by small hyaluronan oligosaccharides. *Clinical and Experimental Metastasis, 28*(2), 113–125. http://dx.doi.org/10.1007/s10585-010-9363-7.

Wei, Q., Galbenus, R., Raza, A., Cerny, R. L., & Simpson, M. A. (2009). Androgen-stimulated UDP-glucose dehydrogenase expression limits prostate androgen availability without impacting hyaluronan levels. *Cancer Research*, *69*(6), 2332–2339. http://dx.doi.org/10.1158/0008-5472.CAN-08-3083.
West, D. C., Hampson, I. N., Arnold, F., & Kumar, S. (1985). Angiogenesis induced by degradation products of hyaluronic acid. *Science*, *228*(4705), 1324–1326.
Yang, C., Cao, M., Liu, H., He, Y., Xu, J., Du, Y., et al. (2012). The high and low molecular weight forms of hyaluronan have distinct effects on CD44 clustering. *Journal of Biological Chemistry*, *287*(51), 43094–43107. http://dx.doi.org/10.1074/jbc.M112.349209.
Yao, J., Zhang, L., Zhou, J., Liu, H., & Zhang, Q. (2013). Efficient simultaneous tumor targeting delivery of all-trans retinoid acid and paclitaxel based on hyaluronic acid-based multifunctional nanocarrier. *Molecular Pharmaceutics*, *10*(3), 1080–1091. http://dx.doi.org/10.1021/mp3005808.
Yoon, H. Y., Koo, H., Choi, K. Y., Lee, S. J., Kim, K., Kwon, I. C., et al. (2012). Tumor-targeting hyaluronic acid nanoparticles for photodynamic imaging and therapy. *Biomaterials*, *33*(15), 3980–3989. http://dx.doi.org/10.1016/j.biomaterials.2012.02.016.
Yoshida, T., Matsuda, Y., Naito, Z., & Ishiwata, T. (2012). CD44 in human glioma correlates with histopathological grade and cell migration. *Pathology International*, *62*(7), 463–470. http://dx.doi.org/10.1111/j.1440-1827.2012.02823.x.
Yoshihara, S., Kon, A., Kudo, D., Nakazawa, H., Kakizaki, I., Sasaki, M., et al. (2005). A hyaluronan synthase suppressor, 4-methylumbelliferone, inhibits liver metastasis of melanoma cells. *FEBS Letters*, *579*(12), 2722–2726. http://dx.doi.org/10.1016/j.febslet.2005.03.079.
Zhang, J., Chang, B., & Liu, J. (2013). CD44 standard form expression is correlated with high-grade and advanced-stage ovarian carcinoma but not prognosis. *Human Pathology*, *44*(9), 1882–1889. http://dx.doi.org/10.1016/j.humpath.2013.02.016.
Zhang, S., Wu, C. C., Fecteau, J. F., Cui, B., Chen, L., Zhang, L., et al. (2013). Targeting chronic lymphocytic leukemia cells with a humanized monoclonal antibody specific for CD44. *Proceedings of the National Academy of Sciences of the United States of America*, *110*(15), 6127–6132. http://dx.doi.org/10.1073/pnas.1221841110.
Zou, L., Song, X., Yi, T., Li, S., Deng, H., Chen, X., et al. (2013). Administration of PLGA nanoparticles carrying shRNA against focal adhesion kinase and CD44 results in enhanced antitumor effects against ovarian cancer. *Cancer Gene Therapy*, *20*(4), 242–250. http://dx.doi.org/10.1038/cgt.2013.12.

CHAPTER THREE

Aberrant Posttranscriptional Processing of Hyaluronan Synthase 1 in Malignant Transformation and Tumor Progression

Sophia Adamia*[,1], Jitra Kriangkum†, Andrew R. Belch†, Linda M. Pilarski†

*Medical Oncology, Dana-Farber Cancer Institute, Harvard Medical School, Boston, Massachusetts, USA
†Department of Oncology, University of Alberta and Cross Cancer Institute, Edmonton, Alberta, Canada
[1]Corresponding author: e-mail address: sophia_adamia@dfci.harvard.edu

Contents

1. Splicing and Cancer 68
   1.1 Splicing in the human genome 68
2. Control of Pre-mRNA Splicing 69
   2.1 Splicing mutations 70
3. Impact on Cancer of Alterations in Splicing Machinery 71
4. Aberrant Splicing of Hyaluronan Synthase 1 72
   4.1 B lineage malignancies 72
   4.2 Aberrant splice variants in hyaluronan synthase 1 73
5. Clinical Impact of Aberrant HAS1 Splicing 75
6. Genetic Variations in HAS1 76
7. Functional Impact of HAS1Vs 76
   7.1 Synthesis of HA 76
   7.2 Functional outcomes of HA synthesis in patients 77
8. Functional Outcomes of HAS1Vs in Transfectants 80
   8.1 HAS1Vs synthesize intracellular HA 80
   8.2 HAS1FL and HAS1Vs form heteromultimers with HAS1Vs as the dominant partners 81
   8.3 Multimers with HAS1Vs stabilize HAS1FL and maintain intracellular HA 82
9. HAS1Vs and Mitotic Catastrophe 83
   9.1 A potential mechanism underlying the oncogenic properties of aberrant HAS1 splice variants: HAS1V-mediated rescue from RHAMM-mediated mitotic catastrophe 83

Advances in Cancer Research, Volume 123
ISSN 0065-230X
http://dx.doi.org/10.1016/B978-0-12-800092-2.00003-4

10. Conclusion 88
Acknowledgments 88
References 88

Abstract

It is becoming increasingly apparent that splicing defects play a key role in cancer, and that alterations in genomic splicing elements promote aberrant splicing. Alternative splicing increases the diversity of the human transcriptome and increases the numbers of functional gene products. However, dysregulation that leads to aberrant pre-mRNA splicing can contribute to cancer. Hyaluronan (HA), known to be an important component of cancer progression, is synthesized by hyaluronan synthases (HASs). In cancer cells, hyaluronan synthase 1 (*HAS1*) pre-mRNA is abnormally spliced to generate a family of aberrant splice variants (HAS1Vs) that synthesize extracellular and intracellular HA. HAS1Vs are clinically relevant, being found almost exclusively in malignant cells. Expression of aberrant HAS1Vs predicts poor survival in multiple myeloma. In this review, we summarize the unusual properties of HAS1Vs and their relationship to cancer. HAS1Vs form heterogeneous multimers with normally spliced HAS1 as well as with each other and with HAS3. Aberrant variants of HAS1 synthesize HA. Extracellular HA synthesized by HAS1Vs is likely to promote malignant spread. We speculate that synthesis of intracellular HA plays a fundamental and early role in oncogenesis by promoting genetic instability and the emergence of viable cancer variants that lead to aggressive disease.

1. SPLICING AND CANCER

1.1. Splicing in the human genome

Pre-RNA splicing is a fine-tuned epigenetic process that provides not only plasticity but also considerable diversity for the human genome (Shkreta et al., 2013). As a result of alternative splicing, one coding sequence is able to encode a large number of functional protein products. The fine machinery of alternative splicing is regulated by the classical, *cis* and *trans* splicing elements, as well as splicing factors and small nuclear proteins that are known as splicing suppressors and enhancers.

The accuracy of splicing is maintained by classical splice elements, including the 5′-splice site known as the splicing donor and 3′-splice site termed the splicing acceptor (Black, 2003; Matera & Wang, 2014; Roy, Haupt, & Griffiths, 2013). Alterations in splicing are dramatically more common in genes associated with susceptibility and/or progression of cancer (Abrahamsson et al., 2009; Adamia et al., 2014; Brown et al., 2011; Lo et al., 2009; Lu et al., 2009; Stark, Wichman, Avivi, & Assaraf, 2009;

Vassen et al., 2009; Yang, Huang, Wu, & Benz, 2008). The four main types of alternative splicing events documented in the human genome thus far include exon skipping, alternative 5′ and 3′ splicing, and intron retention (Roy et al., 2013). Among conserved alternative splicing events, exon skipping is the most prevalent type of alternative splicing while entire intron retention is least frequent. However, partial exon deletion or intron retention appears to be characteristic of genes involved in malignant transformation (Romano, Buratti, & Baralle, 2013). Aberrant splicing has been reported in a number of genes that play a role in cancer (Adamia, Maxwell, & Pilarski, 2005; Adamia, Pilarski, Belch, & Pilarski, 2013; Adamia et al., 2008; David et al., 2002; Kriangkum, Warkentin, Belch, & Pilarski, 2013; Shkreta et al., 2013).

2. CONTROL OF PRE-mRNA SPLICING

The elements that assemble into the spliceosome are splicing factors proteins that bind to pre-mRNAs in a specific order (Black, 2003; Matlin, Clark, & Smith, 2005). The "*trans*" splicing elements are proteins that can potentially bind to any region of the genome holding the target pre-mRNA sequence, as long as the requisite "*cis*" splicing elements located on the pre-mRNA template itself, such as exonic and intronic splicing enhancer and splicing suppressor sequences, are present (Matera & Wang, 2014). Correct splicing can be altered not only by mutations occurring on the classical splice sites but also by genetic variations that take place within the *cis*-splicing elements. Even distant "deep" intronic mutations on *cis* elements can modulate splicing events (Ast, 2004). Thus, any genetic mutations, insertions or deletions that occur in splicing elements themselves are likely to have a genome-wide impact on alternative splicing. In addition, at a much more specific level, the gene sequences of exons and introns in a given gene are responsible for spliceosome binding. Mutations, insertions and deletions in a given gene can dramatically change the splice variants that are generated by the splicing process and in consequence the nature of splice variants that result (Adamia et al., 2008; Wappenschmidt et al., 2012).

Alternative splicing occurs in nearly all genes and is a critical component of normal gene expression and tissue-specific gene regulation. In cancer cells, however, aberrant splicing can be frequent, generating abnormal proteins that may promote disease progression (Fig. 3.1) (Romano et al., 2013).

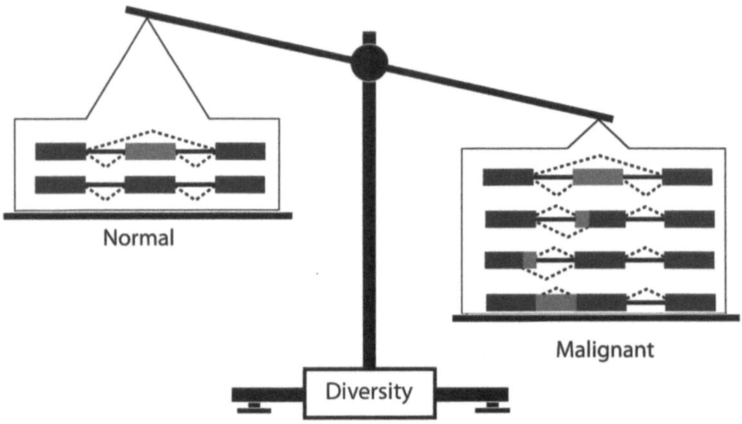

Figure 3.1 Aberrant splicing events increase transcriptome diversity of malignant cells leading to gene products not found in healthy cells. (See the color plate.)

These have the potential to function as oncogenes or cancer promoters. If somatic mutations arise in genes encoding components of splicing machinery such as splicing factors or *trans* elements of splicing, splicing throughout the whole genome may be affected. However, if genetic variations occur in a target gene to enhance or create new splice sites or change the binding of otherwise normal splicing elements, only that gene is likely to be affected.

2.1. Splicing mutations

Highly recurrent alterations in genes encoding components of the spliceosome are the most unexpected and dramatic findings in the cancer genome. These have been discovered recently through use of new high-throughput technologies such as next-generation sequencing. Recent work suggests that somatic mutations in splicing factors and *trans* splicing elements (splicing enhancer and suppressor proteins), presumably having a genome-wide impact, may have a key role in the events that lead to chronic lymphocytic leukemia and myeloid cancers (Adamia, Pilarski, Bar-Natan, Stone, & Griffin, 2013; Cazzola, Rossi, & Malcovati, 2013).

Genetic changes can occur throughout an individual gene and may have an impact on splicing by altering overall spliceosome assembly. Mutations or polymorphisms can also affect distribution patterns of *cis* elements, including splicing enhancers and suppressors, throughout an affected gene and thus could enhance or suppress spliceosomal assembly. As a result of changes to *cis* elements, the "destined" location of spliceosomal assembly will be compromised and new sites for 1st and 2nd splicing reactions will be activated, leading to aberrant splicing of an individual gene. Intronic mutations or other genetic changes within introns appear to play a central role in generation of aberrant splice variants (Adamia, Pilarski, Belch, et al., 2013; Kriangkum et al., 2013).

The same holds true if classical splice sites are affected by genetic variations. Assembly of exons into mRNA through normal alternative splicing requires preferential use of 5′-splice sites at the terminus of each intron. If genetic variations cause an alternate, perhaps otherwise cryptic splice site to become the dominant sequence in a splicing event, this can lead to all or a fragment of intronic sequence incorporated into the mRNA template (Adamia et al., 2008; Spinelli et al., 2013). This in turn leads to the potential generation of an aberrant splice-variant protein with altered functional impact(s) compared to the normally spliced variant. Aberrant intronic splice variants appear to be predominantly found in cancer cells. Thus, any analysis of aberrant splicing in cancer needs to consider intronic as well as exonic sequence variations. In plasma cell dyscrasias (e.g., multiple myeloma (MM) and Waldenstrom macroglobulinemia (WM)), aberrant splicing leading to a family of altered proteins, some incorporating intronic sequences, was identified in cancer cells from a large proportion of patients (Adamia et al., 2003, 2008, 2005; Spinelli et al., 2013).

3. IMPACT ON CANCER OF ALTERATIONS IN SPLICING MACHINERY

Genome-wide studies have revealed the existence of cancer-specific splicing patterns (Romano et al., 2013; Shkreta et al., 2013). Numerous genes possess different splice variants that regulate native proteins in a dominant-negative, mutually antagonistic manner. There is the possibility that novel splice variants could alter function of their native counterpart proteins. Thus, novel splice variants have the potential to function as oncogenes or cancer promoters. Recently, novel splice variants of many oncogenes and tumor suppressor genes were identified by genome-wide analysis of splicing abnormalities in acute myeloid leukemia patients (Adamia et al., 2014). In neuroblastoma cells, a neurotrophin receptor tyrosine kinase TrkA splice-variant receptor (TrkAIII) has oncogenic activity through its regulation of tumor behavior (Farina et al., 2009). BCR–ABL-positive acute lymphoblastic leukemia patients develop primary imatinib resistance due to expression of the Ik6, a splice variant of Ikaros (Iacobucci et al., 2008; Klein et al., 2006). *AIMP2* (aminoacyl-tRNA synthetase-interacting multifunctional protein 2) is a tumor suppressor gene important for controlling cell fate. Expression of AIMP2-DX2A, a splice variant of the AIMP2 is associated with advanced lung cancer stage and shorter overall survival in lung cancer specimens (Choi et al., 2011). Oncogenic V600E *BRAF* mutations in melanoma are the molecular target for the V600E BRAF kinase inhibitor vemurafinib (Hartsough, Basile, & Aplin, 2014). The BRAF splice variant alters effects of vemurafinib, because *BRAF* exons 4–8 encode the Ras-binding domain. Deletion of this domain allows the shorter splice variant of V600E BRAF to undergo constitutive dimerization in a Ras-independent manner.

4. ABERRANT SPLICING OF HYALURONAN SYNTHASE 1

4.1. B lineage malignancies

4.1.1 Multiple myeloma

MM is an incurable cancer that accounts for 1% of all cancers, 10% of all hematological malignancies and 19% of deaths from hematological malignancies. It kills over 15,000 North Americans each year, and the median survival is 3–4 years. MM is characterized by the presence of monoclonal immunoglobulin (Ig) in the blood, lytic bone lesions, and monoclonal plasma cells in the bone marrow (BM). MM may arise through progressive acquisition of diverse genetic abnormalities and is characterized by complex chromosomal abnormalities. The MM clone is definitively identified by expression of a unique IgH VDJ gene rearrangement, termed clonotypic. A variety of work leads to the idea that the generative compartment in MM includes B lineage cells in the BM, the blood, or both, at a stage of differentiation preceding that of plasma cells. Our work shows that MM is a hierarchy of B lineage cells (Debes Marun, Belch, & Pilarski, 2012; Pilarski et al., 2013; Pilarski, Reiman, Pilarski, Orr, & Belch, 2004). Circulating MM-B cells that harbor the clonotypic IgH VDJ, identifying them as members of the malignant clone, have cancer stem cell capabilities as defined by their ability to transfer human disease to immunodeficient mice (Pilarski & Belch, 2002; Pilarski, Hipperson, et al., 2000). In this context, it is provocative that aberrant hyaluronan synthase 1 (*HAS1*) splicing within the circulating B cell compartment correlates with reduced survival of MM patients (Adamia et al., 2005). We speculate that aberrant HAS1 splicing in clonotypic B cells may be a fundamental component of cancer stem cell function in MM.

4.1.2 Waldenstrom macroglobulinemia

WM is characterized by lymphoplasmacytic lymphoma and IgM monoclonal gammopathy. It accounts for 2% of hematological malignancies with a median survival of 5 years. About 50% of patients have predominant, well-differentiated lymphocytes, while 40% of them have more plasmacytic morphology. Malignant WM cells are found in both blood and BM. WM includes multiple compartments of malignant cells spanning multiple differentiation stages in the B lymphocyte lineage.

4.2. Aberrant splice variants in hyaluronan synthase 1

HASs are overexpressed in malignant cells from MM and WM. Hyaluronan synthases are encoded by the *HAS1, HAS2, and HAS3* genes which are evolutionarily conserved, located on, respectively, chromosomes 19, 8, and 16. The enzymes have seven predicted membrane-spanning regions and a large cytoplasmic loop. Although *HAS1* is overexpressed in malignant cells relative to normal cells, these transcripts are hard to detect due to their transient expression. Thus, *HAS1* detection in malignant cells requires use of sensitive technology for careful analysis. While evaluating overexpression of *HASs* in hematopoietic cells (Adamia et al., 2003; Adamia, Reiman, et al., 2005), an unexpected complexity of mRNA species encoding HAS1 was observed in malignant B cells from MM and WM patients. This led to the discovery of a family of aberrant HAS1 splice-variant transcripts (HAS1Vs), HAS1Va, HAS1Vb, and HAS1Vc (Fig. 3.2) (Adamia, Reiman, et al., 2005).

Cloning and sequencing analyses revealed that one variant transcript, HAS1Va, resulted from skipping exon 4, producing a frameshift and a premature stop codon. HAS1Vb and HAS1Vc retained fragments of intron 4 identifying them as intronic splice variants (Adamia, Reiman, et al., 2005), and encoded truncated proteins because of a premature stop codon in exon 5. HAS1Vs appeared to be restricted to mature B lymphocytes and were not detectable in end-stage plasma cells, the antibody-forming cells of the B lineage and a relatively nongenerative compartment that is responsible for the pathology of MM (Pilarski et al., 2013). Aberrantly spliced HAS1Vs were the first abnormal biomarkers to be identified in and restricted to MM-B lymphocytes (Adamia, Reiman, et al., 2005), an earlier stage compartment of the malignancy that appears to include cancer stem cells (Kirshner et al., 2008; Pilarski et al., 2013). Aberrant HAS1 splicing was

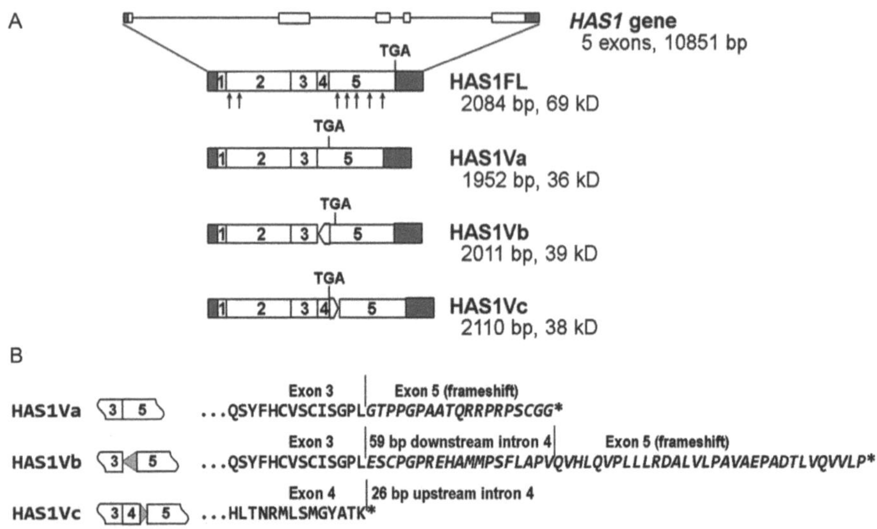

Figure 3.2 Splice isoforms of HAS1 are generated by alternative splicing within exons 3, 4, and 5. Constitutive splicing of HAS1 full length (HAS1FL) and alternative splicing of HAS1Vs are summarized in (A). HAS1Va skips exon 4; HAS1Vb skips exon 4 and retains 59 bp of downstream intron 4; HAS1Vc retains 26 bp of upstream intron 4. Arrow, transmembrane domain; closed box, untranslated region; open box, exon; triangular box, partially retained intron; TGA, termination codon. Splice variants are truncated proteins. Deduced amino acid sequences downstream of alternative splice sites are shown in (B). Italic letters refer to codon insertion and frameshift sequence; *, translation ends.

apparent in over half of MM patients at the time of diagnosis and persisted intermittently in relapse and throughout disease progression (Adamia, Reiman, et al., 2005). Essentially, all MM patients had B cells undergoing aberrant HAS1 splicing at some stage of their disease.

Among malignant cells from WM, nearly all patients harbored B lineage malignant cells that underwent aberrant splicing of HAS1. HAS1Vs were never detected in normal hematopoietic cells, including mature B cells from normal donors, the healthy counterpart of the malignant B cells in MM and WM (Adamia, Reiman, et al., 2005). Additional work by others showed that HAS1Va was expressed in bladder cancer cells suggesting that perhaps aberrant HAS1 splicing is also relevant for solid tumors (Golshani et al., 2007). Expression of HAS1Vs was also found in B cells from patients with monoclonal gammopathy of undetermined significance (MGUS), a "premalignant" condition that is thought to precede development of frank MM or WM. However, some evidence suggests that MGUS may instead be malignancy under control (Almeida et al., 1999; Ocqueteau et al., 1998), an alternate view that may be consistent with the presence of aberrant HAS1Vs in MGUS. These observations raise the provocative speculation that expression of HAS1Vs may be a transforming event that promotes eventual progression to frank plasma cell dyscrasia (particularly MM or WM). Important questions arise regarding the clinical and functional impact of aberrant HAS1 splice variants as addressed below.

5. CLINICAL IMPACT OF ABERRANT HAS1 SPLICING

HAS1Vs gain clinical relevance from the fact that their expression was detected in malignant cells at time of diagnosis or relapse but were absent during remission, suggesting expression of splice variants was associated with clinical course of the disease (Adamia, Reiman, et al., 2005). For a series of individual malignant B cells obtained by cell sorting from WM blood and BM, 51% expressed HAS1FL (HAS1 full length), 78% HAS1Va, 32% HAS1Vb, and 22% HAS1Vc (Adamia, Treon, et al., 2005). In MM malignant B cell populations, 58% expressed HAS1FL, 68% HAS1Va, 46% HAS1Vb, and 33% HAS1Vc (Adamia, Reiman, et al., 2005). Expression patterns for MGUS were similar (Adamia, Reiman, et al., 2005). In WM, aberrant splicing events are found in 86% of analyzed patients, among whom nearly half express intronic splice variants (Adamia, Treon, et al., 2005), suggesting that retention of intronic sequences may be critical to the clinical impact of aberrant HAS1 splicing. Among MM patients, at diagnosis, during progression or in relapse, about half had undergone aberrant splicing to generate intronic HAS1 splice variants (Adamia, Reiman, et al., 2005). Thus, aberrant splicing and the expression of HAS1Vs are frequent in malignant B cells from clinical samples.

Among the HAS1Vs in MM, HAS1Vb is an intronic splice variant with significant prognostic value (Adamia, Reiman, et al., 2005), and there is a trend toward association of HAS1Va with poor outcome. For an unselected cohort of 58 patients with MM, aberrant splicing that generates HAS1Vb was found to be predictive for poor survival (Adamia, Reiman, et al., 2005). Interestingly, there was little overlap between the expression of HAS1Vs and the t(4;14) translocation in MM, a biomarker that also predicts poor outcome (Keats et al., 2005). Of patients with the t(4;14) translocation, only one also expressed HAS1Vs, suggestive of an inverse correlation between these two biomarkers. Nearly, all of the $t(4;14)^{+}$ MM patients were found to be in the HAS1-negative arm. Thus, aberrant HAS1V expression may identify a genetically distinct subgroup of MM patients.

These findings suggest that the *HAS1* gene is involved in early stages of tumorigenesis as well as contributing to disease progression. HAS1 novel variants (HAS1Va, HAS1Vb, and HAS1Vc) are upregulated in the circulating components of the myeloma clone and are biologically relevant markers of a circulating tumor burden.

6. GENETIC VARIATIONS IN HAS1

Genes such as *HAS1* which have a high GC-rich content have a tendency to accumulate mutations (Wolfe, Sharp, & Li, 1989). If mutations accumulate in splicing regions, they might be expected to influence spliceosomal assembly and "cutting and pasting" of pre-mRNA to generate aberrant splice variants. As indicated in Fig. 3.2, aberrant splicing in HAS1 includes exon skipping and intronic inclusion. Segments of the *HAS1* gene were sequenced from MM and WM patients and healthy donors (Adamia, Pilarski, Belch, et al., 2009; Adamia et al., 2008); overall, nearly 50 recurrent mutations were identified in the region from exon 3 to intron 4, with a given mutation being recurrent at least two, but as many as seven, unrelated patients. They were not found in healthy donors. *In silico* analysis predicted that the pattern of mutations and polymorphisms would generate splicing of HAS1Vb (Adamia et al., 2008). Genetic variations in introns 3 and 4 had a strong influence on HAS1 splicing events (Adamia et al., 2008; Kriangkum et al., 2013). *In vitro* mutagenesis experiments indicated that the combination of introduced mutations in HAS1 intron 3 together with introduced deletions in HAS1 intron 4 promoted a shift to the aberrant splicing pattern found in patients (Kriangkum et al., 2013) that is known to be clinically significant (Adamia, Reiman, et al., 2005). Thus, aberrant intronic HAS1 splicing in malignant cells appears to involve intronic HAS1 mutations and deletions that are recurrent in MM patients but absent from healthy donors. Taken together, this suggests that mutational events and inherited polymorphisms play a central role in aberrant HAS1 splicing.

7. FUNCTIONAL IMPACT OF HAS1VS

7.1. Synthesis of HA

In other work, we have speculated that HAS1Vs may contribute to genetic instability in MM (Adamia, Pilarski, Belch, et al., 2013; Pilarski et al., 2005) by contributing to survival from mitotic catastrophe arising from RHAMM (receptor for hyaluronan-mediated motility) toxicity, as discussed below. Levels of HA represent a balance between synthesis and degradation by hyaluronidases (Bharadwaj et al., 2011). Our focus here is on HA synthesis. In this context, the characteristics of B and hematopoietic stem cells in MM may be consistent with the expected impact of expressing HAS1Vs. An analysis of HA production by *ex vivo* malignant B cells from MM patients suggested the individual HAS1Vs have different patterns of HA synthesis (Adamia, Reiman, et al., 2005). HAS1FL arises from the normal alternative splicing patterns that do not incorporate intronic sequences. HAS1FL is overexpressed in MM-B cells but not detectable in MM-plasma cells (MM-PC). HAS3 is expressed in essentially all hematopoietic cells from normal donors and from patients. MM-B cells from populations that harbor only HAS1 and HAS3 or those with only HAS3 were unable to produce any HA in an *in vitro* assay. MM-B cells expressing HAS1Va, or coexpressing HAS1Va and HAS1FL, made only extracellular HA (Fig. 3.3). MM-B cells expressing only HAS1Vb or coexpressing both HAS1Vb and HAS1FL made only intracellular HA (Fig. 3.3). Interestingly MM-B cell populations that expressed both HAS1Va and HAS1Vb synthesized both extracellular and intracellular HA. Unlike HAS1V-expressing MM-B cells, MM-PC were unable to produce any HA.

7.2. Functional outcomes of HA synthesis in patients

The functional outcomes of extracellular or intracellular HA have yet to be determined. However extracellular HA is likely to be an important mediator for motility that underlies malignant spread. MM-B cells which are highly motile synthesize extracellular HA but MM-PC which lack motility do not (Adamia, Reiman, et al., 2005; Adamia, Maxwell, et al., 2005). In the case of intracellular HA, the correlation between splice-variant transcripts and the presence of intracellular HA implicates HAS1Vb in its synthesis. Only those MM-B cell populations with detectable HAS1Vb transcripts also expressed intracellular HA (Adamia, Reiman, et al., 2005). The impact of intracellular HA will be discussed in a later section.

7.2.1 Motility and malignant spread

Among the B lineage malignant cells in MM, only the B cell compartment is motile when evaluated *in vitro* (Masellis-Smith, Belch, Mant, Turley, & Pilarski, 1996). MM-PC, although responsible for most of the pathology in MM, nevertheless lack motility and presumably the ability to mediate malignant spread. MM-PC also fail to synthesize HA *in vitro*, unlike

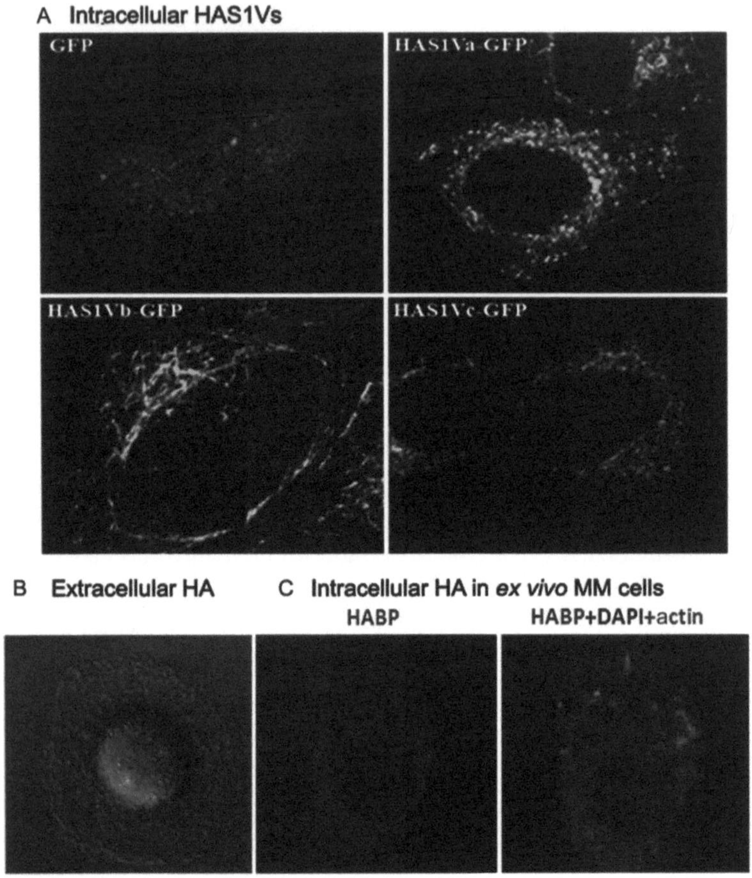

Figure 3.3 Localization of HAS1Vs and synthesis of HA. (A) Transfected HAS1V-GFP constructs localize with cytoskeletal structures (microtubules). (B) Synthesis of extracellular HA by an *ex vivo* MM cell. (C) Synthesis of intracellular HA by *ex vivo* MM cells: blue, DAPI; red, hyaluronan-binding protein (HABP); green, actin. (See the color plate.)

MM-B cells which synthesize a readily detectable extracellular HA matrix (Adamia, Maxwell, et al., 2005; Adamia, Reiman, et al., 2005). Motility, and by extrapolation malignant spread, in MM appears to require HA, presumably synthesized *in vivo* by HASs (Masellis-Smith et al., 1996). MM-B cells express all HAS1 family members as well as HAS2 and HAS3 in various combinations (Adamia, Reiman, et al., 2005). MM-PC express HAS2 and HAS3 but lack HAS1 family members (Adamia, Maxwell, et al., 2005). This, together with the inability of MM-PC to synthesize extracellular HA matrices, suggests that expression of HAS2 and HAS3 is insufficient for HA-dependent motility, at least in the context of MM-PC. Since malignant B cell populations from MM patients often lack expression of HAS1FL but still remain able to synthesize extracellular HA matrices, and those that do express HAS1FL appear unable to synthesize HA (Adamia, Reiman, et al., 2005), the implication is that their HA-dependent motility is likely to involve HA synthesis by HAS1Vs, particularly HAS1Va which was shown to produce extracellular HA (Adamia, Reiman, et al., 2005).

7.2.2 HAS1Vs and oncogenic events

Intriguingly, HAS1Vb, which associates with production of intracellular HA, showed the strongest correlation with poor outcome in MM patients (Adamia, Reiman, et al., 2005), suggesting it may play a very significant role in disease progression. Analysis of HAS1V transfectants indicates that HAS1Vb can confer anchorage independence but does not lead to tumorigenesis *in vivo*. In contrast, HAS1Vc, which shows no significant clinical correlations with patient outcome (Adamia, Reiman, et al., 2005), confers vigorous anchorage-independent growth and is aggressively oncogenic *in vivo* (Ghosh, Kuppusamy, & Pilarski, 2009). HAS1 variants with a significant impact on clinical outcome (HAS1Vb and to a lesser extent HAS1Va) do not appear to be oncogenic as single agents. This leads to the thought that in patients, HAS1Vs may act in concert with each other and potentially with HAS3, to yield an oncogenic outcome in the initial stages of disease. The molecular interactions among HAS1 family members and of HAS1 family members with other partners that lead to oncogenesis may differ from HAS1 interactions that promote disease progression.

7.2.3 Intermolecular interactions of HAS1Vs

The interactions of HAS1 family members with each other and with HAS3 are only partially known (Ghosh et al., 2009). These "inter-HAS" interactions may be different for the synthesis of intracellular HA as compared to those for synthesis of extracellular HA. Bacterial HASs appear to interact with membrane cardiolipins to form pores that extrude HA polymers into the extracellular environment (Tlapak-Simmons, Baggenstoss, Clyne, & Weigel, 1999). There is as yet no evidence to indicate whether or not mammalian HASs form pores, but this is a possibility, given that HAS1 splice variants are expressed in combination with HAS1FL and HAS3 (Ghosh et al., 2009) and these combinations produce extracellular HA (Adamia, Reiman, et al., 2005).

The role of potential HAS pores in intracellular synthesis of HA remains unexplored. However, the ubiquitous presence of HAS3 in all *ex vivo* cell white blood cell populations studied (Adamia et al., 2003; Adamia,

Reiman, et al., 2005), as well as human thymocytes, CD34+ hematopoietic cells and blood cancers (unpublished) raises the possibility that multimolecular interactions among HAS1 family members and HAS3 may be essential components of extracellular or intracellular HA synthesis, at least for nonadherent components of the blood system.

To evaluate coexpression of HAS1FL and HAS1Vs in individual *ex vivo* cells from patient samples, malignant B cells were purified from blood and from BM of patients with WM and analyzed by single-cell RT-PCR for HAS1 family member transcripts (Adamia et al., 2003). All B cells expressed HAS3 and most expressed HAS1Va. Unexpectedly, HAS1Vb was found in 76% of BM-B cells but in only ~10% of blood-B cells in WM (Adamia et al., 2003), indicating an apparent functional difference among malignant B cells in two different compartments of the hematopoietic system from the same patient. Interestingly, HAS1FL and HAS2 were essentially absent from individual B cells in both tissues. This confirms that aberrant splicing to generate HAS1Vs can occur in the absence of the normal splicing events that would otherwise generate HAS1FL. Thus, if pore formation or multimerization occurs in these malignant cells, it appears to involve HAS1Vs and HAS3, but probably does not require HAS1 or HAS2.

7.2.4 HAS1Vs and genetic instability

If HAS1Vs contribute to genetic instability, one would predict that MM-B cells and perhaps the hematopoietic cells from which they arise would have chromosomal abnormalities or abnormal DNA content. The studies by our group (Debes Marun et al., 2012; Pilarski & Belch, 2002; Pilarski, Giannakopoulos, et al., 2000; Pilarski, Mant, & Belch, 1994) suggest that abnormal mitoses leading to chromosomal missegregation may be a frequent process among earlier stage components of the MM clone, possibly driven by HAS1Vs. As discussed below, HAS1Vs may confer rescue from mitotic catastrophe, culminating in aneuploid daughter cells.

8. FUNCTIONAL OUTCOMES OF HAS1VS IN TRANSFECTANTS

8.1. HAS1Vs synthesize intracellular HA

Although it is essential to evaluate molecular impacts in cancer patients and to verify clinical relevance, mechanistic explanations require more defined systems. Use of transfected HAS constructs showed they could encode enzymatically functional protein synthesizing HA *de novo*. HAS1 family members actively synthesized elongating HA, which remained attached to the protein, providing a means to "pull down" HAS1 family proteins undergoing active HA synthesis. This also provided confirmation that such elongating HA was present inside the cell, because all extracellular HA had been removed by hyaluronidase prior to precipitating HAS family members (Ghosh et al., 2009). HAS1Vs localized in the interior of the cell, with a pronounced perinuclear distribution that extended into the cytoplasmic spaces and the Golgi apparatus. HAS1FL and HAS3 showed diffuse staining throughout the cytoplasm with only partial localization to the Golgi. Removal of all extracellular HA by hyaluronidase digestion failed to reduce the HA made by HAS1Vs further confirming its intracellular location. The only *de novo* HA synthesized by HAS1Vs was intracellular (Ghosh et al., 2009). The loss of C-terminal transmembrane domains arising from aberrant splicing may prevent HAS1Vs from "extruding" *de novo* HA molecules into the organelle lumen, thereby retaining HA within the cytoplasm. In contrast, hyaluronidase treatment significantly diminished the HA made by HAS1FL and HAS3 in cells with intact membranes, indicating that as predicted, *de novo* synthesized HA made by HAS1FL and HAS3 had an extracellular membrane localization. Thus, the expression profile of HAS1Vs and other HAS isozymes, with consequent differential pore formation, may determine whether or not *de novo* synthesis of intracellular HA can occur.

8.2. HAS1FL and HAS1Vs form heteromultimers with HAS1Vs as the dominant partners

HAS1Vs localize to intracellular structures (Figs. 3.3 and 3.4). In patient cells, HAS1FL and HAS1Vs appear to be coexpressed in a variety of combinations. To mimic this in transfectants, HAS1Vs were coexpressed alone or in combination with HAS1FL and each other to determine the mechanistic implications of expressing more than one HAS1 family member in the same cell. In cotransfectants, HAS1Vs retained their patchy intracellular distributions. However surprisingly, in cotransfections with HAS1Vs, HAS1FL and HAS3 became colocalized inside transfected cells. These *in vitro* expression experiments demonstrated that HAS1Vs are widely distributed in both cytoskeletal and noncytoskeletal membrane-bound fractions, and that redistribution of HAS1FL occurs in cotransfectants. Although significantly bound to the cytoskeleton, HAS1Vb, in particular, showed a much broader

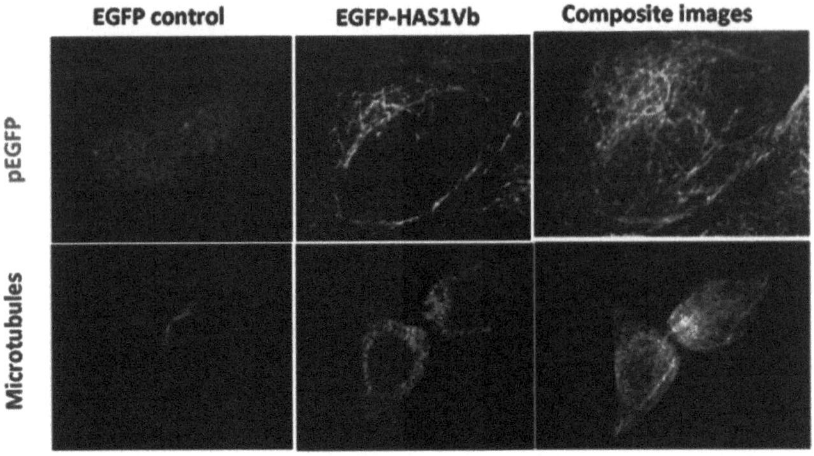

Figure 3.4 HAS1Vb-GFP (green) localizes with microtubules (red). Composite images are assembled z stacks. (See the color plate.)

distribution than did other HAS1 family members, including nuclear localization and possible presence in soluble cytoplasmic regions of the cell, consistent with the observed association between intracellular HA and HAS1Vb (Adamia, Reiman, et al., 2005). This work confirms that in cancer cells, the HAS1 family members are tightly associated with the cellular cytoskeletal network (Figs. 3.4 and 3.5).

8.3. Multimers with HAS1Vs stabilize HAS1FL and maintain intracellular HA

In extracellular matrix, HA has a high turnover rate and membrane-associated HASs appear to have very short half-lives (Bansal & Mason, 1986; Mian, 1986). Because HAS1FL is dominantly relocalized by HAS1Vs, normal turnover rates of these proteins may be altered. In fact, HAS1Vs were shown to have much slower turnover rates than does HAS1FL (Ghosh et al., 2009), the normally spliced variant. Unexpectedly, coexpressed HAS1Vs also protected HAS1FL from degradation, thereby altering turnover rates. This may reflect the fact that HAS1FL and HAS1Vs appear to form a covalently linked "protein-island." Furthermore, the formation of hetero-multimeric complexes of HAS1Vs with or without HAS1FL may be an important precursor to oncogenesis, consistent with their clinical impact. We speculate that this modulation of localization and the increased stability conferred on HAS1FL by association with aberrant HAS1Vs may be

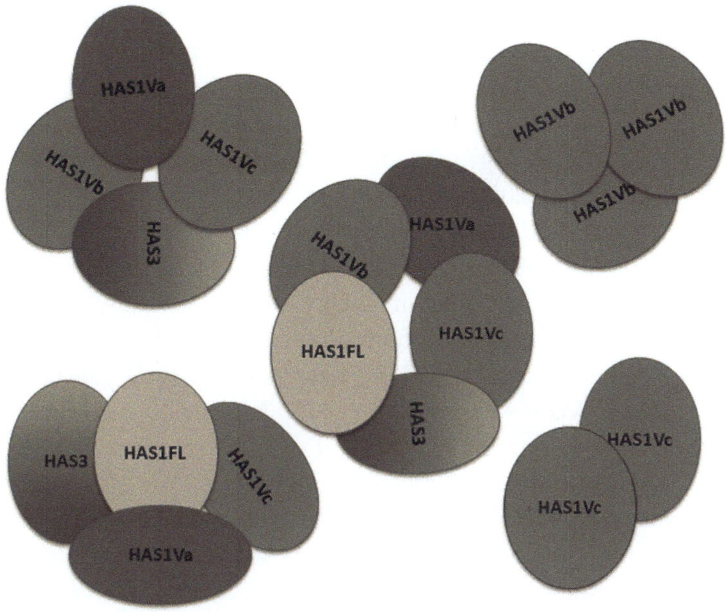

Figure 3.5 Heterogeneity of HAS1 dimers, trimers, and multimers: Potential assemblies of HAS1, HAS1Vs, and HAS3. (See the color plate.)

significant contributors to the malignant process through altered HA dynamics and disrupted cell-cycle regulation.

9. HAS1VS AND MITOTIC CATASTROPHE

9.1. A potential mechanism underlying the oncogenic properties of aberrant HAS1 splice variants: HAS1V-mediated rescue from RHAMM-mediated mitotic catastrophe

9.1.1 Receptor for hyaluronan-mediated motility (RHAMM)

RHAMM (HMMR, CD168) was originally defined as an HA-binding cell-surface protein that allowed cells to migrate (Hardwick et al., 1992). However, RHAMM is a multifunctional protein expressed at multiple cellular and extracellular locales. RHAMM is a cytoplasmic protein that is unconventionally exported to the cell surface (Maxwell, McCarthy, & Turley, 2008). Extracellular RHAMM is a hyaluronan-binding protein that associates with and activates CD44 resulting in stimulation of the Ras/Erk1,2 pathway (Hamilton et al., 2007; Tolg et al., 2010), a cascade that functions in cellular proliferation and is often dysregulated in cancer. Intracellular RHAMM is found in multiple compartments, but its associations with the centrosome and mitotic spindle are the best characterized. RHAMM/centrosome/spindle interactions are required for normal spindle formation and for passage through G2/M (Maxwell, Keats, Belch, Pilarski, & Reiman, 2005; Maxwell et al., 2003; Mohapatra, Yang, Wright, Turley, & Greenberg, 1996). In normal lymphocytes, RHAMM is nearly always intracellular (Pilarski & Belch, 2002). Surface RHAMM is largely absent from B or T cells of healthy donors or healthy cells of cancer patients. RHAMM is expressed both as an intracellular and a cell-surface protein in malignant B and PC in MM and other B lineage cancers (Crainie, Belch, Mant, & Pilarski, 1999; Maxwell et al., 2004; Till, Zuzel, & Cawley, 1999; Turley, Belch, Poppema, & Pilarski, 1993), with important functional roles (Masellis-Smith et al., 1996; Maxwell et al., 2005; Pilarski, Masellis-Smith, et al., 1994). Intracellular RHAMM proteins bind to the mitotic spindle; blocking this RHAMM function results in aberrant mitotic spindle formation (Maxwell et al., 2003, 2005). RHAMM is alternatively spliced to yield three isoforms that are found in healthy cells as well as in malignant cells from patients (Amara, Entwistle, Kuschak, Turley, & Wright, 1996; Assmann, Marshall, Fieber, Hofmann, & Hart, 1998; Crainie et al., 1999; Pilarski & Belch, 2002; Pilarski et al., 1999). Dysregulation of overexpressed RHAMM isoforms correlates with poor outcome in MM (Maxwell et al., 2004), lymphoma (Nagel, Hirschmann, Dirnhofer, Gunthert, & Tzankov, 2010) and colorectal cancers (Lugli et al., 2006; Zlobec, Baker, Terracciano, & Lugli, 2008), and is biologically important in other cancers (Tzankov et al., 2011). Oligonucleotide expression microarray profiling of MM cells identified a significant correlation between RHAMM and reduced overall survival (Maxwell et al., 2004); RHAMM was also identified in exon array screening for activation markers promoting oncogenesis.

9.1.2 Impact of RHAMM overexpression and isoform imbalance

In MM, unifying abnormalities have been elusive. However, overexpression of RHAMM in multiple cellular locales characterizes all MM patients analyzed (Maxwell et al., 2004; Turley et al., 1993). While RHAMM is unlikely to be the only biomarker involved in oncogenesis and disease progression, it is pervasive in its activities, contributing to oncogenic events at multiple points during disease initiation and progression (Maxwell et al., 2008; Telmer, Tolg, McCarthy, & Turley, 2011). It includes intracellular forms that contribute to aberrant mitosis (Assmann, Jenkinson, Marshall, & Hart, 1999; Maxwell et al., 2005, 2003) and surface forms that control migratory events underlying disease spread (Hamilton et al., 2007; Masellis-Smith et al., 1996). Surface RHAMM overexpression may potentiate survival of unanchored malignant cells (e.g., preventing annoikis), while intracellular forms of RHAMM contribute to extensive chromosomal instability by transiently compromising separation of the mitotic spindles. This is consistent with the complex chromosomal abnormalities found in MM (Avet-Loiseau et al., 1997; Debes Marun et al., 2012). Thus, RHAMM has a multifaceted impact on malignancy that is dependent upon its subcellular compartmentalization and at least in part on HA.

Relative imbalance between RHAMM isoforms in MM cells, with predominant expression of the RHAMM isoform lacking exon 4, was found to predict significantly shorter survival of MM patients, as measured by Kaplan Meier analysis (Maxwell et al., 2004), confirming its clinical relevance. In transfectants, too much or too little RHAMM leads to aberrant centrosomes and mitotic spindles, mitotic arrest, and cell death (Maxwell et al., 2005), generating RHAMM "toxicity" through mitotic catastrophe. RHAMM localizes to the centrosome and participates in maintaining mitotic spindle integrity (Maxwell et al., 2003). Extensive mitotic abnormalities and chromosomal missegregation were observed in RHAMM overexpressors. In this context, dysregulation of intracellular RHAMM in malignant cells from clinical samples is associated with mitotic spindle and centrosomal abnormalities.

9.1.3 Interaction between HAS1Vs, intracellular HA, and RHAMM: The Goldilocks hypothesis

We speculate that binding by RHAMM of intracellular HA is a critical component in the early stages of oncogenesis (Adamia, Pilarski, Belch, et al., 2013). Multiple studies link chromosomal instability with the emergence and progression of cancer (Carter, Eklund, Kohane, Harris, & Szallasi, 2006; Ghadimi et al., 2000; Gutierrez et al., 2004; Kops, Weaver, & Cleveland, 2005; Lingle et al., 2002; Perez de Castro, de Carcer, & Malumbres, 2007; Weaver, Silk, Montagna, Verdier-Pinard, & Cleveland, 2007). For proteins involved in cell-cycle regulation, dosage is important. For any given protein, both over- and underexpression can be detrimental, whereas levels that are "just right" allow passage through normal cell division (the "Goldilocks" hypothesis) (Adamia, Pilarski, Belch, et al., 2013; Pilarski et al., 2005). RHAMM overexpression promotes apoptosis and RHAMM silencing hinders apoptosis (Shi et al., 2007). Thus, acquisition by cancer cells of molecular adaptations that can modulate RHAMM overexpression are predicted to result in a biological balance that promotes malignant progression. Aneuploid cells arising from RHAMM-mediated mitotic abnormalities may be encouraged by "dynamic selection" (de Anta et al., 2006) for resistance e.g. (Kuo, Vyas, Jiang, & Hittelman, 1994; Snijders et al., 2008; Swanton, Tomlinson, & Downward, 2006) and emerge as dominant clones (Bahlis, 2012) that mediate relapse and aggressive disease.

In malignant cells that overexpress RHAMM, the levels of intracellular RHAMM appear to be comparable to, or higher than, those leading to RHAMM toxicity in transfectants (unpublished). This suggests that in patients, cancer cells have acquired mechanisms to rescue themselves from the mitotic catastrophe that would otherwise result. The finding of aberrant splice HAS1 variants in cancer cells from patients may provide a mechanism that enables survival of RHAMM overexpressors. The finding that aberrant splicing to generate HAS1Vb leads to significantly reduced survival, suggests that the presence of these aberrant HAS1 splice variants tips the balance in favor of the cancer and against the host.

Intracellular HA is predicted to compete with the mitotic apparatus for binding to RHAMM, thereby modulating the ability of overexpressed RHAMM to cause mitotic catastrophe and cell death. HA-mediated competition may impose control over RHAMM localization patterns inside the cell and on its surface. The HA-binding domains of RHAMM overlap its centrosomal targeting regions (Maxwell et al., 2003), suggesting that intracellular HA could competitively inhibit centrosomal targeting by RHAMM. HAS1 variants can potentially synergize with RHAMM to promote the emergence of increasingly aggressive genetic variants in MM (Fig. 3.6). Abnormalities in RHAMM isoforms and expression levels, together with overexpression of HAS1 variants may determine the ultimate balance between apoptosis/death or chromosomal instability/viability/clonal emergence. High or low levels of RHAMM irreversibly arrest mitosis (Maxwell et al., 2003), leading to cell death unless mitotic rescue permits progression through the cell cycle and potentially viable progeny. Most such cells will die, but those with less compromising abnormalities may survive and emerge as aggressive new cancer variants.

HAS1Vs, particularly HAS1Vb, appear to synthesize intracellular HA which potentially has multiple effects on RHAMM and could modulate the otherwise lethal overexpression of RHAMM. Possible mechanisms

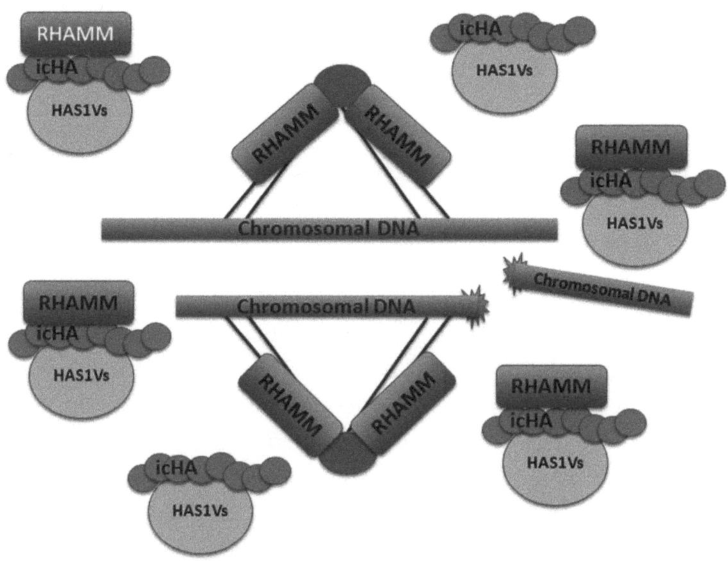

Figure 3.6 Hypothesis: Intracellular HA synthesized by HAS1Vs modulates hyperactive RHAMM toxicity, thereby enabling survival of chromosomally abnormal but viable cancer variants. This model predicts that intracellular HA acts as a competitor to sequester some overexpressed RHAMM away from the mitotic spindle, based on the fact that the HA-binding domain of RHAMM is also the microtubule-binding domain. Thus if RHAMM binds to icHA, it becomes unavailable to the mitotic spindle. ic HA, intracellular HA; brown, mitotic spindle. (See the color plate.)

include (a) competition for RHAMM binding between intracellular HA and components of the mitotic spindle or (b) through redistribution away from their usual subcellular locations of overexpressed RHAMM and its molecular associates in the mitotic apparatus (e.g., dynein; Maxwell et al., 2003). Competitive inhibition or sequestration of intracellular RHAMM by intracellular HA may rescue dividing malignant cells from otherwise lethal effects of RHAMM, leading to more aggressive disease. HAS1Vb thus represents a prognostic biomarker that reflects biologically important properties of a malignant clone as it undergoes stepwise oncogenesis and/or disease progression. At the intracellular level, the impact of HAS1Vs could manifest as increased genetic abnormalities with direct oncogenic potential and/or by predisposing cells to additional oncogenic changes that may enable clonal survival and expansion (Pilarski et al., 2013). Taken together, these observations suggest that in the context of overexpressed and aberrantly spliced HAS1, overexpression of RHAMM isoforms may potentiate malignant transformation and disease progression. Because HAS1V is not expressed by normal blood and BM cells (Adamia, Maxwell, et al., 2005), therapeutic ablation of HAS1V gene expression may selectively attack malignant clones characterized by overexpression of RHAMM and aberrant HAS1 splicing.

10. CONCLUSION

It is becoming increasingly apparent that splicing defects play a key role in cancer, and that genomic changes in splicing elements promote aberrant splicing. Aberrant HAS1 splicing and the consequent production of intracellular HA may contribute to oncogenic events and promote genetic instability in cancer stem cells and their progeny. Aberrant HAS1 splice variants are transforming and in addition to MM and WM, may also characterize colorectal, bladder, lung, breast, and ovarian cancers. Aberrant splicing of HAS1, which predicts for poor survival in MM, appears to arise in part due to inherited predispositions and acquired somatic mutations in HAS1. The centrosomal targeting domain of RHAMM overlaps with its hyaluronan-binding domains, leading to the hypothesis that HAS1Vs modulate overexpressed RHAMM by synthesis of intracellular HA and competition for RHAMM between HA and mitotic spindle components. We speculate that HAS1 splice variants promote emergence of aggressive malignant disease through rescue from RHAMM toxicity, which leads to cell death unless mitotic rescue by intracellular HA permits progression through the cell cycle and survival of viable cancer variants.

ACKNOWLEDGMENTS

We thank Dr. Patrick Pilarski for help with the illustrations. We are grateful for donations from the Myeloma Alberta Support Society and to the American Society for Hematology scholar award program supporting S. A.

REFERENCES

Abrahamsson, A. E., Geron, I., Gotlib, J., Dao, K. H., Barroga, C. F., Newton, I. G., et al. (2009). Glycogen synthase kinase 3beta missplicing contributes to leukemia stem cell generation. *Proceedings of the National Academy of Sciences of the United States of America*, *106*, 3925–3929.

Adamia, S., Crainie, M., Kriangkum, J., Mant, M. J., Belch, A. R., & Pilarski, L. M. (2003). Abnormal expression of hyaluronan synthases in patients with Waldenstrom's macroglobulimenia. *Seminar in Oncology*, *30*, 165–168.

Adamia, S., Haibe-Kains, B., Pilarski, P. M., Bar-Natan, M., Pevzner, S., Avet-Loiseau, H., et al. (2014). A genome-wide aberrant RNA splicing in patients with acute myeloid leukemia identifies novel potential disease markers and therapeutic targets. *Clinical Cancer Research*, *20*(5), 1135–1145.

Adamia, S., Maxwell, C. A., & Pilarski, L. M. (2005). Hyaluronan and hyaluronan synthases: Potential therapeutic targets in cancer. *Current Drug Targets: Cardiovascular & Haematological Disorders*, *5*, 3–14.

Adamia, S., Pilarski, P. M., Bar-Natan, M., Stone, R. M., & Griffin, J. D. (2013). Alternative splicing in chronic myeloid leukemia (CML): A novel therapeutic target? *Current Cancer Drug Targets*, *13*, 735–748.

Adamia, S., Pilarski, P. M., Belch, A. R., & Pilarski, L. M. (2009). Genetic abnormalities in Waldenstrom's macroglobulinemia. *Clinical Lymphoma & Myeloma*, *9*, 30–32.

Adamia, S., Pilarski, P. M., Belch, A. R., & Pilarski, L. M. (2013). Aberrant splicing, hyaluronan synthases and intracellular hyaluronan as drivers of oncogenesis and potential drug targets. *Current Cancer Drug Targets*, *13*, 347–361.

Adamia, S., Reichert, A. A., Kuppusamy, H., Kriangkum, J., Ghosh, A., Hodges, J. J., et al. (2008). Inherited and acquired variations in the hyaluronan synthase 1 (HAS1) gene may contribute to disease progression in multiple myeloma and Waldenstrom macroglobulinemia. *Blood*, *112*, 5111–5121.

Adamia, S., Reiman, T., Crainie, M., Mant, M. J., Belch, A. R., & Pilarski, L. M. (2005). Intronic splicing of hyaluronan synthase 1 (HAS1): A biologically relevant indicator of poor outcome in multiple myeloma. *Blood*, *105*, 4836–4844.

Adamia, S., Treon, S. P., Reiman, T., Tournilhac, O., McQuarrie, C., Mant, M. J., et al. (2005). Potential impact of a single nucleotide polymorphism in the hyaluronan synthase 1 gene in Waldenstrom's macroglobulinemia. *Clinical Lymphoma*, *5*, 253–256.

Almeida, J., Orfao, A., Mateo, G., Ocqueteau, M., Garcia-Sanz, R., Moro, M. J., et al. (1999). Immunophenotypic and DNA content characteristics of plasma cells in multiple myeloma and monoclonal gammopathy of undetermined significance. *Pathologie Biologie (Paris)*, *47*, 119–127.

Amara, F. M., Entwistle, J., Kuschak, T. I., Turley, E. A., & Wright, J. A. (1996). Transforming growth factor-beta1 stimulates multiple protein interactions at a unique cis-element in the 3′-untranslated region of the hyaluronan receptor RHAMM mRNA. *Journal of Biological Chemistry*, *271*, 15279–15284.

Assmann, V., Jenkinson, D., Marshall, J. F., & Hart, I. R. (1999). The intracellular hyaluronan receptor RHAMM/IHABP interacts with microtubules and actin filaments. *Journal of Cell Science*, *112*(Pt. 22), 3943–3954.

Assmann, V., Marshall, J. F., Fieber, C., Hofmann, M., & Hart, I. R. (1998). The human hyaluronan receptor RHAMM is expressed as an intracellular protein in breast cancer cells. *Journal of Cell Science*, *111*(Pt. 12), 1685–1694.

Ast, G. (2004). How did alternative splicing evolve? Nature reviews. *Genetics*, *5*, 773–782.

Avet-Loiseau, H., Andree-Ashley, L. E., Moore, D., Mellerin, M. P., Feusner, J., Bataille, R., et al. (1997). Molecular cytogenetic abnormalities in multiple myeloma and plasma cell leukemia measured using comparative genomic hybridization. *Genes, Chromosomes & Cancer*, *19*, 124–133.

Bahlis, N. J. (2012). Darwinian evolution and tiding clones in multiple myeloma. *Blood*, *120*, 927–928.

Bansal, M. K., & Mason, R. M. (1986). Evidence for rapid metabolic turnover of hyaluronate synthetase in Swarm rat chondrosarcoma chondrocytes. *The Biochemical Journal*, *236*, 515–519.

Bharadwaj, A. G., Goodrich, N. P., McAtee, C. O., Haferbier, K., Oakley, G. G., Wahl, J. K., et al. (2011). Hyaluronan suppresses prostate tumor cell proliferation through diminished expression of N-cadherin and aberrant growth factor receptor signaling. *Experimental Cell Research*, *317*, 1214–1225.

Black, D. L. (2003). Mechanisms of alternative pre-messenger RNA splicing. *Annual Review of Biochemistry, 72*, 291–336.

Brown, R. L., Reinke, L. M., Damerow, M. S., Perez, D., Chodosh, L. A., Yang, J., et al. (2011). CD44 splice isoform switching in human and mouse epithelium is essential for epithelial-mesenchymal transition and breast cancer progression. *The Journal of Clinical Investigation, 121*, 1064–1074.

Carter, S. L., Eklund, A. C., Kohane, I. S., Harris, L. N., & Szallasi, Z. (2006). A signature of chromosomal instability inferred from gene expression profiles predicts clinical outcome in multiple human cancers. *Nature Genetics, 38*, 1043–1048.

Cazzola, M., Rossi, M., & Malcovati, L. (2013). Biologic and clinical significance of somatic mutations of SF3B1 in myeloid and lymphoid neoplasms. *Blood, 121*, 260–269.

Choi, J. W., Kim, D. G., Lee, A.-E., Kim, H. R., Lee, J. Y., Kwon, N. H., et al. (2011). Cancer-associated splicing variant of tumor suppressor AIMP2/p38: Pathological implication in tumorigenesis. *PLoS Genetics*, 7, e1001351.

Crainie, M., Belch, A. R., Mant, M. J., & Pilarski, L. M. (1999). Overexpression of the receptor for hyaluronan-mediated motility (RHAMM) characterizes the malignant clone in multiple myeloma: Identification of three distinct RHAMM variants. *Blood, 93*, 1684–1696.

David, A., Mabjeesh, N., Azar, I., Biton, S., Engel, S., Bernstein, J., et al. (2002). Unusual alternative splicing within the human kallikrein genes KLK2 and KLK3 gives rise to novel prostate-specific proteins. *Journal of Biological Chemistry, 277*, 18084–18090.

de Anta, J. M., Mayo, C., Sole, F., Salido, M., Espinet, B., Corzo, C., et al. (2006). Methotrexate resistance in vitro is achieved by a dynamic selection process of tumor cell variants emerging during treatment. *International Journal of Cancer, 119*, 1607–1615.

Debes Marun, C. S., Belch, A. R., & Pilarski, L. M. (2012). In multiple myeloma, bone-marrow lymphocytes harboring the same chromosomal abnormalities as autologous plasma cells predict poor survival. *American Journal of Hematology, 87*, 579–587.

Farina, A. R., Tacconelli, A., Cappabianca, L., Cea, G., Chioda, A., Romanelli, A., et al. (2009). The neuroblastoma tumour-suppressor TrkAI and its oncogenic alternative TrkAIII splice variant exhibit geldanamycin-sensitive interactions with Hsp90 in human neuroblastoma cells. *Oncogene, 28*, 4075–4094.

Ghadimi, B. M., Sackett, D. L., Difilippantonio, M. J., Schrock, E., Neumann, T., Jauho, A., et al. (2000). Centrosome amplification and instability occurs exclusively in aneuploid, but not in diploid colorectal cancer cell lines, and correlates with numerical chromosomal aberrations. *Genes, Chromosomes & Cancer, 27*, 183–190.

Ghosh, A., Kuppusamy, H., & Pilarski, L. M. (2009). Aberrant splice variants of HAS1 (Hyaluronan Synthase 1) multimerize with and modulate normally spliced HAS1 protein: A potential mechanism promoting human cancer. *Journal of Biological Chemistry, 284*, 18840–18850.

Golshani, R., Hautmann, S. H., Estrella, V., Cohen, B. L., Kyle, C. C., Manoharan, M., et al. (2007). HAS1 expression in bladder cancer and its relation to urinary HA test. *International Journal of Cancer, 120*, 1712–1720.

Gutierrez, N. C., Garcia, J. L., Hernandez, J. M., Lumbreras, E., Castellanos, M., Rasillo, A., et al. (2004). Prognostic and biologic significance of chromosomal imbalances assessed by comparative genomic hybridization in multiple myeloma. *Blood, 104*, 2661–2666.

Hamilton, S. R., Fard, S. F., Paiwand, F. F., Tolg, C., Veiseh, M., Wang, C., et al. (2007). The hyaluronan receptors CD44 and RHAMM (CD168) form complexes with ERK1,2 that sustain high basal motility in breast cancer cells. *Journal of Biological Chemistry, 282*, 16667–16680.

Hardwick, C., Hoare, K., Owens, R., Hohn, H. P., Hook, M., Moore, D., et al. (1992). Molecular cloning of a novel hyaluronan receptor that mediates tumor cell motility. *Journal of Cell Biology, 117*, 1343–1350.

Hartsough, E. J., Basile, K., & Aplin, A. E. (2014). Beneficial effects of raf inhibitor in mutant BRAF splice variant-expressing melanoma. *Molecular Cancer Research*, *12*(5), 795–802.

Iacobucci, I., Lonetti, A., Messa, F., Cilloni, D., Arruga, F., Ottaviani, E., et al. (2008). Expression of spliced oncogenic Ikaros isoforms in Philadelphia-positive acute lymphoblastic leukemia patients treated with tyrosine kinase inhibitors: Implications for a new mechanism of resistance. *Blood*, *112*, 3847–3855.

Keats, J. J., Maxwell, C. A., Taylor, B. J., Hendzel, M. J., Chesi, M., Bergsagel, P. L., et al. (2005). Overexpression of transcripts originating from the MMSET locus characterizes all t(4;14)(p16;q32)-positive multiple myeloma patients. *Blood*, *105*, 4060–4069.

Kirshner, J., Thulien, K. J., Martin, L. D., Debes Marun, C., Reiman, T., Belch, A. R., et al. (2008). A unique three-dimensional model for evaluating the impact of therapy on multiple myeloma. *Blood*, *112*, 2935–2945.

Klein, F., Feldhahn, N., Herzog, S., Sprangers, M., Mooster, J. L., Jumaa, H., et al. (2006). BCR-ABL1 induces aberrant splicing of IKAROS and lineage infidelity in pre-B lymphoblastic leukemia cells. *Oncogene*, *25*, 1118–1124.

Kops, G. J. P. L., Weaver, B. A. A., & Cleveland, D. W. (2005). On the road to cancer: Aneuploidy and the mitotic checkpoint. *Nature Reviews Cancer*, *5*, 773–785.

Kriangkum, J., Warkentin, A., Belch, A. R., & Pilarski, L. M. (2013). Alteration of introns in a hyaluronan synthase 1 (HAS1) minigene convert Pre-mRNA [corrected] splicing to the aberrant pattern in multiple myeloma (MM): MM patients harbor similar changes. *PLoS One*, *8*, e53469.

Kuo, M. T., Vyas, R. C., Jiang, L. X., & Hittelman, W. N. (1994). Chromosome breakage at a major fragile site associated with P-glycoprotein gene amplification in multidrug-resistant CHO cells. *Molecular and Cellular Biology*, *14*, 5202–5211.

Lingle, W. L., Barrett, S. L., Negron, V. C., D'Assoro, A. B., Boeneman, K., Liu, W., et al. (2002). Centrosome amplification drives chromosomal instability in breast tumor development. *Proceedings of the National Academy of Sciences of the United States of America*, *99*, 1978–1983.

Lo, T. C., Barnhill, L. M., Kim, Y., Nakae, E. A., Yu, A. L., & Diccianni, M. B. (2009). Inactivation of SHIP1· in T-cell acute lymphoblastic leukemia due to mutation and extensive alternative splicing. *Leukemia Research*, *33*, 1562–1566.

Lu, X., Feng, X., Man, X., Yang, G., Tang, L., Du, D., et al. (2009). Aberrant splicing of Hugl-1 is associated with hepatocellular carcinoma progression. *Clinical Cancer Research*, *15*, 3287–3296.

Lugli, A., Zlobec, I., Gunthert, U., Minoo, P., Baker, K., Tornillo, L., et al. (2006). Overexpression of the receptor for hyaluronic acid mediated motility is an independent adverse prognostic factor in colorectal cancer. *Modern Pathology*, *19*, 1302–1309.

Masellis-Smith, A., Belch, A. R., Mant, M. J., Turley, E. A., & Pilarski, L. M. (1996). Hyaluronan-dependent motility of B cells and leukemic plasma cells in blood, but not of bone marrow plasma cells, in multiple myeloma: Alternate use of receptor for hyaluronan-mediated motility (RHAMM) and CD44. *Blood*, *87*, 1891–1899.

Matera, A. G., & Wang, Z. (2014). A day in the life of the spliceosome. *Nature Reviews Molecular Cell Biology*, *15*, 108–121.

Matlin, A. J., Clark, F., & Smith, C. W. (2005). Understanding alternative splicing: Towards a cellular code. *Nature Reviews Molecular Cell Biology*, *6*, 386–398.

Maxwell, C. A., Keats, J. J., Belch, A. R., Pilarski, L. M., & Reiman, T. (2005). Receptor for hyaluronan-mediated motility correlates with centrosome abnormalities in multiple myeloma and maintains mitotic integrity. *Cancer Research*, *65*, 850–860.

Maxwell, C. A., Keats, J. J., Crainie, M., Sun, X., Yen, T., Shibuya, E., et al. (2003). RHAMM is a centrosomal protein that interacts with dynein and maintains spindle pole stability. *Molecular Biology of the Cell*, *14*, 2262–2276.

Maxwell, C. A., McCarthy, J., & Turley, E. (2008). Cell-surface and mitotic-spindle RHAMM: Moonlighting or dual oncogenic functions? *Journal of Cell Science, 121*, 925–932.

Maxwell, C. A., Rasmussen, E., Zhan, F., Keats, J. J., Adamia, S., Strachan, E., et al. (2004). RHAMM expression and isoform balance predict aggressive disease and poor survival in multiple myeloma. *Blood, 104*, 1151–1158.

Mian, N. (1986). Analysis of cell-growth-phase-related variations in hyaluronate synthase activity of isolated plasma-membrane fractions of cultured human skin fibroblasts. *The Biochemical Journal, 237*, 333–342.

Mohapatra, S., Yang, X., Wright, J. A., Turley, E. A., & Greenberg, A. H. (1996). Soluble hyaluronan receptor RHAMM induces mitotic arrest by suppressing Cdc2 and cyclin B1 expression. *Journal of Experimental Medicine, 183*, 1663–1668.

Nagel, S., Hirschmann, P., Dirnhofer, S., Gunthert, U., & Tzankov, A. (2010). Coexpression of CD44 variant isoforms and receptor for hyaluronic acid-mediated motility (RHAMM, CD168) is an International Prognostic Index and C-MYC gene status-independent predictor of poor outcome in diffuse large B-cell lymphomas. *Experimental Hematology, 38*, 38–45.

Ocqueteau, M., Orfao, A., Almeida, J., Blade, J., Gonzalez, M., Garcia-Sanz, R., et al. (1998). Immunophenotypic characterization of plasma cells from monoclonal gammopathy of undetermined significance patients. Implications for the differential diagnosis between MGUS and multiple myeloma. *The American Journal of Pathology, 152*, 1655–1665.

Perez de Castro, I., de Carcer, G., & Malumbres, M. (2007). A census of mitotic cancer genes: New insights into tumor cell biology and cancer therapy. *Carcinogenesis, 28*, 899–912.

Pilarski, L. M., Adamia, S., Maxwell, C. A., Pilarski, P. M., Reiman, T., & Belch, A. R. (2005). Hyaluronan synthases and RHAMM as synergistic mediators of malignancy in B lineage cancers. In E. A. Balazs & V. C. Hascall (Eds.), *Hyaluronan structure, metabolism, biological activities, therapeutic applications* (pp. 329–338). Edgewater, New Jersey, USA: Matrix Biology Institute.

Pilarski, L. M., & Belch, A. R. (2002). Clonotypic myeloma cells able to xenograft myeloma to nonobese diabetic severe combined immunodeficient mice copurify with CD34 (+) hematopoietic progenitors. *Clinical Cancer Research, 8*, 3198–3204.

Pilarski, L. M., Debes Marun, C. S., Martin, L. D., Venner, C. P., Pilarski, P. M., & Belch, A. R. (2013). B lymphocytes as cancer stem cells in multiple myeloma. *The Journal of OncoPathology, 1*, 10–21.

Pilarski, L. M., Giannakopoulos, N. V., Szczepek, A. J., Masellis, A. M., Mant, M. J., & Belch, A. R. (2000). In multiple myeloma, circulating hyperdiploid B cells have clonotypic immunoglobulin heavy chain rearrangements and may mediate spread of disease. *Clinical Cancer Research, 6*, 585–596.

Pilarski, L. M., Hipperson, G., Seeberger, K., Pruski, E., Coupland, R. W., & Belch, A. R. (2000). Myeloma progenitors in the blood of patients with aggressive or minimal disease: Engraftment and self-renewal of primary human myeloma in the bone marrow of NOD SCID mice. *Blood, 95*, 1056–1065.

Pilarski, L. M., Mant, M. J., & Belch, A. R. (1994). Circulating monoclonal multi-drug resistant B cells may comprise the malignant stem cells population in multiple myeloma. In F. Dammaco & B. Barlogie (Eds.), *Challenges in modern medicine: Vol 4* (pp. 31–39). Rome, Italy: Ares Serono Symposia.

Pilarski, L. M., Masellis-Smith, A., Belch, A. R., Yang, B., Savani, R. C., & Turley, E. A. (1994). RHAMM, a receptor for hyaluronan-mediated motility, on normal human lymphocytes, thymocytes and malignant B cells: A mediator in B cell malignancy? *Leukemia and Lymphoma, 14*, 363–374.

Pilarski, L. M., Pruski, E., Wizniak, J., Paine, D., Seeberger, K., Mant, M. J., et al. (1999). Potential role for hyaluronan and the hyaluronan receptor RHAMM in mobilization and trafficking of hematopoietic progenitor cells. *Blood, 93*, 2918–2927.
Pilarski, L. M., Reiman, T., Pilarski, P. M., Orr, F. W., & Belch, A. R. (2004). The malignant hierarchy in multiple myeloma: Relationships between malignant cells and bone disease. In G. Singh, & F. W. Orr (Eds.), *Bone metastasis and molecular mechanisms: Pathophysiology* (pp. 109–138). The Netherlands: Kluwer Academic Publishers.
Romano, M., Buratti, E., & Baralle, D. (2013). Role of pseudoexons and pseudointrons in human cancer. *International Journal of Cell Biology, 2013*, 810572.
Roy, B., Haupt, L. M., & Griffiths, L. R. (2013). Review: Alternative splicing (AS) of genes as an approach for generating protein complexity. *Current Genomics, 14*, 182–194.
Shi, Y., Reiman, T., Li, W., Maxwell, C. A., Sen, S., Pilarski, L., et al. (2007). Targeting aurora kinases as therapy in multiple myeloma. *Blood, 109*, 3915–3921.
Shkreta, L., Bell, B., Revil, T., Venables, J. P., Prinos, P., Elela, S. A., et al. (2013). Cancer-associated perturbations in alternative pre-messenger RNA splicing. *Cancer Treatment and Research, 158*, 41–94.
Snijders, A. M., Hermsen, M. A., Baughman, J., Buffart, T. E., Huey, B., Gajduskova, P., et al. (2008). Acquired genomic aberrations associated with methotrexate resistance vary with background genomic instability. *Genes, Chromosomes & Cancer, 47*, 71–83.
Spinelli, R., Pirola, A., Redaelli, S., Sharma, N., Raman, H., Valletta, S., et al. (2013). Identification of novel point mutations in splicing sites integrating whole-exome and RNA-seq data in myeloproliferative diseases. *Molecular Genetics and Genomic Medicine, 1*, 246–259.
Stark, M., Wichman, C., Avivi, I., & Assaraf, Y. G. (2009). Aberrant splicing of folylpolyglutamate synthetase as a novel mechanism of antifolate resistance in leukemia. *Blood, 113*, 4362–4369.
Swanton, C., Tomlinson, I., & Downward, J. (2006). Chromosomal instability, colorectal cancer and taxane resistance. *Cell Cycle, 5*, 818–823.
Telmer, P. G., Tolg, C., McCarthy, J. B., & Turley, E. A. (2011). How does a protein with dual mitotic spindle and extracellular matrix receptor functions affect tumor susceptibility and progression? *Communicative and Integrative Biology, 4*, 182–185.
Till, K. J., Zuzel, M., & Cawley, J. C. (1999). The role of hyaluronan and interleukin 8 in the migration of chronic lymphocytic leukemia cells within lymphoreticular tissues. *Cancer Research, 59*, 4419–4426.
Tlapak-Simmons, V. L., Baggenstoss, B. A., Clyne, T., & Weigel, P. H. (1999). Purification and lipid dependence of the recombinant hyaluronan synthases from Streptococcus pyogenes and Streptococcus equisimilis. *Journal of Biological Chemistry, 274*, 4239–4245.
Tolg, C., Hamilton, S. R., Morningstar, L., Zhang, J., Zhang, S., Esguerra, K. V., et al. (2010). RHAMM promotes interphase microtubule instability and mitotic spindle integrity through MEK1/ERK1/2 activity. *Journal of Biological Chemistry, 285*, 26461–26474.
Turley, E. A., Belch, A. J., Poppema, S., & Pilarski, L. M. (1993). Expression and function of a receptor for hyaluronan-mediated motility on normal and malignant B lymphocytes. *Blood, 81*, 446–453.
Tzankov, A., Strasser, U., Dirnhofer, S., Menter, T., Arber, C., Jotterand, M., et al. (2011). In situ RHAMM protein expression in acute myeloid leukemia blasts suggests poor overall survival. *Annals of Hematology, 90*, 901–909.
Vassen, L., Khandanpour, C., Ebeling, P., van der Reijden, B. A., Jansen, J. H., Mahlmann, S., et al. (2009). Growth factor independent 1b (Gfi1b) and a new splice variant of Gfi1b are highly expressed in patients with acute and chronic leukemia. *International Journal of Hematology, 89*, 422–430.

Wappenschmidt, B., Becker, A. A., Hauke, J., Weber, U., Engert, S., Kohler, J., et al. (2012). Analysis of 30 putative BRCA1 splicing mutations in hereditary breast and ovarian cancer families identifies exonic splice site mutations that escape in silico prediction. *PLoS One*, *7*, e50800.

Weaver, B. A. A., Silk, A. D., Montagna, C., Verdier-Pinard, P., & Cleveland, D. W. (2007). Aneuploidy acts both oncogenically and as a tumor suppressor. *Cancer Cell*, *11*, 25–36.

Wolfe, K. H., Sharp, P. M., & Li, W. H. (1989). Mutation rates differ among regions of the mammalian genome. *Nature*, *337*, 283–285.

Yang, G., Huang, S. C., Wu, J. Y., & Benz, E. J., Jr. (2008). Regulated Fox-2 isoform expression mediates protein 4.1R splicing during erythroid differentiation. *Blood*, *111*, 392–401.

Zlobec, I., Baker, K., Terracciano, L. M., & Lugli, A. (2008). RHAMM, p21 combined phenotype identifies microsatellite instability-high colorectal cancers with a highly adverse prognosis. *Clinical Cancer Research*, *14*, 3798–3806.

CHAPTER FOUR

Hyaluronan Synthases Posttranslational Regulation in Cancer

Davide Vigetti, Alberto Passi[1]

Dipartimento di Scienze Chirurgiche e Morfologiche, Università degli Studi dell'Insubria, Varese, Italy

[1]Corresponding author: e-mail address: alberto.passi@uninsubria.it

Contents

1. Introduction 96
2. HA Synthesis 98
3. HA Catabolism 101
4. HAS2 Phosphorylation, AMPK, and Its Dual Effects on Tumors 103
5. O-GlcNAcylation and Cancer 108
   5.1 HAS2 O-GlcNAcylation 112
6. Conclusions 113
Acknowledgment 113
References 113

Abstract

Hyaluronan (HA) is a critical component of cancer microenvironment that is known to increase tumor progression and aggressiveness. The synthesis of HA starts from the cytosolic precursors UDP-*N*-acetylglucosamine and UDP-glucuronic acid. These two sugar nucleotides have several functions in addition to glycoconjugate synthesis and glucuronidation reactions, each of which can have a critical role in cancer. HA is synthesized by a family of three HA synthase (HAS) enzymes and, in this review, we described the main posttranslational modifications that are known to regulate HA metabolism. In particular, as the main HAS in adult tissues is HAS2, we focused on the role of AMPK-mediated phosphorylation and glycosylation by O-linked *N*-acetylglucosamine (O-GlcNAcylation) of HAS2 which mediate HAS2 inactivation and activation, respectively. HA catabolism, furnishing glucuronic acid and *N*-acetylglucosamine, can represent for a cancer cell a valid source of substrates to sustain complex tumor metabolism, and we highlight a presumable metabolic fate of such sugars in tumor cells.

Advances in Cancer Research, Volume 123
ISSN 0065-230X
http://dx.doi.org/10.1016/B978-0-12-800092-2.00004-6

1. INTRODUCTION

Communication among cells is fundamental in multicellular organisms and can regulate several aspects of cell biology not only in physiological but also in pathological conditions. Cell microenvironment is critical to regulate signals that reach the cell surface. Not only nutrients and waste products diffuse in and out of the surroundings of the cells; of equal significance are all signal molecules that can modulate the survival, death, proliferation, differentiation, and migration of the cells. Moreover, in this extracellular space take place all the critical molecular recognition reactions that permit immune competent cells to trigger specific response to damaged components or foreign structures. The extracellular matrix (ECM) components can interact with such signal molecules and, hence, have an active role in the regulation of many cellular events (Vigetti, Gotte, Pavao, & Theocharis, 2013).

Hyaluronan (HA) is one of the most abundant and ubiquitous components of vertebrate ECM (Laurent & Fraser, 1992). This glycosaminoglycan (GAG) is a linear unsulfated polysaccharide that is composed of repeating units of D-glucuronic (GlcUA) acid and *N*-acetylglucosamine (GlcNAc) linked together by alternating beta-1,4 and beta-1,3 glycosidic bonds. HA length can greatly vary from the disaccharide unit to thousands of disaccharides, determining the polydispersity of HA *in vivo* (Jiang, Liang, & Noble, 2011). HA is synthesized as a high-molecular-weight polymer by three HA synthases (HAS1, 2, and, 3) that use cytosolic UDP-GlcUA and UDP-GlcNAc as precursors (Weigel & DeAngelis, 2007). The degradation of HA is due to a family of enzymes named hyaluronidases (HYALs) (Stern, 2008). HYALs produce low-molecular-weight HA fragments in the extracellular space that are further internalized in the cell by receptor-mediated endocytosis and degraded by other enzymes in endosomes and lysosomes (Yoshida et al., 2013).

HA has many biological activities that depend on the polysaccharide molecular weight and, interestingly, some of these properties appear to be contradictory (Stern, Asari, & Sugahara, 2006). High-molecular-weight HA is a space-filling molecule with hydration capability, but it also has several effects on cell biology. In fact, high-molecular-weight HA has anti-angiogenic, immunosuppressive, and anti-inflammatory properties and promotes tissue integrity and quiescence. On the other hand, small HA fragments are proangiogenic, proinflammatory, and immunostimulatory

(Stern et al., 2006; Chapter 1). In general, these short oligosaccharides tend to be involved in the body alarm system, transmitting various modes of "danger signals" (Powell & Horton, 2005). Interestingly, some of the smaller HA oligosaccharides or monodisperse HA appear to ameliorate the effects of these stress signals (Asari, Kanemitsu, & Kurihara, 2010; Hill, Kessler, Rho, Cowman, & de la Motte, 2012; Hill et al., 2013). Further, HA can modulate other physiological processes in development, differentiation, and wound healing, as well as pathological issues having a role in tumor biology and cardiovascular pathologies (Hascall & Karamanos, 2011; Jiang, Liang, & Noble, 2007; Kashima et al., 2013; Toole, Wight, & Tammi, 2002; Vigetti, Viola, Karousou, Genasetti, et al., 2008).

This multiplicity of effects can be ascribed to the presence of several HA receptors on the cell surface that can trigger specific signaling cascades. These receptors are CD44, RHAMM, Lyve-1, HA receptor for endocytosis, and Toll-like receptors 4 and 2 (TLR4 and 2) (Jiang et al., 2011). The latter two receptors are turned on in the presence of short fragments of HA and can mediate many of the proinflammatory effects of low-molecular-weight HA (Fieber et al., 2004; Jiang et al., 2005). On the other hand, very low-molecular-weight HA can prevent HA receptor clustering and reduce signaling (Toole, 2004; Vigetti, Viola, Karousou, Rizzi, et al., 2008). This aspect could open new possibilities for the use of HA oligosaccharides as pharmacological agents.

The role of HA in the tumor microenvironment has been investigated for a long time and it is now clear that HA accumulates in the stroma of various tumors, activating signaling pathways that modulate cell proliferation, motility, and invasiveness (Iijima, Konno, & Itano, 2011; Misra et al., 2011). The presence of HA is in many cancers associated with malignant progression and poor survival as in the case of breast, colorectal, prostate, lung, ovarian, and gastric tumors (Tammi et al., 2008). HA is able to induce epithelial–mesenchymal transition (EMT), a critical event in the initiation of metastasis (Itano & Kimata, 2008). HA also modulates angiogenesis, which is an absolute requirement for tumor growth and metastasis. HA dimensions are critical to angiogenesis; high- and low-molecular-weight HA have opposite effects on neovascularization, inhibiting, and promoting angiogenesis, respectively (Sironen et al., 2011).

Another important issue in cancer is inflammation. In fact, chronic inflammation is a risk factor for cancer development and it contributes to the malignant microenvironment (Coussens & Werb, 2002; Hanahan & Weinberg, 2011; Mantovani, Allavena, Sica, & Balkwill, 2008).

Inflammation promotes tumorigenesis and cancer progression by supplying growth factors that sustain cancer cell proliferation and/or survival, proangiogenic factors, ECM-modifying enzymes that promote invasion and metastasis, and signals that lead to EMT (Bauer et al., 2012). HA has a critical role in the modulation of inflammation, and presence of this polysaccharide facilitates immune cell infiltration in the tumor stroma, influencing processes such as macrophage differentiation and/or chemokines release (Acquati et al., 2011; Fuchs et al., 2013; Kobayashi et al., 2010).

From this brief list of HA effects on tumors, it is clear that HA has a pivotal role in tumor cell modulation and in tumor-associated processes. However, it is not trivial to identify which cell synthesizes HA to increase the tumor microenvironment content of this GAG. If tumor cell metabolism is focused to produce energy and cellular components that permit rapid growth (Hanahan & Weinberg, 2011), then HA synthesis would be expected to be negligible for cancer cells. However, alteration of the expression of HA-synthetic enzymes is one of the most common events in cell transformation (Auvinen et al., 2014; Itano, Sawai, Miyaishi, & Kimata, 1999; Kosaki, Watanabe, & Yamaguchi, 1999; Li & Heldin, 2001; Nishida, Knudson, Knudson, & Ishiguro, 2005; Simpson, Wilson, Furcht, et al., 2002; Simpson, Wilson, & McCarthy, 2002; Udabage et al., 2005) and the ability of cancer cells to produce HA may be important for autocrine stimulation of receptors such as CD44 and RHAMM (Hall & Turley, 1995; Misra et al., 2011; Slevin et al., 2007). HA may create a well-hydrated microenvironment in which nutrients will easily diffuse and cells efficiently divide. A HA-rich microenvironment could result from crosstalk between tumor and adjacent stromal cells, which then produce HA to favor tumor growth (Knudson, Biswas, & Toole, 1984). Moreover, cancer cells, through the action of HA-degrading enzymes, could take advantage of low-molecular-weight HA to promote angiogenesis. The complete degradation of HA could provide sugars for metabolic requirements of cancer cells. This point could be crucial in early phases of tumorigenesis when new blood vessels are not yet formed. In this review, we focus on HA synthesis and we will discuss key mechanisms to control HA production and their involvement in tumor biology.

2. HA SYNTHESIS

It is now well accepted that cancer is in part a metabolic disorder (Vander Heiden, Cantley, & Thompson, 2009). Here, we describe the metabolism of HA, from the precursors to the final degradation products to highlight the advantage that such metabolism could bring to tumor cells. HA production depends on the availability of UDP-sugars precursors (i.e., UDP-GlcUA and UDP-GlcNAc) in the cells. This implies that cell metabolism must support anabolic pathways, and therefore the cells producing HA must have a positive energy balance. In fact, UDP-sugar precursors represent molecules with a high cost in terms of energy, as UDP-sugars compete with glycolysis and other pathways (i.e., pentose phosphate pathway) for their synthesis (Fig. 4.1).

The synthesis of UDP-GlcUA requires the activity of the UDP-glucose dehydrogenase (UGDH), which produces UDP-GlcUA from UDP-glucose (Vigetti et al., 2006). This reaction is possible in the presence of NAD^+, which is converted to NADH during the twofold oxidation of the C-6 of UDP-glucose. This double oxidation is very uncommon in nature. Moreover, this aspect of HA synthesis introduces the importance of UDP-GlcUA in terms of energy balance. When all the steps of the anabolic pathway from glucose to UDP-GlcUA are considered, it is evident that the costs of UDP-GlcUA synthesis are completely repaid by the reoxidation in mitochondria of NADH to produce ATP molecules. Moreover, the UGDH reaction can contribute to the NAD^+/NADH ratio that is critical to regulate the activity of NAD-dependent histone deacetylases (i.e., sirtuins), which regulate enzyme activity and gene expression, and are critical in cancer (Yuan, Su, & Chen, 2013).

The synthesis of UDP-GlcUA is a critical step not only for GAG synthesis. In fact, UDP-GlcUA is the donor substrate for glucuronidation. This is a critical detoxification reaction that is catalyzed by a family of isoenzymes called uridine diphosphoglucuronosyltransferases (UGT). Glucuronidation reactions take place mainly in liver, where they serve in part to reduce the cytotoxicity of antiproliferative drugs and may enhance chemoresistance (Nagar & Remmel, 2006). Moreover, as several hormones (i.e., estrogens) are glucuronidated, UGT can interfere with hormone-induced growth of tumors, as described for breast cancer (Guillemette, Millikan, Newman, & Housman, 2000).

As mentioned above, HA is synthesized by HAS1, 2, and, 3 (Itano & Kimata, 2002; Vigetti, Viola, Karousou, De Luca, & Passi, 2014), which localize to the cellular plasma membrane as discussed in several chapters within this volume. In mammals, genes encoding HA precursors and synthesis are probably regulated to coordinate the production of substrates with the necessity to synthesize HA (Tammi et al., 2011). In fact, overexpression of UGDH with

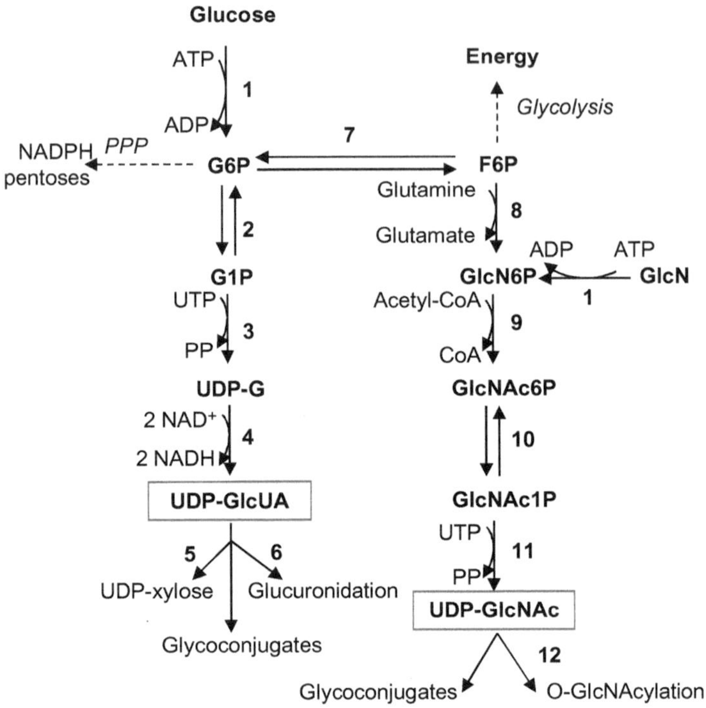

Figure 4.1 HA precursor synthesis pathways. The two UDP-sugars precursors for HA synthesis are boxed and indicated within the context of the biochemical pathways in which they are involved. Numbers represent the specific enzyme that catalyzes the reaction. G6P, glucose 6-phosphate; G1P, glucose 1-phosphate; UDP-G, UDP-glucose; UDP-GlcUA, UDP-glucuronic acid; F6P, fructose 6-phosphate; Glc6P, glucose 6-phosphate; GlcNAc6P, *N*-acetylglucosamine 6-phosphate; GlcNAc1P, *N*-acetylglucosamine 1-phosphate; UDP-GlcNAc, UDP-*N*-acetylglucosamine; GlcN, glucosamine; PPP, pentose phosphate pathway. 1: hexokinase/glucokinase; 2: phosphoglucomutase; 3: UDP-glucose pyrophosphorylase; 4: UDP-glucose dehydrogenase; 5: UDP-GlcUA decarboxylase (a.k.a. UDP-xylose synthase); 6: UDP glucuronosyltransferases; 7: phosphoglucose isomerase; 8: glutamine:fructose 6-phosphate transaminase; 9: glucosamine-phosphate *N*-acetyltransferase; 10: phosphoacetylglucosamine mutase; 11: UDP-*N*-acetylglucosamine pyrophosphorylase; 12: O-GlcNAc transferase.

concomitant increase of UDP-GlcUA availability induces HAS2 and HAS3 expression (Vigetti et al., 2006). Although the molecular mechanism of such regulation is not known, this point could be important in cancers that have altered UGDH expression (Huang et al., 2010). HASes, and HAS2, in particular, are also regulated at transcriptional level and many signaling molecules induce variations in HAS mRNA content (Karousou et al., 2013; Tammi et al., 2011; Vigetti, Genasetti, et al., 2009). Recently, our research group discovered that oxysterols and their receptors (i.e., liver X receptors) can modulate HAS2 and 3 gene expression in smooth muscle cells (Viola et al., 2013). Interestingly, HAS2 has a natural antisense transcript (HAS2-AS1) that has been described to stabilize HAS2 mRNA (Michael et al., 2011). These long noncoding RNAs are known to regulate gene expression by epigenetic modifications (Rinn & Chang, 2012) and this point opens the fascinating possibility that epigenetics contribute to the cell microenvironment.

In recent years, the cytosolic portion of the HAS2 isoenzyme has been found to contain posttranslational modifications (Table 4.1) that are critical to regulate the catalytic activity (Tammi et al., 2011) and may modulate HA polymer length (Fig. 4.2).

The critical role of HA in tumorigenesis and cancer progression is highlighted by the effects of the HAS inhibitor 4-methylumbelliferone (4-MU) (Nakamura et al., 1995). This compound lowers HA deposition by reducing the cellular concentration of UDP-GlcUA (Kakizaki et al., 2004), decreasing the expression of HASes (Vigetti, Rizzi, et al., 2009), and is able to induce apoptosis (Vigetti, Rizzi, et al., 2011). Interestingly, 4-MU has been shown to inhibit proliferation and motility in esophageal squamous cell carcinoma or prostate cancer cells and to decrease liver metastases in a mouse model for melanoma (Lokeshwar et al., 2010; Twarock et al., 2011; Whatcott, Han, Posner, Hostetter, & Von Hoff, 2011; Yoshihara et al., 2005).

3. HA CATABOLISM

Degradation of HA and balance of HA levels are mainly mediated by the HYALs as discussed in other chapters in this volume. The mechanism of HA degradation is complex and involves different cellular compartments (Stern, 2004). At the cell surface, HYAL2 is thought to degrade high-molecular-mass HA to an approximately 20 kDa product, whereas HYAL1 can degrade high-molecular-mass HA to small oligomers, primarily to tetrasaccharides. HYAL1 has a lysosomal localization and the 20 kDa HA fragments are internalized in the cells via CD44-mediated endocytosis as depicted in Fig. 4.3. Recently, a new HA-degrading enzyme (KIAA1199) distinct from HYALs and CD44 was described (Yoshida et al., 2013) (Fig. 4.3A and B).

HA degradation yields GlcUA and GlcNAc. These two sugars could be used by cancer cells to provide energy and building blocks to sustain the rapid growth of tumor cells as ribose, hexosamine, and sialic acid as shown in Fig. 4.3C and D.

Table 4.1 HAS2 posttranslational modifications

Modification	Enzyme	Stimuli	Residue	Effects	Reference
Phosphorylation	ERK 1/2	Undetermined	Unknown	Increased activity	Bourguignon et al. (2007)
Phosphorylation	AMPK	Low ATP:AMP ratio; metformin; and 2-deoxyglucose	T110	Decreased activity	Vigetti, Clerici, et al. (2011)
Ubiquitination			K190	Dimerization	Karousou et al. (2010)
O-GlcNAcylation	OGT	Hyperglycemia, increased HBP flux	S221	Increased stability and activity	Vigetti et al. (2012)

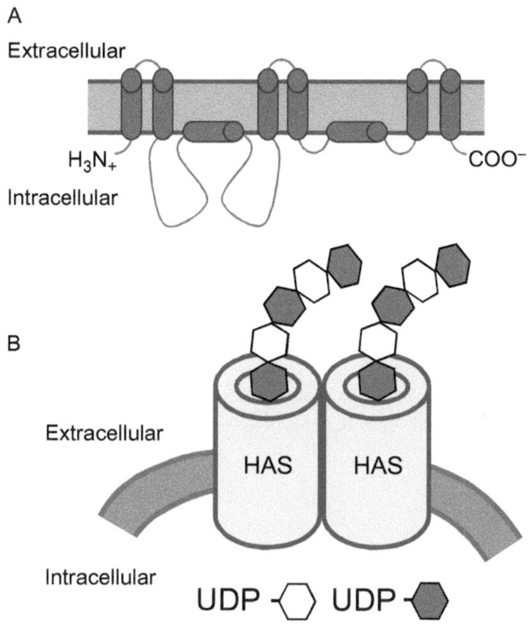

Figure 4.2 Mammalian HAS topology models. (A) Putative topological representation of a mammalian HAS protein in which are highlighted: six transmembrane domains, two membrane-associated domains, and the large predicted cytosolic loop. (B) Tridimensional representation of HAS. As HASes are capable of extruding the HA nascent chain to the extracellular space, the enzyme is represented as a hollow cylinder, with an extracellular domain, hydrophobic domain, and an internal cytoplasmic domain. *In vivo* it is thought that HAS2 and HAS3 proteins can complex with other HAS enzymes dependent upon ubiquitination status. The scheme shows a HAS dimer. White and gray hexagons represent the two UDP-sugar precursors.

4. HAS2 PHOSPHORYLATION, AMPK, AND ITS DUAL EFFECTS ON TUMORS

It is well known that *HAS* genes are regulated at the expression level and alterations of HAS transcripts are found to be related to cell transformation and tumor progression (Adamia, Pilarski, Belch, & Pilarski, 2013; Itano & Kimata, 2008; Sironen et al., 2011). Among the three HAS enzymes, HAS2 seems to be the most intriguing not only because knockout mice die early with severe heart malformations (Camenisch, Schroeder, Bradley, Klewer, & McDonald, 2002; Camenisch et al., 2000) but also because HAS2 protein is strictly regulated by several posttranslational

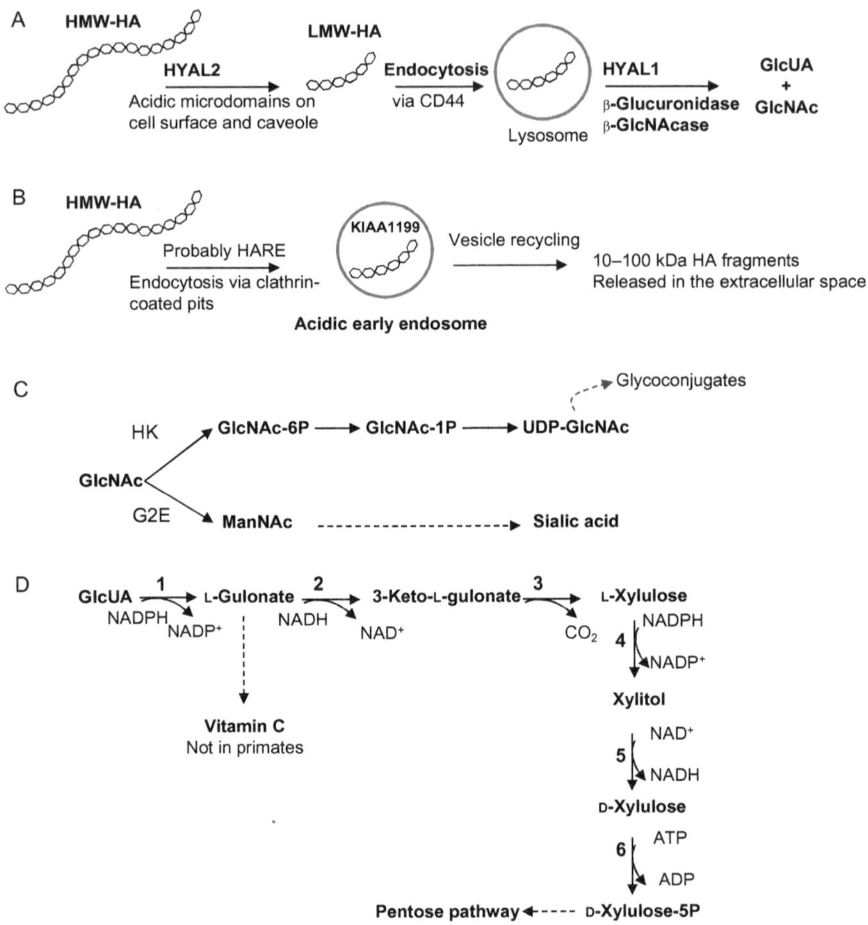

Figure 4.3 HA catabolism. (A) Model of the HA degradation pathway via HYAL2, receptor-mediated endocytosis of ~20 kDa HA fragments, and lysosomal degradation to monosaccharides. (B) Recently proposed model of HA fragmentation independent of HYAL enzyme by KIAA1199. (C) GlcNAc derived from GAG degradation can be recycled and incorporated back into UDP-GlcNAc (using hexokinase (HK)) or sialic acid (in several steps not shown here, using GlcNac 2-epimerase (G2E) to synthesize *N*-acetylmannosamine (ManNAc)). (D) GlcUA derived from GAG degradation can be converted into pentose phosphate pathway intermediates. 1: glucuronate reductase; 2: gulonate 3-dehydrogenase; 3: dehydro-L-gulonate decarboxylase; 4: L-xylulose reductase; 5: xylitol dehydrogenase; 6: D-xylulose kinase.

modifications that alter activity (Bourguignon, Gilad, & Peyrollier, 2007; Vigetti, Clerici, et al., 2011; Vigetti et al., 2012), half-life (Vigetti et al., 2012), structure (Karousou et al., 2010), and probably localization (Hascall et al., 2014). In tumor biology, HAS2 was shown to suppress tissue metalloproteinase inhibitor 1 (TIMP-1) (Bernert, Porsch, & Heldin, 2011), which has a critical role in the remodeling of cell microenvironment by regulating metalloproteinases. Recently, HAS2 has been described to favor EMT independently of its enzymatic activity, suggesting a new and not yet investigated function (Porsch et al., 2013). This opens the possibility that HAS2 could be not only an enzyme but also a structural protein linking the cell surface with cytoskeleton or a coreceptor able to modulate signaling cascades. These possibilities remain a hypothesis until a molecular partner for HAS2 is identified.

HAS localization is a critical point for their proper function. In fact, these enzymes follow the secretory pathway to reach the plasma membrane, where HA synthesis takes place. This process must be finely tuned, but there is little information about the HAS maturation process. HA is present not only in the extracellular space but also inside the cells (Evanko & Wight, 1999). Although HA-containing vesicles can derive from extracellular HA catabolism (which precedes endocytosis), the possibility that HASes could be active in the ER and Golgi is intriguing. HA cable-like structures with high-adhesive capacity for immune cells have been observed to start not from the cell surface, but from inside vesicles (de La Motte, Hascall, Calabro, Yen-Lieberman, & Strong, 1999; de la Motte, Hascall, Drazba, Bandyopadhyay, & Strong, 2003; Majors et al., 2003), which could be due to HAS deregulation in these intracellular vesicles (Hascall et al., 2004). *In vitro* experiments on purified microsomal membranes (containing ER and Golgi) revealed that HASes are active in these subcellular compartments and such activity can be regulated by several stimuli (Vigetti, Genasetti, et al., 2009). Interestingly, HASes are quickly removed from the plasma membrane by endocytosis or vesicle recycling indicating that HA synthesis is a very regulated process (Rilla et al., 2005).

In the plasma membrane, HASes can form dimers (Fig. 4.2B). In particular, ectopic overexpression of tagged HAS2 and HAS3 in COS-1 cells revealed the formation of homo- and heterodimers. Interestingly, the ubiquitination of lysine 190 of HAS2-mediated HAS2 homodimer formation and the mutation of this lysine residue reduced HAS2 wild-type activity (Karousou et al., 2010). This mutated HAS2 represents a powerful tool to study this protein, since it functions as a dominant-negative enzyme.

HA synthesis is rapidly regulated by external stimuli suggesting posttranslational regulation of HASes (Vigetti, Genasetti, et al., 2009). Several treatments (phorbol esters, interleukin 1beta, and PDGF) that induce protein phosphorylation also increase HA secretion in the medium, and phosphatase treatment of the membrane reduces HA synthesis. Also, the activation of ERK1/2 (Bourguignon et al., 2007) and PKC (Wang & Hascall, 2004) was found to lead to activation of HASes. These observations suggested that one of the main posttranslational modifications (i.e., phosphorylation) was acting to increase HA deposition via HAS activation. To complicate this picture, in 2011 our research group found that threonine 110 of HAS2 can be phosphorylated by the adenosine monophosphate-activated protein kinase (AMPK) (Vigetti, Clerici, et al., 2011), which drastically reduced HA deposition without altering other GAGs synthesis in human aortic smooth muscle cells. This effect was due to the inhibition of HAS2 activity by AMPK phosphorylation, leading to reduced cell proliferation, motility, and adhesion of inflammatory cells. This discovery shed light on the complexity of the regulation of HAS2 by phosphorylation indicating HAS2 as one of the most important enzymes in HA synthesis.

AMPK is a master regulator of metabolism and its activity is mainly dependent on ATP:AMP ratio in the cells (Hardie & Alessi, 2013). In conditions of low-cellular energy, AMPK phosphorylates several target proteins with the aim of turning on catabolism and turning off anabolism as shown in Fig. 4.4.

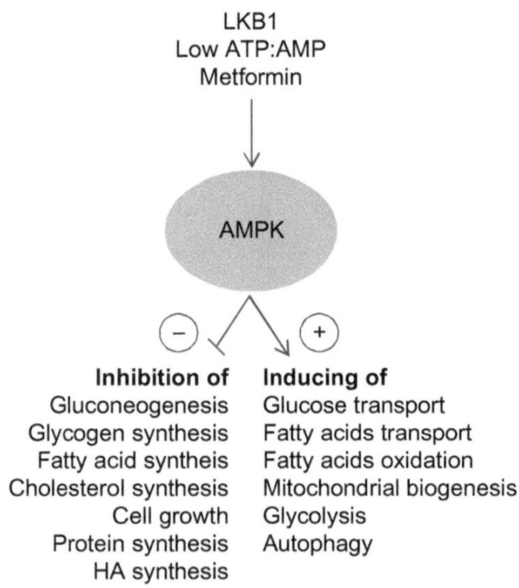

Figure 4.4 Main effects of AMPK on metabolism. Activation of AMPK by various stimuli, including AMP or kinases (e.g., LKB1), blocks anabolism, whereas AMPK activity increases flux through catabolic pathways to increase ATP levels.

AMPK is a heterotrimeric protein with a catalytic α subunit and two regulatory subunits, β and γ that, when the ATP:AMP ratio decreases, inhibits anabolic processes and induces catabolic pathways in order to restore ATP levels (Towler & Hardie, 2007). The mechanism for activating AMPK is complex and involves phosphorylation of a critical threonine-172 residue on the α-catalytic subunit (Liang & Mills, 2013). Among the AMPK upstream kinases, LKB1 is critical in tumor biology. In fact, LKB1 is a well-known tumor suppressor gene and many antitumor effects are mediated by AMPK (Hardie & Alessi, 2013). In fact, AMPK activation alters cell growth by blocking cell cycle-stabilizing p53 and cyclin-dependent kinase inhibitors $p21^{WAF}$ and $p27^{CIP1}$, inhibiting the biosynthesis of most important cellular macromolecule including glucose, fatty acids, triglycerides, cholesterol, glycogen, ribosomal RNA, and proteins. Acting on the mechanistic target of rapamycin complex 1 (mTORC1), Raptor, and TSC2, AMPK decreases translation of many factors including HIF-1α, which is required in many tumors to contribute to the Warburg effect (Hardie & Alessi, 2013; Liang & Mills, 2013).

Whether the tumor suppressor function of LKB1 is actually mediated through AMPK is still under investigation. Interestingly, the AMPK activating drug metformin (a well-known antidiabetic agent) significantly reduces the incidence of cancer to a greater extent than other hypoglycemic treatments (Evans, Donnelly, Emslie-Smith, Alessi, & Morris, 2005). Our data indicated that metformin is able to drastically reduce HA synthesis (Vigetti, Clerici, et al., 2011) probably contributing to prevention of cancer progression and development.

On the other hand, AMPK can also favor tumor growth. In many human cancers, AMPK subunits are significantly elevated and probably sustain tumor metabolism (Liang & Mills, 2013). The aim of AMPK activation is to produce energy, which is accomplished partly through AMPK-stimulated mitochondrial biogenesis and expression of nuclear-encoded mitochondrial genes by upregulating PGC-1α. Glucose transporter GLUT1, glycolytic enzymes, and lactate transporters are upregulated by AMPK (Liang & Mills, 2013). Moreover, AMPK can confer on cancer cells the flexibility to use nutrients other than glucose, facilitating fatty acid catabolism and favoring the release of fatty acids by adjacent adipocytes in the tumor microenvironment (Nieman et al., 2011). Many tumors increase mitochondrial glutaminolysis, which essentially sustains the Krebs cycle by using glutamine and preventing glucose catabolism (Gao et al., 2009). This point is crucial as AMPK, by inducing mitochondrial biogenesis, promotes glutaminolysis and permits the anabolic use of glucose to produce ribose for nucleotide synthesis, and NADPH for protection against oxidative damage. Moreover, glucose can also be available in sustaining glycoconjugate synthesis (by entering the hexosamine biosynthetic pathway (HBP) and GlcUA-biosynthetic pathways) favoring cell growth and division.

5. O-GLCNACYLATION AND CANCER

The O-GlcNAcylation of the proteins, reported for the first time by Torres and Hart (1984), is a dynamic posttranslational modification that regulates several biological processes and in human diseases. O-GlcNAcylation is a reaction in which β-*N*-acetylglucosamine (O-GlcNAc) is linked to the side chain hydroxyl group of a serine or threonine residue, and it is catalyzed by O-GlcNAc transferase (OGT). The removal of the O-GlcNAc is catalyzed by O-GlcNAcase (OGA) (Fig. 4.5A). This reaction is often in a dynamic equilibrium with protein phosphorylation, competing for alternate modification of the same amino acid residue, which plays a critical role in protein structure and activity (Hart, Housley, & Slawson, 2007). Compared to *N*-glycosylation, this reaction is more dynamic and is reversible. Moreover, it is also reported that this protein modification is a critical key regulator of diverse cellular processes, such as signal transduction, transcription, translation, and proteasomal degradation (Hanover, Krause, & Love, 2012; Hart, Slawson, Ramirez-Correa, & Lagerlof, 2011).

An intriguing aspect of this protein machinery underlying these biochemical pathways is that in the human genome to date, 428 serine/threonine kinases and about 40 phosphatases are described (Alonso et al., 2004; Moorhead, Trinkle-Mulcahy, & Ulke-Lemee, 2007), but only two single genes encode the enzymes that are responsible for cyclic O-GlcNAcylation, OGT, and OGA (also known as MGEA5). So far it has been reported that more than 3000 proteins can be O-GlcNAcylated in all tissues. Modified proteins may be mainly located in the nucleus, but an increasing body of literature reports O-GlcNAcylated proteins in the cytoplasm, and such modifications are present in all metazoans (Ma, Vocadlo, & Vosseller, 2013), all cells and tissues, and are often reported in tumors (Ma & Vosseller, 2013). For instance, almost all glycolytic enzymes are O-GlcNAcylated and therefore can be regulated by UDP-GlcNAc levels (Yi et al., 2012). UDP-GlcNAc is derived from extracellular glucose through the HBP. Because UDP-GlcNAc and protein O-GlcNAc

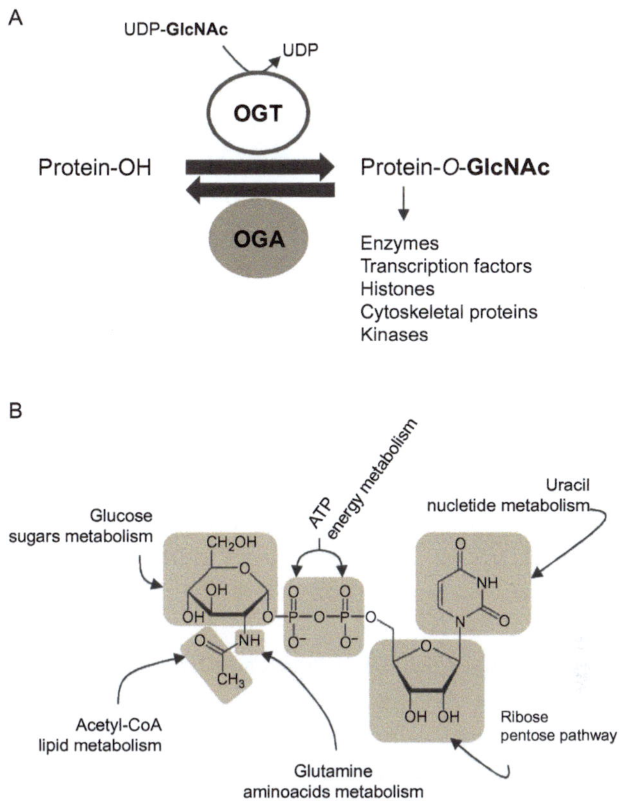

Figure 4.5 O-GlcNAcylation and UDP-GlcNAc. (A) UDP-GlcNAc is a substrate of OGT, which catalyzes the O-GlcNAcylation of hydroxyl residues on target proteins. OGA hydrolyzes the GlcNac moiety to restore –OH functional groups on affected residues. Many different proteins can be modified by O-GlcNAc. (B) UDP-GlcNAc synthesis is regulated by crosstalk with sugar, lipid, amino acid, energetic, and nucleotide metabolic pathways and is therefore considered a nutrient sensor.

levels in the cell fluctuate with the availability of glucose, free fatty acids, uridine, and the amino acid glutamine, it has long been speculated that O-GlcNAc serves as a nutrient sensor (Zachara & Hart, 2004; Zachara et al., 2004) (Fig. 4.5B). UDP-sugar availability is also critical for HA synthesis influencing HAS expression (Vigetti et al., 2006).

In aerobic glycolysis, a phenomenon also known as the "Warburg effect" (reviewed in Hanahan & Weinberg, 2011), cancer cells preferentially hyperactivate glycolysis rather than depending on oxidative phosphorylation as the main energy source, even in the presence of high-oxygen tension.

This condition seems to be related to the necessity of the cancer cells to produce molecules necessary for cancer growth, such as nucleotides, lipids, and pentose sugars. The switch to glycolysis in cancer cells is driven in part by tumor cell hypoxia, oncogenes (e.g., Ras and Myc), and mutant tumor suppressors (e.g., TP53), which activate HIF-1α, PI3K/Akt/mTOR, and c-Myc pathways to promote Warburg-like tumor metabolism (DeBerardinis, 2008). This great influx of glucose into the cancer induces not only the glycolytic pathway but also the pentose phosphate pathway for nucleotide production and HBP. In fact, about 2–5% of glucose entering a cell fluxes through the HBP. Thus, increased cancer cell glucose uptake may drive increased HBP flux.

In recent years, the roles of O-GlcNAcylation in several human malignant tumors have been investigated, and O-GlcNAcylation was found to be linked to cellular features relevant to metastasis. Hence, a large number of cancer-associated proteins, including p53, c-myc, IKK, and Snail1 (Lee, Jan, & Jan, 2009; Park et al., 2010), are modified by O-GlcNAc and increased O-GlcNAcylation is correlated to tumorigenesis (Caldwell et al., 2010).

Compared to normal cells, OGT transcripts and O-GlcNAc content are increased in a variety of cancers including prostate cancer, leukemia, colorectal cancer, bladder cancer, sarcoma, and leukemia, while OGA transcripts are decreased in lymphoma, brain and ovarian cancer, and leukemia (http://www.oncomine.org) (Ma & Vosseller, 2013). O-GlcNAc levels are also increased in hepatocellular carcinoma tissues and, interestingly, low expression of OGA is suggested as a prognostic marker for tumor recurrence (Zhu et al., 2012). Another study suggests that urinary content of OGT and OGA may be used as a prognostic tool for bladder cancer (Rozanski et al., 2012). Several studies support the concept that O-GlcNAcylation plays a critical role in breast cancer and in lung (Caldwell et al., 2010; Lynch et al., 2012). It has been reported that cell adhesion mediated by E-cadherin, beta-catenin, and P120 is influenced by O-GlcNAcylation, which increases cell motility, and may thereby impact metastasis (Zhu, Leber, & Andrews, 2001).

Certain oncogenes such as Kras not only upregulate levels of glucose transporters and glycolytic enzymes but also upregulate glutamine: fructose-6-phosphate amidotransferase 1 (GFAT1), the rate-limiting enzyme in the HBP (Ying et al., 2012). Thus, both excess glucose uptake and upregulation of HBP enzymes likely drive increased HBP flux in cancer. In cancer cells, it is the upregulated glucose transport and glutamine uptake that is regulated by the oncogene c-Myc. In addition to the other metabolic pathways where glutamine is involved as anaplerotic agent or in glutaminolysis, the increased glutamine can be also used as substrate for HBP by GFAT, so excess glutamine uptake in cancer cells could contribute to increased flux through the HBP, and increased levels of UDP-GlcNAc. In support of these findings is the recent study which demonstrated a high level of UDP-GlcNAc in pancreatic cancer (Ma & Vosseller, 2013). Therefore, increased HBP flux is linked to increased levels of O-GlcNAc (Wells, Vosseller, & Hart, 2003). Hence, it appears that increased HBP flux and elevated UDP-GlcNAc is a general feature of cancer cells that contributes to hyper-O-GlcNAcylation, since UDP-GlcNAc availability is critical for OGT activity (Lazarus, Nam, Jiang, Sliz, & Walker, 2011).

The changes in patterns of OGT and OGA expression are consistent with feedback signals to "normalize" increases or decreases in O-GlcNAc levels, although the mechanisms underlying such a feedback mechanism are not known. Despite these negative feedback mechanisms, which attempt to "normalize" changing O-GlcNAc levels, cancer cells maintain a strikingly elevated level of O-GlcNAcylation (hyper-O-GlcNAcylation), elevated OGT, and decreased OGA compared to nontransformed cells. The control of UDP-GlcNAc content is therefore an intriguing target in cancer studies, and knocking down OGT inhibits many hallmarks of cancer, but does not affect the cell proliferation of immortalized nontransformed cells derived from breast, prostate, and pancreas (Caldwell et al., 2010; Lynch et al., 2012; Ma & Vosseller, 2013), which suggests that OGT may be a cancer-specific target. To support this hypothesis, the use of inhibitors of key enzymes of HBP reduces cancer cell growth *in vitro* (Gloster et al., 2011; Ma & Vosseller, 2013).

Hyper-O-GlcNAcylation contributes to transformed phenotypes, as reduction of hyper-O-GlcNAcylation in breast, prostate, and pancreatic cancer models inhibits cancer cell proliferation and tumor growth *in vitro* and *in vivo*. Recent evidence indicates that tumor angiogenesis may in part be supported by hyper-O-GlcNAcylation, which is related to VEGF expression. Eventually, the hyper-O-GlcNAcylation in tumors may contribute to angiogenesis through upregulation of VEGF, MMPs, and FGF signaling. Increasing hyper-O-GlcNAcylation enhances the migration and invasion of breast and liver cancer cells, and conversely, lowering hyper-O-GlcNAcylation by knockdown of OGT inhibits tumor invasion and metastasis *in vivo* and *in vitro* in breast and prostate cancer cells (Caldwell et al., 2010).

5.1. HAS2 O-GlcNAcylation

In our recent study (Vigetti et al., 2012), we described that HAS2 is O-GlcNAcylated after glucosamine treatments and this covalent modification increased its activity both in isolated membranes and in cell culture. In this work, we described an increased protein stability and a reduction of proteasome degradation of HAS2 and a consequent increase in HA synthesis. Treatment with glucosamine mimics the effect of glucose and this would maintain high-ATP level, blocking AMPK activity, and favoring HA synthesis. In other set of studies, it was found that IL-1β stimulates the synthesis of HA, triggering the NF-kB pathway (Vigetti et al., 2010), and in cancer this signaling is controlled by O-GlcNAcylation (Gerritsen et al., 1997; Ozcan, Andrali, & Cantrell, 2010). In this case, the results showed that mRNA of HAS2 increased markedly as well as CD44, as expected for an increased protein O-GlcNAcylation and in accordance with many prior cancer characterizations, where CD44 is overexpressed.

O-GlcNAcylation represents a general regulatory mechanism (like phosphorylation) that may adapt metabolism to an excess of nutrients. The increased synthesis of HA stimulated by O-GlcNAcylation could represent a way to reduce intracellular UDP-GlcNAc. The cell is able to drive sugar nucleotides inside the ER and Golgi, but in cancer conditions, these transporters may be easily saturated and, therefore, they would not be able to diminish UDP-sugar concentration efficiently. This hypothesis is supported by the observation that O-GlcNAcylation regulates only HA synthesis without affecting other GAG production in the Golgi (Vigetti et al., 2012). Similarly, AMPK regulates only HA synthesis without altering that of GAGs (Vigetti, Clerici, et al., 2011), suggesting that HA synthesis may be more dynamic than that of other glycoconjugates. The other GAGs are strictly required for cell viability and their regulation may be more complex because it would control not only the GAG synthetic machinery (GAGosome) (Presto et al., 2008) but also the core protein expression. HA has no core protein and its synthesis involves just the HASes. Thus, the synthesis of HA could represent a valve to quickly adjust the UDP-sugar availability in the cells: in the case of low-nutrients and low-energy HA synthesis can be switched off to preserve building blocks for vital functions, whereas in the case of excess of nutrients and energy HA can be synthesized and secreted in the cell microenvironment. Therefore, HA could be a signal of optimal cell conditions to sustain cell growth.

6. CONCLUSIONS

Aberrant HA synthesis in cancer accelerates tumor growth by triggering signaling cascades from HA receptors (i.e., CD44 and RHAMM), promoting angiogenesis, favoring EMT, and modulating immune cell recruitment and inflammation in the stroma. HAS2 is a critical enzyme-producing HA in adult tissues and our review highlights the complexity of HAS2 regulation at posttranslational levels through phosphorylation (AMPK) and O-GlcNAcylation (OGT). Therefore, we propose that HAS2 is the main regulatory point in HA synthesis, able to directly and quickly integrate the metabolic stimuli and consequently adapt its enzymatic activity to cell necessity.

ACKNOWLEDGMENT

This work was supported by FAR and EU grant IRSES INFLAMA to A. P.

REFERENCES

Acquati, F., Bertilaccio, S., Grimaldi, A., Monti, L., Cinquetti, R., Bonetti, P., et al. (2011). Microenvironmental control of malignancy exerted by RNASET2, a widely conserved extracellular RNase. *Proceedings of the National Academy of Sciences of the United States of America, 108*, 1104–1109.

Adamia, S., Pilarski, P. M., Belch, A. R., & Pilarski, L. M. (2013). Aberrant splicing, hyaluronan synthases and intracellular hyaluronan as drivers of oncogenesis and potential drug targets. *Current Cancer Drug Targets, 13*, 347–361.

Alonso, A., Sasin, J., Bottini, N., Friedberg, I., Friedberg, I., Osterman, A., et al. (2004). Protein tyrosine phosphatases in the human genome. *Cell, 117*, 699–711.

Asari, A., Kanemitsu, T., & Kurihara, H. (2010). Oral administration of high molecular weight hyaluronan (900 kDa) controls immune system via Toll-like receptor 4 in the intestinal epithelium. *The Journal of Biological Chemistry, 285*, 24751–24758.

Auvinen, P., Rilla, K., Tumelius, R., Tammi, M., Sironen, R., Soini, Y., et al. (2014). Hyaluronan synthases (HAS1–3) in stromal and malignant cells correlate with breast cancer grade and predict patient survival. *Breast Cancer Research and Treatment, 143*, 277–286.

Bauer, I., Grozio, A., Lasiglie, D., Basile, G., Sturla, L., Magnone, M., et al. (2012). The NAD^+-dependent histone deacetylase SIRT6 promotes cytokine production and migration in pancreatic cancer cells by regulating Ca^{2+} responses. *The Journal of Biological Chemistry, 287*, 40924–40937.

Bernert, B., Porsch, H., & Heldin, P. (2011). Hyaluronan synthase 2 (HAS2) promotes breast cancer cell invasion by suppression of tissue metalloproteinase inhibitor 1 (TIMP-1). *The Journal of Biological Chemistry, 286*, 42349–42359.

Bourguignon, L. Y., Gilad, E., & Peyrollier, K. (2007). Heregulin-mediated ErbB2-ERK signaling activates hyaluronan synthases leading to CD44-dependent ovarian tumor cell growth and migration. *The Journal of Biological Chemistry, 282*, 19426–19441.

Caldwell, S. A., Jackson, S. R., Shahriari, K. S., Lynch, T. P., Sethi, G., Walker, S., et al. (2010). Nutrient sensor O-GlcNAc transferase regulates breast cancer tumorigenesis through targeting of the oncogenic transcription factor FoxM1. *Oncogene, 29*, 2831–2842.

Camenisch, T. D., Schroeder, J. A., Bradley, J., Klewer, S. E., & McDonald, J. A. (2002). Heart-valve mesenchyme formation is dependent on hyaluronan-augmented activation of ErbB2-ErbB3 receptors. *Nature Medicine, 8*, 850–855.

Camenisch, T. D., Spicer, A. P., Brehm-Gibson, T., Biesterfeldt, J., Augustine, M. L., Calabro, A., Jr., et al. (2000). Disruption of hyaluronan synthase-2 abrogates normal cardiac morphogenesis and hyaluronan-mediated transformation of epithelium to mesenchyme. *The Journal of Clinical Investigation, 106*, 349–360.

Coussens, L. M., & Werb, Z. (2002). Inflammation and cancer. *Nature, 420*, 860–867.

de La Motte, C. A., Hascall, V. C., Calabro, A., Yen-Lieberman, B., & Strong, S. A. (1999). Mononuclear leukocytes preferentially bind via CD44 to hyaluronan on human intestinal mucosal smooth muscle cells after virus infection or treatment with poly(I.C). *The Journal of Biological Chemistry, 274*, 30747–30755.

de la Motte, C. A., Hascall, V. C., Drazba, J., Bandyopadhyay, S. K., & Strong, S. A. (2003). Mononuclear leukocytes bind to specific hyaluronan structures on colon mucosal smooth muscle cells treated with polyinosinic acid:polycytidylic acid: Inter-alpha-trypsin inhibitor is crucial to structure and function. *The American Journal of Pathology, 163*, 121–133.

DeBerardinis, R. J. (2008). Is cancer a disease of abnormal cellular metabolism? New angles on an old idea. *Genetics in Medicine, 10*, 767–777.

Evanko, S. P., & Wight, T. N. (1999). Intracellular localization of hyaluronan in proliferating cells. *The Journal of Histochemistry and Cytochemistry, 47*, 1331–1342.

Evans, J. M., Donnelly, L. A., Emslie-Smith, A. M., Alessi, D. R., & Morris, A. D. (2005). Metformin and reduced risk of cancer in diabetic patients. *BMJ, 330*, 1304–1305.

Fieber, C., Baumann, P., Vallon, R., Termeer, C., Simon, J. C., Hofmann, M., et al. (2004). Hyaluronan-oligosaccharide-induced transcription of metalloproteases. *Journal of Cell Science, 117*, 359–367.

Fuchs, K., Hippe, A., Schmaus, A., Homey, B., Sleeman, J. P., & Orian-Rousseau, V. (2013). Opposing effects of high- and low-molecular weight hyaluronan on CXCL12-induced CXCR4 signaling depend on CD44. *Cell Death & Disease, 4*, e819.

Gao, P., Tchernyshyov, I., Chang, T. C., Lee, Y. S., Kita, K., Ochi, T., et al. (2009). c-Myc suppression of miR-23a/b enhances mitochondrial glutaminase expression and glutamine metabolism. *Nature, 458*, 762–765.

Gerritsen, M. E., Williams, A. J., Neish, A. S., Moore, S., Shi, Y., & Collins, T. (1997). CREB-binding protein/p300 are transcriptional coactivators of p65. *Proceedings of the National Academy of Sciences of the United States of America, 94*, 2927–2932.

Gloster, T. M., Zandberg, W. F., Heinonen, J. E., Shen, D. L., Deng, L., & Vocadlo, D. J. (2011). Hijacking a biosynthetic pathway yields a glycosyltransferase inhibitor within cells. *Nature Chemical Biology, 7*, 174–181.

Guillemette, C., Millikan, R. C., Newman, B., & Housman, D. E. (2000). Genetic polymorphisms in uridine diphospho-glucuronosyltransferase 1A1 and association with breast cancer among African Americans. *Cancer Research, 60*, 950–956.

Hall, C. L., & Turley, E. A. (1995). Hyaluronan: RHAMM mediated cell locomotion and signaling in tumorigenesis. *Journal of Neuro-Oncology, 26*, 221–229.

Hanahan, D., & Weinberg, R. A. (2011). Hallmarks of cancer: The next generation. *Cell, 144*, 646–674.

Hanover, J. A., Krause, M. W., & Love, D. C. (2012). Bittersweet memories: Linking metabolism to epigenetics through O-GlcNAcylation. *Nature Reviews Molecular Cell Biology, 13*, 312–321.

Hardie, D. G., & Alessi, D. R. (2013). LKB1 and AMPK and the cancer-metabolism link—Ten years after. *BMC Biology, 11*, 36.

Hart, G. W., Housley, M. P., & Slawson, C. (2007). Cycling of O-linked beta-N-acetylglucosamine on nucleocytoplasmic proteins. *Nature, 446*, 1017–1022.

Hart, G. W., Slawson, C., Ramirez-Correa, G., & Lagerlof, O. (2011). Cross talk between O-GlcNAcylation and phosphorylation: Roles in signaling, transcription, and chronic disease. *Annual Review of Biochemistry, 80*, 825–858.

Hascall, V., & Karamanos, N. (2011). Regulatory roles of hyaluronan in health and disease. *The FEBS Journal, 278*, 1411.

Hascall, V. C., Majors, A. K., De La Motte, C. A., Evanko, S. P., Wang, A., Drazba, J. A., et al. (2004). Intracellular hyaluronan: A new frontier for inflammation? *Biochimica et Biophysica Acta, 1673*, 3–12.

Hascall, V. C., Wang, A., Tammi, M., Oikari, S., Tammi, R., Passi, A., et al. (2014). The dynamic metabolism of hyaluronan regulates the cytosolic concentration of UDP-GlcNAc. *Matrix Biology, 35*, 14–17.

Hill, D. R., Kessler, S. P., Rho, H. K., Cowman, M. K., & de la Motte, C. A. (2012). Specific-sized hyaluronan fragments promote expression of human beta-defensin 2 in intestinal epithelium. *The Journal of Biological Chemistry, 287*, 30610–30624.

Hill, D. R., Rho, H. K., Kessler, S. P., Amin, R., Homer, C. R., McDonald, C., et al. (2013). Human milk hyaluronan enhances innate defense of the intestinal epithelium. *The Journal of Biological Chemistry, 288*, 29090–29104.

Huang, D., Casale, G. P., Tian, J., Lele, S. M., Pisarev, V. M., Simpson, M. A., et al. (2010). UDP-glucose dehydrogenase as a novel field-specific candidate biomarker of prostate cancer. *International Journal of Cancer, 126*, 315–327.

Iijima, J., Konno, K., & Itano, N. (2011). Inflammatory alterations of the extracellular matrix in the tumor microenvironment. *Cancers (Basel), 3*, 3189–3205.

Itano, N., & Kimata, K. (2002). Mammalian hyaluronan synthases. *IUBMB Life, 54*, 195–199.

Itano, N., & Kimata, K. (2008). Altered hyaluronan biosynthesis in cancer progression. *Seminars in Cancer Biology, 18*, 268–274.

Itano, N., Sawai, T., Miyaishi, O., & Kimata, K. (1999). Relationship between hyaluronan production and metastatic potential of mouse mammary carcinoma cells. *Cancer Research, 59*, 2499–2504.

Jiang, D., Liang, J., Fan, J., Yu, S., Chen, S., Luo, Y., et al. (2005). Regulation of lung injury and repair by Toll-like receptors and hyaluronan. *Nature Medicine, 11*, 1173–1179.

Jiang, D., Liang, J., & Noble, P. W. (2007). Hyaluronan in tissue injury and repair. *Annual Review of Cell and Developmental Biology, 23*, 435–461.

Jiang, D., Liang, J., & Noble, P. W. (2011). Hyaluronan as an immune regulator in human diseases. *Physiological Reviews, 91*, 221–264.

Kakizaki, I., Kojima, K., Takagaki, K., Endo, M., Kannagi, R., Ito, M., et al. (2004). A novel mechanism for the inhibition of hyaluronan biosynthesis by 4-methylumbelliferone. *The Journal of Biological Chemistry, 279*, 33281–33289.

Karousou, E., Kamiryo, M., Skandalis, S. S., Ruusala, A., Asteriou, T., Passi, A., et al. (2010). The activity of hyaluronan synthase 2 is regulated by dimerization and ubiquitination. *The Journal of Biological Chemistry, 285*, 23647–23654.

Karousou, E., Stachtea, X., Moretto, P., Viola, M., Vigetti, D., D'Angelo, M. L., et al. (2013). New insights into the pathobiology of Down syndrome—Hyaluronan synthase-2 overexpression is regulated by collagen VI alpha2 chain. *The FEBS Journal, 280*, 2418–2430.

Kashima, Y., Takahashi, M., Shiba, Y., Itano, N., Izawa, A., Koyama, J., et al. (2013). Crucial role of hyaluronan in neointimal formation after vascular injury. *PLoS One, 8*, e58760.

Knudson, W., Biswas, C., & Toole, B. P. (1984). Interactions between human tumor cells and fibroblasts stimulate hyaluronate synthesis. *Proceedings of the National Academy of Sciences of the United States of America, 81*, 6767–6771.

Kobayashi, N., Miyoshi, S., Mikami, T., Koyama, H., Kitazawa, M., Takeoka, M., et al. (2010). Hyaluronan deficiency in tumor stroma impairs macrophage trafficking and tumor neovascularization. *Cancer Research, 70*, 7073–7083.

Kosaki, R., Watanabe, K., & Yamaguchi, Y. (1999). Overproduction of hyaluronan by expression of the hyaluronan synthase Has2 enhances anchorage-independent growth and tumorigenicity. *Cancer Research, 59*, 1141–1145.

Laurent, T. C., & Fraser, J. R. (1992). Hyaluronan. *FASEB Journal, 6*, 2397–2404.

Lazarus, M. B., Nam, Y., Jiang, J., Sliz, P., & Walker, S. (2011). Structure of human O-GlcNAc transferase and its complex with a peptide substrate. *Nature, 469*, 564–567.

Lee, H. H., Jan, L. Y., & Jan, Y. N. (2009). Drosophila IKK-related kinase Ik2 and Katanin p60-like 1 regulate dendrite pruning of sensory neuron during metamorphosis. *Proceedings of the National Academy of Sciences of the United States of America, 106*, 6363–6368.

Li, Y., & Heldin, P. (2001). Hyaluronan production increases the malignant properties of mesothelioma cells. *British Journal of Cancer, 85*, 600–607.

Liang, J., & Mills, G. B. (2013). AMPK: A contextual oncogene or tumor suppressor? *Cancer Research, 73*, 2929–2935.

Lokeshwar, V. B., Lopez, L. E., Munoz, D., Chi, A., Shirodkar, S. P., Lokeshwar, S. D., et al. (2010). Antitumor activity of hyaluronic acid synthesis inhibitor 4-methylumbelliferone in prostate cancer cells. *Cancer Research, 70*, 2613–2623.

Lynch, T. P., Ferrer, C. M., Jackson, S. R., Shahriari, K. S., Vosseller, K., & Reginato, M. J. (2012). Critical role of O-linked beta-*N*-acetylglucosamine transferase in prostate cancer invasion, angiogenesis, and metastasis. *The Journal of Biological Chemistry, 287*, 11070–11081.

Ma, Z., Vocadlo, D. J., & Vosseller, K. (2013). Hyper-O-GlcNAcylation is anti-apoptotic and maintains constitutive NF-kappaB activity in pancreatic cancer cells. *The Journal of Biological Chemistry, 288*, 15121–15130.

Ma, Z., & Vosseller, K. (2013). O-GlcNAc in cancer biology. *Amino Acids, 45*, 719–733.

Majors, A. K., Austin, R. C., de la Motte, C. A., Pyeritz, R. E., Hascall, V. C., Kessler, S. P., et al. (2003). Endoplasmic reticulum stress induces hyaluronan deposition and leukocyte adhesion. *The Journal of Biological Chemistry, 278*, 47223–47231.

Mantovani, A., Allavena, P., Sica, A., & Balkwill, F. (2008). Cancer-related inflammation. *Nature, 454*, 436–444.

Michael, D. R., Phillips, A. O., Krupa, A., Martin, J., Redman, J. E., Altaher, A., et al. (2011). The human hyaluronan synthase 2 (HAS2) gene and its natural antisense RNA exhibit coordinated expression in the renal proximal tubular epithelial cell. *The Journal of Biological Chemistry, 286*, 19523–19532.

Misra, S., Heldin, P., Hascall, V. C., Karamanos, N. K., Skandalis, S. S., Markwald, R. R., et al. (2011). Hyaluronan-CD44 interactions as potential targets for cancer therapy. *The FEBS Journal, 278*, 1429–1443.

Moorhead, G. B., Trinkle-Mulcahy, L., & Ulke-Lemee, A. (2007). Emerging roles of nuclear protein phosphatases. *Nature Reviews Molecular Cell Biology, 8*, 234–244.

Nagar, S., & Remmel, R. P. (2006). Uridine diphosphoglucuronosyltransferase pharmacogenetics and cancer. *Oncogene, 25*, 1659–1672.

Nakamura, T., Takagaki, K., Shibata, S., Tanaka, K., Higuchi, T., & Endo, M. (1995). Hyaluronic-acid-deficient extracellular matrix induced by addition of 4-methylumbelliferone to the medium of cultured human skin fibroblasts. *Biochemical and Biophysical Research Communications, 208*, 470–475.

Nieman, K. M., Kenny, H. A., Penicka, C. V., Ladanyi, A., Buell-Gutbrod, R., Zillhardt, M. R., et al. (2011). Adipocytes promote ovarian cancer metastasis and provide energy for rapid tumor growth. *Nature Medicine*, *17*, 1498–1503.

Nishida, Y., Knudson, W., Knudson, C. B., & Ishiguro, N. (2005). Antisense inhibition of hyaluronan synthase-2 in human osteosarcoma cells inhibits hyaluronan retention and tumorigenicity. *Experimental Cell Research*, *307*, 194–203.

Ozcan, S., Andrali, S. S., & Cantrell, J. E. (2010). Modulation of transcription factor function by O-GlcNAc modification. *Biochimica et Biophysica Acta*, *1799*, 353–364.

Park, S. Y., Kim, H. S., Kim, N. H., Ji, S., Cha, S. Y., Kang, J. G., et al. (2010). Snail1 is stabilized by O-GlcNAc modification in hyperglycaemic condition. *The EMBO Journal*, *29*, 3787–3796.

Porsch, H., Bernert, B., Mehic, M., Theocharis, A. D., Heldin, C. H., & Heldin, P. (2013). Efficient TGFbeta-induced epithelial-mesenchymal transition depends on hyaluronan synthase HAS2. *Oncogene*, *32*, 4355–4365.

Powell, J. D., & Horton, M. R. (2005). Threat matrix: Low-molecular-weight hyaluronan (HA) as a danger signal. *Immunologic Research*, *31*, 207–218.

Presto, J., Thuveson, M., Carlsson, P., Busse, M., Wilen, M., Eriksson, I., et al. (2008). Heparan sulfate biosynthesis enzymes EXT1 and EXT2 affect NDST1 expression and heparan sulfate sulfation. *Proceedings of the National Academy of Sciences of the United States of America*, *105*, 4751–4756.

Rilla, K., Siiskonen, H., Spicer, A. P., Hyttinen, J. M., Tammi, M. I., & Tammi, R. H. (2005). Plasma membrane residence of hyaluronan synthase is coupled to its enzymatic activity. *The Journal of Biological Chemistry*, *280*, 31890–31897.

Rinn, J. L., & Chang, H. Y. (2012). Genome regulation by long noncoding RNAs. *Annual Review of Biochemistry*, *81*, 145–166.

Rozanski, W., Krzeslak, A., Forma, E., Brys, M., Blewniewski, M., Wozniak, P., et al. (2012). Prediction of bladder cancer based on urinary content of MGEA5 and OGT mRNA level. *Clinical Laboratory*, *58*, 579–583.

Simpson, M. A., Wilson, C. M., Furcht, L. T., Spicer, A. P., Oegema, T. R., Jr., & McCarthy, J. B. (2002). Manipulation of hyaluronan synthase expression in prostate adenocarcinoma cells alters pericellular matrix retention and adhesion to bone marrow endothelial cells. *The Journal of Biological Chemistry*, *277*, 10050–10057.

Simpson, M. A., Wilson, C. M., & McCarthy, J. B. (2002). Inhibition of prostate tumor cell hyaluronan synthesis impairs subcutaneous growth and vascularization in immunocompromised mice. *The American Journal of Pathology*, *161*, 849–857.

Sironen, R. K., Tammi, M., Tammi, R., Auvinen, P. K., Anttila, M., & Kosma, V. M. (2011). Hyaluronan in human malignancies. *Experimental Cell Research*, *317*, 383–391.

Slevin, M., Krupinski, J., Gaffney, J., Matou, S., West, D., Delisser, H., et al. (2007). Hyaluronan-mediated angiogenesis in vascular disease: Uncovering RHAMM and CD44 receptor signaling pathways. *Matrix Biology*, *26*, 58–68.

Stern, R. (2004). Hyaluronan catabolism: A new metabolic pathway. *European Journal of Cell Biology*, *83*, 317–325.

Stern, R. (2008). Hyaluronidases in cancer biology. *Seminars in Cancer Biology*, *18*, 275–280.

Stern, R., Asari, A. A., & Sugahara, K. N. (2006). Hyaluronan fragments: An information-rich system. *European Journal of Cell Biology*, *85*, 699–715.

Tammi, R. H., Kultti, A., Kosma, V. M., Pirinen, R., Auvinen, P., & Tammi, M. I. (2008). Hyaluronan in human tumors: Pathobiological and prognostic messages from cell-associated and stromal hyaluronan. *Seminars in Cancer Biology*, *18*, 288–295.

Tammi, R. H., Passi, A. G., Rilla, K., Karousou, E., Vigetti, D., Makkonen, K., et al. (2011). Transcriptional and post-translational regulation of hyaluronan synthesis. *The FEBS Journal*, *278*, 1419–1428.

Toole, B. P. (2004). Hyaluronan: From extracellular glue to pericellular cue. *Nature Reviews Cancer*, *4*, 528–539.

Toole, B. P., Wight, T. N., & Tammi, M. I. (2002). Hyaluronan-cell interactions in cancer and vascular disease. *The Journal of Biological Chemistry*, *277*, 4593–4596.

Torres, C. R., & Hart, G. W. (1984). Topography and polypeptide distribution of terminal *N*-acetylglucosamine residues on the surfaces of intact lymphocytes. Evidence for *O*-linked GlcNAc. *The Journal of Biological Chemistry*, *259*, 3308–3317.

Towler, M. C., & Hardie, D. G. (2007). AMP-activated protein kinase in metabolic control and insulin signaling. *Circulation Research*, *100*, 328–341.

Twarock, S., Freudenberger, T., Poscher, E., Dai, G., Jannasch, K., Dullin, C., et al. (2011). Inhibition of oesophageal squamous cell carcinoma progression by in vivo targeting of hyaluronan synthesis. *Molecular Cancer*, *10*, 30.

Udabage, L., Brownlee, G. R., Waltham, M., Blick, T., Walker, E. C., Heldin, P., et al. (2005). Antisense-mediated suppression of hyaluronan synthase 2 inhibits the tumorigenesis and progression of breast cancer. *Cancer Research*, *65*, 6139–6150.

Vander Heiden, M. G., Cantley, L. C., & Thompson, C. B. (2009). Understanding the Warburg effect: The metabolic requirements of cell proliferation. *Science*, *324*, 1029–1033.

Vigetti, D., Clerici, M., Deleonibus, S., Karousou, E., Viola, M., Moretto, P., et al. (2011). Hyaluronan synthesis is inhibited by adenosine monophosphate-activated protein kinase through the regulation of HAS2 activity in human aortic smooth muscle cells. *The Journal of Biological Chemistry*, *286*, 7917–7924.

Vigetti, D., Deleonibus, S., Moretto, P., Karousou, E., Viola, M., Bartolini, B., et al. (2012). Role of UDP-*N*-acetylglucosamine (GlcNAc) and *O*-GlcNAcylation of hyaluronan synthase 2 in the control of chondroitin sulfate and hyaluronan synthesis. *The Journal of Biological Chemistry*, *287*, 35544–35555.

Vigetti, D., Genasetti, A., Karousou, E., Viola, M., Clerici, M., Bartolini, B., et al. (2009). Modulation of hyaluronan synthase activity in cellular membrane fractions. *The Journal of Biological Chemistry*, *284*, 30684–30694.

Vigetti, D., Genasetti, A., Karousou, E., Viola, M., Moretto, P., Clerici, M., et al. (2010). Proinflammatory cytokines induce hyaluronan synthesis and monocyte adhesion in human endothelial cells through hyaluronan synthase 2 (HAS2) and the nuclear factor-kappaB (NF-kappaB) pathway. *The Journal of Biological Chemistry*, *285*, 24639–24645.

Vigetti, D., Gotte, M., Pavao, M. S., & Theocharis, A. D. (2013). Cellular microenvironment in human pathologies. *BioMed Research International*, *2013*, 946–958.

Vigetti, D., Ori, M., Viola, M., Genasetti, A., Karousou, E., Rizzi, M., et al. (2006). Molecular cloning and characterization of UDP-glucose dehydrogenase from the amphibian Xenopus laevis and its involvement in hyaluronan synthesis. *The Journal of Biological Chemistry*, *281*, 8254–8263.

Vigetti, D., Rizzi, M., Moretto, P., Deleonibus, S., Dreyfuss, J. M., Karousou, E., et al. (2011). Glycosaminoglycans and glucose prevent apoptosis in 4-methylumbelliferone-treated human aortic smooth muscle cells. *The Journal of Biological Chemistry*, *286*, 34497–34503.

Vigetti, D., Rizzi, M., Viola, M., Karousou, E., Genasetti, A., Clerici, M., et al. (2009). The effects of 4-methylumbelliferone on hyaluronan synthesis, MMP2 activity, proliferation, and motility of human aortic smooth muscle cells. *Glycobiology*, *19*, 537–546.

Vigetti, D., Viola, M., Karousou, E., De Luca, G., & Passi, A. (2014). Metabolic control of hyaluronan synthases. *Matrix Biology*, *35*, 8–13.

Vigetti, D., Viola, M., Karousou, E., Genasetti, A., Rizzi, M., Clerici, M., et al. (2008). Vascular pathology and the role of hyaluronan. *ScientificWorldJournal*, *8*, 1116–1118.

Vigetti, D., Viola, M., Karousou, E., Rizzi, M., Moretto, P., Genasetti, A., et al. (2008). Hyaluronan-CD44-ERK1/2 regulate human aortic smooth muscle cell motility during aging. *The Journal of Biological Chemistry*, *283*, 4448–4458.

Viola, M., Bartolini, B., Vigetti, D., Karousou, E., Moretto, P., Deleonibus, S., et al. (2013). Oxidized low density lipoprotein (LDL) affects hyaluronan synthesis in human aortic smooth muscle cells. *The Journal of Biological Chemistry*, *288*, 29595–29603.

Wang, A., & Hascall, V. C. (2004). Hyaluronan structures synthesized by rat mesangial cells in response to hyperglycemia induce monocyte adhesion. *The Journal of Biological Chemistry*, *279*, 10279–10285.

Weigel, P. H., & DeAngelis, P. L. (2007). Hyaluronan synthases: A decade-plus of novel glycosyltransferases. *The Journal of Biological Chemistry*, *282*, 36777–36781.

Wells, L., Vosseller, K., & Hart, G. W. (2003). A role for *N*-acetylglucosamine as a nutrient sensor and mediator of insulin resistance. *Cellular and Molecular Life Sciences*, *60*, 222–228.

Whatcott, C. J., Han, H., Posner, R. G., Hostetter, G., & Von Hoff, D. D. (2011). Targeting the tumor microenvironment in cancer: Why hyaluronidase deserves a second look. *Cancer Discovery*, *1*, 291–296.

Yi, W., Clark, P. M., Mason, D. E., Keenan, M. C., Hill, C., Goddard, W. A., 3rd., et al. (2012). Phosphofructokinase 1 glycosylation regulates cell growth and metabolism. *Science*, *337*, 975–980.

Ying, H., Kimmelman, A. C., Lyssiotis, C. A., Hua, S., Chu, G. C., Fletcher-Sananikone, E., et al. (2012). Oncogenic Kras maintains pancreatic tumors through regulation of anabolic glucose metabolism. *Cell*, *149*, 656–670.

Yoshida, H., Nagaoka, A., Kusaka-Kikushima, A., Tobiishi, M., Kawabata, K., Sayo, T., et al. (2013). KIAA1199, a deafness gene of unknown function, is a new hyaluronan binding protein involved in hyaluronan depolymerization. *Proceedings of the National Academy of Sciences of the United States of America*, *110*, 5612–5617.

Yoshihara, S., Kon, A., Kudo, D., Nakazawa, H., Kakizaki, I., Sasaki, M., et al. (2005). A hyaluronan synthase suppressor, 4-methylumbelliferone, inhibits liver metastasis of melanoma cells. *FEBS Letters*, *579*, 2722–2726.

Yuan, H., Su, L., & Chen, W. Y. (2013). The emerging and diverse roles of sirtuins in cancer: A clinical perspective. *Onco Targets and Therapy*, *6*, 1399–1416.

Zachara, N. E., & Hart, G. W. (2004). *O*-GlcNAc a sensor of cellular state: The role of nucleocytoplasmic glycosylation in modulating cellular function in response to nutrition and stress. *Biochimica et Biophysica Acta*, *1673*, 13–28.

Zachara, N. E., O'Donnell, N., Cheung, W. D., Mercer, J. J., Marth, J. D., & Hart, G. W. (2004). Dynamic *O*-GlcNAc modification of nucleocytoplasmic proteins in response to stress. A survival response of mammalian cells. *The Journal of Biological Chemistry*, *279*, 30133–30142.

Zhu, W., Leber, B., & Andrews, D. W. (2001). Cytoplasmic *O*-glycosylation prevents cell surface transport of E-cadherin during apoptosis. *The EMBO Journal*, *20*, 5999–6007.

Zhu, Q., Zhou, L., Yang, Z., Lai, M., Xie, H., Wu, L., et al. (2012). *O*-GlcNAcylation plays a role in tumor recurrence of hepatocellular carcinoma following liver transplantation. *Medical Oncology*, *29*, 985–993.

CHAPTER FIVE

Hyaluronan-Coated Extracellular Vesicles—A Novel Link Between Hyaluronan and Cancer

Kirsi Rilla[*,1], **Hanna Siiskonen**[*,†], **Markku Tammi**[*], **Raija Tammi**[*]
[*]Institute of Biomedicine, University of Eastern Finland, Kuopio, Finland
[†]Department of Dermatology, Kuopio University Hospital and University of Eastern Finland, Kuopio, Finland
[1]Corresponding author: e-mail address: kirsi.rilla@uef.fi

Contents

1. Introduction 122
2. Extracellular Vesicles as Novel Communicators Between Cells 124
   2.1 EVs—A quickly expanding field of research 124
   2.2 Variable nomenclature of EVs 124
   2.3 Origin and formation of EVs 125
   2.4 Methods for the analysis of EVs 125
   2.5 EVs as carriers of regulatory molecules into target cells 126
3. EVs Promote Tumor Progression 126
   3.1 EVs associate with cancers 126
   3.2 Promotion of invasion and metastasis by EVs 127
   3.3 Thrombosis and EVs 128
   3.4 Angiogenesis is promoted by EVs 128
   3.5 EVs participate in antitumor immunity 128
4. HA Synthesis Enhances Shedding of Extracellular Vesicles 129
   4.1 HA–EVs are common in cells with active HA secretion 129
   4.2 HAS-induced protrusions as a platform for vesicle shedding 129
   4.3 Regulation of HA synthesis and EV shedding by cellular glucose supply? 131
   4.4 HA–EVs in tissue regeneration 133
   4.5 Body fluids rich in HA are also rich in EVs 133
   4.6 HA–EVs provide the missing link between HA and cancer? 134
5. HA–EVs as Predictors, Targets, and Carriers of Therapy 137
   5.1 HA–EVs as potential biomarkers for HA-rich tumors 137
   5.2 Utilization of HA–EVs as targets and carriers for therapy 138
6. Conclusions 139
Acknowledgments 139
References 140

Advances in Cancer Research, Volume 123
ISSN 0065-230X
http://dx.doi.org/10.1016/B978-0-12-800092-2.00005-8

Abstract

The synthesis of hyaluronan (HA) on the plasma membrane is a unique and still partly mysterious way of macromolecular biosynthesis. HA forms pericellular coats around many cell types and accumulates in the extracellular matrix (ECM) of growing and renewing tissues. It is secreted to high concentrations in body fluids with antifriction properties like pleural, peritoneal, and synovial fluids, but is also detectable in plasma, saliva, and urine. In pathological states, like cancer and inflammation, the amount of HA is increased around cells, in the ECM, and in the body fluids. HA is an indicator of poor prognosis for cancer patients and creates a favorable environment for cellular growth and motility.

The recent finding that HA-coated extracellular vesicles act both as a product of HA synthase activity and as special vehicles for HA, and perhaps carry signals important for malignant growth, provides a novel link between HA and cancer. HA could be carried on the surface of these vesicles in tissues and body fluids, creating beneficial environments by itself, or by associated molecules, for the invasion and metastasis of cancer cells. The HA-coated plasma membrane protrusions and vesicles shed from them are potential biomarkers in cancer and other HA-associated disease states.

1. INTRODUCTION

Until lately, it was believed that the communication between cells not in direct contact with each other depends on the secretion of soluble factors and their interaction with specific cellular receptors. The discovery of secreted extracellular vesicles (EVs) has revolutionized our knowledge of cellular signaling events. EVs are detected in body fluids as well as in the interstitial space between cells. They have been found to modify both normal tissue homeostasis as well as pathological states via paracrine and endocrine pathways. In cancer patients, they can travel into surrounding tissues, lymph nodes, or even distant tissue sites to create a favorable niche for metastasis by influencing cellular interactions and signaling events (Camussi et al., 2011; Tetta, Ghigo, Silengo, Deregibus, & Camussi, 2013; Thery, Zitvogel, & Amigorena, 2002).

Accumulation of hyaluronan (HA) correlates with the aggressiveness of cancers by enhancing processes involved in malignant growth, like cell proliferation, invasion, metastasis, and tumor–stroma interactions (Fig. 5.1). Many epithelial cancers, like colon (Ropponen et al., 1998), breast (Auvinen et al., 2000), prostate (Lipponen et al., 2001), and ovarian cancers (Anttila et al., 2000; Hiltunen et al., 2002) contain increased levels of HA. Accumulation of HA both around the tumor cells and in the surrounding

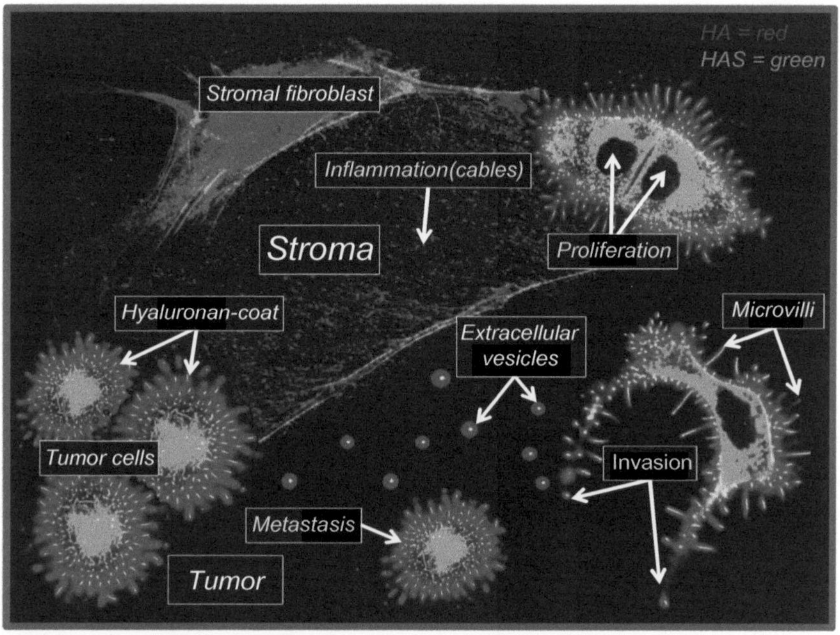

Figure 5.1 A collage of (manipulated) images illustrating possible roles of HA in cancer progression (arrows). HA produced by HA synthases forms a tight coat around tumor cells, inducing growth of microvilli and shedding of EVs, promoting cell proliferation, invasion, and metastasis. Stromal fibroblasts surrounding the tumor cells also form a coat of HA and, together with proteoglycans, form cable-like structures promoting a cancer-supporting type of inflammation. Cancer cells and stromal cells communicate via secreted EVs, which also promote metastasis. (See the color plate.)

stroma acts as a strong, independent, unfavorable prognostic factor (Anttila et al., 2000; Auvinen et al., 2000). Immunostaining of HA synthases (HASs) has shown that HAS proteins are increased in endometrial (Nykopp et al., 2010), ovarian (Nykopp et al., 2009), and breast (Auvinen et al., 2014) cancers, suggesting that the increased HAS activity is responsible for the accumulation of HA in these cancer types. HA and HASs could act as biomarkers for cancer and noninvasive methods to detect them in cancer patients would be valuable in diagnostics.

We have recently shown that overexpression of HAS3 in epithelial cultures disturbs the organization of normal epithelium, resulting in intraluminal invasion, enhanced growth (Rilla et al., 2012), and increased secretion of EVs (Rilla, Pasonen-Seppänen, et al., 2013), offering a potential explanation for the contribution of HA to the premalignant changes of epithelial cells and surrounding stroma. This review focuses on the characteristics and putative roles of the newly discovered HA-coated EVs as general carriers of HA, potential diagnostic tools, and targets of therapy in HA-rich cancers.

2. EXTRACELLULAR VESICLES AS NOVEL COMMUNICATORS BETWEEN CELLS

2.1. EVs—A quickly expanding field of research

EVs are tiny plasma membrane-derived extracellular particles, ranging from 30 up to 1000 nm in diameter, secreted into the extracellular environment by many cell types. They transfer compounds, such as growth factors, and even DNA between cells in embryonic development, wound healing and other tissue injuries, inflammation, and cancer (Camussi, Deregibus, Bruno, Cantaluppi, & Biancone, 2010). Studies on EVs form a quickly expanding field of research that has attracted much interest because of their potential use as disease biomarkers and therapeutic tools. The International Society for Extracellular Vesicles http://www.isev.org/ was founded in 2011, and the first volume of the Society's official Journal, "Journal of Extracellular Vesicles (JEV)" was launched in 2012 (Lötvall et al., 2012). Several integrated databases for vesicle data sharing have been developed, like vesiclepedia (http://microvesicles.org/; Kalra et al., 2012), Exocarta (http://exocarta.org/; Mathivanan & Simpson, 2009), and EVpedia (http://evpedia.info/; Kim et al., 2013). Clear evidence for the recent research activity and rapid development of the field for EVs is that in PubMed the term "extracellular vesicle" results in more than 1600 matches, and one-third of the studies were published within five years.

2.2. Variable nomenclature of EVs

Researchers have invented dozens of different names for secreted vesicles, reflecting their specific functions, their cell of origin or the process used for vesicle collection. The most popular among the terms are microvesicles, microparticles, nanoparticles, exosomes, ectosomes, oncosomes, and circulating microvesicles. The vesicles originating via exocytosis of multivesicular bodies are called exosomes (30–100 nm), while vesicles derived from direct budding of the plasma membranes are called microvesicles (100–1000 nm) and vesicles released during apoptosis represent apoptotic bodies (800 nm to 5 μm). However, complete discrimination between these subpopulations is not possible with the currently available purification methods, and there is a significant overlap between particles termed exosomes and microvesicles in many studies. Therefore, the preparations studied may consist of mixtures of vesicles with different size, composition, cellular origin, and mechanism of secretion. Common features of all these vesicles, independent of their origin, function, or sedimentation are that they are extracellular, delimited by phospholipid bilayer, and independent "extracellular organelles," that offer completely novel aspects both in cell biological and physiological point of view. For simplicity, we will use "EV" as a general term in this review, as recently suggested by Gould and Raposo (2013). When referring to the newly discovered HA-coated EVs, we use an abbreviation HA–EV.

2.3. Origin and formation of EVs

Shedding of EVs has been observed in many normal cell types, including macrophages (Bhatnagar, Shinagawa, Castellino, & Schorey, 2007), epithelial cells (van Niel et al., 2001), neurons (Faure et al., 2006), platelets (Aatonen, Grönholm, & Siljander, 2012; Heijnen, Schiel, Fijnheer, Geuze, & Sixma, 1999), mesenchymal stem cells (Sabin & Kikyo, 2014), and several cancer cell types (reviewed in Martins, Dias, & Hainaut, 2013).

The formation of EVs is a continuous process, but their numbers are increased especially during inflammation (Hulsmans & Holvoet, 2013), tissue regeneration (Sabin & Kikyo, 2014), and cancer progression (Kim et al., 2003). Once formed, they can stay in the vicinity of the site of their release, but they can also migrate significant distances by diffusion, eventually ending up in different body fluids.

2.4. Methods for the analysis of EVs

The small and varied size and differences in their origin and composition have raised many challenges for the analysis and quantification of EVs. EVs are usually too small to be seen under a standard light microscope, but utilization of fluorescent labels has enabled their visualization even in live cell cultures (Tian, Wang, Wang, Zhu, & Xiao, 2010). EVs have been isolated from cell culture media and biological fluids, like plasma, urine, and ascites using ultracentrifugation (Raposo et al., 1996), often in sucrose gradients, or serial filtration methods (Lamparski et al., 2002). Conventional FACS analysis cannot distinguish vesicles smaller than 200–300 nm, but novel techniques to count nanosized vesicles, like nanoparticle tracking analysis (Dragovic et al., 2011) have been recently developed. Electron microscopy has been used to confirm the purity of the preparations, and characterize the ultrastructure and size of the EVs (Raposo et al., 1996).

Even atomic force microscopy can be used for this purpose (Tian et al., 2010). The molecular contents of purified EV populations have been analyzed by immunoblotting, mass spectrometry, qPCR, and different microarrays (Graner et al., 2009; Mathivanan & Simpson, 2009; Xiao et al., 2012).

2.5. EVs as carriers of regulatory molecules into target cells

EVs represent their cellular origin and contain both cytosolic and membrane-associated proteins derived from their parental cells, a "molecular signature" of the cell of origin. However, due to currently unknown sorting mechanisms, the contents of the vesicles may differ from the average contents of their parental cells. Membranes of EVs have been shown to be enriched in cholesterol, sphingolipids, and ceramides, typical components of lipid rafts (Mathivanan & Simpson, 2009), supposed to be involved in EV fusion with target cells (Vickers & Remaley, 2012). EVs interact with their target cells by specific interactions, initiating fusion of the vesicle and target cell membranes. This results in the release of the EV contents into the cytosol and inclusion of the EV membrane components to the target cell plasma membrane. Another route for internalization is receptor-mediated endocytosis, which results in the release of their cargo into multivesicular bodies or degradation in lysosomes. Internalized vesicles can also be re-released into the extracellular space via transcytosis (Schneider & Simons, 2013).

EVs contain intracellular proteins, second messengers and genetic material. They are rich in membrane trafficking regulators (Rab proteins), cytoskeletal proteins like actin and tubulin, regulators of cytoskeleton and cell–extracellular matrix (ECM) interactions like myosins and integrins (Thery et al., 2002). EVs carry and transfer functional mRNA and miRNAs (Valadi et al., 2007) to target cells where they are translated into, or regulate the synthesis of proteins that modulate signaling pathways. A comprehensive list of EV-associated mRNAs, proteins, and lipids from more than 140 studies are provided in the ExoCarta database (http://exocarta.org/).

3. EVs PROMOTE TUMOR PROGRESSION

3.1. EVs associate with cancers

EVs have been shown to participate in many processes during the pathogenesis of cancer. The connection of EVs with cancer was observed as early as the 1970s (Lee, Laurie, & Townsend, 1975), when EVs were found in osteogenic sarcoma (Lee et al., 1975). Thereafter, EVs have been found to associate with tumor progression in various malignancies in bone, (Yoshida, Miyazaki, & Yumoto, 1991), fibrosarcoma, brain tumors (Al-Nedawi et al., 2008), and oligodendrogliomas (Ginestra et al., 1997). EVs are secreted also by ovarian (Dolo et al., 1999; Ginestra, Miceli, Dolo, Romano, & Vittorelli, 1999; Graves et al., 2004; Millimaggi et al., 2007), breast (Cassara et al., 1998; Galindo-Hernandez et al., 2013; Lau & Wong, 2012), prostate (Angelucci et al., 2000; Morello et al., 2013; Renzulli et al., 2010), melanoma (Lima, Chammas, Monteiro, Moreira, & Barcinski, 2009; Peinado et al., 2012), and colorectal (Hong et al., 2009) cancer cells.

The exchange of bioactive molecules via EVs may occur between cancer cells as shown for the oncogenic growth factor receptors in brain tumors (Al-Nedawi et al., 2008), or between cancer cells and nonmalignant recipient cells (Skog et al., 2008). Cancer cells may also themselves be the targets of EVs secreted by other cells. For example, breast cancer cells have been shown to increase invasiveness *in vitro* when incubated with EVs released by platelets (Janowska-Wieczorek, Marquez-Curtis, Wysoczynski, & Ratajczak, 2006) or macrophages (Yang et al., 2011).

3.2. Promotion of invasion and metastasis by EVs

In murine models, tumor cell-derived EVs enhance the metastasis of melanoma cells (Lima et al., 2009), while platelet-derived EVs promote the metastasis of lung cancer cells (Janowska-Wieczorek et al., 2006). During cancer progression the surrounding tumor–stroma and even the more distant tissues are modified and changed by partly unknown mechanisms. EVs released by the cancer cells into the interstitial space of surrounding tissues, or into peripheral circulation (Iero et al., 2008; Kim et al., 2003), provide a plausible mechanism to enhance cancer progression in several ways. EVs are rich in proteolytic enzymes that facilitate matrix remodeling around the tumor, and thus directly help invasion of the malignant cells (Angelucci et al., 2000; Ginestra et al., 1998, 1997; Graves et al., 2004; Lo Cicero, Majkowska, Nagase, Di Liegro, & Troeberg, 2012). The bioactive materials in EVs may lead to enhanced migration of cells (Akao et al., 2011; Li et al., 2013; Morello et al., 2013) through transfer of active molecules into the target cells, or by signaling through cell surface receptors. EVs released by cancer cells are able to educate the surrounding stromal cells (Josefsson et al., 2011), and cells at distant sites, like the progenitors in bone marrow (Peinado et al., 2012), toward a prometastatic phenotype.

Chromosome segregation 1-like protein at the EV membrane has been proposed to be important for the metastasis of melanoma cells (Liao et al., 2012). EVs may also be involved in multidrug resistance by expulsion of anticancer drugs, as demonstrated with doxorubicin (Shedden, Xie, Chandaroy, Chang, & Rosania, 2003) and cisplatin (Safaei et al., 2005).

3.3. Thrombosis and EVs

Several cancer cells, for example melanoma (Lima et al., 2011) and breast cancer cells (Lima, Leal, Vargas, Porto-Carreiro, & Monteiro, 2013) secrete EVs, which contain the tissue factor protein. As this factor can initiate the coagulation cascade, it has been proposed that the prothrombotic state often found in malignancies could be due to EVs secreted by the cancer cells (Del Conde et al., 2007). In melanocytes the production of procoagulant EVs is associated with malignant transformation (Lima et al., 2011), and in colorectal cancer cells the activity of the tissue factor in EVs seems to reflect the genetic status of the cancer cells (Yu et al., 2005).

3.4. Angiogenesis is promoted by EVs

EVs have been suggested to enhance angiogenesis by multiple mechanisms. Sphingomyelin in EVs derived from fibrosarcoma cells induces angiogenesis by stimulating endothelial cell migration and neovascularization *in vivo* (Kim et al., 2002). In glioblastoma multiforme cells, hypoxia triggers the release of EVs containing tissue factor, which leads to activation of endothelial cells and angiogenic signaling (Svensson et al., 2011). In human lung cancer cells, platelet-derived EVs stimulate mRNA production of angiogenic factors such as MMP-9, VEGF, IL-8, and HGF (Janowska-Wieczorek et al., 2005). In addition, cell cycle-related mRNA in EVs secreted by colorectal cancer cells increases endothelial cell proliferation (Hong et al., 2009).

3.5. EVs participate in antitumor immunity

EVs participate also in antitumor immunity, one of the key features for the survival of tumor cells under the host's immune surveillance. Melanoma cells have been reported to produce large quantities of phosphatidylserine-containing EVs, which stimulate anti-inflammatory responses during melanoma progression (Lima et al., 2009). In ovarian cancer, EVs containing the Fas ligand are toxic to activate CD4+ T-lymphocytes, thus suppressing antitumor immune responses (Meng, Kang, & Fishman, 2005). Similar results have been obtained in melanoma (Andreola et al., 2002) and colorectal cancer cells (Huber et al., 2005). In addition, melanoma and colorectal carcinoma cells release EVs that inhibit the differentiation of monocytes to antigen-presenting cells, both *in vitro* and *in vivo*, and the monocytes then suppress T-cell functions (Valenti et al., 2006).

4. HA SYNTHESIS ENHANCES SHEDDING OF EXTRACELLULAR VESICLES

4.1. HA–EVs are common in cells with active HA secretion

While studying EGFP-HAS3 overexpressing cells we found that they released EVs covered with HAS and HA (Rilla, Pasonen-Seppänen, et al., 2013). Even more importantly, their shedding was induced by HA synthesis in many cell types when HAS2 or HAS3 was overexpressed (Rilla, Pasonen-Seppänen, et al., 2013). Active EV secretion seems to be a typical feature of human mesenchymal stem cells (Qu et al., 2014) and mesothelial cells (Rilla, Pasonen-Seppänen, et al., 2013), which naturally secrete large amounts of HA. Similar association between high-HA levels (Hascall et al., 2004; Jiang, Liang, & Noble, 2007; Toole & Hascall, 2002) and active EV release (Hulsmans & Holvoet, 2013; Martins et al., 2013; Sabin & Kikyo, 2014) is found in conditions like cancer, inflammation, embryonic development, diabetes, atherosclerosis, and healing tissues, suggesting that HA metabolism is associated with EV secretion also *in vivo*. The HA–EVs may carry HA on their surface even to distant sites, but at the moment we do not know if HA–EVs are even able to produce new HA. At least they contain HAS on their membranes (Rilla, Oikari, et al., 2013; Rilla, Pasonen-Seppänen, et al., 2013). Interestingly, Has2 and Has3 mRNAs have been detected in the EVs secreted by colorectal cancer cells (Hong et al., 2009).

4.2. HAS-induced protrusions as a platform for vesicle shedding

Active HAS accumulates on special plasma membrane protrusions (Fig. 5.2), and induces their growth and maintenance (Rilla et al., 2005; Kultti et al., 2006). Probing live cell cultures with a fluorescent HA-binding probe has shown that these HA-induced cell protrusions are an inherent feature of cells with high-HAS activity. They scaffold the traditional HA glycocalyx in many cell types (Rilla, Tiihonen, Kultti, Tammi, & Tammi, 2008), and increase the plasma membrane area for active HA secretion. The average diameter of the HAS-induced protrusions is around 100 nm, and can reach

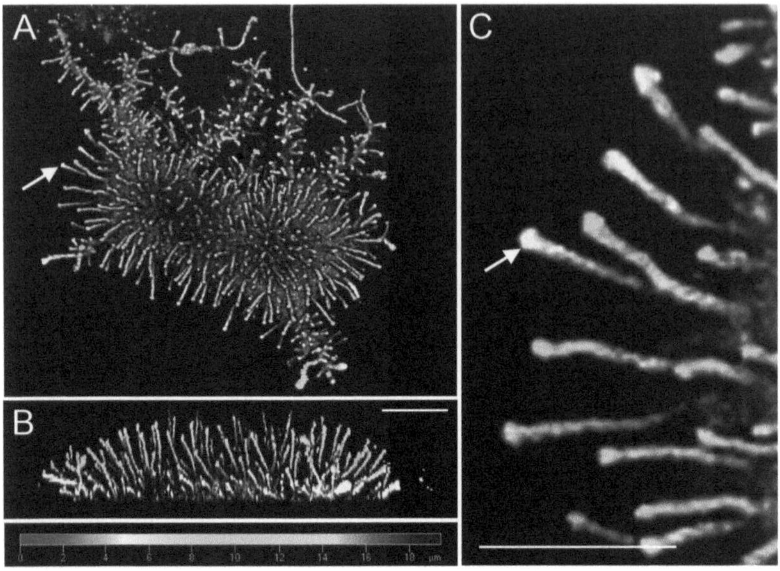

Figure 5.2 HAS3 overexpression induces formation of plasma membrane protrusions. A 3D projection created from a compressed stack of confocal optical sections of an MCF-7 human breast adenocarcinoma cell overexpressing GFP-HAS3 (horizontal view in A and C and vertical view in B). Depth color coding has been used to indicate the various positions and lengths of the protrusions. Note the bulbous extensions at the distal tips of the tentacles (arrows in A and C). Magnification bars 10 μm in B and 5 μm in C. (See the color plate.)

a length of 20 μm, or even more (Kultti et al., 2006). A typical morphological feature of the HAS-induced protrusion is a bulbous enlargement at the distal tip, which is the putative point of EV shedding (Figs. 5.2 and 5.3). The diameter of the HA-generated EVs is around 100 nm (Fig. 5.3C and D), similar to the diameter of the protrusions (Kultti et al., 2006) supporting the function of protrusions as platforms for the shedding of EVs. HAS-induced protrusions contain CD44 and actin (Kultti et al., 2006), proteins detected also in HAS-induced EVs (Rilla, Pasonen-Seppänen, et al., 2013). Furthermore, factors like EGF, known to induce HA secretion and filopodia formation (Pienimäki et al., 2001), also induce EV shedding in HER2-positive breast cancer cells (Ciravolo et al., 2012). The synthesis of HA on the long protrusions, and shedding of the HA-coated EVs could facilitate the transport of concentrated HA away from the cell body. This may be important especially in tissues with a large extracellular space, like cartilage and bone, and in the connective tissue compartments of renewing, healing, or malignant tissues.

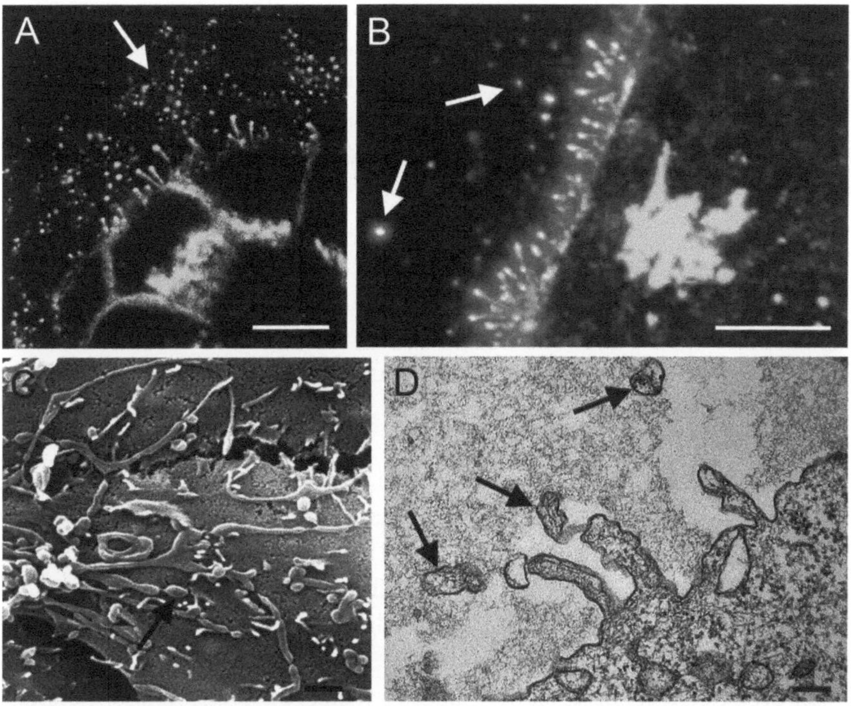

Figure 5.3 Extracellular vesicles originate from the tips of HAS-induced microvilli. Confocal (A and B), SEM (C), and TEM (D) images of GFP-HAS3 overexpressing MDCK cells. Budding of an EV from the tips of microvilli is shown in D. Red color indicates HA, blue shows nuclei in B, and green indicates GFP-HAS3 in A and B. Arrows indicate EVs or tips of microvilli in all panels. (A) is a single confocal optical section and (B) is a compressed stack of optical sections. Magnification bars 10 μm in A and B, 1 μm in C, and 100 nm in D. (See the color plate.)

Interestingly, as early as 1967, Wolf et al. described the budding of lipid-rich material from the enlarged termini of the extensions of activated platelets, called "platelet dust" (Polasek, 1982; Wolf, 1967), and a similar phenomenon was found in ionophore-treated erythrocytes (White, 1974). More recently, shedding of vesicles has been shown to occur from the tips of the microvilli in intestinal enterocytes (McConnell et al., 2009). These findings suggest that a significant proportion of circulating vesicles may originate from various plasma membrane extensions.

4.3. Regulation of HA synthesis and EV shedding by cellular glucose supply?

One of the most important factors regulating HAS activity is the availability of the UDP-sugar substrates. Depletion of the intracellular UDP-GlcUA and

UDP-GlcNAc with 4-methylumbelliferone (Kultti et al., 2009; Rilla et al., 2005) and mannose (Jokela et al., 2008), respectively, inhibits the traffic of GFP-HAS to the plasma membrane, and HA synthesis. In cell cultures, HA synthesis correlates with the concentration of glucose in the culture medium (Rilla, Oikari, et al., 2013; Rilla, Pasonen-Seppänen, et al., 2013) and synthesis is induced by high-glucose availability (Sainio, Jokela, Tammi, & Järveläinen, 2010), which secures the supply of the UDP-sugars. Without sufficient glucose supply, UDP-sugars are depleted and HAS escapes from the plasma membrane into intracellular compartments, and the HAS-induced plasma membrane protrusions disappear (Rilla, Oikari, et al., 2013; Tammi et al., 2011). As a consequence, HA–EV shedding is also reduced. In line with the *in vitro* observations on the effects of glucose, increased levels of HA are found in diabetic patients (Mine, Okada, Kawahara, Tabata, & Tanaka, 2006).

The above observations suggest that HAS activity is regulated also under other conditions where sugar levels are altered. Indeed, cancer cells rely on high-glucose uptake (Semenza, 2008) and the increased intracellular glucose level probably contributes to the increased production of HA and the activity of HASs, found in malignant tumors like breast cancer (Auvinen et al., 2014, 2000).

In breast cancer, high blood glucose predicts progression and metastasis of the tumor (Friedl & Wolf, 2003). High blood glucose and type 2 diabetes are often associated with obesity, and obesity is also an unfavorable prognostic factor for breast cancer patients (Contiero et al., 2013). Interestingly, HA is increased in obesity (Kang et al., 2013), and our recent data show that increased body mass associates with HA accumulation in the tumor–stroma of breast cancers (Auvinen et al., 2014). Moreover, obesity correlates with HAS expression in breast cancer patients (Auvinen et al., 2014).

The fact that the amount of EVs in the plasma is also increased in diabetic patients (Diamant et al., 2002; Sabatier et al., 2002) and in obesity (Müller, 2012), suggests that there is a strong correlation between high glucose, HA synthesis, and EV shedding. Our data in cell culture models show that glucose concentration in the culture medium influence the release of EVs, correlating with HA synthesis (Rilla, Oikari, et al., 2013; Rilla, Pasonen-Seppänen, et al., 2013). Formerly, EVs were considered a way of waste elimination. This may be partly true, because enhanced HA synthesis and HA–EV shedding provides a pathway for cells to get rid of an overload of glucose and other metabolites.

4.4. HA–EVs in tissue regeneration

The role of stem cells in tissue regeneration, their recruitment into injured tissues, and use for the treatment of various diseases is currently under investigation, and subject to clinical trials. Interestingly, the EVs secreted from stem cells have been shown as essential factors for tissue healing and regeneration (Sabin & Kikyo, 2014). They can transfer information between bone marrow stem cells and damaged tissue in acute injuries (Reis et al., 2012), suggesting that stem cells can repair damaged tissues without directly replacing the parenchymal cells. HA has been known for a long time as one of the most important constituents of the body's natural biological toolbox for healing (Jiang et al., 2007). The high levels of HA secreted by human mesenchymal stem cells (Qu et al., 2014), and the abundance of the HA–EVs shed by them (Fig. 5.4), suggest that HA is an important constituent of the stem cell-derived EVs and supports stem cell functions in tissue remodeling and regeneration.

4.5. Body fluids rich in HA are also rich in EVs

EVs have been isolated from almost all of our normal body fluids like plasma, urine, saliva, breast milk, synovial fluid, amniotic fluid, and cerebrospinal fluid (Raposo & Stoorvogel, 2013). All these fluids contain HA, but the fluids particularly rich in HA like synovial fluid, malignant pleural effusion and plasma of patients of malignant mesothelioma (Creaney, Dick, Segal, Musk, & Robinson, 2013), and ascites of ovarian cancer patients (Yeo, Nagy, Yeo, Dvorak, & Toole, 1996), are especially rich in EVs (Bard et al., 2004; Peng, Yan, & Keng, 2011). At present, there is no direct evidence linking HA synthesis to the formation of EVs in these fluids or showing EVs act as specific carriers of HA in them. However, it is interesting to note that the protrusions are a characteristic feature of mesothelium (Mutsaers, 2004) and synovial membrane (Lukoschek & Addicks, 1991), both active in HA synthesis and a source of EVs. Likewise, chondrocytes present microvilli and cartilage contains EVs (Hale & Wuthier, 1987). Tumor fibroblasts contain microvilli (Rilla et al., 2008), the formation of which correlates with malignant transformation (Hamada, Ren, & Hosokawa, 1999) and invasive behavior of cancer cells (Zu, Sidhu, Wieczorek, & Cassai, 2002).

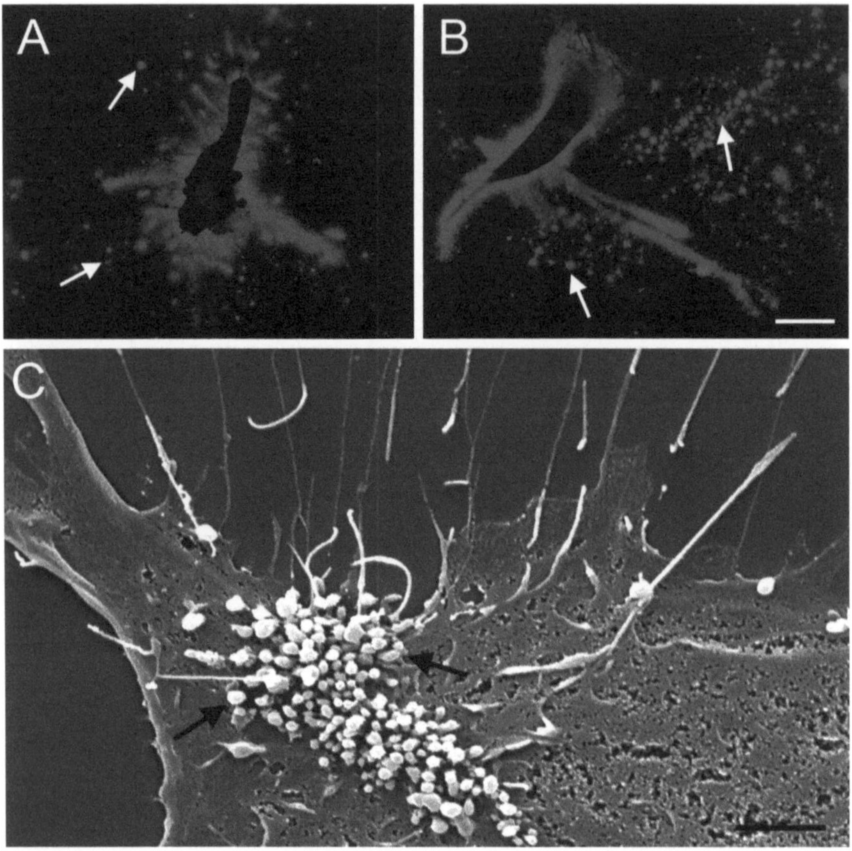

Figure 5.4 HA-positive extracellular vesicles are secreted from cell types with active HA synthesis. A cell stained with fHABC indicates the HA coat (red) and DRAQ5 the nucleus (blue) in a live human mesenchymal stem cell (A) and a rat primary mesothelial cell (B). Human mesothelial cell (LP-9) imaged with SEM is shown in C. Arrows indicate secreted EVs. Magnification bars 10 μm in B and 2 μm in C. (See the color plate.)

4.6. HA–EVs provide the missing link between HA and cancer?

HA is accumulated in the stroma and around tumor cells in many epithelial cancer types like breast, ovarian, and prostate cancers (Anttila et al., 2000; Auvinen et al., 2013, 2000; Josefsson et al., 2011; Lipponen et al., 2001), and promotes the malignant properties of the tumor cells via tumor–stroma interactions and by increasing tumor vasculature (Sironen et al., 2011). However, the detailed mechanisms behind these events are still unknown. Tumor-associated HA reduces the apoptotic efficiency of chemotherapic drugs (Ricciardelli et al., 2013; Toole & Slomiany, 2008; Varadi et al., 2012) while removal of HA, or inhibition of its synthesis, potentiates the effects of cytostatic drugs and restricts tumor growth (Nakazawa et al., 2006; Provenzano et al., 2012). The HER2-positive breast cancers are enriched in HA, and the tumors with the highest level of HA show a tendency to resist trastuzumab therapy (Auvinen et al., 2013).

Most cancer types with increased HA content have enhanced capability to produce EVs, as summarized in Table 5.1. Interestingly, HER2-positive tumors secrete EVs displaying HER2 on their surfaces. The HER2-positive EVs are suggested to compromise the therapy by binding trastuzumab (Ciravolo et al., 2012), offering a mechanism for resistance to HER2-targeted therapy.

Table 5.1 Many tumors that have increased HA content have also increased secretion of EVs, suggesting a relationship between HA synthesis and EV shedding

Cancer type	Extracellular vesicles	HA in tissue
Melanoma	↑ Protein concentration of EVs in plasma (Peinado et al., 2012)	↓ (Karjalainen et al., 2000; Siiskonen, Poukka, Tyynelä-Korhonen, Sironen, & Pasonen-Seppänen, 2013)
Squamous cell carcinoma of mouth	↑ In serum (Kim et al., 2005)	↑ (Kosunen et al., 2004)
Adenocarcinoma of lung	↑ In plasma (Rabinowits, Gercel-Taylor, Day, Taylor, & Kloecker, 2009)	↑ (Pirinen et al., 2001)
Adenocarcinoma of breast	↑ In plasma (Galindo-Hernandez et al., 2013)	↑ (Auvinen et al., 1997, 2000)
Serous ovarian carcinoma	↑ In ascites (Ginestra et al., 1999; Graves et al., 2004)	↑ (Anttila et al., 2000; Hiltunen et al., 2002)
Prostate carcinoma	↑ In plasma (Lance, Drake, & Troyer, 2011; Morello et al., 2013; Tavoosidana et al., 2011)	↑ (Josefsson et al., 2011; Lipponen et al., 2001; Lokeshwar et al., 2001)
Colon carcinoma	↑ In plasma (Silva et al., 2012)	↑ (Ropponen et al., 1998)
Acute myeloid leukemia	↑ In serum (Szczepanski, Szajnik, Welsh, Whiteside, & Boyiadzis, 2011)	↑ (Sundström et al., 2005)

Immunostaining of HASs is typically intense in tumor cells (Auvinen et al., 2014; Nykopp et al., 2009, 2010), while the HA produced is accumulated in the stroma rather than tumor cells (Fig. 5.5). Transport and possible secretion of HA by EVs released by tumor cells, and the exchange of HASs between tumor and stromal cells could explain these apparently conflicting observations. Thus, since EVs can carry functional *Has* mRNAs (Hong et al., 2009), and even HAS proteins (Kim et al., 2013), the target cell could receive a functional HA synthesis machinery from the donor cell. HA synthesis could also continue extracellularly on the membrane of the HAS-positive EVs. It is possible that the HA–EVs also recruit specific proteins, lipids, and RNAs that convey important signals to the target cells.

EVs can interact with the cells they recognize by cell-type specific, membrane-bound receptors. Secretion of EVs by tumor cells may give an explanation for the accumulation of HA in the premetastatic mesentery, and the subsequent attachment of CD44-positive tumor cells to the peritoneal wall (Yeo et al., 1996).

The massive shedding of HA–EVs from the basal side of the MDCK cell-derived epithelial cysts into the surrounding matrix indicates that EVs can carry HA into adjacent stroma, or tissues at more distant sites (Fig. 5.6). The HA–EVs contain also CD44 (Fig. 5.6F), which may help to retain HA on the EVs. EVs may interact with stromal cells by specific receptor interactions, like HA binding to CD44 on the target cell, resulting in the fusion and release of the cargo in the target cells to modify their functions. This idea is supported by the recent suggestion that EVs promote tumor cell

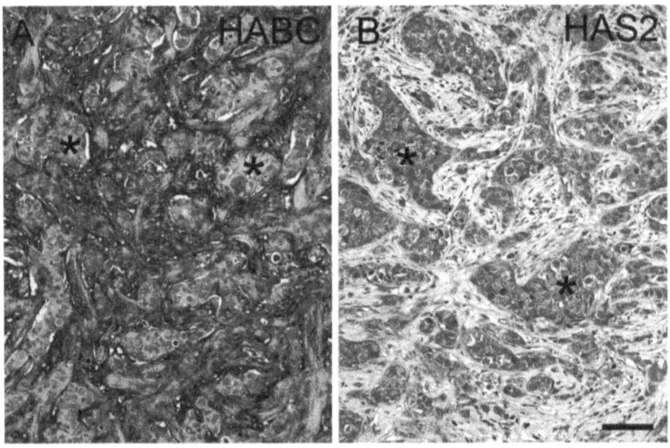

Figure 5.5 HA and HAS2 in breast carcinoma sections. HA staining is most intense in the stroma, shown in A, while HAS2 immunostaining is strongest in tumor cells, but also seen in stromal fibroblasts (B). Asterisks indicate tumor cells. Magnification bar 50 μm. (See the color plate.)

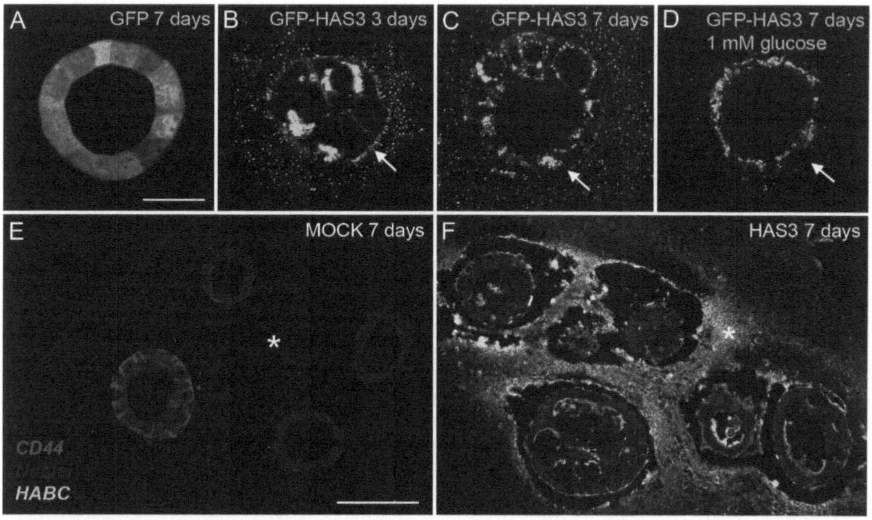

Figure 5.6 Extracellular vesicle secretion and HA accumulation in the extracellular matrix of epithelial organotypic cultures of MDCK cells. The GFP-HAS-positive plasma membrane protrusions exist on the basal face of the epithelial cysts during the early stages of the morphogenesis in a basement membrane extract (B, 3 days and C, 7 days). High numbers of EVs are present in all stages, as compared to control cultures expressing GFP only (A, 7 days). Low-glucose concentration (1 mM) results in a decreased secretion of EVs (D, 7 days) when compared with the normal glucose (5 mM) in the other images. Live cultures are presented in (A)–(D). Paraffin sections of 7-day-old cultures stained for nuclei (blue), CD44 (red), and HA (green) show that HA was secreted by the GFP-HAS3 overexpressing epithelial cells, and accumulated in the ECM (F), while MOCK-transfected cells do not produce detectable amounts of HA (E). All panels are confocal optical sections. Magnification bars 20 μm in A and 50 μm in E. (See the color plate.)

motility and metastasis via CD44–HA interactions (Mu, Rana, & Zoller, 2013), and the finding that the induction of the premetastatic niche by exosomes is dependent on CD44v6 (Jung et al., 2009). These direct and indirect ways to modify target tissues would explain many unanswered questions concerning HA relationship with cancer.

5. HA–EVs AS PREDICTORS, TARGETS, AND CARRIERS OF THERAPY

5.1. HA–EVs as potential biomarkers for HA-rich tumors

In many areas of medicine, biomarkers are sought to monitor the state of diseases. Biomarkers can also lead the way toward personalized medicine.

An ideal biomarker is identified from an easily obtainable sample like blood, urine, or saliva, and detectable with fast, accurate, and inexpensive methods. EVs fulfill these criteria, and have the potential for prediction, diagnosis, prognosis, and testing the efficacy of treatments in diseases such as cancer, type II diabetes, and obesity (Müller, 2012).

Solid tumors are often difficult to reach and detect, but the appearance of specific, tumor-derived EVs in body fluids, and their quantification could reveal the presence of tumors, even in their early stages. EVs could also predict tumor aggressiveness, and indicate possible recurrence after treatment (Müller, 2012). Analysis of EVs might also aid in the planning of treatments for individual patients (Skog et al., 2008).

Since the role of HA in cancers is well acknowledged, noninvasive methods to detect HA in cancer patients would be valuable in diagnostics, and real-time disease monitoring. If the amount of HA–EVs in body fluids correlates with the HA content of the original tumor, it can be easily and rapidly detected with a highly specific, fluorescently labeled probe for HA (Rilla et al., 2008). HA–EVs might be used to predict disease progression, and help in the selection of therapeutic measures, for example to decide if the therapy should include HA depletion to enhance the efficacy of other therapies (Provenzano et al., 2012).

5.2. Utilization of HA–EVs as targets and carriers for therapy

In addition to indicators of disease state, HA–EVs could act as potential targets of therapy, or even harnessed as specific carriers for targeted therapy (Collet, Grillon, Nadim, & Kieda, 2013). HA-coated liposomes (Arpicco et al., 2013), HA-decorated PLGA-PEG nanoparticles (Vangara, Liu, & Palakurthi, 2013), and HA-coated magnetic nanoparticles (El-Dakdouki et al., 2013) have been tested to target drugs to CD44-rich tumors. In the same way, artificially formed HA–EVs could be used as vehicles to deliver anticancer factors to CD44 expressing cancers.

Selective capturing and removal of harmful EVs delivering oncogenic, thrombogenic, or inflammatory signals from the entire circulation system have been tested as a therapy for cancer. These methods are based on the recognition of EV surface proteins by utilizing affinity agents fitted for dialysis machines (Marleau, Chen, Joyce, & Tullis, 2012), and could also be applied to eliminate HA–EVs from the circulation of patients.

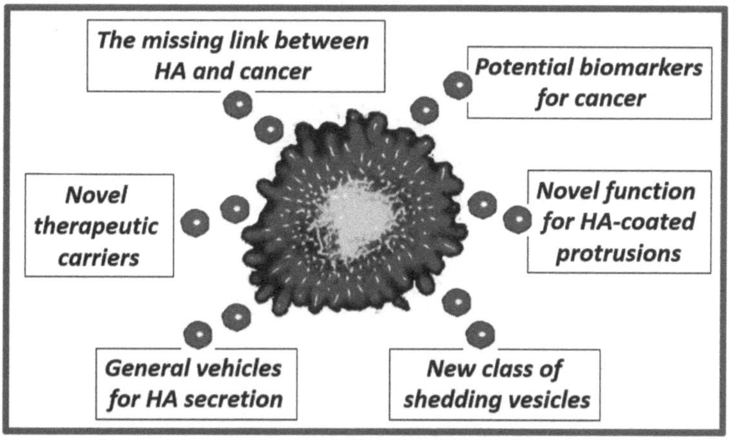

Figure 5.7 Putative roles of the HA-induced extracellular vesicles. Schematic presentation of the potential connections between HA, cancer, and HA-induced EVs. HA-coated extracellular vesicles act as special carriers for HA and may give an answer to many open questions regarding HA and its role in cancer. The HA–EVs may also act as cancer biomarkers as well as a platform to deliver therapeutic drugs. (See the color plate.)

6. CONCLUSIONS

The very active research during the recent years has raised EVs as a new tissue compartment, in addition to cells and the surrounding ECM. Careful characterization of the molecular composition of the newly discovered HA–EVs and of their functional effects on target cells are needed to find out if they represent a unique species among EVs. If so, it will be important to study in more detail how their shedding is regulated and what are the mechanisms of their interaction with target cells. Figure 5.7 summarizes the novel aspects emerged in this review: HA–EVs shed from HA-coated protrusions can potentially act as vehicles for HA secretion, and potential biomarkers, therapeutic vehicles, and a previously unknown link between HA content and cancer promotion.

ACKNOWLEDGMENTS

The original studies of the authors, forming the basis of the review, were financially supported by the Spearhead Funds from the University of Eastern Finland (Cancer Center of Eastern Finland) (R. T. and M. T.), Sigrid Juselius foundation (M. T. and R. T.), Academy of Finland (M. T.), Kuopio University Hospital (EVO and VTR)

(M. T.), and personal grants for H. S. (from Northern Savo Cultural Foundation) and K. R. (from North Savo Cancer Foundation).

REFERENCES

Aatonen, M., Grönholm, M., & Siljander, P. R. (2012). Platelet-derived microvesicles: Multitalented participants in intercellular communication. *Seminars in Thrombosis and Hemostasis, 38*, 102–113.

Akao, Y., Iio, A., Itoh, T., Noguchi, S., Itoh, Y., Ohtsuki, Y., et al. (2011). Microvesicle-mediated RNA molecule delivery system using monocytes/macrophages. *Molecular Therapy, 19*, 395–399.

Al-Nedawi, K., Meehan, B., Micallef, J., Lhotak, V., May, L., Guha, A., et al. (2008). Intercellular transfer of the oncogenic receptor EGFRvIII by microvesicles derived from tumour cells. *Nature Cell Biology, 10*, 619–624.

Andreola, G., Rivoltini, L., Castelli, C., Huber, V., Perego, P., Deho, P., et al. (2002). Induction of lymphocyte apoptosis by tumor cell secretion of FasL-bearing microvesicles. *Journal of Experimental Medicine, 195*, 1303–1316.

Angelucci, A., D'Ascenzo, S., Festuccia, C., Gravina, G. L., Bologna, M., Dolo, V., et al. (2000). Vesicle-associated urokinase plasminogen activator promotes invasion in prostate cancer cell lines. *Clinical and Experimental Metastasis, 18*, 163–170.

Anttila, M. A., Tammi, R. H., Tammi, M. I., Syrjänen, K. J., Saarikoski, S. V., & Kosma, V. M. (2000). High levels of stromal HA predict poor disease outcome in epithelial ovarian cancer. *Cancer Research, 60*, 150–155.

Arpicco, S., Lerda, C., Dalla Pozza, E., Costanzo, C., Tsapis, N., Stella, B., et al. (2013). Hyaluronic acid-coated liposomes for active targeting of gemcitabine. *European Journal of Pharmaceutics and Biopharmaceutics, 85*, 373–380.

Auvinen, P. K., Parkkinen, J. J., Johansson, R. T., Ågren, U. M., Tammi, R. H., Eskelinen, M. J., et al. (1997). Expression of HA in benign and malignant breast lesions. *International Journal of Cancer, 74*, 477–481.

Auvinen, P., Rilla, K., Tumelius, R., Tammi, M., Sironen, R., Soini, Y., et al. (2014). HA synthases (HAS1-3) in stromal and malignant cells correlate with breast cancer grade and predict patient survival. *Breast Cancer Research and Treatment, 143*, 277–286.

Auvinen, P., Tammi, R., Kosma, V. M., Sironen, R., Soini, Y., Mannermaa, A., et al. (2013). Increased HA content and stromal cell CD44 associate with HER2 positivity and poor prognosis in human breast cancer. *International Journal of Cancer, 132*, 531–539.

Auvinen, P., Tammi, R., Parkkinen, J., Tammi, M., Ågren, U., Johansson, R., et al. (2000). HA in peritumoral stroma and malignant cells associates with breast cancer spreading and predicts survival. *American Journal of Pathology, 156*, 529–536.

Bard, M. P., Hegmans, J. P., Hemmes, A., Luider, T. M., Willemsen, R., Severijnen, L. A., et al. (2004). Proteomic analysis of exosomes isolated from human malignant pleural effusions. *American Journal of Respiratory Cell and Molecular Biology, 31*, 114–121.

Bhatnagar, S., Shinagawa, K., Castellino, F. J., & Schorey, J. S. (2007). Exosomes released from macrophages infected with intracellular pathogens stimulate a proinflammatory response in vitro and in vivo. *Blood, 110*, 3234–3244.

Camussi, G., Deregibus, M. C., Bruno, S., Cantaluppi, V., & Biancone, L. (2010). Exosomes/microvesicles as a mechanism of cell-to-cell communication. *Kidney International, 78*, 838–848.

Camussi, G., Deregibus, M. C., Bruno, S., Grange, C., Fonsato, V., & Tetta, C. (2011). Exosome/microvesicle-mediated epigenetic reprogramming of cells. *American Journal of Cancer Research, 1*, 98–110.

Cassara, D., Ginestra, A., Dolo, V., Miele, M., Caruso, G., Lucania, G., et al. (1998). Modulation of vesicle shedding in 8701 BC human breast carcinoma cells. *Journal of Submicroscopic Cytology and Pathology*, *30*, 45–53.

Ciravolo, V., Huber, V., Ghedini, G. C., Venturelli, E., Bianchi, F., Campiglio, M., et al. (2012). Potential role of HER2-overexpressing exosomes in countering trastuzumab-based therapy. *Journal of Cellular Physiology*, *227*, 658–667.

Collet, G., Grillon, C., Nadim, M., & Kieda, C. (2013). Trojan horse at cellular level for tumor gene therapies. *Gene*, *525*, 208–216.

Contiero, P., Berrino, F., Tagliabue, G., Mastroianni, A., Di Mauro, M. G., Fabiano, S., et al. (2013). Fasting blood glucose and long-term prognosis of non-metastatic breast cancer: A cohort study. *Breast Cancer Research and Treatment*, *138*, 951–959.

Creaney, J., Dick, I. M., Segal, A., Musk, A. W., & Robinson, B. W. (2013). Pleural effusion hyaluronic acid as a prognostic marker in pleural malignant mesothelioma. *Lung Cancer*, *82*, 491–498.

Del Conde, I., Bharwani, L. D., Dietzen, D. J., Pendurthi, U., Thiagarajan, P., & Lopez, J. A. (2007). Microvesicle-associated tissue factor and Trousseau's syndrome. *Journal of Thrombosis and Haemostasis*, *5*, 70–74.

Diamant, M., Nieuwland, R., Pablo, R. F., Sturk, A., Smit, J. W., & Radder, J. K. (2002). Elevated numbers of tissue-factor exposing microparticles correlate with components of the metabolic syndrome in uncomplicated type 2 diabetes mellitus. *Circulation*, *106*, 2442–2447.

Dolo, V., D'Ascenzo, S., Violini, S., Pompucci, L., Festuccia, C., Ginestra, A., et al. (1999). Matrix-degrading proteinases are shed in membrane vesicles by ovarian cancer cells in vivo and in vitro. *Clinical and Experimental Metastasis*, *17*, 131–140.

Dragovic, R. A., Gardiner, C., Brooks, A. S., Tannetta, D. S., Ferguson, D. J., Hole, P., et al. (2011). Sizing and phenotyping of cellular vesicles using nanoparticle tracking analysis. *Nanomedicine*, *7*, 780–788.

El-Dakdouki, M. H., Xia, J., Zhu, D. C., Kavunja, H., Grieshaber, J., O'Reilly, S., et al. (2013). Assessing the in vivo efficacy of doxorubicin loaded HA nanoparticles. *ACS Applied Materials & Interfaces*, *8*, 697–705.

Faure, J., Lachenal, G., Court, M., Hirrlinger, J., Chatellard-Causse, C., Blot, B., et al. (2006). Exosomes are released by cultured cortical neurones. *Molecular and Cellular Neuroscience*, *31*, 642–648.

Friedl, P., & Wolf, K. (2003). Tumour-cell invasion and migration: Diversity and escape mechanisms. *Nature Reviews. Cancer*, *3*, 362–374.

Galindo-Hernandez, O., Villegas-Comonfort, S., Candanedo, F., Gonzalez-Vazquez, M. C., Chavez-Ocana, S., Jimenez-Villanueva, X., et al. (2013). Elevated concentration of microvesicles isolated from peripheral blood in breast cancer patients. *Archives of Medical Research*, *44*, 208–214.

Ginestra, A., La Placa, M. D., Saladino, F., Cassara, D., Nagase, H., & Vittorelli, M. L. (1998). The amount and proteolytic content of vesicles shed by human cancer cell lines correlates with their in vitro invasiveness. *Anticancer Research*, *18*, 3433–3437.

Ginestra, A., Miceli, D., Dolo, V., Romano, F. M., & Vittorelli, M. L. (1999). Membrane vesicles in ovarian cancer fluids: A new potential marker. *Anticancer Research*, *19*, 3439–3445.

Ginestra, A., Monea, S., Seghezzi, G., Dolo, V., Nagase, H., Mignatti, P., et al. (1997). Urokinase plasminogen activator and gelatinases are associated with membrane vesicles shed by human HT1080 fibrosarcoma cells. *Journal of Biological Chemistry*, *272*, 17216–17222.

Gould, S. J., & Raposo, G. (2013). As we wait: Coping with an imperfect nomenclature for extracellular vesicles. *Journal of Extracellular Vesicles*, *2*, 1–3, http://dx.doi.org/10.3402/jev.v2i0.20389.

Graner, M. W., Alzate, O., Dechkovskaia, A. M., Keene, J. D., Sampson, J. H., Mitchell, D. A., et al. (2009). Proteomic and immunologic analyses of brain tumor exosomes. *FASEB Journal, 23*, 1541–1557.

Graves, L. E., Ariztia, E. V., Navari, J. R., Matzel, H. J., Stack, M. S., & Fishman, D. A. (2004). Proinvasive properties of ovarian cancer ascites-derived membrane vesicles. *Cancer Research, 64*, 7045–7049.

Hale, J. E., & Wuthier, R. E. (1987). The mechanism of matrix vesicle formation. Studies on the composition of chondrocyte microvilli and on the effects of microfilament-perturbing agents on cellular vesiculation. *Journal of Biological Chemistry, 262*, 1916–1925.

Hamada, J., Ren, J., & Hosokawa, M. (1999). Malignant progression of a weakly malignant rat mammary tumor cell line, ER-1, by tumor microenvironmental factors. *Medical Electron Microscopy, 32*, 87–93.

Hascall, V. C., Majors, A. K., De La Motte, C. A., Evanko, S. P., Wang, A., Drazba, J. A., et al. (2004). Intracellular HA: A new frontier for inflammation? *Biochimica et Biophysica Acta, 1673*, 3–12.

Heijnen, H. F., Schiel, A. E., Fijnheer, R., Geuze, H. J., & Sixma, J. J. (1999). Activated platelets release two types of membrane vesicles: Microvesicles by surface shedding and exosomes derived from exocytosis of multivesicular bodies and alpha-granules. *Blood, 94*, 3791–3799.

Hiltunen, E. L., Anttila, M., Kultti, A., Ropponen, K., Penttinen, J., Yliskoski, M., et al. (2002). Elevated HA concentration without hyaluronidase activation in malignant epithelial ovarian tumors. *Cancer Research, 62*, 6410–6413.

Hong, B. S., Cho, J. H., Kim, H., Choi, E. J., Rho, S., Kim, J., et al. (2009). Colorectal cancer cell-derived microvesicles are enriched in cell cycle-related mRNAs that promote proliferation of endothelial cells. *BMC Genomics, 10*, 556.

Huber, V., Fais, S., Iero, M., Lugini, L., Canese, P., Squarcina, P., et al. (2005). Human colorectal cancer cells induce T-cell death through release of proapoptotic microvesicles: Role in immune escape. *Gastroenterology, 128*, 1796–1804.

Hulsmans, M., & Holvoet, P. (2013). MicroRNA-containing microvesicles regulating inflammation in association with atherosclerotic disease. *Cardiovascular Research, 100*, 7–18.

Iero, M., Valenti, R., Huber, V., Filipazzi, P., Parmiani, G., Fais, S., et al. (2008). Tumour-released exosomes and their implications in cancer immunity. *Cell Death and Differentiation, 15*, 80–88.

Janowska-Wieczorek, A., Marquez-Curtis, L. A., Wysoczynski, M., & Ratajczak, M. Z. (2006). Enhancing effect of platelet-derived microvesicles on the invasive potential of breast cancer cells. *Transfusion, 46*, 1199–1209.

Janowska-Wieczorek, A., Wysoczynski, M., Kijowski, J., Marquez-Curtis, L., Machalinski, B., Ratajczak, J., et al. (2005). Microvesicles derived from activated platelets induce metastasis and angiogenesis in lung cancer. *International Journal of Cancer, 113*, 752–760.

Jiang, D., Liang, J., & Noble, P. W. (2007). HA in tissue injury and repair. *Annual Review of Cell and Developmental Biology, 23*, 435–461.

Jokela, T. A., Jauhiainen, M., Auriola, S., Kauhanen, M., Tiihonen, R., Tammi, M. I., et al. (2008). Mannose inhibits HA synthesis by down-regulation of the cellular pool of UDP-N-acetylhexosamines. *Journal of Biological Chemistry, 283*, 7666–7673.

Josefsson, A., Adamo, H., Hammarsten, P., Granfors, T., Stattin, P., Egevad, L., et al. (2011). Prostate cancer increases HA in surrounding nonmalignant stroma, and this response is associated with tumor growth and an unfavorable outcome. *American Journal of Pathology, 179*, 1961–1968.

Jung, T., Castellana, D., Klingbeil, P., Cuesta Hernandez, I., Vitacolonna, M., Orlicky, D. J., et al. (2009). CD44v6 dependence of premetastatic niche preparation by exosomes. *Neoplasia, 11*, 1093–1105.

Kalra, H., Simpson, R. J., Ji, H., Aikawa, E., Altevogt, P., Askenase, P., et al. (2012). Vesiclepedia: A compendium for extracellular vesicles with continuous community annotation. *PLoS Biology, 10*, e1001450.

Kang, L., Lantier, L., Kennedy, A., Bonner, J. S., Mayes, W. H., Bracy, D. P., et al. (2013). HA accumulates with high-fat feeding and contributes to insulin resistance. *Diabetes, 62*, 1888–1896.

Karjalainen, J. M., Tammi, R. H., Tammi, M. I., Eskelinen, M. J., Ågren, U. M., Parkkinen, J. J., et al. (2000). Reduced level of CD44 and HA associated with unfavorable prognosis in clinical stage I cutaneous melanoma. *American Journal of Pathology, 157*, 957–965.

Kim, D. K., Kang, B., Kim, O. Y., Choi, D. S., Lee, J., Kim, S. R., et al. (2013). EVpedia: An integrated database of high-throughput data for systemic analyses of extracellular vesicles. *Journal of Extracellular Vesicles, 2*, 1–7, http://dx.doi.org/10.3402/jev.v2i0.20384.

Kim, C. W., Lee, H. M., Lee, T. H., Kang, C., Kleinman, H. K., & Gho, Y. S. (2002). Extracellular membrane vesicles from tumor cells promote angiogenesis via sphingomyelin. *Cancer Research, 62*, 6312–6317.

Kim, H. K., Song, K. S., Park, Y. S., Kang, Y. H., Lee, Y. J., Lee, K. R., et al. (2003). Elevated levels of circulating platelet microparticles, VEGF, IL-6 and RANTES in patients with gastric cancer: Possible role of a metastasis predictor. *European Journal of Cancer, 39*, 184–191.

Kim, J. W., Wieckowski, E., Taylor, D. D., Reichert, T. E., Watkins, S., & Whiteside, T. L. (2005). Fas ligand-positive membranous vesicles isolated from sera of patients with oral cancer induce apoptosis of activated T lymphocytes. *Clinical Cancer Research, 11*, 1010–1020.

Kosunen, A., Ropponen, K., Kellokoski, J., Pukkila, M., Virtaniemi, J., Valtonen, H., et al. (2004). Reduced expression of HA is a strong indicator of poor survival in oral squamous cell carcinoma. *Oral Oncology, 40*, 257–263.

Kultti, A., Pasonen-Seppänen, S., Jauhiainen, M., Rilla, K. J., Kärnä, R., Pyoriä, E., et al. (2009). 4-Methylumbelliferone inhibits HA synthesis by depletion of cellular UDP-glucuronic acid and downregulation of HA synthase 2 and 3. *Experimental Cell Research, 315*, 1914–1923.

Kultti, A., Rilla, K., Tiihonen, R., Spicer, A. P., Tammi, R. H., & Tammi, M. I. (2006). HA synthesis induces microvillus-like cell surface protrusions. *Journal of Biological Chemistry, 281*, 15821–15828.

Lamparski, H. G., Metha-Damani, A., Yao, J. Y., Patel, S., Hsu, D. H., Ruegg, C., et al. (2002). Production and characterization of clinical grade exosomes derived from dendritic cells. *Journal of Immunological Methods, 270*, 211–226.

Lance, R. S., Drake, R. R., & Troyer, D. A. (2011). Multiple recognition assay reveals prostasomes as promising plasma biomarkers for prostate cancer. *Expert Review of Anticancer Therapy, 11*, 1341–1343.

Lau, C. S., & Wong, D. T. (2012). Breast cancer exosome-like microvesicles and salivary gland cells interplay alters salivary gland cell-derived exosome-like microvesicles in vitro. *PLoS One, 7*, e33037.

Lee, W. R., Laurie, J., & Townsend, A. L. (1975). Fine structure of a radiation-induced osteogenic sarcoma. *Cancer, 36*, 1414–1425.

Li, J., Zhang, Y., Liu, Y., Dai, X., Li, W., Cai, X., et al. (2013). Microvesicle-mediated transfer of microRNA-150 from monocytes to endothelial cells promotes angiogenesis. *Journal of Biological Chemistry, 288*, 23586–23596.

Liao, C. F., Lin, S. H., Chen, H. C., Tai, C. J., Chang, C. C., Li, L. T., et al. (2012). CSE1L, a novel microvesicle membrane protein, mediates Ras-triggered microvesicle generation and metastasis of tumor cells. *Molecular Medicine*, *18*, 1269–1280.

Lima, L. G., Chammas, R., Monteiro, R. Q., Moreira, M. E., & Barcinski, M. A. (2009). Tumor-derived microvesicles modulate the establishment of metastatic melanoma in a phosphatidylserine-dependent manner. *Cancer Letters*, *283*, 168–175.

Lima, L. G., Leal, A. C., Vargas, G., Porto-Carreiro, I., & Monteiro, R. Q. (2013). Intercellular transfer of tissue factor via the uptake of tumor-derived microvesicles. *Thrombosis Research*, *132*, 450–456.

Lima, L. G., Oliveira, A. S., Campos, L. C., Bonamino, M., Chammas, R., Werneck, C., et al. (2011). Malignant transformation in melanocytes is associated with increased production of procoagulant microvesicles. *Thrombosis and Haemostasis*, *106*, 712–723.

Lipponen, P., Aaltomaa, S., Tammi, R., Tammi, M., Ågren, U., & Kosma, V. M. (2001). High stromal HA level is associated with poor differentiation and metastasis in prostate cancer. *European Journal of Cancer*, *37*, 849–856.

Lo Cicero, A., Majkowska, I., Nagase, H., Di Liegro, I., & Troeberg, L. (2012). Microvesicles shed by oligodendroglioma cells and rheumatoid synovial fibroblasts contain aggrecanase activity. *Matrix Biology*, *31*, 229–233.

Lokeshwar, V. B., Rubinowicz, D., Schroeder, G. L., Forgacs, E., Minna, J. D., Block, N. L., et al. (2001). Stromal and epithelial expression of tumor markers hyaluronic acid and HYAL1 hyaluronidase in prostate cancer. *Journal of Biological Chemistry*, *276*, 11922–11932.

Lötvall, J., Rajendran, L., Gho, Y. S., Thery, C., Wauben, M., Raposo, G., et al. (2012). The launch of Journal of Extracellular Vesicles (JEV), the official journal of the International Society for Extracellular Vesicles—About microvesicles, exosomes, ectosomes and other extracellular vesicles. *Journal of Extracellular Vesicles*, *1*, 1–2, http://dx.doi.org/10.3402/jev.v1i0.18514.

Lukoschek, M., & Addicks, K. (1991). Histologic, morphometric study of the human synovial membrane. *Zeitschrift für Orthopädie und Ihre Grenzgebiete*, *129*, 136–140.

Marleau, A. M., Chen, C. S., Joyce, J. A., & Tullis, R. H. (2012). Exosome removal as a therapeutic adjuvant in cancer. *Journal of Translational Medicine*, *10*, 134.

Martins, V. R., Dias, M. S., & Hainaut, P. (2013). Tumor-cell-derived microvesicles as carriers of molecular information in cancer. *Current Opinion in Oncology*, *25*, 66–75.

Mathivanan, S., & Simpson, R. J. (2009). ExoCarta: A compendium of exosomal proteins and RNA. *Proteomics*, *9*, 4997–5000.

McConnell, R. E., Higginbotham, J. N., Shifrin, D. A., Jr., Tabb, D. L., Coffey, R. J., & Tyska, M. J. (2009). The enterocyte microvillus is a vesicle-generating organelle. *Journal of Cell Biology*, *185*, 1285–1298.

Meng, Y., Kang, S., & Fishman, D. A. (2005). Lysophosphatidic acid stimulates Fas ligand microvesicle release from ovarian cancer cells. *Cancer Immunology, Immunotherapy*, *54*, 807–814.

Millimaggi, D., Mari, M., D'Ascenzo, S., Carosa, E., Jannini, E. A., Zucker, S., et al. (2007). Tumor vesicle-associated CD147 modulates the angiogenic capability of endothelial cells. *Neoplasia*, *9*, 349–357.

Mine, S., Okada, Y., Kawahara, C., Tabata, T., & Tanaka, Y. (2006). Serum HA concentration as a marker of angiopathy in patients with diabetes mellitus. *Endocrine Journal*, *53*, 761–766.

Morello, M., Minciacchi, V. R., de Candia, P., Yang, J., Posadas, E., Kim, H., et al. (2013). Large oncosomes mediate intercellular transfer of functional microRNA. *Cell Cycle*, *12*, 3526–3536.

Mu, W., Rana, S., & Zoller, M. (2013). Host matrix modulation by tumor exosomes promotes motility and invasiveness. *Neoplasia*, *15*, 875–887.

Müller, G. (2012). Microvesicles/exosomes as potential novel biomarkers of metabolic diseases. *Diabetes, Metabolic Syndrome and Obesity, 5*, 247–282.

Mutsaers, S. E. (2004). The mesothelial cell. *International Journal of Biochemistry and Cell Biology, 36*, 9–16.

Nakazawa, H., Yoshihara, S., Kudo, D., Morohashi, H., Kakizaki, I., Kon, A., et al. (2006). 4-Methylumbelliferone, a HA synthase suppressor, enhances the anticancer activity of gemcitabine in human pancreatic cancer cells. *Cancer Chemotherapy and Pharmacology, 57*, 165–170.

Nykopp, T. K., Rilla, K., Sironen, R., Tammi, M. I., Tammi, R. H., Hämäläinen, K., et al. (2009). Expression of HA synthases (HAS1-3) and hyaluronidases (HYAL1-2) in serous ovarian carcinomas: Inverse correlation between HYAL1 and HA content. *BMC Cancer, 9*, 143.

Nykopp, T. K., Rilla, K., Tammi, M. I., Tammi, R. H., Sironen, R., Hämäläinen, K., et al. (2010). HA synthases (HAS1-3) and hyaluronidases (HYAL1-2) in the accumulation of HA in endometrioid endometrial carcinoma. *BMC Cancer, 10*, 512.

Peinado, H., Aleckovic, M., Lavotshkin, S., Matei, I., Costa-Silva, B., Moreno-Bueno, G., et al. (2012). Melanoma exosomes educate bone marrow progenitor cells toward a pro-metastatic phenotype through MET. *Nature Medicine, 18*, 883–891.

Peng, P., Yan, Y., & Keng, S. (2011). Exosomes in the ascites of ovarian cancer patients: Origin and effects on anti-tumor immunity. *Oncology Reports, 25*, 749–762.

Pienimäki, J. P., Rilla, K., Fülop, C., Sironen, R. K., Karvinen, S., Pasonen, S., et al. (2001). Epidermal growth factor activates HA synthase 2 in epidermal keratinocytes and increases pericellular and intracellular HA. *Journal of Biological Chemistry, 276*, 20428–20435.

Pirinen, R., Tammi, R., Tammi, M., Hirvikoski, P., Parkkinen, J. J., Johansson, R., et al. (2001). Prognostic value of HA expression in non-small-cell lung cancer: Increased stromal expression indicates unfavorable outcome in patients with adenocarcinoma. *International Journal of Cancer, 95*, 12–17.

Polasek, J. (1982). The appearance of multivesicular structures during platelet activation as observed by scanning electron microscopy. *Thrombosis Research, 28*, 433–437.

Provenzano, P. P., Cuevas, C., Chang, A. E., Goel, V. K., Von Hoff, D. D., & Hingorani, S. R. (2012). Enzymatic targeting of the stroma ablates physical barriers to treatment of pancreatic ductal adenocarcinoma. *Cancer Cell, 21*, 418–429.

Qu, C., Rilla, K., Tammi, R., Tammi, M., Kröger, H., & Lammi, M. J. (2014). Extensive CD44-dependent HA coats on human bone marrow-derived mesenchymal stem cells produced by HA synthases HAS1, HAS2 and HAS3. *International Journal of Biochemistry and Cell Biology, 48C*, 45–54.

Rabinowits, G., Gercel-Taylor, C., Day, J. M., Taylor, D. D., & Kloecker, G. H. (2009). Exosomal microRNA: A diagnostic marker for lung cancer. *Clinical Lung Cancer, 10*, 42–46.

Raposo, G., Nijman, H. W., Stoorvogel, W., Liejendekker, R., Harding, C. V., Melief, C. J., et al. (1996). B lymphocytes secrete antigen-presenting vesicles. *Journal of Experimental Medicine, 183*, 1161–1172.

Raposo, G., & Stoorvogel, W. (2013). Extracellular vesicles: Exosomes, microvesicles, and friends. *Journal of Cell Biology, 200*, 373–383.

Reis, L. A., Borges, F. T., Simoes, M. J., Borges, A. A., Sinigaglia-Coimbra, R., & Schor, N. (2012). Bone marrow-derived mesenchymal stem cells repaired but did not prevent gentamicin-induced acute kidney injury through paracrine effects in rats. *PLoS One*, 7, e44092.

Renzulli, J. F., 2nd., Del Tatto, M., Dooner, G., Aliotta, J., Goldstein, L., Dooner, M., et al. (2010). Microvesicle induction of prostate specific gene expression in normal human bone marrow cells. *Journal of Urology, 184*, 2165–2171.

Ricciardelli, C., Ween, M. P., Lokman, N. A., Tan, I. A., Pyragius, C. E., & Oehler, M. K. (2013). Chemotherapy-induced HA production: A novel chemoresistance mechanism in ovarian cancer. *BMC Cancer, 13*, 476.

Rilla, K., Oikari, S., Jokela, T. A., Hyttinen, J. M., Kärnä, R., Tammi, R. H., et al. (2013). HA synthase 1 (HAS1) requires higher cellular UDP-GlcNAc concentration than HAS2 and HAS3. *Journal of Biological Chemistry, 288*, 5973–5983.

Rilla, K., Pasonen-Seppänen, S., Deen, A. J., Koistinen, V. V., Wojciechowski, S., Oikari, S., et al. (2013). HA production enhances shedding of plasma membrane-derived microvesicles. *Experimental Cell Research, 319*, 2006–2018.

Rilla, K., Pasonen-Seppänen, S., Kärnä, R., Karjalainen, H. M., Törrönen, K., Koistinen, V., et al. (2012). HAS3-induced accumulation of HA in 3D MDCK cultures results in mitotic spindle misorientation and disturbed organization of epithelium. *Histochemistry and Cell Biology, 137*, 153–164.

Rilla, K., Siiskonen, H., Spicer, A. P., Hyttinen, J. M., Tammi, M. I., & Tammi, R. H. (2005). Plasma membrane residence of HA synthase is coupled to its enzymatic activity. *Journal of Biological Chemistry, 280*, 31890–31897.

Rilla, K., Tiihonen, R., Kultti, A., Tammi, M., & Tammi, R. (2008). Pericellular HA coat visualized in live cells with a fluorescent probe is scaffolded by plasma membrane protrusions. *Journal of Histochemistry and Cytochemistry, 56*, 901–910.

Ropponen, K., Tammi, M., Parkkinen, J., Eskelinen, M., Tammi, R., Lipponen, P., et al. (1998). Tumor cell-associated HA as an unfavorable prognostic factor in colorectal cancer. *Cancer Research, 58*, 342–347.

Sabatier, F., Darmon, P., Hugel, B., Combes, V., Sanmarco, M., Velut, J. G., et al. (2002). Type 1 and type 2 diabetic patients display different patterns of cellular microparticles. *Diabetes, 51*, 2840–2845.

Sabin, K., & Kikyo, N. (2014). Microvesicles as mediators of tissue regeneration. *Translational Research, 163*(4), 286–295.

Safaei, R., Larson, B. J., Cheng, T. C., Gibson, M. A., Otani, S., Naerdemann, W., et al. (2005). Abnormal lysosomal trafficking and enhanced exosomal export of cisplatin in drug-resistant human ovarian carcinoma cells. *Molecular Cancer Therapeutics, 4*, 1595–1604.

Sainio, A., Jokela, T., Tammi, M. I., & Järveläinen, H. (2010). Hyperglycemic conditions modulate connective tissue reorganization by human vascular smooth muscle cells through stimulation of HA synthesis. *Glycobiology, 20*, 1117–1126.

Schneider, A., & Simons, M. (2013). Exosomes: Vesicular carriers for intercellular communication in neurodegenerative disorders. *Cell and Tissue Research, 352*, 33–47.

Semenza, G. L. (2008). Tumor metabolism: Cancer cells give and take lactate. *Journal of Clinical Investigation, 118*, 3835–3837.

Shedden, K., Xie, X. T., Chandaroy, P., Chang, Y. T., & Rosania, G. R. (2003). Expulsion of small molecules in vesicles shed by cancer cells: Association with gene expression and chemosensitivity profiles. *Cancer Research, 63*, 4331–4337.

Siiskonen, H., Poukka, M., Tyynelä-Korhonen, K., Sironen, R., & Pasonen-Seppänen, S. (2013). Inverse expression of hyaluronidase 2 and HA synthases 1–3 is associated with reduced HA content in malignant cutaneous melanoma. *BMC Cancer, 13*, 181.

Silva, J., Garcia, V., Rodriguez, M., Compte, M., Cisneros, E., Veguillas, P., et al. (2012). Analysis of exosome release and its prognostic value in human colorectal cancer. *Genes, Chromosomes & Cancer, 51*, 409–418.

Sironen, R. K., Tammi, M., Tammi, R., Auvinen, P. K., Anttila, M., & Kosma, V. M. (2011). HA in human malignancies. *Experimental Cell Research, 317*, 383–391.

Skog, J., Wurdinger, T., van Rijn, S., Meijer, D. H., Gainche, L., Sena-Esteves, M., et al. (2008). Glioblastoma microvesicles transport RNA and proteins that promote tumour growth and provide diagnostic biomarkers. *Nature Cell Biology, 10*, 1470–1476.

Sundström, G., Dahl, I. M., Hultdin, M., Lundström, B., Wahlin, A., & Engström-Laurent, A. (2005). Bone marrow HA distribution in patients with acute myeloid leukemia. *Medical Oncology, 22*, 71–78.

Svensson, K. J., Kucharzewska, P., Christianson, H. C., Skold, S., Lofstedt, T., Johansson, M. C., et al. (2011). Hypoxia triggers a proangiogenic pathway involving cancer cell microvesicles and PAR-2-mediated heparin-binding EGF signaling in endothelial cells. *Proceedings of the National Academy of Sciences of the United States of America, 108*, 13147–13152.

Szczepanski, M. J., Szajnik, M., Welsh, A., Whiteside, T. L., & Boyiadzis, M. (2011). Blast-derived microvesicles in sera from patients with acute myeloid leukemia suppress natural killer cell function via membrane-associated transforming growth factor-beta1. *Haematologica, 96*, 1302–1309.

Tammi, R. H., Passi, A. G., Rilla, K., Karousou, E., Vigetti, D., Makkonen, K., et al. (2011). Transcriptional and post-translational regulation of HA synthesis. *FEBS Journal, 278*, 1419–1428.

Tavoosidana, G., Ronquist, G., Darmanis, S., Yan, J., Carlsson, L., Wu, D., et al. (2011). Multiple recognition assay reveals prostasomes as promising plasma biomarkers for prostate cancer. *Proceedings of the National Academy of Sciences of the United States of America, 108*, 8809–8814.

Tetta, C., Ghigo, E., Silengo, L., Deregibus, M. C., & Camussi, G. (2013). Extracellular vesicles as an emerging mechanism of cell-to-cell communication. *Endocrine, 44*(1), 11–19. http://dx.doi.org/10.1007/s12020-012-9839-0.

Thery, C., Zitvogel, L., & Amigorena, S. (2002). Exosomes: Composition, biogenesis and function. *Nature Reviews Immunology, 2*, 569–579.

Tian, T., Wang, Y., Wang, H., Zhu, Z., & Xiao, Z. (2010). Visualizing of the cellular uptake and intracellular trafficking of exosomes by live-cell microscopy. *Journal of Cellular Biochemistry, 111*, 488–496.

Toole, B. P., & Hascall, V. C. (2002). HA and tumor growth. *American Journal of Pathology, 161*, 745–747.

Toole, B. P., & Slomiany, M. G. (2008). HA: A constitutive regulator of chemoresistance and malignancy in cancer cells. *Seminars in Cancer Biology, 18*, 244–250.

Valadi, H., Ekstrom, K., Bossios, A., Sjostrand, M., Lee, J. J., & Lotvall, J. O. (2007). Exosome-mediated transfer of mRNAs and microRNAs is a novel mechanism of genetic exchange between cells. *Nature Cell Biology, 9*, 654–659.

Valenti, R., Huber, V., Filipazzi, P., Pilla, L., Sovena, G., Villa, A., et al. (2006). Human tumor-released microvesicles promote the differentiation of myeloid cells with transforming growth factor-beta-mediated suppressive activity on T lymphocytes. *Cancer Research, 66*, 9290–9298.

van Niel, G., Raposo, G., Candalh, C., Boussac, M., Hershberg, R., Cerf-Bensussan, N., et al. (2001). Intestinal epithelial cells secrete exosome-like vesicles. *Gastroenterology, 121*, 337–349.

Vangara, K. K., Liu, J. L., & Palakurthi, S. (2013). Hyaluronic acid-decorated PLGA-PEG nanoparticles for targeted delivery of SN-38 to ovarian cancer. *Anticancer Research, 33*, 2425–2434.

Varadi, T., Mersich, T., Auvinen, P., Tammi, R., Tammi, M., Salamon, F., et al. (2012). Binding of trastuzumab to ErbB2 is inhibited by a high pericellular density of HA. *Journal of Histochemistry and Cytochemistry, 60*, 567–575.

Vickers, K. C., & Remaley, A. T. (2012). Lipid-based carriers of microRNAs and intercellular communication. *Current Opinion in Lipidology, 23*, 91–97.

White, J. G. (1974). Effects of an ionophore, A23187, on the surface morphology of normal erythrocytes. *American Journal of Pathology, 77*, 507–518.

Wolf, P. (1967). The nature and significance of platelet products in human plasma. *British Journal of Haematology*, *13*, 269–288.

Xiao, D., Ohlendorf, J., Chen, Y., Taylor, D. D., Rai, S. N., Waigel, S., et al. (2012). Identifying mRNA, microRNA and protein profiles of melanoma exosomes. *PLoS One*, *7*, e46874.

Yang, M., Chen, J., Su, F., Yu, B., Su, F., Lin, L., et al. (2011). Microvesicles secreted by macrophages shuttle invasion-potentiating microRNAs into breast cancer cells. *Molecular Cancer*, *10*, 117.

Yeo, T. K., Nagy, J. A., Yeo, K. T., Dvorak, H. F., & Toole, B. P. (1996). Increased HA at sites of attachment to mesentery by CD44-positive mouse ovarian and breast tumor cells. *American Journal of Pathology*, *148*, 1733–1740.

Yoshida, H., Miyazaki, S., & Yumoto, T. (1991). Matrix vesicles in bone tumors. Ultrastructural analysis and their significance in neoplastic bone formation. *Acta Pathologica Japonica*, *41*, 610–617.

Yu, J. L., May, L., Lhotak, V., Shahrzad, S., Shirasawa, S., Weitz, J. I., et al. (2005). Oncogenic events regulate tissue factor expression in colorectal cancer cells: Implications for tumor progression and angiogenesis. *Blood*, *105*, 1734–1741.

Zu, Y., Sidhu, G. S., Wieczorek, R., & Cassai, N. D. (2002). Ultrastructurally "invasive" microvilli in an aggressively metastasizing biphasic malignant mesothelioma. *Ultrastructural Pathology*, *26*, 403–409.

CHAPTER SIX

Hyaluronan in the Healthy and Malignant Hematopoietic Microenvironment

Sophia K. Khaldoyanidi[1], Valentina Goncharova, Barbara Mueller, Ingrid U. Schraufstatter
Torrey Pines Institute for Molecular Studies, San Diego, California, USA
[1]Corresponding author: e-mail address: skhaldoyanidi@tpims.org

Contents

1. Introduction 151
  1.1 Multipotent stem cells in bone marrow 151
  1.2 The BM stem cell microenvironment 152
  1.3 Basic aspects of HA biology 152
2. BM MSC, Their Derivatives, and HA 153
  2.1 MSC and HA in BM 153
  2.2 Fibroblasts and HA in BM 159
  2.3 Osteoblasts and HA in BM 161
3. Macrophages and HA in the BM 163
  3.1 Role of macrophages in BM 163
  3.2 Effect of HA on macrophages 163
  3.3 Effect of HA on osteoclasts 166
4. Endothelial Cells and HA in BM 166
  4.1 Role of endothelial cells in BM 166
  4.2 Expression of HA in EC 167
  4.3 Functions of HA in BM EC: Cell recruitment 170
  4.4 Functions of HA in BM EC: Angiogenesis 171
  4.5 Functions of HA in BM EC: Vascular integrity 172
5. Nerve Cells and HA in BM 173
6. Role of HA in the BM Microenvironment in Hematological Malignancies 174
7. Conclusion 177
References 178

Abstract

The fate of both endogenous and transplanted stem cells is dependent on the functional status of the regulatory local microenvironment, which is compromised by disease and therapeutic intervention. The glycosaminoglycan hyaluronan (HA) is a critical component of the hematopoietic microenvironment. We summarize recent advances

Advances in Cancer Research, Volume 123
ISSN 0065-230X
http://dx.doi.org/10.1016/B978-0-12-800092-2.00006-X

in our understanding of the role of HA in regulating mesenchymal stem cells, osteoblasts, fibroblasts, macrophages, and endothelium in bone marrow (BM) and their crosstalk within the hematopoietic microenvironment. HA not only determines the volume, hydration, and microfluidics of the BM interstitial space, but also, via interactions with specific receptors, regulates multiple cell functions including differentiation, migration, and production of regulatory factors. The effects of HA are dependent on the polymer size and are influenced by the formation of complexes with other molecules. In healthy BM, HA synthases and hyaluronidases form a molecular network that maintains extracellular HA levels within a discrete physiological window, but HA homeostasis is often perturbed in pathological conditions, including hematological malignancies. Recent studies have suggested that HA synthases may have functions beyond HA production and contribute to the intracellular regulatory machinery. We discuss a possible role for HA synthases, intracellular and extracellular HA in the malignant BM microenvironment, and resistance to therapy.

ABBREVIATIONS

3D three-dimensional
4MU 4-methylumbelliferone
5FU 5-fluorouracil
AML acute myeloid leukemia
AMPK adenosine monophosphate-activated protein kinase
bFGF basic fibroblast growth factor
BM bone marrow
CEM caveolin-enriched microdomains
CLL chronic lymphatic leukemia
CML chronic myeloid leukemia
COL1A1 collagen, type 1, alpha 1
EC endothelial cells
ECM extracellular matrix
GC glucocorticoids
HA hyaluronan
HA-HC complex of HA with heavy chain of inter-α-inhibitor
HARE hyaluronic acid receptor for endocytosis (also designated Stabilin-2)
HAS HA synthase
HMW high-molecular weight
HSC hematopoietic stem cells
HSPC hematopoietic stem/progenitor cells
HUVEC human umbilical vein endothelial cells
IBSP integrin-binding sialoprotein
IFNγ interferon gamma
IL interleukin
IP-10 interferon gamma-induced protein 10
KO knockout mice
LIF leukemia inhibitory factor
LMW low-molecular weight

LPS lipopolysaccharide
LYVE-1 lymphatic vessel endothelial receptor 1
MAFIA macrophage Fas-induced apoptosis mouse
MAPK mitogen-activated protein kinase
MCP1 monocyte chemotactic protein-1
M-CSF macrophage colony-stimulating factor
MIP-1α macrophage inflammatory protein-1 alpha
MM multiple myeloma
MSC mesenchymal stem cells
PDGF platelet-derived growth factor
PGE2 prostaglandin E2
PTH parathyroid hormone
RANKL receptor activator of the NF-kB ligand
RHAMM hyaluronan-mediated motility receptor
ROS reactive oxygen species
SDF-1 stromal cell-derived factor 1
TGFβ transforming growth factor beta
TLR-4 toll-like receptor 4
TNFα tumor necrosis factor alpha
TSG-6 tumor necrosis factor-α-stimulated gene 6
VCAM-1 vascular cell adhesion molecule-1

1. INTRODUCTION

1.1. Multipotent stem cells in bone marrow

The current literature on stem cell biology contains inaccuracies; therefore, the purpose of this paragraph is to introduce and clarify some basic concepts for the benefit of scientists specialized in other research fields. Stem cells residing in bone marrow (BM) are termed multipotent because their potency is restricted to specific tissue types, in contrast to pluripotent stem cells that are capable of generating all tissues of a fetus. Two types of multipotent stem cells are present in BM: hematopoietic stem cells (HSC) and mesenchymal stem cells (MSC). HSC generate all blood cells, some of which can further mature into highly specialized cells (e.g., monocyte maturation to osteoclasts), whereas MSC differentiate into all connective tissue cells including bone, cartilage, ligament, tendon, stromal fibroblasts, and adipocytes. Although there are reports that stem cells (or mature cells) are able to differentiate into cells outside of their specialization without genetic manipulation, these studies challenge the existing dogma and will require further proof and verification. In addition to their multipotency, stem cells are distinguished from all other cells by their ability to self-renew, which means that during division, one daughter cell remains a stem cell, while the other may undergo lineage commitment and differentiation. HSC and MSC that lose their multipotency or capacity to self-renew become progenitors and are no longer stem cells. Stem cells can remain in a quiescent state or proliferate, depending on signals from the microenvironment. Both HSC and MSC have been studied intensively over the past few decades, and many aspects of their biology have been reviewed, including their cell surface signatures and the molecular pathways regulating their functions (e.g., Levi & Morrison, 2008; Seita & Weissman, 2010).

1.2. The BM stem cell microenvironment

The ability of stem cells to maintain their unique properties is supported by the local microenvironment and changes in this environment can result in loss of multipotency and/or self-renewal and subsequent exhaustion of the stem cell pool. Understanding how stem cells respond to specific microenvironments (healthy or diseased niches or sites of stem cell transplantation) is critical for the effective treatment of cancer and the development of stem cell-based therapies. In BM, the microenvironments for HSC and MSC overlap. Although MSC require a unique microenvironment to maintain their own properties, they also actively contribute to the cellular network that makes up the hematopoietic microenvironment for HSC. The BM microenvironment is a complex structure to which many different cell types contribute. Like a well-tuned orchestra, these cells must work together to provide a healthy microenvironment that not only maintains the stem cell pool under steady-state conditions but also allows a prompt response to stresses and demands without stem cell exhaustion. In this review, we discuss selected cell types in detail, and although many molecules contribute to crosstalk between the cellular elements of the niche (selected reviews Canaani, Kollet, & Lapidot, 2011; Fecteau & Kipps, 2012; Kaur-Bollinger, Gotze, & Oostendorp, 2012; Park, Sykes, & Scadden, 2012; Schraufstatter, Discipio, & Khaldoyanidi, 2011), this review is focused on hyaluronan (HA).

1.3. Basic aspects of HA biology

Here, we briefly touch on key aspects of the molecular structure, biochemistry, receptors, and functions of HA. The reader is referred to the wealth of primary research and review articles for further information.

HA is a glycosaminoglycan consisting of repeating disaccharides (*N*-acetyl-D-glucosamine and glucuronic acid) that form a negatively charged hydrophilic polymer. The biological effects of HA are dependent on the size of this polymer. In contrast to recombinant or synthetic polymers that have precisely defined sizes, naturally occurring HA polymers exist in a range of sizes from low-molecular weight (LMW) HA of <200 kDa to high-molecular weight (HMW) HA of >1000 kDa. HA oligosaccharides consist of only a few disaccharide units (<5 kDa). HA is synthesized by three HA synthases (HAS1, HAS2, and HAS3) and is degraded by a family of hyaluronidases. Under normal homeostatic conditions, the balance between HA synthesis and degradation is carefully controlled to maintain physiological concentration in tissues. Indeed, fluctuations in HA levels may reflect an ongoing pathological process (Fraser, Laurent, & Laurent, 1997; Laurent, Laurent, & Fraser, 1996a, 1996b; Stern, 2008; Weigel & DeAngelis, 2007).

HA binds to several receptors, including CD44, HARE, hyaluronan-mediated motility receptor (RHAMM), lymphatic vessel endothelial receptor 1 (LYVE-1), and toll-like receptor 4 (TLR-4). The structure, regulation, and functions of these receptors are well established. In addition, HA binds several extracellular molecules including growth factors and extracellular matrix (ECM) components. The formation of such complexes significantly affects the biology of HA and may be responsible for some of its tissue-specific effects. HA has been shown in a variety of models to influence numerous cellular functions, including regulation of cytokine production and cell proliferation, differentiation, and migration. Given these functions, it is perhaps not surprising that HA has been implicated in pathologies ranging from oncological, dermatological, and vascular diseases to ophthalmological and orthopedic disorders (Hascall et al., 2004; Jackson, 2009; Sherman, Sleeman, Herrlich, & Ponta, 1994; Sironen et al., 2011).

Despite the many well-characterized functions of HA, relatively little is known about the roles of HA in the hematopoietic system and the hematopoietic microenvironment in particular. The goal of this review is to discuss recent advances in the field and to highlight the many gaps in our knowledge that merit further attention.

2. BM MSC, THEIR DERIVATIVES, AND HA

2.1. MSC and HA in BM

MSC play a dual role in the hematopoietic microenvironment: (1) due to their multipotent nature, MSC can differentiate into monopotent progenitors and generate specialized connective tissue cells, such as osteoblasts and fibroblasts, that actively participate in the regulation of hematopoiesis; and (2) MSC are capable of paracrine signaling (the so-called "chaperone effects") through the production of soluble factors and ECM and the expression of cell surface molecules that participate in cellular crosstalk. MSC have been intensively studied since their discovery in 1976 (Friedenstein, Gorskaja, & Kulagina, 1976). They are stromal-like cells characterized by a $CD105^{+}/CD73^{+}/CD90^{+}/CD45^{-}/CD34^{-}/CD11b^{-}/CD19^{-}/HLA\text{-}DR^{-}$ cell surface signature (Dominici et al., 2006) and are often located perivascularly (for reviews, see Phinney & Prockop, 2007; Schraufstatter et al., 2011). Even though the molecular mechanisms regulating MSC differentiation, self-renewal, trophic factor production, and anti-inflammatory properties have been intensively investigated (Kolf, Cho, & Tuan, 2007), our understanding of the role of HA in MSC biology is limited.

Adult MSC express all three HA synthase (HAS) and produce HA (Calabro, Oken, Hascall, & Masellis, 2002; Caralla, Boehm, Hascall, & Muschler, 2012; Goncharova et al., 2012). Mouse and human MSC differ in their patterns of HAS expression in that human MSC express significantly lower levels of HAS1 than their murine counterparts (Fig. 6.1). HAS1 expression is dramatically increased in MSC from patients with multiple myeloma (MM), but can be downregulated by coculture of MSC with plasma cells (Calabro et al., 2002).

Inhibition of HAS activity and HA production significantly affects the hematopoiesis-supportive function of MSC. We have previously reported that BM cells isolated from HAS1/2/3 triple knockout (tHAS1/2/3 knockout (KO)) mice (*Prx1-Cre;Has2*$^{flox/flox}$*;Has1*$^{-/-}$*;Has3*$^{-/-}$) fail to establish hematopoiesis *in vitro*, in contrast to BM cells from HAS1/3 double KO

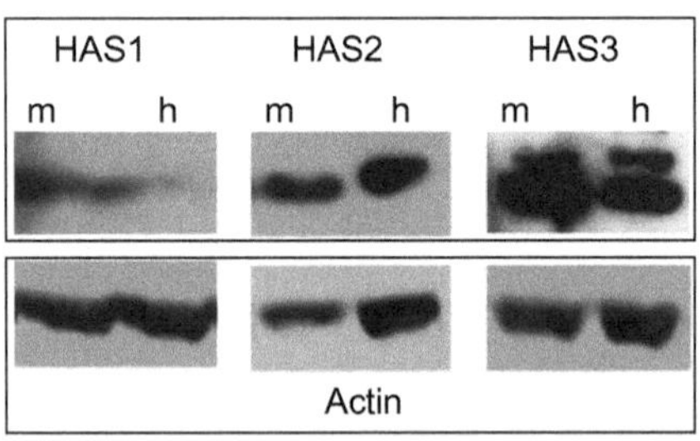

Figure 6.1 HAS expression in MSC. The expression of HAS proteins in mouse and human MSC was compared by Western blotting. Actin was used as a loading control. The positions of HAS1, HAS2, HAS3, and actin are indicated on a representative gel.

and WT mice, which showed similar hematopoietic abilities (Goncharova et al., 2012). Hematopoietic activity of BM from tHAS1/2/3 KO mice was rescued by culture of these cells on the HA-expressing hematopoiesis-supportive S17 stromal cell line (Fig. 6.2). These findings suggest that hematopoietic stem/progenitor cells (HSPC) in the BM of tHAS1/2/3 KO mice require HA-expressing stromal cells to establish hematopoietic foci and produce hematopoietic cells *in vitro*.

In many HA-deficient experimental models, addition of exogenous HA at least partially rescues cell function. The addition of exogenous HMW or LMW HA to tHAS1/2/3 KO BM cultures did not improve hematopoietic activity, as demonstrated by a failure to produce nonadherent hematopoietic cells at any point during the culture period (representative parameters collected on week 3 of culture are shown in Fig. 6.3A). Pharmacological inhibition of HAS expression and HA synthesis by 4-methylumbelliferone (4MU) also profoundly inhibited hematopoiesis *in vitro* (Goncharova et al., 2012). We found that addition of HMW HA to 4MU-treated WT

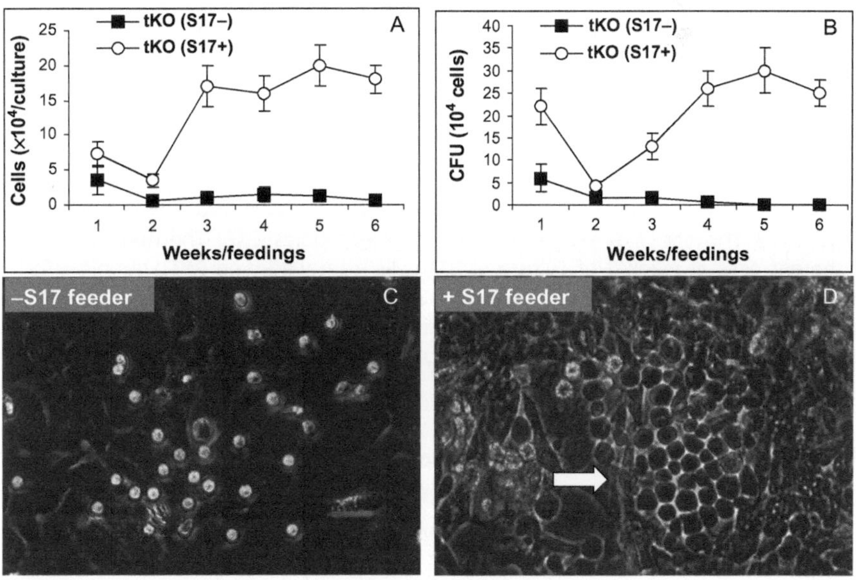

Figure 6.2 Hematopoietic activity in tHAS1/2/3 KO bone marrow cultures is rescued by hematopoiesis-supportive S17 stromal cells. BM cells from tHAS1/2/3 KO mice were cultured on a preestablished S17 feeder layer. The number of nonadherent cells (A) and progenitors (B) in the cultures was measured weekly during feedings. (C and D) Images of the adherent layers taken on week 4 of culture. Hematopoietic foci ("cobblestone" areas) were observed when tHAS1/2/3 KO BM cells were cocultured with the S17 feeder.

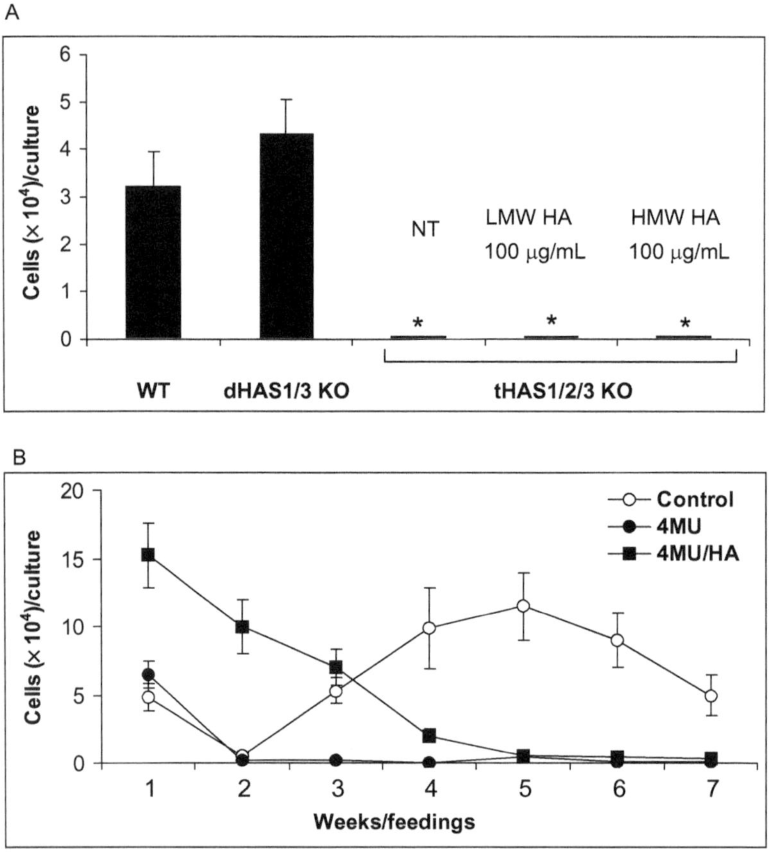

Figure 6.3 The effect of exogenous HA on hematopoiesis. (A) The number of non-adherent cells at week 3 of culture of BM cells from wild-type (WT), double HAS1/3 knockout (dHAS1/3 KO), and triple HAS1/2/3 KO (tHAS1/2/3 KO) mice. LMW HA or HMW HA was added as indicated. (B) Addition of HMW HA (100 μg/mL) to 4MU-treated LTBMC transiently increases the number of nonadherent cells from week 1 through week 3 of culture.

BM cultures transiently improved hematopoietic activity during the first 3 weeks of culture but was followed by a rapid decline in cell production (Fig. 6.3B). These findings were surprising considering that HA–CD44 interactions have previously been reported to regulate hematopoietic activity *in vitro* (Khaldoyanidi, Denzel, & Zoller, 1996; Khaldoyanidi, Karakhanova, Sleeman, Herrlich, & Ponta, 2002; Khaldoyanidi, Moll, Karakhanova, Herrlich, & Ponta, 1999; Khaldoyanidi, Schnabel, Fohr, & Zoller, 1997; Miyake et al., 1990) and suggest that HAS2 may serve cell functions other than synthesis of extracellular HA. This possibility is supported by studies using other cellular models. For example, depletion of HAS2 in the C2C12 myoblast cell line resulted in a loss of pericellular matrix and inhibition of differentiation; addition of exogenous HA restored the pericellular matrix but not cell differentiation (Hunt et al., 2013). Similar results were obtained with the NMuMG cell line (immortalized mammary epithelial cells), which undergo cell differentiation and migration when treated with transforming growth factor beta (TGFβ). In these cells, exogenous HA failed to rescue the effects of HAS2 depletion on TGFβ-induced cell migration and gene expression (Porsch et al., 2013). In another study, human aortic smooth muscle cells were treated with AMPK activators, resulting in phosphorylation of HAS2 and subsequent inhibition of HA synthesis, cell proliferation, and cell motility. Here too, addition of either LMW or HMW HA to the cultures did not overcome the effects of HAS2 dysfunction (Vigetti et al., 2011). Together, these findings suggest that the inability of exogenous HA to overcome HAS2 deficiency is not cell-type specific. Indeed, MSC isolated from tHAS1/2/3 KO mice were morphologically different from control MSC (Fig. 6.4) and exhibited low rates of proliferation and/or self-renewal. This could explain our inability to expand these cells beyond passage 2. The addition of exogenous HMW or LMW HA did not affect either the shape (not shown) or expansion (Fig. 6.4) of these cells *in vitro*. Collectively, these findings suggest that HAS2 may serve critical cellular functions beyond HA production, such as in intracellular regulatory networks, in a variety of cell types. These aspects of HAS2 biology, as well as the origin and function of intracellular HA, remain severely understudied.

Components of the HA pathway, including HAS2, Hyal2, CD44, and RHAMM, are transiently upregulated in mesenchymal cells of the regenerative bud. Inhibition of HAS activity by 4MU reduced mesenchymal cell proliferation and interfered with the regenerative cascade (Contreras, Gaete, Sanchez, Carrasco, & Larrain, 2009). The importance of CD44 in MSC-mediated tissue regeneration has been demonstrated in another study that reported CD44-dependent recruitment of MSC to sites of tissue injury (Zhu et al., 2006). One of several factors that influence the regenerative capacity of MSC is the number of passages before cell transplantation. Interestingly, MSC senescence in culture correlates with a reduction in the expression of HASs and the production of HA (Jung et al., 2011). This observation suggests that HA hydrogel compositions that preserve the stem cell properties of MSC (Mohand-Kaci, Assoul, Martelly, Allaire, & Zidi, 2013), including their ability to differentiate (Zavan et al., 2007), might be of use in regenerative medicine. However, given our recent findings that

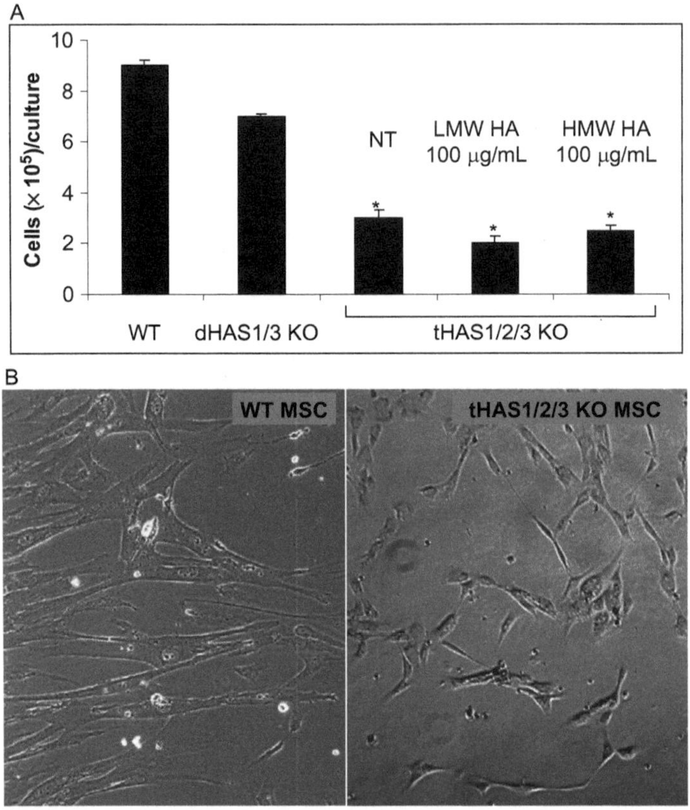

Figure 6.4 Defective function of MSC from tHAS1/2/3 KO mice. (A) The number of MSC harvested after the first passage is shown. LMW HA or HMW HA was added as indicated. (B) Images of MSC from WT or tHAS1/2/3 KO mice (magnification ×20).

HAS2 might play other important roles in MSC homeostasis (Figs. 6.1–6.4), further studies are warranted to better understand HA biology in these cells. Still outstanding is an accurate determination of the physiological HA levels in BM. Decreases (Goncharova et al., 2012) and increases (Jadin et al., 2008) in BM HA levels both cause hematological abnormalities, suggesting that an imbalance in HA synthesis, production, retention, and degradation in the BM may have undesired consequences on hematopoietic homeostasis. Although the physiological levels of HA produced by MSC in BM remain unknown, other mouse models of human disease suggest that HA synthesis requires exquisite fine tuning (Chowdhury, Hemming, Hombach-Klonisch, Flamion, & Triggs-Raine, 2013; Li et al., 2011).

2.2. Fibroblasts and HA in BM

Fibroblasts, also known as stromal cells, are direct MSC descendants and play a critical role in the homeostasis of the hematopoietic microenvironment. Experimentally, it is often not clear whether primary fibroblast cultures are free of MSC. Fibroblasts and MSC cells have morphological similarities and require similar culture conditions, and in many cases, cell surface characterization is either not performed or not reported. Thus, it is possible that some of the findings reported for primary fibroblasts might be relevant for MSC.

Although the effects of HA on fibroblasts isolated from many tissues has been studied in detail, this is not the case for BM-derived fibroblasts. Several studies have identified tissue-specific differences in fibroblast behavior that may preclude extrapolation of the findings to BM fibroblasts. Fibroblasts isolated from different tissues not only respond differently to cytokine stimulation but also display different basal levels of HAS expression and HA production (Li, Asteriou, Bernert, Heldin, & Heldin, 2007; Meran et al., 2007). This is in line with our recent observation that HAS2 expression differs among various stromal fibroblast cell lines established from murine BM, with the highest level being found in the S17 stromal cell line that possesses hematopoiesis-supportive functions (Goncharova et al., 2012).

HA has been suggested to regulate several fibroblast functions, including proliferation and differentiation (Meran et al., 2011, 2007). In particular, human biopsy-derived dermal fibroblasts undergo TGFβ_1-induced differentiation associated with induction of HAS1 and HAS2 expression and assembly of pericellular HA coats, whereas these responses to TGFβ_1 were not observed with oral fibroblasts. Moreover, 4MU treatment significantly attenuated the TGFβ_1-mediated differentiation of dermal fibroblasts, confirming the importance of HA in this process (Meran et al., 2007). Further work by this group showed that the TGFβ_1-induced responses of the dermal fibroblasts, including the changes in HAS2 and HA, did not occur under conditions of *in vitro* senescence (Simpson et al., 2009). TGFβ_1 is also involved in regulating fibroblast proliferation, and, as was observed in the differentiation studies, this occurs in a tissue-specific manner. Whereas dermal fibroblasts proliferate in response to TGFβ_1 through interactions between HA, CD44, and EGFR, oral fibroblasts show an antiproliferative response (Meran et al., 2011). Interestingly, TGFβ_1 failed to increase HAS activity in dermal fibroblasts obtained from patients undergoing breast reduction surgery, but did increase the activity of hyaluronidases

(Li et al., 2007). One possible explanation for the discrepant observations with dermal fibroblasts is that HA biology may be further influenced by intersubject variations that affect the local microenvironment, such as vascularization, hormone levels, inflammation, age, and gender. Of note, since these studies did not provide cell surface signatures, the primary fibroblast cultures may have been contaminated with other cell types, such as MSC, that respond differently to the various treatments. Additional factors that have been reported to influence HA production in fibroblasts are bFGF (Ohno et al., 2009), platelet-derived growth factor (PDGF)-BB (Li et al., 2007), PDGF-AA, and PDGF-CC (Willenberg, Saalbach, Simon, & Anderegg, 2012). It is interesting to note that PDGF secreted by melanoma cells stimulates formation of a thick pericellular HA coat and HA cables in neighboring fibroblasts (Pasonen-Seppanen et al., 2012), which in turn stimulate melanoma cell proliferation in a paracrine manner.

Hormones may also influence HA production in fibroblasts. Dexamethasone treatment decreased the expression of HAS2 and downregulated HA production in human dermal fibroblasts *in vitro* and *in vivo* (Averbeck, Gebhardt, Anderegg, & Simon, 2010; Gebhardt et al., 2010). However, this finding could be limited to skin fibroblasts because glucocorticoids (GC) have the opposite effect in smooth muscle cells, which also originate from MSC. Thus, in airway smooth muscle cells from asthma patients, GC-stimulated HMW HA secretion and deposition through increased expression of HAS1 and reduced expression of hyaluronidase 1 (Papakonstantinou et al., 2012). The effect of GC on HA production has not been investigated in BM fibroblasts, although GC are known to influence hematopoietic homeostasis in healthy BM and they are also used to treat hematological malignancies (Gisslinger & Kees, 2003; Hicsonmez, 2006).

Other chemotherapies are known to influence the BM microenvironment. We have previously reported that 5-fluorouracil (5FU) decreases HA levels in murine BM (Matrosova, Orlovskaya, Serobyan, & Khaldoyanidi, 2004; Matrosova, Orlovskaya, Serobyan, McClelland, & Khaldoyanidi, 2005), and similar results were obtained in the brain of 5FU-treated mice, specifically in the neural stem cell microenvironment (Sherman et al., manuscript in preparation). However, other chemotherapy drugs show contrasting effects on HA production in fibroblast cell lines. In particular, cytarabine treatment upregulated HAS1 expression and increased both secreted and cell surface HA levels in the M210B4 BM-derived fibroblast cell line (Zweegman et al., 2007). Whether other cell lines or primary cells respond similarly to cytarabine treatment remains unclear. This is of particular interest because baseline HAS expression in fibroblast cell lines differs markedly, ranging from barely detectable expression in M10B4 cells to very high levels in the hematopoiesis-supportive S17 cell line (Goncharova et al., 2012). Thus, the effect of chemotherapeutic drugs on HA metabolism in BM stromal fibroblasts remains largely unknown and requires more extensive investigation.

High levels of HAS2 expression and subsequent HA production in dermal fibroblasts is associated with thickened and folded skin in dogs (Docampo et al., 2011; Olsson et al., 2011; Zanna et al., 2009) and in humans (Ramsden et al., 2000). HA appeared to localize not only on the cell membrane but also intracellularly in both the cytoplasm and nucleus. Dermal fibroblasts from subjects with Down syndrome also showed increased HAS2 expression and HA production. Interestingly, silencing of the COL6A2 gene in these cells downregulated HAS2 and decreased secreted HA. Conversely, overexpression of COL6A2 increased expression of HAS2, which may contribute to Down syndrome phenotypes (Karousou et al., 2013). It will be of interest to determine whether the changes in HA levels associated with these conditions are limited to dermal fibroblasts or also occur in BM fibroblasts, and if so, whether the changes correlate with hematopoietic abnormalities.

2.3. Osteoblasts and HA in BM

Osteoblasts play a very important role in the hematopoietic microenvironment. Early studies demonstrated that adhesive interactions between HSC and osteoblasts are required for HSC self-renewal (Taichman & Emerson, 1998), and later confirmation of these studies (Scadden, 2006) led to the concept of the "osteoblastic niche" (Ema & Suda, 2012). Although current knowledge suggests that several cell types are required to build an effective niche for HSC, the role of osteoblasts remains highly appreciated. Osteoblasts are produced by osteogenic differentiation of MSC and in turn undergo differentiation to osteocytes, which create bone. Bone is resorbed by osteoclasts, which are a product of HSC differentiation and maturation. Thus, crosstalk between both BM-resident stem cells and their progeny contributes to bone homeostasis.

There is clear evidence that HA is involved in osteogenic differentiation of MSC and bone formation (Matsumoto et al., 2009) *in vitro* and *in vivo*. HA from the umbilical cord, which is composed of different-sized polymers but is predominantly HMW, has been reported to provide favorable conditions for osteoblast differentiation of BM-derived MSC and bone matrix synthesis. Osteoblast differentiation-associated genes such as osterix, osteocalcin, integrin-binding sialoprotein, osteonectin, and COL1A1 were significantly upregulated when MSC were cultured on immobilized HA (Mathews, Mathew, Gupta, Bhonde, & Totey, 2012). Similarly, osteogenic differentiation of MSC was supported by HA from *Streptococcus equi* incorporated into immobilized films (Serra Moreno, Sabbieti, Agas, Marchetti, & Panero, 2012) and poly-L-lysine/HA multilayer coatings cross-linked with fibronectin (Semenov, Malek, Bittermann, Voros, & Zisch, 2009). In a different experimental setting, HMW HA had the opposite effect on MSC differentiation. Leukemia inhibitory factor (LIF) treatment inhibited osteogenic differentiation of MSC from rat calvaria, and this correlated with increased expression of HAS2 and production of HMW HA (Falconi & Aubin, 2007). In this model, the addition of HMW HA to cultures decreased the number of bone nodules formed. The variable effects of HA on osteogenic differentiation may be explained, at least in part, by the finding that high expression of endogenous HAS2 and production of HMW HA, or the addition of exogenous HMW HA, inhibit osteoblast differentiation at the late osteoprogenitor/early osteoblast stage, such that osteogenic genes would be upregulated but mineralization would be defective in the presence of high concentrations of HA. Thus, the exact mechanisms by which HA contributes to osteoblast generation remains elusive, and further investigation and validation using *in vivo* models is required.

HAS expression and HA production in osteoblasts can be regulated by local microenvironmental factors. Although LIF inhibited HAS (Falconi & Aubin, 2007), parathyroid hormone (PTH) stimulated HAS2-mediated HA synthesis in osteoblasts (Midura, Su, Morcuende, Tammi, & Tammi, 2003). PTH also increased osteoblast numbers and augmented hematopoiesis (Calvi et al., 2003; Stier et al., 2005). Interestingly, PTH peptides are able to stimulate HAS2 expression and HA synthesis and inhibit hyaluronidase 2 expression in osteosarcoma cell lines, and this was associated with increased motility (Berdiaki et al., 2010). Thus, understanding how HAS2 expression is regulated in osteoblasts may help increase our understanding not only of normal hematopoiesis but also of the progression of relevant malignancies.

HA has also been shown to be involved in MSC differentiation toward the chondrocytic lineage (reviewed in Kim, Mauck, & Burdick, 2011); however, since a role for chondrocytes in the hematopoietic microenvironment has not been established, this topic is not discussed here.

3. MACROPHAGES AND HA IN THE BM

3.1. Role of macrophages in BM

BM macrophages play an important role in the complex cell–cell network within the hematopoietic microenvironment. Their roles include production of cytokines and chemokines, clearance of cellular debris from the extracellular space, production and remodeling of the ECM, and communication with neighboring cells. A critical role for BM macrophages in maintaining and regulating the hematopoietic microenvironment was first postulated in the 1980s (Lichtman, 1981; Weiss, 1980, 1981), but further studies were neglected for at least a decade after the role of osteoblasts was proposed (Taichman & Emerson, 1994). Interest in the function of macrophages in the BM microenvironment underwent a resurgence in 2004 following creation of the macrophage Fas-induced apoptosis (MAFIA) mouse (Burnett et al., 2004), in which ablation of BM macrophages leads to extramedullary hematopoiesis. Subsequent studies have shown that BM macrophages are required to maintain osteoblasts (Chang et al., 2008) and work performed during recent years has further strengthened the evidence that macrophages play a critical role in structuring the hematopoietic niche (reviewed in Ehninger & Trumpp, 2011). The ability of macrophages to respond to HA stimulation is a topic of great interest because, in addition to other important functions in BM, macrophages produce soluble factors that are critically involved in homeostasis of the hematopoietic microenvironment (Kaur-Bollinger et al., 2012).

3.2. Effect of HA on macrophages

Cultured murine BM-derived macrophages express the standard form and splice variants of the HA receptor CD44 (Khaldoyanidi et al., 1999). The expression level of CD44 splice variants is especially important because the ligand-binding activity of CD44 depends not only on glycosylation of CD44 (Sackstein, 2012; Sackstein et al., 2008) but also on the presence of alternatively spliced variants (Sleeman et al., 1996). CD44 expression and HA binding were increased by treatment of mouse primary BM-derived macrophages with tumor necrosis factor alpha (TNFα) or lipopolysaccharide (LPS) and interferon gamma (IFNγ) (Ruffell et al., 2011). IFNγ stimulation of human macrophages differentiated from primary monocytes enhanced HMW HA-induced secretion of TNFα, interleukin

(IL)-1β, IL-12p70, and interferon gamma-induced protein 10 (IP-10), but suppressed the secretion of IL-10 (Wallet, Wallet, Guiulfo, Sleasman, & Goodenow, 2010), a factor involved in HSC self-renewal (Kang et al., 2007). Stimulation of BM macrophages with a human umbilical cord-derived HA preparation containing different-sized polymers also increased production of IL-1 and IL-6, both of which are involved in the regulation of hematopoiesis (Khaldoyanidi et al., 1999). Interestingly, HMW HA (2700 kDa) suppressed LPS-stimulated production of prostaglandin E2 (PGE2) (Yasuda, 2010) and IL-6 (Yasuda, 2011) in the U937 human monocytic cell line. The anti-inflammatory effect of HMW HA was also reported in microglial cells, which are the product of monocyte differentiation in the brain. Thus, addition of HMW HA to LPS-treated rat primary microglial cultures decreased IL-1β, IL-6, and TNFα gene expression and IL-6 and nitric oxide production (Austin, Gilchrist, & Fehlings, 2012). In addition, a naturally occurring HMW HA and α-trypsin inhibitor complex (HA-HC) showed strong anti-inflammatory effects on LPS-stimulated macrophages via multiple mechanisms, suppressing the M1 while promoting M2 phenotype (He, Zhang, Tighe, Son, & Tseng, 2013).

As in other cell types, the molecular weight of HA strongly influences its effects on macrophage activity. Current understanding suggests that LMW HA has a proinflammatory effect on macrophages. Specifically, LMW HA induced the expression of chemokines such as MIP-1α, MIP-1β, KC/CXCL1, MCP1, IL-8, and IP-10, as well as proinflammatory cytokines such as IL-12 and TNFα (Osterholt et al., 2012; Powell & Horton, 2005). Reactive oxygen species (ROS) can synergize with LMW HA to induce inflammatory gene expression in the mouse alveolar macrophage cell line MH-S and in thioglycolate-elicited peritoneal macrophages (Eberlein et al., 2008). The effect of LMW HA on macrophages can be inhibited by antioxidants. Further work is necessary to determine whether BM macrophages respond similarly to LMW HA, but if so, this finding would be highly relevant to hematology because ROS strongly influence HSC and their microenvironment and contribute to disease development (Ishdorj, Li, & Gibson, 2012; Urao & Ushio-Fukai, 2013; Watson, Mortensen, & Simon, 2011).

Chronic inflammation negatively influences hematopoietic homeostasis and, similarly to the solid tumor microenvironment (Iijima, Konno, & Itano, 2011), can aggravate pathological conditions in BM (Hasselbalch, 2013). In addition to promoting inflammatory cytokine production, LMW HA can contribute to the pathophysiologic "vicious cycle" of inflammation by inhibiting the adenosine A2a receptor in macrophages and thus the anti-inflammatory function of adenosine (Collins et al., 2011).

LMW HA stimulated COX2 expression, arachidonic acid release, and PGE2 production in primary human monocytes, primary BM-derived macrophages, and RAW264.7 cells (a murine leukemic monocyte/macrophage cell line) (Sokolowska et al., 2013). LMW HA also polarized human macrophages toward an M1 phenotype (Sokolowska et al., 2013). The effects of LMW HA on macrophages are also important during tumor growth. Cancer cells often release LMW HA polymers that cause monocyte dysfunction and induce formation of immunosuppressive macrophages (Kuang et al., 2007).

LMW HA can also induce IFNβ production via activation of TLR-4 in murine macrophages, including MH-S cells (a murine alveolar macrophage cell line), RAW264.7 cells, and primary peritoneal macrophages (Black et al., 2013). Macrophages of all origins tested in this study responded similarly to LMW HA, suggesting that primary BM-derived macrophages may also produce type I IFNs (IFNα, IFNβ) in response to LMW HA. In BM, signaling of these IFNs via the IFNα/β receptor played an important role in modulating HSC and leukemic stem cell proliferation and differentiation (Schurch, Riether, & Ochsenbein, 2013), suggesting that HA may contribute to a milieu supportive of hematological malignancies.

Macrophages not only bind and respond to HA, they also produce HA via HAS1 and HAS2, which are upregulated in response to stress (Calve, Isaac, Gumucio, & Mendias, 2012). TGFβ is a positive regulator of HA synthesis and participates in the chemotactic response of BM macrophages (Foley et al., 2012). Of note, serum amyloid A3 and HA form a complex that increases monocyte recruitment, adhesion, and retention in tissues (Han et al., 2007).

Monocytes recruited to the BM differentiate into macrophages, but the role of HA in this process remains unclear. In the human monocytic leukemia cell line THP-1, HAS1 was essentially undetectable before, during, and after differentiation into macrophages, whereas HAS2 and HAS3 expression were changed. HAS2 was present at extremely low levels in non-differentiated THP-1 cells but was rapidly and markedly upregulated (800-fold) upon induction of differentiation and then became undetectable in the fully differentiated macrophages. In contrast, HAS3 was present in THP-1 monocytes, but its expression decreased upon induction of differentiation (Chang et al., 2012). Whether the transient upregulation of HAS2 observed here is accompanied by increased production of HA remains to be investigated. In addition, it will be of interest to determine whether HA and/or HAS are involved in differentiation of primary blood-derived monocytes into tissue-specific macrophages.

3.3. Effect of HA on osteoclasts

In BM, macrophages can differentiate into bone-resorbing osteoclasts. Differentiation of osteoclast precursors into mature osteoclasts requires their interaction with osteoblasts/stromal cells (Suda et al., 1999), which express the receptor activator of the NF-kB ligand (RANKL), a member of the TNF superfamily. RANKL interacts with the osteoclast cell surface receptor RANK, which in turn recruits TNF receptor-associated factors (Darnay, Haridas, Ni, Moore, & Aggarwal, 1998), and plays a crucial role in the osteoclast differentiation axis. HMW HA downregulated RANKL expression in stromal cells and inhibited stromal cell-supported osteoclast formation, whereas the HAS synthesis inhibitor 4MU increased RANKL expression (Ariyoshi, Okinaga, Knudson, Knudson, & Nishihara, 2014). In addition to RANKL, differentiation of osteoclasts is also governed by macrophage colony-stimulating factor (M-CSF). Interestingly, HMW HA has potent anti-osteoclastogenic effects in primary mouse and human osteoclast precursor cell cultures (Chang et al., 2007). HMW HA inhibited M-CSF-dependent signaling pathways involving ROS, Rac, and MAPK, resulting in suppression of AP-1 and MITF transcriptional activity and subsequent RANK expression. Furthermore, in an *in vivo* mouse model of calvarial bone resorption, HA reduced RANKL-induced bone erosion and osteoclastogenesis (Chang et al., 2007). However, stimulation of osteoblasts by hormones and inflammatory mediators increased M-CSF and RANKL levels, which triggered efficient osteoclastogenesis and overcame the negative influence by HA.

These findings substantiate a role for HA in regulation of all macrophage functions. However, most studies have been performed using macrophage cell lines or macrophages from tissues other than BM, including alveolar and peritoneal macrophages. Although BM macrophages may respond to HA in a similar fashion, further investigation is warranted to determine how HA specifically regulates BM macrophages in normal and diseased microenvironments.

4. ENDOTHELIAL CELLS AND HA IN BM

4.1. Role of endothelial cells in BM

Endothelial cells (EC) contribute to the structure of the hematopoietic microenvironment in numerous ways, including separation of the BM extravascular space from the blood flow, regulation of liquid exchange, maintenance of interstitial fluid flow, support of cell retention/mobilization (including that of HSC), production of cytokines, chemokines, and ECM, and participation in intercellular crosstalk via cell surface molecules. A distinct type of EC in BM makes up the sinusoids, which are distinguished by sieve plates of fenestrae without diaphragms and noncontinuous or absent basement membranes (Kashimura, 1982; Tavassoli & Shaklai, 1979; Wisse, 1970). In contrast to most endothelia, BM sinusoidal EC constitutively express VCAM-1 and stromal cell-derived factor 1 (SDF-1), which contribute to HSC migration to the BM (Sipkins et al., 2005). Although most molecular pathways regulating the functions of BM EC have been examined in depth, studies on the influence of HA have mostly been performed on ECs from other tissues.

4.2. Expression of HA in EC

There have been no systematic studies of HA function in BM vasculature, but the available information suggests that HA plays a key role in BM EC biology. The level of HA expressed and retained on EC varies with the tissue of origin (Szczepanek, Kieda, & Cichy, 2008), although little is known of HAS or HA expression in BM EC. We recently reported that STR-12 cells, a line established from murine BM EC, express high levels of HAS1-3 and cell surface-associated HA when cultured under static conditions (Goncharova et al., 2012).

Interestingly, the expression of HA on EC is influenced by physiological shear stress. In the stable EC-RF24 cell line established from human umbilical vein endothelial cells (HUVEC), shear stress significantly increased the amount of HA incorporated in the endothelial glycocalyx and the amount shed from the surface into the supernatant (Gouverneur, Spaan, Pannekoek, Fontijn, & Vink, 2006). Another study confirmed this observation and reported that while HAS2 is detected in primary HUVEC cultured in static conditions, shear stress conditions increased both HAS2 expression and HA production (Maroski et al., 2011). The authors expanded on previous studies and showed that several parameters associated with shear flow influence not only the production of HA but also whether it is released into the supernatant or incorporated into the glycocalyx. These parameters included the level of wall shear stress, constant laminar flow versus pulsatile conditions, and duration of exposure to shear stress (Maroski et al., 2011). Most studies on HA in vascular biology have aimed to better understand the protective role of HA on a vascular wall during the development of atherosclerosis because HA is a component of the glycocalyx and is essential for its size and structure (Pahakis, Kosky, Dull, & Tarbell, 2007). However, these findings are also highly relevant to the BM microenvironment where sinusoidal and vascular EC are exposed to different levels of shear stress. The study of marrow vasculature *in vivo* is technically difficult because of the surrounding bone. To overcome this challenge, we developed a "bone and marrow" transplant approach using dorsal skin chambers (Sikora & Khaldoyanidi, 2010). In this approach, a piece of bone with the attached marrow is placed in the chamber window and becomes revascularized within 10–14 days. Notably, a slow pulsatile flow in the marrow vasculature and sinusoids can be observed for several weeks. This approach may be useful to address some remaining questions about the role of HA in BM EC, such as how HAS expression, HA synthesis, and HA retention versus shedding into the extracellular space are regulated, during both normal steady-state hematopoiesis and pathological conditions affecting the BM.

The physiological blood flow contributes to all vital processes in the body, including maintenance of the semi-liquid interstitial fluid in BM, which depends critically on the amount of water retained in the extravascular space. While it is well established that HA binds and retains water in tissues, there are only a few studies examining the function of HA in maintaining the BM interstitial fluid and in regulating its flow. Glycocalyx-mediated mechanotransduction of interstitial flow shear stress is thought to significantly influence cell motility and play a role in cell migration (Qazi, Palomino, Shi, Munn, & Tarbell, 2013). A study using a three-dimensional (3D) interstitial flow model (Qazi, Shi, & Tarbell, 2011) showed that CD44 expression on SN12L1 cells was significantly increased by interstitial flow compared to static conditions, and blocking CD44 inhibited flow-induced cell migration. Interestingly, enzymatic degradation of HA inhibited flow-enhanced SN12L1 migration, supporting earlier observations on the role of HA in fluid shear stress mechanotransduction in other cell types (Pahakis et al., 2007). No studies have yet been performed examining HA and interstitial fluid flow in the context of the hematopoietic microenvironment. This is surprising considering that the ECM and interstitial fluid flow have been shown to affect proliferation and differentiation of MSC (Kim, You, Yellowley, & Jacobs, 2006; Knippenberg et al., 2005; Riddle, Taylor, Genetos, & Donahue, 2006) and osteogenesis (Kim et al., 2006). Furthermore, the importance of the volume and fluid pressure characteristics of the BM has been demonstrated not only during normal hematopoiesis but also in pathological conditions associated with acute myeloid leukemia (AML), anemia, and therapy with hematopoietic growth factor (Iversen, Berggreen, Nicolaysen, & Heyeraas, 2001; Iversen, Nicolaysen, & Benestad, 1992, 1993; Iversen, Thing-Mortensen, Nicolaysen, & Benestad, 1993; Mortensen et al., 1998). Given the significant role played by HA in interstitial fluid regulation, it is possible that HA may be involved in the development of pathological conditions in the BM as a part of the mechanotransduction machinery.

A variety of additional factors have been shown to contribute to the regulation of HAS expression and HA production in EC, although no information is available on BM-derived EC. Specifically, TNFα treatment of human intestinal EC increased HAS3 expression, HA production, and the accumulation of HA cable-like structures (Kessler et al., 2008). A follow-up study demonstrated that IL-1β, TNFα, and TNFβ, but not TGFα and TGFβ, strongly induced HAS2 expression and HA synthesis in HUVEC via the NF-kB pathway (Vigetti et al., 2010). Other inflammatory factors, such as C-reactive protein, induced HA shedding from EC, leading to endothelial dysfunction (Devaraj, Yun, Adamson, Galvez, & Jialal, 2009). The association between HA and inflammation is highly relevant to BM biology since an inflammatory element is present in marrow during pathological conditions. Specifically, chemotherapy leads to an inflammatory reaction reflected by increased levels of C-reactive protein in the marrow of MM patients (Harada et al., 2013).

Several disease states are correlated with vasculature-associated deposition of HA. A profound increase in HA deposition within blood vessels of the colon was observed prior to significant inflammatory infiltration in mice with experimental colitis and in humans with inflammatory bowel disease (Kessler et al., 2008). The pattern of HA deposition in lungs was species specific: in humans, HA was deposited in patches in the microvasculature of the alveolar septum, whereas it was present only in the periarterial space in the mouse lung (Richter, Wicklein, Geleff, & Schumacher, 2012). Increased intracellular HA and deposition of HA in blood vessels were reported in human brains after stroke, and correlated with increased expression of HAS1 and 2, likely as a response to ischemia (Al'Qteishat et al., 2006). Interestingly, hyaluronidase 1 was also upregulated in the microvessels of the stroke patients and was most likely responsible for the increased levels of LMW HA breakdown products in the serum. RHAMM was observed intracellularly and in the nuclei in the tunica media of larger blood vessels and in EC of microvessels in the stroke-affected tissue, whereas CD44 and tumor necrosis factor-α-stimulated gene 6 (TSG-6) were mainly increased in infiltrating mononuclear cells (Al'Qteishat et al., 2006).

Interestingly, postischemic shedding of HA in EC from other tissues (e.g., coronary vessels) could be inhibited by hydrocortisone, which preserved the endothelial glycocalyx and sustained the vascular barrier (Chappell et al., 2007). The perturbations in HA homeostasis observed in ischemic brain may reflect autocrine mechanisms by which the microvasculature responds to stress and could also occur in other tissues, including BM.

4.3. Functions of HA in BM EC: Cell recruitment

HA clearly plays an important role in mediating EC integrity and functions. A large body of evidence suggests that HA present on the EC cell surface is responsible for cell attachment and recruitment. CD44/HA interactions are involved in regulation of homing of normal HSC and malignant cells into BM (Avigdor et al., 2004; Ellis et al., 2011; Khaldoyanidi et al., 1996; Krause, Lazarides, von Andrian, & Van Etten, 2006). We recently reported that 4MU-mediated inhibition of HA synthesis in STR-12 EC cells decreased their ability to support rolling and adhesion of the HSPC line FDCP-mix (Goncharova et al., 2012), confirming that the initial steps of the homing cascade are mediated by endothelial HA. These *in vitro* findings are consistent with the observed decrease in homing of transplanted HSC in 4MU-treated mice (Goncharova et al., 2012). Interactions of other hematopoietic cells with EC also depend on HA. U937 monocyte adhesion to IL-1β-stimulated HUVEC required endothelial HA and was decreased by hyaluronidase treatment (Vigetti et al., 2010). Similarly, the removal of HA from the surface of HUVEC by hyaluronidase treatment inhibited their interactions with neutrophils (Butler, Rainger, & Nash, 2009). EC isolated from various tissues require HA to support the recruitment of circulating cells. For example, slow rolling lymphocyte interactions with EC isolated from mouse brain were inhibited by EC treatment with PH20 hyaluronidase (Winkler et al., 2012). In addition to direct binding between EC HA and CD44 expressed on circulating cells, CD44 expressed on EC may function as an anchor for endogenous HA. This is supported by the observation that TNFα-stimulated CD44−/− EC expressed lower levels of HA and subsequently had a reduced ability to support slow rolling of activated lymphocytes (Winkler et al., 2012). Interestingly, pretreatment of lymphocytes, but not EC, with LMW HA (dodecasaccharides) impaired rolling of activated lymphocytes on brain EC (Winkler et al., 2013). In larger vessels, vascular EC are shielded from flowing blood by the glycocalyx, and the incorporation of HA into the glycocalyx is essential to maintain its permeability barrier properties. Under these circumstances, endothelial HA could also function to protect EC from aberrant cell attachment.

4.4. Functions of HA in BM EC: Angiogenesis

HA has been implicated in the regulation of angiogenesis. Inhibition of endogenous HA synthesis by 4MU in HUVEC (RF24 cell line) and human microvascular EC impaired angiogenesis at several steps, including proliferation, adhesion, tube formation, and ECM remodeling (Garcia-Vilas, Quesada, & Medina, 2013). The size and origin of exogenous HA strongly influences its effect on EC growth and the formation of vessels. In early studies, HMW HA was reported to be an inhibitor of angiogenesis (Feinberg & Beebe, 1983; West & Kumar, 1989). Recent studies supported these observations and demonstrated that HMW HA inhibits HUVEC proliferation via merlin activation (Mo et al., 2011). Interestingly, a covalent complex of HMW HA and the heavy chain of inter-α-inhibitor (HA-HC) purified from amniotic membrane had multiple effects on HUVEC, including suppression of HUVEC viability and proliferation, alterations of cell shape, and impairment of VEGF-stimulated cell migration and tube formation (Shay, He, Sakurai, & Tseng, 2011). However, other studies suggest that the biological effects of HA–protein complexes depend on the protein. In particular, HMW HA alone had no effect on lymphangiogenesis *in vivo*, whereas HA–versican complexes stimulated stromal support and lymphangiogenesis (Koyama et al., 2008). An important consideration here is that *in vivo*, HA acts in combination with tissue-specific microenvironmental signals. The concept that glycosaminoglycans can participate in chemokine and cytokine presentation was proposed decades ago (Gordon, Riley, Watt, & Greaves, 1987; Roberts et al., 1988), and HA is known to contribute to the sequestration and presentation of soluble factors. In agreement with this concept, HMW HA was shown to enhance HUVEC proliferation in the presence of SDF-1 (Fuchs et al., 2013). Stimulatory effects of HMW HA alone on angiogenesis have also been reported; for example, HMW HA induced tube formation by HUVEC, which was prevented by hyaluronidase treatment (Park, Kim, et al., 2012). Restoration of blood perfusion and salvaging of the ischemic limb, angiogenesis, and cell retention was also reported to be significantly increased upon supplementation of transplanted HUVEC with HMW HA (Tang, Liao, Tang, Tsai, & Hsieh, 2011).

The effects of LMW HA *in vitro* are influenced by the culture conditions. LMW HA inhibits HUVEC proliferation in the presence of SDF-1 (Fuchs et al., 2013) but was reported to stimulate HUVEC proliferation in the absence of SDF-1 (Mo et al., 2011; Park, Kim, et al., 2012). These effects were found to be mediated by the HA receptors CD44 and RHAMM, because knockdown of either receptor inhibited LMW HA-induced EC tube formation in Matrigel (Matou-Nasri, Gaffney, Kumar, & Slevin, 2009). CD44 was suggested to mediate the effects of LMW HA on angiogenesis through activation of tyrosine phosphorylation of Src, FAK, and ERK1/2 and upregulation of c-Jun and c-Fos expression (Wang et al., 2011). Notably, CD44 deficiency impairs angiogenesis. Nascent vessels in Matrigel implants from CD44−/− mice demonstrated irregular luminal surfaces characterized by retracted cells and thinned endothelia (Cao et al., 2005).

Thus, the effects of HA on angiogenesis appear to depend not only on the size of the HA polymers but also on the HA-binding proteins present in the local microenvironment and the HA receptors expressed on EC. Each of these vary with the tissue, suggesting that whether HA influences angiogenesis or serves as a stabilizing molecule to maintain steady-state conditions might also be tissue specific. This is illustrated by a study demonstrating that HA promotes BM neovascularization *in vivo* (Raines et al., 2011). In that study, a rat tibial marrow ablation model was used to assess neovascularization during bone formation and regeneration of marrow using HA of different sizes. Whereas total blood vessel volume and vessel connectivity in the marrow cavity were increased only in LMW HA-treated animals, the vessel number was increased in animals treated with HMW HA, LMW HA, or a mixture of polymers (Raines et al., 2011). Unfortunately, little is known of the effects of HA on the biology of EC within the HSC niche in BM. Given that the cell surface signatures and biology of BM-derived EC differ significantly from those of HUVEC or EC isolated from other organs and tissues, progress in this field would be significantly enhanced if BM-derived EC were included in the *in vitro* and *in vivo* studies.

4.5. Functions of HA in BM EC: Vascular integrity

The effect of HA on vascular integrity in the BM remains largely unknown, although this information is highly relevant to our understanding of the pathogenesis of hematologic diseases and the development of novel therapeutic approaches. Based on studies performed with EC from other organs and tissues, removal of endogenous HA from EC appears to be associated with reduced vascular dilatation properties and increased vascular vulnerability (Nieuwdorp et al., 2007). In human pulmonary EC, binding of HMW HA to CD44 within caveolin-enriched microdomains (CEM) enhanced the EC barrier function, promoted redistribution of CEM to areas of cell-cell contact, and mediated recruitment of cytoskeletal regulatory proteins (Singleton et al., 2010). The importance of CD44 in this process is supported by studies with *in vivo* models. HMW HA-mediated protection from LPS-induced pulmonary vascular hyperpermeability was significantly reduced in CD44−/− mice (Singleton et al., 2007). In contrast, HA contributed to the glycocalyx of glomerular microvascular EC but had a limited effect on vascular permeability (Foster et al., 2013). These observations support the concept that the role of HA in EC biology is tissue specific and, therefore, knowledge gained in different experimental models cannot be extrapolated *a priori* to BM physiology and pathology.

Vascular integrity is critical for preventing thrombus generation, and not surprisingly, EC dysfunction increases the thromboembolic risk. Interestingly, inhibition of HA synthesis by 4MU results in a prothrombotic state, but because 4MU does not influence thrombin activity, this effect may be caused by a reduction in HA in the glycocalyx (Nagy et al., 2010). Although EC-associated HA may be important for blood coagulation, additional mechanisms are likely to be involved. During our studies with HAS-deficient mice (Goncharova et al., 2012), we observed increased coagulation of blood harvested from tHAS1/2/3 KO mice. Whereas 6 mg/mL EDTA was necessary to prevent coagulation of blood drawn from wild-type and dHAS1/3 KO mice, twice that concentration was necessary to prevent thrombus formation by blood from tHAS1/2/3 KO mice. Given that HA deficiency does not affect thrombin activity (Nagy et al., 2010), this observation suggests that the involvement of HA in blood coagulation occurs via additional mechanisms, including platelet activation.

5. NERVE CELLS AND HA IN BM

As is the case for all tissues, BM is innervated (Afan, Broome, Nicholls, Whetton, & Miyan, 1997; Artico et al., 2002). Innervation plays an important role in regulating hematopoiesis in the BM and contributes to the complex microenvironment (Mendez-Ferrer, Lucas, Battista, & Frenette, 2008). Elements of the cholinergic system are expressed in BM cells and participate in the regulation of hematopoietic activity (Khaldoyanidi et al., 2001; Serobyan et al., 2007). However, whether and how HA might play a role in regulating the nervous system in BM is completely unknown. This is perhaps surprising given our extensive knowledge of the role of HA in the brain (Preston & Sherman, 2011).

6. ROLE OF HA IN THE BM MICROENVIRONMENT IN HEMATOLOGICAL MALIGNANCIES

The BM microenvironment plays a key role in leukemia stem cell survival and drug resistance (Dalton, Hazlehurst, Shain, Landowski, & Alsina, 2004; Fecteau & Kipps, 2012). There is growing interest in the function of HA as an element of the malignant microenvironment in BM and its involvement in the pathogenesis of hematopoietic malignancies.

The possibility that serum HA could serve as a disease marker for patients with liver cirrhosis, rheumatoid arthritis, sepsis, and cancer was first suggested decades ago (Engstrom-Laurent & Hallgren, 1987; Engstrom-Laurent, Laurent, Lilja, & Laurent, 1985; Laurent et al., 1996a). Serum HA levels vary with age and this should be taken into consideration when evaluating human samples. Concentrations of HA in the serum of healthy individuals increase with age from 29 ± 17 μg/L at 16 years of age to 177 ± 133 μg/L at age 75 and older (Lindqvist & Laurent, 1992). In middle-aged subjects, the average concentration is 30–40 μg/L (Laurent et al., 1996a). Serum HA concentrations also fluctuate with the time of day (Lindqvist et al., 1988).

Analysis of serum samples from a large number of MM patients revealed that abnormally low or high HA values were associated with advanced stage of disease and with shorter median survival (Dahl, Turesson, Holmberg, & Lilja, 1999). Since myeloma cells migrate to the BM (Cook, Dumbar, & Franklin, 1997), their interactions with the microenvironment may significantly influence disease progression. In another study, BM plasma cells and peripheral B cells from MM patients were analyzed for HAS expression (Adamia et al., 2005). HAS3 was expressed in both cell types, HAS2 expression was restricted to plasma cells, and HAS1 was predominantly expressed in B cells. While B cells from healthy donors failed to establish a pericellular coat, B cells from MM patients showed HA matrix around the cells. Interestingly, expression of the HAS1Vb splice variant correlated both with the presence of HA in intracellular compartments (cytoskeletal and perinuclear) and with reduced survival of MM patients (Adamia et al., 2005). This intriguing observation raises several questions about the mechanism(s) controlling the change in HAS expression pattern and HA accumulation (including intracellular location) during maturation of

B cells to plasma cells in MM patients, and the role of the BM microenvironment in this process. Understanding the biology of HA in MM cells is also important for the development of novel therapeutic applications because MM B cells are drug resistant (Pilarski, Szczepek, & Belch, 1997) and HA has been shown to contribute to multidrug resistance (Misra, Ghatak, Zoltan-Jones, & Toole, 2003). In contrast to BM-derived MSC from healthy donors, which express HAS2, MSC from MM patients express HAS1, similarly to B cells, which could be downregulated by culture with plasma cells (Calabro et al., 2002). Since both synthases produce HMW HA, the relevance of the switch in HAS form to the pathogenesis of MM remains to be understood. Notably, HA acted as a survival factor against dexamethasone-induced apoptosis in MM cell lines (Vincent, Jourdan, Sy, Klein, & Mechti, 2001; Vincent, Molina, Espert, & Mechti, 2003), and dexamethasone treatment has been shown to decrease the expression of HAS2 and production of HA in human dermal fibroblasts (Averbeck et al., 2010; Gebhardt et al., 2010). The effect of dexamethasone on the BM microenvironment, either healthy or malignant, is not yet known.

AML patients also show abnormalities in HA accumulation. HA levels in BM biopsies of AML patients were reported to be significantly increased (Sundstrom et al., 2005; Sundstrom, Hultdin, Engstrom-Laurent, & Dahl, 2010), and serum HA levels were also increased in a limited number of patients (Zweegman et al., 2007). It should be noted that most studies of human samples do not differentiate between LMW and HMW HA due to technical limitations. Because injection of HMW HA results in accumulation of LMW HA in the BM of mice (Courel et al., 2004), it is likely that HA size is strictly regulated through local hyaluronidases in normal BM. Furthermore, under certain pathological conditions, such as inflammation, HA is organized into novel, cable-like structures that appear as thin HA strands arising from cell surfaces (Itano et al., 2004; Knudson & Knudson, 1993). Cable structures can be formed by the coalescence of LMW HA polymers or cross-linking with other proteins, such as the protein product of TSG-6. However, the mechanism by which these structures are formed and whether they are present in BM is not yet known. Similarly, we know little about how the balance between HMW versus LMW HA is regulated in human BM in health or disease. This is an important point because the effect of HA in BM, as in all other organs, is determined by its size. For example, *in vitro* studies with the myeloid leukemic cell line KG1 showed that LMW HA could reduce drug-induced apoptosis, whereas HMW HA had no effect (Onoda, Nakaseko, Yokota, & Saito, 2009). Thus, local hyaluronidases might contribute to the accumulation of LMW HA in the leukemic microenvironment, thereby providing protection for leukemic cells (including leukemic stem cells) and contributing to drug resistance. HSC from healthy donors and AML blasts can both interact with HA in a CD44-dependent manner (Crosby, Lalor, Ross, Newsome, & Adams, 2009; Hanke et al., 2014; White, Nash, Kavanagh, Savage, & Kalia, 2013). Ligation of CD44 on AML cells can induce either cell differentiation (Charrad et al., 1999) or apoptosis (Qian, Xia, Ling, Waxman, & Jing, 2012). In a human/mouse xenograft model, a CD44-specific antibody was shown to eradicate AML cells in BM (Jin, Hope, Zhai, Smadja-Joffe, & Dick, 2006), providing the rationale for an ongoing clinical trial. However, it is important to note that CD44s is also expressed by normal HSC and other healthy tissues, raising concerns about deleterious side effects from CD44-targeted therapies. Nevertheless, there is reason to believe that highly specific anti-CD44 therapies could be developed, considering the heterogeneity of the CD44 family, including differences in glycosylation patterns and splice variants, as well as the tendency of CD44 to form complexes with other (often tissue specific) molecules. Similarly, RHAMM has shown promise as a leukemia-associated antigen, leading to the development of a peptide vaccination approach (Greiner et al., 2010; Tzankov et al., 2011) and subsequent clinical trials. However, RHAMM is also expressed on healthy HSC and activated T cells and does not fulfill the criteria for an ideal target antigen for immunotherapy of AML (Snauwaert et al., 2012). In addition to CD44 and RHAMM, leukemic cells express CD38, a cyclic ADP ribose hydrolase, which can interact with HA, thereby allowing leukemic cells to be entrapped within the BM microenvironment (Gallay et al., 2007). The cellular and molecular mechanisms involved in this process are unknown.

Recent studies demonstrate that AML and chronic myeloid leukemia (CML) cells respond differently to changes in the hematopoietic microenvironment. Activation of the PTH receptor on osteoblastic cells and subsequent increases in TGFβ_1 create a microenvironment favorable for AML but inhospitable to CML stem cells in Col1-caPPR mice (Krause et al., 2013), suggesting that TGFβ_1 has opposing effects on these cells. Interestingly, PTH also stimulates HAS2-mediated HA synthesis in osteoblasts (Midura et al., 2003). TGFβ may also influence HA levels, although the effects depend on experimental conditions. In fibroblasts, TGFβ_1 was reported to increase the activity of hyaluronidases (Li et al., 2007), but it has also been shown to augment the expression of HAS1 and HAS2 and the assembly of pericellular HA coats (Meran et al., 2007). Thus, it is likely that the HA matrix is modified in mice with the activated PTH receptor (Col1-caPPR mouse model) (Calvi et al., 2003) and also contributes to the selective environment for leukemic cells. However, the expression and activation of the HA pathway elements in BM, including HAS, hyaluronidases, and HA size distribution, have not yet been examined in these mice.

The differential response of leukemic cells to the microenvironment and CD44 activation is also illustrated by recent studies on chronic lymphatic leukemias (CLL). In human CLL cells, CD44 expression can be induced by stimulation with IL-4 or contact with stromal cells. In addition, and in contrast to AML cells, signaling through CD44-protected CLL cells from apoptosis (Fedorchenko et al., 2013).

An investigation of the expression of HAS and HA in primary effusion lymphoma cell lines reported that HAS expression and HMW HA secretion was higher in chemoresistant than in chemosensitive cell lines. Of note, addition of exogenous LMW HA enhanced the sensitivity of cells to chemotherapy drugs by disrupting interactions between HMW HA and LYVE-1, CD44, and possibly other HA receptors (Qin et al., 2011). This interesting finding supports the notion that a better understanding of the HA matrix in BM could be of great importance for the development of new therapeutic approaches.

Because CD44 is overexpressed in most human cancers, HA is recognized as a promising drug delivery vehicle. Some of the key properties of HA for this application are its high chemical versatility, biocompatibility, and affinity for CD44, which provides an efficient means of internalization of HA (Brown, 2008). Indeed, HA esterified with butyric acid and retinoic acid has been shown to induce apoptosis in retinoic acid-resistant leukemic cell lines (Coradini et al., 2006) supporting the concept that HA may be used for drug delivery in hematopoietic malignancies.

7. CONCLUSION

While the role of HA in many tissues and organ systems has been studied extensively, it is clear that our understanding of its role in BM physiology remains limited. One major challenge remaining is to understand how elements of the HA pathway maintain HA levels within a narrow physiological window. Similarly, we do not know how HA synthesis by HAS and degradation by hyaluronidases are balanced to provide an optimal composition of HA polymers in healthy BM, or how this balance shifts during BM diseases such as hematological malignancies. Another question to be addressed is whether there are BM-specific or disease-specific HA-binding proteins, and if so, how these proteins influence HA functions in the hematopoietic microenvironment. The enzymatic-independent functions of HAS, particularly HAS2, in the hematopoietic microenvironment also remain unknown.

A thorough understanding of the basic biology of the HA pathway in the BM, especially the differences between normal and diseased BM, is essential to ensure the safety and efficacy of novel drug candidates, not only for treatment of hematological disorders but also for stem cell-based tissue regeneration.

REFERENCES

Adamia, S., Reiman, T., Crainie, M., Mant, M. J., Belch, A. R., & Pilarski, L. M. (2005). Intronic splicing of hyaluronan synthase 1 (HAS1): A biologically relevant indicator of poor outcome in multiple myeloma. *Blood, 105*, 4836–4844.

Afan, A. M., Broome, C. S., Nicholls, S. E., Whetton, A. D., & Miyan, J. A. (1997). Bone marrow innervation regulates cellular retention in the murine haemopoietic system. *British Journal of Haematology, 98*, 569–577.

Al'Qteishat, A., Gaffney, J., Krupinski, J., Rubio, F., West, D., Kumar, S., et al. (2006). Changes in hyaluronan production and metabolism following ischaemic stroke in man. *Brain, 129*, 2158–2176.

Ariyoshi, W., Okinaga, T., Knudson, C. B., Knudson, W., & Nishihara, T. (2014). High molecular weight hyaluronic acid regulates osteoclast formation by inhibiting receptor activator of NF-kappaB ligand through Rho kinase. *Osteoarthritis and Cartilage, 22*, 111–120.

Artico, M., Bosco, S., Cavallotti, C., Agostinelli, E., Giuliani-Piccari, G., Sciorio, S., et al. (2002). Noradrenergic and cholinergic innervation of the bone marrow. *International Journal of Molecular Medicine, 10*, 77–80.

Austin, J. W., Gilchrist, C., & Fehlings, M. G. (2012). High molecular weight hyaluronan reduces lipopolysaccharide mediated microglial activation. *Journal of Neurochemistry, 122*, 344–355.

Averbeck, M., Gebhardt, C., Anderegg, U., & Simon, J. C. (2010). Suppression of hyaluronan synthase 2 expression reflects the atrophogenic potential of glucocorticoids. *Experimental Dermatology, 19*, 757–759.

Avigdor, A., Goichberg, P., Shivtiel, S., Dar, A., Peled, A., Samira, S., et al. (2004). CD44 and hyaluronic acid cooperate with SDF-1 in the trafficking of human CD34+ stem/progenitor cells to bone marrow. *Blood, 103*, 2981–2989.

Berdiaki, A., Datsis, G. A., Nikitovic, D., Tsatsakis, A., Katonis, P., Karamanos, N. K., et al. (2010). Parathyroid hormone (PTH) peptides through the regulation of hyaluronan metabolism affect osteosarcoma cell migration. *IUBMB Life, 62*, 377–386.

Black, K. E., Collins, S. L., Hagan, R. S., Hamblin, M. J., Chan-Li, Y., Hallowell, R. W., et al. (2013). Hyaluronan fragments induce IFNbeta via a novel TLR4-TRIF-TBK1-IRF3-dependent pathway. *Journal of Inflammation (London), 10*, 23.

Brown, T. J. (2008). The development of hyaluronan as a drug transporter and excipient for chemotherapeutic drugs. *Current Pharmaceutical Biotechnology, 9*, 253–260.

Burnett, S. H., Kershen, E. J., Zhang, J., Zeng, L., Straley, S. C., Kaplan, A. M., et al. (2004). Conditional macrophage ablation in transgenic mice expressing a Fas-based suicide gene. *Journal of Leukocyte Biology, 75*, 612–623.

Butler, L. M., Rainger, G. E., & Nash, G. B. (2009). A role for the endothelial glycosaminoglycan hyaluronan in neutrophil recruitment by endothelial cells cultured for prolonged periods. *Experimental Cell Research, 315*, 3433–3441.

Calabro, A., Oken, M. M., Hascall, V. C., & Masellis, A. M. (2002). Characterization of hyaluronan synthase expression and hyaluronan synthesis in bone marrow mesenchymal progenitor cells: Predominant expression of HAS1 mRNA and up-regulated hyaluronan synthesis in bone marrow cells derived from multiple myeloma patients. *Blood, 100*, 2578–2585.

Calve, S., Isaac, J., Gumucio, J. P., & Mendias, C. L. (2012). Hyaluronic acid, HAS1, and HAS2 are significantly upregulated during muscle hypertrophy. *American Journal of Physiology. Cell Physiology, 303*, C577–C588.

Calvi, L. M., Adams, G. B., Weibrecht, K. W., Weber, J. M., Olson, D. P., Knight, M. C., et al. (2003). Osteoblastic cells regulate the haematopoietic stem cell niche. *Nature, 425*, 841–846.

Canaani, J., Kollet, O., & Lapidot, T. (2011). Neural regulation of bone, marrow, and the microenvironment. *Frontiers in Bioscience (Scholar Edition), 3*, 1021–1031.

Cao, J. J., Singleton, P. A., Majumdar, S., Boudignon, B., Burghardt, A., Kurimoto, P., et al. (2005). Hyaluronan increases RANKL expression in bone marrow stromal cells through CD44. *Journal of Bone and Mineral Research, 20*, 30–40.

Caralla, T., Boehm, C., Hascall, V., & Muschler, G. (2012). Hyaluronan as a novel marker for rapid selection of connective tissue progenitors. *Annals of Biomedical Engineering, 40*, 2559–2567.

Chang, M. Y., Chan, C. K., Braun, K. R., Green, P. S., O'Brien, K. D., Chait, A., et al. (2012). Monocyte-to-macrophage differentiation: Synthesis and secretion of a complex extracellular matrix. *The Journal of Biological Chemistry, 287*, 14122–14135.

Chang, E. J., Kim, H. J., Ha, J., Kim, H. J., Ryu, J., Park, K. H., et al. (2007). Hyaluronan inhibits osteoclast differentiation via Toll-like receptor 4. *Journal of Cell Science, 120*, 166–176.

Chang, M. K., Raggatt, L. J., Alexander, K. A., Kuliwaba, J. S., Fazzalari, N. L., Schroder, K., et al. (2008). Osteal tissue macrophages are intercalated throughout human and mouse bone lining tissues and regulate osteoblast function in vitro and in vivo. *The Journal of Immunology, 181*, 1232–1244.

Chappell, D., Jacob, M., Hofmann-Kiefer, K., Bruegger, D., Rehm, M., Conzen, P., et al. (2007). Hydrocortisone preserves the vascular barrier by protecting the endothelial glycocalyx. *Anesthesiology, 107*, 776–784.

Charrad, R. S., Li, Y., Delpech, B., Balitrand, N., Clay, D., Jasmin, C., et al. (1999). Ligation of the CD44 adhesion molecule reverses blockage of differentiation in human acute myeloid leukemia. *Nature Medicine, 5*, 669–676.

Chowdhury, B., Hemming, R., Hombach-Klonisch, S., Flamion, B., & Triggs-Raine, B. (2013). Murine hyaluronidase 2 deficiency results in extracellular hyaluronan accumulation and severe cardiopulmonary dysfunction. *The Journal of Biological Chemistry, 288*, 520–528.

Collins, S. L., Black, K. E., Chan-Li, Y., Ahn, Y. H., Cole, P. A., Powell, J. D., et al. (2011). Hyaluronan fragments promote inflammation by down-regulating the anti-inflammatory A2a receptor. *American Journal of Respiratory Cell and Molecular Biology, 45*, 675–683.

Contreras, E. G., Gaete, M., Sanchez, N., Carrasco, H., & Larrain, J. (2009). Early requirement of hyaluronan for tail regeneration in Xenopus tadpoles. *Development, 136*, 2987–2996.

Cook, G., Dumbar, M., & Franklin, I. M. (1997). The role of adhesion molecules in multiple myeloma. *Acta Haematologica, 97*, 81–89.

Coradini, D., Pellizzaro, C., Scarlata, I., Zorzet, S., Garrovo, C., Abolafio, G., et al. (2006). A novel retinoic/butyric hyaluronan ester for the treatment of acute promyelocytic leukemia: Preliminary preclinical results. *Leukemia, 20*, 785–792.

Courel, M. N., Maingonnat, C., Bertrand, P., Chauzy, C., Smadja-Joffe, F., & Delpech, B. (2004). Biodistribution of injected tritiated hyaluronic acid in mice: A comparison between macromolecules and hyaluronic acid-derived oligosaccharides. *In Vivo, 18*, 181–187.

Crosby, H. A., Lalor, P. F., Ross, E., Newsome, P. N., & Adams, D. H. (2009). Adhesion of human haematopoietic (CD34+) stem cells to human liver compartments is integrin and CD44 dependent and modulated by CXCR3 and CXCR4. *Journal of Hepatology, 51*, 734–749.

Dahl, I. M., Turesson, I., Holmberg, E., & Lilja, K. (1999). Serum hyaluronan in patients with multiple myeloma: Correlation with survival and Ig concentration. *Blood, 93*, 4144–4148.

Dalton, W. S., Hazlehurst, L., Shain, K., Landowski, T., & Alsina, M. (2004). Targeting the bone marrow microenvironment in hematologic malignancies. *Seminars in Hematology, 41*, 1–5.

Darnay, B. G., Haridas, V., Ni, J., Moore, P. A., & Aggarwal, B. B. (1998). Characterization of the intracellular domain of receptor activator of NF-kappaB (RANK). Interaction with tumor necrosis factor receptor-associated factors and activation of NF-kappab and c-Jun N-terminal kinase. *The Journal of Biological Chemistry, 273*, 20551–20555.

Devaraj, S., Yun, J. M., Adamson, G., Galvez, J., & Jialal, I. (2009). C-reactive protein impairs the endothelial glycocalyx resulting in endothelial dysfunction. *Cardiovascular Research, 84*, 479–484.

Docampo, M. J., Zanna, G., Fondevila, D., Cabrera, J., Lopez-Iglesias, C., Carvalho, A., et al. (2011). Increased HAS2-driven hyaluronic acid synthesis in shar-pei dogs with hereditary cutaneous hyaluronosis (mucinosis). *Veterinary Dermatology, 22*, 535–545.

Dominici, M., Le Blanc, K., Mueller, I., Slaper-Cortenbach, I., Marini, F., Krause, D., et al. (2006). Minimal criteria for defining multipotent mesenchymal stromal cells. The International Society for Cellular Therapy position statement. *Cytotherapy, 8*, 315–317.

Eberlein, M., Scheibner, K. A., Black, K. E., Collins, S. L., Chan-Li, Y., Powell, J. D., et al. (2008). Anti-oxidant inhibition of hyaluronan fragment-induced inflammatory gene expression. *Journal of Inflammation (London), 5*, 20.

Ehninger, A., & Trumpp, A. (2011). The bone marrow stem cell niche grows up: Mesenchymal stem cells and macrophages move in. *The Journal of Experimental Medicine, 208*, 421–428.

Ellis, S. L., Grassinger, J., Jones, A., Borg, J., Camenisch, T., Haylock, D., et al. (2011). The relationship between bone, hemopoietic stem cells, and vasculature. *Blood, 118*, 1516–1524.

Ema, H., & Suda, T. (2012). Two anatomically distinct niches regulate stem cell activity. *Blood, 120*, 2174–2181.

Engstrom-Laurent, A., & Hallgren, R. (1987). Circulating hyaluronic acid levels vary with physical activity in healthy subjects and in rheumatoid arthritis patients. Relationship to synovitis mass and morning stiffness. *Arthritis and Rheumatism, 30*, 1333–1338.

Engstrom-Laurent, A., Laurent, U. B., Lilja, K., & Laurent, T. C. (1985). Concentration of sodium hyaluronate in serum. *Scandinavian Journal of Clinical and Laboratory Investigation, 45*, 497–504.

Falconi, D., & Aubin, J. E. (2007). LIF inhibits osteoblast differentiation at least in part by regulation of HAS2 and its product hyaluronan. *Journal of Bone and Mineral Research, 22*, 1289–1300.

Fecteau, J. F., & Kipps, T. J. (2012). Structure and function of the hematopoietic cancer niche: Focus on chronic lymphocytic leukemia. *Frontiers in Bioscience (Scholar Edition), 4*, 61–73.

Fedorchenko, O., Stiefelhagen, M., Peer-Zada, A. A., Barthel, R., Mayer, P., Eckei, L., et al. (2013). CD44 regulates the apoptotic response and promotes disease development in chronic lymphocytic leukemia. *Blood, 121*, 4126–4136.

Feinberg, R. N., & Beebe, D. C. (1983). Hyaluronate in vasculogenesis. *Science, 220*, 1177–1179.

Foley, J. P., Lam, D., Jiang, H., Liao, J., Cheong, N., McDevitt, T. M., et al. (2012). Toll-like receptor 2 (TLR2), transforming growth factor-beta, hyaluronan (HA), and receptor for HA-mediated motility (RHAMM) are required for surfactant protein A-stimulated macrophage chemotaxis. *The Journal of Biological Chemistry, 287*, 37406–37419.

Foster, R. R., Armstrong, L., Baker, S., Wong, D. W., Wylie, E. C., Ramnath, R., et al. (2013). Glycosaminoglycan regulation by VEGFA and VEGFC of the glomerular microvascular endothelial cell glycocalyx in vitro. *The American Journal of Pathology, 183*, 604–616.

Fraser, J. R., Laurent, T. C., & Laurent, U. B. (1997). Hyaluronan: Its nature, distribution, functions and turnover. *Journal of Internal Medicine, 242*, 27–33.

Friedenstein, A. J., Gorskaja, J. F., & Kulagina, N. N. (1976). Fibroblast precursors in normal and irradiated mouse hematopoietic organs. *Experimental Hematology, 4*, 267–274.

Fuchs, K., Hippe, A., Schmaus, A., Homey, B., Sleeman, J. P., & Orian-Rousseau, V. (2013). Opposing effects of high- and low-molecular weight hyaluronan on CXCL12-induced CXCR4 signaling depend on CD44. *Cell Death and Disease, 4*, e819.

Gallay, N., Anani, L., Lopez, A., Colombat, P., Binet, C., Domenech, J., et al. (2007). The role of platelet/endothelial cell adhesion molecule 1 (CD31) and CD38 antigens in marrow microenvironmental retention of acute myelogenous leukemia cells. *Cancer Research, 67*, 8624–8632.

Garcia-Vilas, J. A., Quesada, A. R., & Medina, M. A. (2013). 4-Methylumbelliferone inhibits angiogenesis in vitro and in vivo. *Journal of Agricultural and Food Chemistry, 61*, 4063–4071.

Gebhardt, C., Averbeck, M., Diedenhofen, N., Willenberg, A., Anderegg, U., Sleeman, J. P., et al. (2010). Dermal hyaluronan is rapidly reduced by topical treatment with glucocorticoids. *The Journal of Investigative Dermatology, 130*, 141–149.

Gisslinger, H., & Kees, M. (2003). Therapy strategies for multiple myeloma: Current status. *Wiener Klinische Wochenschrift, 115*, 451–461.

Goncharova, V., Serobyan, N., Iizuka, S., Schraufstatter, I., de Ridder, A., Povaliy, T., et al. (2012). Hyaluronan expressed by the hematopoietic microenvironment is required for bone marrow hematopoiesis. *The Journal of Biological Chemistry, 287*, 25419–25433.

Gordon, M. Y., Riley, G. P., Watt, S. M., & Greaves, M. F. (1987). Compartmentalization of a haematopoietic growth factor (GM-CSF) by glycosaminoglycans in the bone marrow microenvironment. *Nature, 326*, 403–405.

Gouverneur, M., Spaan, J. A., Pannekoek, H., Fontijn, R. D., & Vink, H. (2006). Fluid shear stress stimulates incorporation of hyaluronan into endothelial cell glycocalyx. *American Journal of Physiology. Heart and Circulatory Physiology, 290*, H458–H462.

Greiner, J., Schmitt, A., Giannopoulos, K., Rojewski, M. T., Gotz, M., Funk, I., et al. (2010). High-dose RHAMM-R3 peptide vaccination for patients with acute myeloid leukemia, myelodysplastic syndrome and multiple myeloma. *Haematologica, 95*, 1191–1197.

Han, C. Y., Subramanian, S., Chan, C. K., Omer, M., Chiba, T., Wight, T. N., et al. (2007). Adipocyte-derived serum amyloid A3 and hyaluronan play a role in monocyte recruitment and adhesion. *Diabetes, 56*, 2260–2273.

Hanke, M., Hoffmann, I., Christophis, C., Schubert, M., Hoang, V. T., Zepeda-Moreno, A., et al. (2014). Differences between healthy hematopoietic progenitors and leukemia cells with respect to CD44 mediated rolling versus adherence behavior on hyaluronic acid coated surfaces. *Biomaterials, 35*, 1411–1419.

Harada, T., Ozaki, S., Oda, A., Fujii, S., Nakamura, S., Miki, H., et al. (2013). Association of Th1 and Th2 cytokines with transient inflammatory reaction during lenalidomide plus

dexamethasone therapy in multiple myeloma. *International Journal of Hematology, 97*, 743–748.
Hascall, V. C., Majors, A. K., De La Motte, C. A., Evanko, S. P., Wang, A., Drazba, J. A., et al. (2004). Intracellular hyaluronan: A new frontier for inflammation? *Biochimica et Biophysica Acta, 1673*, 3–12.
Hasselbalch, H. C. (2013). Chronic inflammation as a promotor of mutagenesis in essential thrombocythemia, polycythemia vera and myelofibrosis. A human inflammation model for cancer development? *Leukemia Research, 37*, 214–220.
He, H., Zhang, S., Tighe, S., Son, J., & Tseng, S. C. (2013). Immobilized heavy chain-hyaluronic acid polarizes lipopolysaccharide-activated macrophages toward M2 phenotype. *The Journal of Biological Chemistry, 288*, 25792–25803.
Hicsonmez, G. (2006). The effect of steroid on myeloid leukemic cells: The potential of short-course high-dose methylprednisolone treatment in inducing differentiation, apoptosis and in stimulating myelopoiesis. *Leukemia Research, 30*, 60–68.
Hunt, L. C., Gorman, C., Kintakas, C., McCulloch, D. R., Mackie, E. J., & White, J. D. (2013). Hyaluronan synthesis and myogenesis: A requirement for hyaluronan synthesis during myogenic differentiation independent of pericellular matrix formation. *The Journal of Biological Chemistry, 288*, 13006–13021.
Iijima, J., Konno, K., & Itano, N. (2011). Inflammatory alterations of the extracellular matrix in the tumor microenvironment. *Cancers (Basel), 3*, 3189–3205.
Ishdorj, G., Li, L., & Gibson, S. B. (2012). Regulation of autophagy in hematological malignancies: Role of reactive oxygen species. *Leukemia and Lymphoma, 53*, 26–33.
Itano, N., Sawai, T., Atsumi, F., Miyaishi, O., Taniguchi, S., Kannagi, R., et al. (2004). Selective expression and functional characteristics of three mammalian hyaluronan synthases in oncogenic malignant transformation. *The Journal of Biological Chemistry, 279*, 18679–18687.
Iversen, P. O., Berggreen, E., Nicolaysen, G., & Heyeraas, K. (2001). Regulation of extracellular volume and interstitial fluid pressure in rat bone marrow. *American Journal of Physiology. Heart and Circulatory Physiology, 280*, H1807–H1813.
Iversen, P. O., Nicolaysen, G., & Benestad, H. B. (1992). Blood flow to bone marrow during development of anemia or polycythemia in the rat. *Blood, 79*, 594–601.
Iversen, P. O., Nicolaysen, G., & Benestad, H. B. (1993). The leukopoietic cytokine granulocyte colony-stimulating factor increases blood flow to rat bone marrow. *Experimental Hematology, 21*, 231–235.
Iversen, P. O., Thing-Mortensen, B., Nicolaysen, G., & Benestad, H. B. (1993). Decreased blood flow to rat bone marrow, bone, spleen, and liver in acute leukemia. *Leukemia Research, 17*, 663–668.
Jackson, D. G. (2009). Immunological functions of hyaluronan and its receptors in the lymphatics. *Immunological Reviews, 230*, 216–231.
Jadin, L., Wu, X., Ding, H., Frost, G. I., Onclinx, C., Triggs-Raine, B., et al. (2008). Skeletal and hematological anomalies in HYAL2-deficient mice: A second type of mucopolysaccharidosis IX? *The FASEB Journal, 22*, 4316–4326.
Jin, L., Hope, K. J., Zhai, Q., Smadja-Joffe, F., & Dick, J. E. (2006). Targeting of CD44 eradicates human acute myeloid leukemic stem cells. *Nature Medicine, 12*, 1167–1174.
Jung, E. M., Kwon, O., Kwon, K. S., Cho, Y. S., Rhee, S. K., Min, J. K., et al. (2011). Evidences for correlation between the reduced VCAM-1 expression and hyaluronan synthesis during cellular senescence of human mesenchymal stem cells. *Biochemical and Biophysical Research Communications, 404*, 463–469.
Kang, Y. J., Yang, S. J., Park, G., Cho, B., Min, C. K., Kim, T. Y., et al. (2007). A novel function of interleukin-10 promoting self-renewal of hematopoietic stem cells. *Stem Cells, 25*, 1814–1822.

Karousou, E., Stachtea, X., Moretto, P., Viola, M., Vigetti, D., D'Angelo, M. L., et al. (2013). New insights into the pathobiology of Down syndrome–hyaluronan synthase-2 overexpression is regulated by collagen VI alpha2 chain. *The FEBS Journal, 280*, 2418–2430.

Kashimura, M. (1982). Scanning electron microscopy studies of bone marrow. *Scanning Electron Microscopy, Pt 1*, 445–453.

Kaur-Bollinger, P., Gotze, K. S., & Oostendorp, R. A. (2012). Role of secreted factors in the regulation of hematopoietic stem cells by the bone marrow microenvironment. *Frontiers in Bioscience (Landmark Edition), 17*, 876–891.

Kessler, S., Rho, H., West, G., Fiocchi, C., Drazba, J., & de la Motte, C. (2008). Hyaluronan (HA) deposition precedes and promotes leukocyte recruitment in intestinal inflammation. *Clinical and Translational Science, 1*, 57–61.

Khaldoyanidi, S., Denzel, A., & Zoller, M. (1996). Requirement for CD44 in proliferation and homing of hematopoietic precursor cells. *Journal of Leukocyte Biology, 60*, 579–592.

Khaldoyanidi, S., Karakhanova, S., Sleeman, J., Herrlich, P., & Ponta, H. (2002). CD44 variant-specific antibodies trigger hemopoiesis by selective release of cytokines from bone marrow macrophages. *Blood, 99*, 3955–3961.

Khaldoyanidi, S., Moll, J., Karakhanova, S., Herrlich, P., & Ponta, H. (1999). Hyaluronate-enhanced hematopoiesis: Two different receptors trigger the release of interleukin-1beta and interleukin-6 from bone marrow macrophages. *Blood, 94*, 940–949.

Khaldoyanidi, S., Schnabel, D., Fohr, N., & Zoller, M. (1997). Functional activity of CD44 isoforms in haemopoiesis of the rat. *British Journal of Haematology, 96*, 31–45.

Khaldoyanidi, S., Sikora, L., Orlovskaya, I., Matrosova, V., Kozlov, V., & Sriramarao, P. (2001). Correlation between nicotine-induced inhibition of hematopoiesis and decreased CD44 expression on bone marrow stromal cells. *Blood, 98*, 303–312.

Kim, I. L., Mauck, R. L., & Burdick, J. A. (2011). Hydrogel design for cartilage tissue engineering: A case study with hyaluronic acid. *Biomaterials, 32*, 8771–8782.

Kim, C. H., You, L., Yellowley, C. E., & Jacobs, C. R. (2006). Oscillatory fluid flow-induced shear stress decreases osteoclastogenesis through RANKL and OPG signaling. *Bone, 39*, 1043–1047.

Knippenberg, M., Helder, M. N., Doulabi, B. Z., Semeins, C. M., Wuisman, P. I., & Klein-Nulend, J. (2005). Adipose tissue-derived mesenchymal stem cells acquire bone cell-like responsiveness to fluid shear stress on osteogenic stimulation. *Tissue Engineering, 11*, 1780–1788.

Knudson, C. B., & Knudson, W. (1993). Hyaluronan-binding proteins in development, tissue homeostasis, and disease. *The FASEB Journal, 7*, 1233–1241.

Kolf, C. M., Cho, E., & Tuan, R. S. (2007). Mesenchymal stromal cells. Biology of adult mesenchymal stem cells: Regulation of niche, self-renewal and differentiation. *Arthritis Research and Therapy, 9*, 204.

Koyama, H., Kobayashi, N., Harada, M., Takeoka, M., Kawai, Y., Sano, K., et al. (2008). Significance of tumor-associated stroma in promotion of intratumoral lymphangiogenesis: Pivotal role of a hyaluronan-rich tumor microenvironment. *The American Journal of Pathology, 172*, 179–193.

Krause, D. S., Fulzele, K., Catic, A., Sun, C. C., Dombkowski, D., Hurley, M. P., et al. (2013). Differential regulation of myeloid leukemias by the bone marrow microenvironment. *Nature Medicine, 19*, 1513–1517.

Krause, D. S., Lazarides, K., von Andrian, U. H., & Van Etten, R. A. (2006). Requirement for CD44 in homing and engraftment of BCR-ABL-expressing leukemic stem cells. *Nature Medicine, 12*, 1175–1180.

Kuang, D. M., Wu, Y., Chen, N., Cheng, J., Zhuang, S. M., & Zheng, L. (2007). Tumor-derived hyaluronan induces formation of immunosuppressive macrophages through transient early activation of monocytes. *Blood, 110*, 587–595.

Laurent, T. C., Laurent, U. B., & Fraser, J. R. (1996a). Serum hyaluronan as a disease marker. *Annals of Medicine, 28*, 241–253.
Laurent, T. C., Laurent, U. B., & Fraser, J. R. (1996b). The structure and function of hyaluronan: An overview. *Immunology and Cell Biology, 74*, A1–A7.
Levi, B. P., & Morrison, S. J. (2008). Stem cells use distinct self-renewal programs at different ages. *Cold Spring Harbor Symposia on Quantitative Biology, 73*, 539–553.
Li, L., Asteriou, T., Bernert, B., Heldin, C. H., & Heldin, P. (2007). Growth factor regulation of hyaluronan synthesis and degradation in human dermal fibroblasts: Importance of hyaluronan for the mitogenic response of PDGF-BB. *The Biochemical Journal, 404*, 327–336.
Li, Y., Jiang, D., Liang, J., Meltzer, E. B., Gray, A., Miura, R., et al. (2011). Severe lung fibrosis requires an invasive fibroblast phenotype regulated by hyaluronan and CD44. *The Journal of Experimental Medicine, 208*, 1459–1471.
Lichtman, M. A. (1981). The ultrastructure of the hemopoietic environment of the marrow: A review. *Experimental Hematology, 9*, 391–410.
Lindqvist, U., Engstrom-Laurent, A., Laurent, U., Nyberg, A., Bjorklund, U., Eriksson, H., et al. (1988). The diurnal variation of serum hyaluronan in health and disease. *Scandinavian Journal of Clinical and Laboratory Investigation, 48*, 765–770.
Lindqvist, U., & Laurent, T. C. (1992). Serum hyaluronan and aminoterminal propeptide of type III procollagen: Variation with age. *Scandinavian Journal of Clinical and Laboratory Investigation, 52*, 613–621.
Maroski, J., Vorderwulbecke, B. J., Fiedorowicz, K., Da Silva-Azevedo, L., Siegel, G., Marki, A., et al. (2011). Shear stress increases endothelial hyaluronan synthase 2 and hyaluronan synthesis especially in regard to an atheroprotective flow profile. *Experimental Physiology, 96*, 977–986.
Mathews, S., Mathew, S. A., Gupta, P. K., Bhonde, R., & Totey, S. (2012). Glycosaminoglycans enhance osteoblast differentiation of bone marrow derived human mesenchymal stem cells. *Journal of Tissue Engineering and Regenerative Medicine, 8*(2), 143–152.
Matou-Nasri, S., Gaffney, J., Kumar, S., & Slevin, M. (2009). Oligosaccharides of hyaluronan induce angiogenesis through distinct CD44 and RHAMM-mediated signalling pathways involving Cdc2 and gamma-adducin. *International Journal of Oncology, 35*, 761–773.
Matrosova, V. Y., Orlovskaya, I. A., Serobyan, N., & Khaldoyanidi, S. K. (2004). Hyaluronic acid facilitates the recovery of hematopoiesis following 5-fluorouracil administration. *Stem Cells, 22*, 544–555.
Matrosova, V., Orlovskaya, I., Serobyan, N., McClelland, M., & Khaldoyanidi, S. (2005). Hyaluronan facilitates recovery of hematopoiesis after therapy-induced myelosuppression. In E. A. Balazs, & V. C. Hascall (Eds.), *Hyaluronan, structure, metabolism, biological activities, therapeutic applications: Vol. 2.* (pp. 813–820). Edgewater, NJ 07020 (USA): Matrix Biology Institute.
Matsumoto, K., Li, Y., Jakuba, C., Sugiyama, Y., Sayo, T., Okuno, M., et al. (2009). Conditional inactivation of Has2 reveals a crucial role for hyaluronan in skeletal growth, patterning, chondrocyte maturation and joint formation in the developing limb. *Development, 136*, 2825–2835.
Mendez-Ferrer, S., Lucas, D., Battista, M., & Frenette, P. S. (2008). Haematopoietic stem cell release is regulated by circadian oscillations. *Nature, 452*, 442–447.
Meran, S., Luo, D. D., Simpson, R., Martin, J., Wells, A., Steadman, R., et al. (2011). Hyaluronan facilitates transforming growth factor-beta1-dependent proliferation via CD44 and epidermal growth factor receptor interaction. *The Journal of Biological Chemistry, 286*, 17618–17630.
Meran, S., Thomas, D., Stephens, P., Martin, J., Bowen, T., Phillips, A., et al. (2007). Involvement of hyaluronan in regulation of fibroblast phenotype. *The Journal of Biological Chemistry, 282*, 25687–25697.

Midura, R. J., Su, X., Morcuende, J. A., Tammi, M., & Tammi, R. (2003). Parathyroid hormone rapidly stimulates hyaluronan synthesis by periosteal osteoblasts in the tibial diaphysis of the growing rat. *The Journal of Biological Chemistry*, *278*, 51462–51468.

Misra, S., Ghatak, S., Zoltan-Jones, A., & Toole, B. P. (2003). Regulation of multidrug resistance in cancer cells by hyaluronan. *The Journal of Biological Chemistry*, *278*, 25285–25288.

Miyake, K., Medina, K. L., Hayashi, S., Ono, S., Hamaoka, T., & Kincade, P. W. (1990). Monoclonal antibodies to Pgp-1/CD44 block lympho-hemopoiesis in long-term bone marrow cultures. *The Journal of Experimental Medicine*, *171*, 477–488.

Mo, W., Yang, C., Liu, Y., He, Y., Wang, Y., & Gao, F. (2011). The influence of hyaluronic acid on vascular endothelial cell proliferation and the relationship with ezrin/merlin expression. *Acta Biochimica et Biophysica Sinica (Shanghai)*, *43*, 930–939.

Mohand-Kaci, F., Assoul, N., Martelly, I., Allaire, E., & Zidi, M. (2013). Optimized hyaluronic acid-hydrogel design and culture conditions for preservation of mesenchymal stem cell properties. *Tissue Engineering Part C: Methods*, *19*, 288–298.

Mortensen, B. T., Jensen, P. O., Helledie, N., Iversen, P. O., Ralfkiaer, E., Larsen, J. K., et al. (1998). Changing bone marrow micro-environment during development of acute myeloid leukaemia in rats. *British Journal of Haematology*, *102*, 458–464.

Nagy, N., Freudenberger, T., Melchior-Becker, A., Rock, K., Ter Braak, M., Jastrow, H., et al. (2010). Inhibition of hyaluronan synthesis accelerates murine atherosclerosis: Novel insights into the role of hyaluronan synthesis. *Circulation*, *122*, 2313–2322.

Nieuwdorp, M., Holleman, F., de Groot, E., Vink, H., Gort, J., Kontush, A., et al. (2007). Perturbation of hyaluronan metabolism predisposes patients with type 1 diabetes mellitus to atherosclerosis. *Diabetologia*, *50*, 1288–1293.

Ohno, T., Yoo, M. J., Swanson, E. R., Hirano, S., Ossoff, R. H., & Rousseau, B. (2009). Regenerative effects of basic fibroblast growth factor on extracellular matrix production in aged rat vocal folds. *The Annals of Otology, Rhinology, and Laryngology*, *118*, 559–564.

Olsson, M., Meadows, J. R., Truve, K., Rosengren Pielberg, G., Puppo, F., Mauceli, E., et al. (2011). A novel unstable duplication upstream of HAS2 predisposes to a breed-defining skin phenotype and a periodic fever syndrome in Chinese Shar-Pei dogs. *PLoS Genetics*, *7*, e1001332.

Onoda, M., Nakaseko, C., Yokota, A., & Saito, Y. (2009). Ligation of CD44 with low-molecular-weight hyaluronan and a monoclonal antibody leads to inhibition of drug-induced apoptosis in a human myeloid cell line. *Hematology*, *14*, 213–219.

Osterholt, H. C., Dannevig, I., Wyckoff, M. H., Liao, J., Akgul, Y., Ramgopal, M., et al. (2012). Antioxidant protects against increases in low molecular weight hyaluronan and inflammation in asphyxiated newborn pigs resuscitated with 100% oxygen. *PLoS One*, *7*, e38839.

Pahakis, M. Y., Kosky, J. R., Dull, R. O., & Tarbell, J. M. (2007). The role of endothelial glycocalyx components in mechanotransduction of fluid shear stress. *Biochemical and Biophysical Research Communications*, *355*, 228–233.

Papakonstantinou, E., Klagas, I., Karakiulakis, G., Hostettler, K., S'ng, C. T., Kotoula, V., et al. (2012). Steroids and beta2-agonists regulate hyaluronan metabolism in asthmatic airway smooth muscle cells. *American Journal of Respiratory Cell and Molecular Biology*, *47*, 759–767.

Park, D., Kim, Y., Kim, H., Kim, K., Lee, Y. S., Choe, J., et al. (2012). Hyaluronic acid promotes angiogenesis by inducing RHAMM-TGFbeta receptor interaction via CD44-PKCdelta. *Molecules and Cells*, *33*, 563–574.

Park, D., Sykes, D. B., & Scadden, D. T. (2012). The hematopoietic stem cell niche. *Frontiers in Bioscience*, *17*, 30–39.

Pasonen-Seppanen, S., Takabe, P., Edward, M., Rauhala, L., Rilla, K., Tammi, M., et al. (2012). Melanoma cell-derived factors stimulate hyaluronan synthesis in dermal fibroblasts by upregulating HAS2 through PDGFR-PI3K-AKT and p38 signaling. *Histochemistry and Cell Biology*, *138*, 895–911.

Phinney, D. G., & Prockop, D. J. (2007). Concise review: Mesenchymal stem/multipotent stromal cells: The state of transdifferentiation and modes of tissue repair—Current views. *Stem Cells*, *25*, 2896–2902.

Pilarski, L. M., Szczepek, A. J., & Belch, A. R. (1997). Deficient drug transporter function of bone marrow-localized and leukemic plasma cells in multiple myeloma. *Blood*, *90*, 3751–3759.

Porsch, H., Bernert, B., Mehic, M., Theocharis, A. D., Heldin, C. H., & Heldin, P. (2013). Efficient TGFbeta-induced epithelial-mesenchymal transition depends on hyaluronan synthase HAS2. *Oncogene*, *32*, 4355–4365.

Powell, J. D., & Horton, M. R. (2005). Threat matrix: Low-molecular-weight hyaluronan (HA) as a danger signal. *Immunologic Research*, *31*, 207–218.

Preston, M., & Sherman, L. S. (2011). Neural stem cell niches: Roles for the hyaluronan-based extracellular matrix. *Frontiers in Bioscience (Scholar Edition)*, *3*, 1165–1179.

Qazi, H., Palomino, R., Shi, Z. D., Munn, L. L., & Tarbell, J. M. (2013). Cancer cell glycocalyx mediates mechanotransduction and flow-regulated invasion. *Integrative Biology (Cambridge)*, *5*, 1334–1343.

Qazi, H., Shi, Z. D., & Tarbell, J. M. (2011). Fluid shear stress regulates the invasive potential of glioma cells via modulation of migratory activity and matrix metalloproteinase expression. *PLoS One*, *6*, e20348.

Qian, H., Xia, L., Ling, P., Waxman, S., & Jing, Y. (2012). CD44 ligation with A3D8 antibody induces apoptosis in acute myeloid leukemia cells through binding to CD44s and clustering lipid rafts. *Cancer Biology and Therapy*, *13*, 1276–1283.

Qin, Z., Dai, L., Bratoeva, M., Slomiany, M. G., Toole, B. P., & Parsons, C. (2011). Cooperative roles for emmprin and LYVE-1 in the regulation of chemoresistance for primary effusion lymphoma. *Leukemia*, *25*, 1598–1609.

Raines, A. L., Sunwoo, M., Gertzman, A. A., Thacker, K., Guldberg, R. E., Schwartz, Z., et al. (2011). Hyaluronic acid stimulates neovascularization during the regeneration of bone marrow after ablation. *Journal of Biomedical Materials Research Part A*, *96*, 575–583.

Ramsden, C. A., Bankier, A., Brown, T. J., Cowen, P. S., Frost, G. I., McCallum, D. D., et al. (2000). A new disorder of hyaluronan metabolism associated with generalized folding and thickening of the skin. *The Journal of Pediatrics*, *136*, 62–68.

Richter, U., Wicklein, D., Geleff, S., & Schumacher, U. (2012). The interaction between CD44 on tumour cells and hyaluronan under physiologic flow conditions: Implications for metastasis formation. *Histochemistry and Cell Biology*, *137*, 687–695.

Riddle, R. C., Taylor, A. F., Genetos, D. C., & Donahue, H. J. (2006). MAP kinase and calcium signaling mediate fluid flow-induced human mesenchymal stem cell proliferation. *American Journal of Physiology. Cell Physiology*, *290*, C776–C784.

Roberts, R., Gallagher, J., Spooncer, E., Allen, T. D., Bloomfield, F., & Dexter, T. M. (1988). Heparan sulphate bound growth factors: A mechanism for stromal cell mediated haemopoiesis. *Nature*, *332*, 376–378.

Ruffell, B., Poon, G. F., Lee, S. S., Brown, K. L., Tjew, S. L., Cooper, J., et al. (2011). Differential use of chondroitin sulfate to regulate hyaluronan binding by receptor CD44 in inflammatory and interleukin 4-activated macrophages. *The Journal of Biological Chemistry*, *286*, 19179–19190.

Sackstein, R. (2012). Glycoengineering of HCELL, the human bone marrow homing receptor: Sweetly programming cell migration. *Annals of Biomedical Engineering*, *40*, 766–776.

Sackstein, R., Merzaban, J. S., Cain, D. W., Dagia, N. M., Spencer, J. A., Lin, C. P., et al. (2008). Ex vivo glycan engineering of CD44 programs human multipotent mesenchymal stromal cell trafficking to bone. *Nature Medicine*, *14*, 181–187.

Scadden, D. T. (2006). The stem-cell niche as an entity of action. *Nature*, *441*, 1075–1079.

Schraufstatter, I. U., Discipio, R. G., & Khaldoyanidi, S. (2011). Mesenchymal stem cells and their microenvironment. *Frontiers in Bioscience, 17*, 2271–2288.

Schurch, C. M., Riether, C., & Ochsenbein, A. F. (2013). Interferons in hematopoiesis and leukemia. *Oncoimmunology, 2*, e24572.

Seita, J., & Weissman, I. L. (2010). Hematopoietic stem cell: Self-renewal versus differentiation. *Wiley Interdisciplinary Reviews: Systems Biology and Medicine, 2*, 640–653.

Semenov, O. V., Malek, A., Bittermann, A. G., Voros, J., & Zisch, A. H. (2009). Engineered polyelectrolyte multilayer substrates for adhesion, proliferation, and differentiation of human mesenchymal stem cells. *Tissue Engineering Part A, 15*, 2977–2990.

Serobyan, N., Jagannathan, S., Orlovskaya, I., Schraufstatter, I., Skok, M., Loring, J., et al. (2007). The cholinergic system is involved in regulation of the development of the hematopoietic system. *Life Sciences, 80*, 2352–2360.

Serra Moreno, J., Sabbieti, M. G., Agas, D., Marchetti, L., & Panero, S. (2012). Polysaccharides immobilized in polypyrrole matrices are able to induce osteogenic differentiation in mouse mesenchymal stem cells. *Journal of Tissue Engineering and Regenerative Medicine*. http://dx.doi.org/10.1002/term.1601, wileyonlinelibrary.com.

Shay, E., He, H., Sakurai, S., & Tseng, S. C. (2011). Inhibition of angiogenesis by HC.HA, a complex of hyaluronan and the heavy chain of inter-alpha-inhibitor, purified from human amniotic membrane. *Investigative Ophthalmology & Visual Science, 52*, 2669–2678.

Sherman, L., Sleeman, J., Herrlich, P., & Ponta, H. (1994). Hyaluronate receptors: Key players in growth, differentiation, migration and tumor progression. *Current Opinion in Cell Biology, 6*, 726–733.

Sikora, L., & Khaldoyanidi, S. (2010). Visualization of the hematopoietic microenvironment: An alternative approach using the dorsal skinfold chamber model. *Journal of Stem Cells, 5*, 113–116.

Simpson, R. M., Meran, S., Thomas, D., Stephens, P., Bowen, T., Steadman, R., et al. (2009). Age-related changes in pericellular hyaluronan organization leads to impaired dermal fibroblast to myofibroblast differentiation. *The American Journal of Pathology, 175*, 1915–1928.

Singleton, P. A., Mirzapoiazova, T., Guo, Y., Sammani, S., Mambetsariev, N., Lennon, F. E., et al. (2010). High-molecular-weight hyaluronan is a novel inhibitor of pulmonary vascular leakiness. *American Journal of Physiology. Lung Cellular and Molecular Physiology, 299*, L639–L651.

Singleton, P. A., Salgia, R., Moreno-Vinasco, L., Moitra, J., Sammani, S., Mirzapoiazova, T., et al. (2007). CD44 regulates hepatocyte growth factor-mediated vascular integrity. Role of c-Met, Tiam1/Rac1, dynamin 2, and cortactin. *The Journal of Biological Chemistry, 282*, 30643–30657.

Sipkins, D. A., Wei, X., Wu, J. W., Runnels, J. M., Cote, D., Means, T. K., et al. (2005). In vivo imaging of specialized bone marrow endothelial microdomains for tumour engraftment. *Nature, 435*, 969–973.

Sironen, R. K., Tammi, M., Tammi, R., Auvinen, P. K., Anttila, M., & Kosma, V. M. (2011). Hyaluronan in human malignancies. *Experimental Cell Research, 317*, 383–391.

Sleeman, J., Rudy, W., Hofmann, M., Moll, J., Herrlich, P., & Ponta, H. (1996). Regulated clustering of variant CD44 proteins increases their hyaluronate binding capacity. *The Journal of Cell Biology, 135*, 1139–1150.

Snauwaert, S., Vanhee, S., Goetgeluk, G., Verstichel, G., Van Caeneghem, Y., Velghe, I., et al. (2012). RHAMM/HMMR (CD168) is not an ideal target antigen for immunotherapy of acute myeloid leukemia. *Haematologica, 97*, 1539–1547.

Sokolowska, M., Chen, L. Y., Eberlein, M., Martinez-Anton, A., Liu, Y., Alsaaty, S., et al. (2013). Low molecular weight hyaluronan activates cytosolic phospholipase A2alpha and eicosanoid production in monocytes and macrophages. *The Journal of Biological Chemistry, 289*(7), 4470–4488.

Stern, R. (2008). Hyaluronidases in cancer biology. *Seminars in Cancer Biology, 18*, 275–280.

Stier, S., Ko, Y., Forkert, R., Lutz, C., Neuhaus, T., Grunewald, E., et al. (2005). Osteopontin is a hematopoietic stem cell niche component that negatively regulates stem cell pool size. *The Journal of Experimental Medicine, 201*, 1781–1791.

Suda, T., Takahashi, N., Udagawa, N., Jimi, E., Gillespie, M. T., & Martin, T. J. (1999). Modulation of osteoclast differentiation and function by the new members of the tumor necrosis factor receptor and ligand families. *Endocrine Reviews, 20*, 345–357.

Sundstrom, G., Dahl, I. M., Hultdin, M., Lundstrom, B., Wahlin, A., & Engstrom-Laurent, A. (2005). Bone marrow hyaluronan distribution in patients with acute myeloid leukemia. *Medical Oncology, 22*, 71–78.

Sundstrom, G., Hultdin, M., Engstrom-Laurent, A., & Dahl, I. M. (2010). Bone marrow hyaluronan and reticulin in patients with malignant disorders. *Medical Oncology, 27*, 618–623.

Szczepanek, K., Kieda, C., & Cichy, J. (2008). Differential binding of hyaluronan on the surface of tissue-specific endothelial cell lines. *Acta Biochimica Polonica, 55*, 35–42.

Taichman, R. S., & Emerson, S. G. (1994). Human osteoblasts support hematopoiesis through the production of granulocyte colony-stimulating factor. *The Journal of Experimental Medicine, 179*, 1677–1682.

Taichman, R. S., & Emerson, S. G. (1998). The role of osteoblasts in the hematopoietic microenvironment. *Stem Cells, 16*, 7–15.

Tang, Z. C., Liao, W. Y., Tang, A. C., Tsai, S. J., & Hsieh, P. C. (2011). The enhancement of endothelial cell therapy for angiogenesis in hindlimb ischemia using hyaluronan. *Biomaterials, 32*, 75–86.

Tavassoli, M., & Shaklai, M. (1979). Absence of tight junctions in endothelium of marrow sinuses: Possible significance for marrow cell egress. *British Journal of Haematology, 41*, 303–307.

Tzankov, A., Strasser, U., Dirnhofer, S., Menter, T., Arber, C., Jotterand, M., et al. (2011). In situ RHAMM protein expression in acute myeloid leukemia blasts suggests poor overall survival. *Annals of Hematology, 90*, 901–909.

Urao, N., & Ushio-Fukai, M. (2013). Redox regulation of stem/progenitor cells and bone marrow niche. *Free Radical Biology and Medicine, 54*, 26–39.

Vigetti, D., Clerici, M., Deleonibus, S., Karousou, E., Viola, M., Moretto, P., et al. (2011). Hyaluronan synthesis is inhibited by adenosine monophosphate-activated protein kinase through the regulation of HAS2 activity in human aortic smooth muscle cells. *The Journal of Biological Chemistry, 286*, 7917–7924.

Vigetti, D., Genasetti, A., Karousou, E., Viola, M., Moretto, P., Clerici, M., et al. (2010). Proinflammatory cytokines induce hyaluronan synthesis and monocyte adhesion in human endothelial cells through hyaluronan synthase 2 (HAS2) and the nuclear factor-kappaB (NF-kappaB) pathway. *The Journal of Biological Chemistry, 285*, 24639–24645.

Vincent, T., Jourdan, M., Sy, M. S., Klein, B., & Mechti, N. (2001). Hyaluronic acid induces survival and proliferation of human myeloma cells through an interleukin-6-mediated pathway involving the phosphorylation of retinoblastoma protein. *The Journal of Biological Chemistry, 276*, 14728–14736.

Vincent, T., Molina, L., Espert, L., & Mechti, N. (2003). Hyaluronan, a major non-protein glycosaminoglycan component of the extracellular matrix in human bone marrow, mediates dexamethasone resistance in multiple myeloma. *British Journal of Haematology, 121*, 259–269.

Wallet, M. A., Wallet, S. M., Guiulfo, G., Sleasman, J. W., & Goodenow, M. M. (2010). IFNgamma primes macrophages for inflammatory activation by high molecular weight hyaluronan. *Cellular Immunology, 262*, 84–88.

Wang, Y. Z., Cao, M. L., Liu, Y. W., He, Y. Q., Yang, C. X., & Gao, F. (2011). CD44 mediates oligosaccharides of hyaluronan-induced proliferation, tube formation and signal transduction in endothelial cells. *Experimental Biology and Medicine (Maywood, N.J.)*, *236*, 84–90.

Watson, A. S., Mortensen, M., & Simon, A. K. (2011). Autophagy in the pathogenesis of myelodysplastic syndrome and acute myeloid leukemia. *Cell Cycle*, *10*, 1719–1725.

Weigel, P. H., & DeAngelis, P. L. (2007). Hyaluronan synthases: A decade-plus of novel glycosyltransferases. *The Journal of Biological Chemistry*, *282*, 36777–36781.

Weiss, L. (1980). The haemopoietic microenvironment of bone marrow: An ultrastructural study of the interactions of blood cells, stroma and blood vessels. *Ciba Foundation Symposium*, *71*, 3–19.

Weiss, L. (1981). Haemopoiesis in mammalian bone marrow. *Ciba Foundation Symposium*, *84*, 5–21.

West, D. C., & Kumar, S. (1989). Hyaluronan and angiogenesis. *Ciba Foundation Symposium*, *143*, 187–201 discussion 201-7, 281-5.

White, R. L., Nash, G., Kavanagh, D. P., Savage, C. O., & Kalia, N. (2013). Modulating the adhesion of haematopoietic stem cells with chemokines to enhance their recruitment to the ischaemically injured murine kidney. *PLoS One*, *8*, e66489.

Willenberg, A., Saalbach, A., Simon, J. C., & Anderegg, U. (2012). Melanoma cells control HA synthesis in peritumoral fibroblasts via PDGF-AA and PDGF-CC: Impact on melanoma cell proliferation. *The Journal of Investigative Dermatology*, *132*, 385–393.

Winkler, C. W., Foster, S. C., Itakura, A., Matsumoto, S. G., Asari, A., McCarty, O. J., et al. (2013). Hyaluronan oligosaccharides perturb lymphocyte slow rolling on brain vascular endothelial cells: Implications for inflammatory demyelinating disease. *Matrix Biology*, *32*, 160–168.

Winkler, C. W., Foster, S. C., Matsumoto, S. G., Preston, M. A., Xing, R., Bebo, B. F., et al. (2012). Hyaluronan anchored to activated CD44 on central nervous system vascular endothelial cells promotes lymphocyte extravasation in experimental autoimmune encephalomyelitis. *The Journal of Biological Chemistry*, *287*, 33237–33251.

Wisse, E. (1970). An electron microscopic study of the fenestrated endothelial lining of rat liver sinusoids. *Journal of Ultrastructure Research*, *31*, 125–150.

Yasuda, T. (2010). Hyaluronan inhibits prostaglandin E2 production via CD44 in U937 human macrophages. *The Tohoku Journal of Experimental Medicine*, *220*, 229–235.

Yasuda, T. (2011). Hyaluronan inhibits Akt, leading to nuclear factor-kappaB down-regulation in lipopolysaccharide-stimulated U937 macrophages. *Journal of Pharmacological Sciences*, *115*, 509–515.

Zanna, G., Docampo, M. J., Fondevila, D., Bardagi, M., Bassols, A., & Ferrer, L. (2009). Hereditary cutaneous mucinosis in shar pei dogs is associated with increased hyaluronan synthase-2 mRNA transcription by cultured dermal fibroblasts. *Veterinary Dermatology*, *20*, 377–382.

Zavan, B., Giorgi, C., Bagnara, G. P., Vindigni, V., Abatangelo, G., & Cortivo, R. (2007). Osteogenic and chondrogenic differentiation: Comparison of human and rat bone marrow mesenchymal stem cells cultured into polymeric scaffolds. *European Journal of Histochemistry*, *51*(Suppl. 1), 1–8.

Zhu, H., Mitsuhashi, N., Klein, A., Barsky, L. W., Weinberg, K., Barr, M. L., et al. (2006). The role of the hyaluronan receptor CD44 in mesenchymal stem cell migration in the extracellular matrix. *Stem Cells*, *24*, 928–935.

Zweegman, S., Kessler, F. L., Kerkhoven, R. M., Heimerikx, M., Celie, J. W., Janssen, J. J., et al. (2007). Reduced supportive capacity of bone marrow stroma upon chemotherapy is mediated via changes in glycosaminoglycan profile. *Matrix Biology*, *26*, 561–571.

CHAPTER SEVEN

Hyaluronan Regulation of Endothelial Barrier Function in Cancer

Patrick A. Singleton*,†,1
*Department of Medicine, Section of Pulmonary and Critical Care, Chicago, Illinois, USA
†Department of Anesthesia and Critical Care, The University of Chicago, Chicago, Illinois, USA
[1]Corresponding author: e-mail address: psinglet@medicine.bsd.uchicago.edu

Contents

1. Introduction 192
2. HA Regulation of Vascular Integrity 194
3. HA Regulation of Endothelial Barrier Function During Tumor Angiogenesis 197
4. HA Regulation of Endothelial Barrier Function During Cancer Metastasis 198
5. Potential Therapeutic Effects of HMW-HA in Inhibiting Endothelial Barrier Disruption During Cancer Progression 200
6. Conclusions 201
Acknowledgments 203
References 203

Abstract

Vascular integrity or the maintenance of blood vessel continuity is a fundamental process regulated by endothelial cell–cell junctions. Defects in endothelial barrier function are an initiating factor in several disease processes including tumor angiogenesis and metastasis. The glycosaminoglycan, hyaluronan (HA), maintains vascular integrity through specific mechanisms including HA-binding protein signaling in caveolin-enriched microdomains, a subset of lipid rafts. Certain disease states, including cancer, increase enzymatic hyaluronidase activity and reactive oxygen species generation, which break down high molecular weight HA (HMW-HA) to low molecular weight fragments (LMW-HA). LMW-HA can activate specific HA-binding proteins during tumor progression to promote disruption of endothelial cell–cell contacts. In contrast, exogenous administration of HMW-HA promotes enhancement of vascular integrity. This review focuses on the roles of HA in regulating angiogenic and metastatic processes based on its size and the HA-binding proteins present. Further, potential therapeutic applications of HMW-HA in treating cancer are discussed.

Advances in Cancer Research, Volume 123
ISSN 0065-230X
http://dx.doi.org/10.1016/B978-0-12-800092-2.00007-1

1. INTRODUCTION

The vascular endothelium, which lines the inner surface of blood vessels, acts as a selectively permeable barrier to regulate the movement of liquid and solutes between blood and the surrounding tissue, particularly in the microvasculature (Curry, 2005; Dejana, Tournier-Lasserve, & Weinstein, 2009; Vandenbroucke, Mehta, Minshall, & Malik, 2008). The semipermeable nature of the endothelium allows plasma fluid, nutrients, and even cells to move out of the blood and into the tissues, while metabolic products may be taken up by the circulation. This exchange between the blood and the tissues is vital for organ function and tissue viability by maintaining fluid and metabolic homeostasis. Vascular barrier function is dependent on the integrity of the endothelial cell (EC) layer (Lennon & Singleton, 2011a, 2011b). Several mechanisms regulate basal vascular integrity including the endothelial glycocalyx, a meshwork of hyaluronan (HA), proteoglycans, glycolipids, and proteins between the vascular luminal space and the EC surface, endothelial cell–cell junctions which are controlled by tight junctions, adherens junctions, and caveolin-enriched microdomains (CEM) (Lennon & Singleton, 2011a, 2011b). Disruption or dysregulation of the endothelial layer can lead to altered permeability resulting in leakage of fluid, solutes and proteins from the blood into the underlying tissue resulting in edema. Dysregulation of endothelial barrier function can occur in a wide range of human pathologies including tumor angiogenesis and metastasis (Pardue, Ibrahim, & Ramamurthi, 2008; Reymond, d'Agua, & Ridley, 2013; Singleton & Bourguignon, 2002, 2004; Slevin et al., 2007).

The major nonsulfated glycosaminoglycan in most tissues, HA, plays a fundamental role in the maintenance of vascular integrity (Singleton, Dudek, Ma, & Garcia, 2006; Singleton et al., 2010). HA is composed of a linear repeat of disaccharide units consisting of D-glucuronic acid and *N*-acetylglucosamine (Almond, 2007; Cantor, 2007; Cantor & Nadkarni, 2006; Fraser, Laurent, & Laurent, 1997; Gaffney, Matou-Nasri, Grau-Olivares, & Slevin, 2010; Girish & Kemparaju, 2007; Jiang, Liang, & Noble, 2007, 2011; Liao, Jones, Forbes, Martin, & Brown, 2005; Olczyk, Komosinska-Vassev, Winsz-Szczotka, Kuznik-Trocha, & Olczyk, 2008; Scott & Heatley, 2002; Singleton et al., 2010; Tammi, Day, & Turley, 2002; Toole, 2004; Wang, de la Motte, Lauer, & Hascall, 2011). The major form of HA *in vivo*, high molecular weight HA

(HMW-HA), has a molecular weight >1 million Da. HMW-HA exhibits a random coil structure that can expand in aqueous solutions (Scott & Heatley, 2002). Aqueous HMW-HA is highly viscous and elastic, properties which contribute to its filtering functions in the glycocalyx (Furlan, La Penna, Perico, & Cesaro, 2005; Henry & Duling, 1999). HA is a dynamic molecule with a high rate of metabolism. In humans, up to one third of total HA is turned over in the body everyday (Fraser et al., 1997). The majority of HA within the vasculature is incorporated into the endothelial glycocalyx and the extracellular matrix of the underlying tissue (Nandi, Estess, & Siegelman, 2000; Olczyk et al., 2008; Wheeler-Jones, Farrar, & Pitsillides, 2010). The levels of soluble HA are low in normal human plasma due to rapid removal by the liver and kidneys (Fraser et al., 1997).

In several types of cancer, there is enhanced production of HA in the tumor stroma, which correlates with angiogenesis, metastasis, and poor patient outcome (Sironen et al., 2011; Tammi et al., 2008). HA is synthesized by hyaluronan synthases (HAS) (Weigel & DeAngelis, 2007). The three main HAS (HAS1, HAS2, and HAS3) differ in the Km values for their substrates (D-glucuronic acid and *N*-acetylglucosamine) leading to differential rates of hyaluronan synthesis and secretion from the plasma membrane (Weigel & DeAngelis, 2007). HAS1 and HAS2 produce HA with a molecular weight >500 kDa and HAS3 produces <500 kDa HA (Itano & Kimata, 2002). HAS2 upregulation is correlated with breast cancer invasion (Bernert, Porsch, & Heldin, 2011). Inhibition of HA synthesis with 4-methylumbelliferone has been shown to promote growth arrest, apoptosis, and metastasis in tumor cells (Arai et al., 2011; Futamura et al., 2013; Saito, Dai, & Asano, 2013). HA is degraded in cancer by hyaluronidases and reactive oxygen species (ROS) to produce lower molecular weight fragments (<500 kDa) (Girish & Kemparaju, 2007). HA fragments in the ~10–15 disaccharide range promote angiogenesis and have been observed in several cancer types. There are six hyaluronidase genes encoding HYAL-1,2,3,4, PHYAL1 (a pseudogene), and PH-20 (Csoka, Frost, & Stern, 2001; Stern, 2003; Stern, Asari, & Sugahara, 2006; Stern, Kogan, Jedrzejas, & Soltes, 2007). HYAL enzymes have different cellular localization and optimal pH activity which can lead to generation of different sized HA fragments (Bourguignon, Singleton, Diedrich, Stern, & Gilad, 2004; Csoka et al., 2001; Stern, 2003, 2008; Stern et al., 2006, 2007). Although the correlation of specific hyaluronidases with tumor progression varies, several aggressive cancer types have upregulated Hyal-1 and/or Hyal-2 expression

(Tan, Wang, Li, et al., 2011; Tan, Wang, Su, et al., 2011; Udabage, Brownlee, Nilsson, & Brown, 2005). The differential mechanisms of HA's regulation of vascular integrity during cancer are discussed below.

2. HA REGULATION OF VASCULAR INTEGRITY

Vascular integrity or the maintenance of blood vessel continuity is a fundamental process in which HA plays numerous roles including regulation of endothelial cell–cell junctions (Lennon & Singleton, 2011a, 2011b; Singleton & Lennon, 2011). Our previous published data indicate that HMW-HA (~1 million Da) enhances vascular integrity both *in vitro* and *in vivo* (Singleton et al., 2006, 2007, 2010). Using a murine model of LPS-induced acute lung injury with pulmonary vascular hyperpermeability, we observed that intravenous administration of HMW-HA protected against impaired pulmonary vascular integrity (Singleton et al., 2007, 2010). This effect was also observed in a ventilator-induced model of acute lung injury (Lennon & Singleton, 2011b; Liu et al., 2008). The protective effects of HMW-HA were abrogated in CD44 or caveolin-1 knockout mice indicating the importance of these molecules in endothelial barrier enhancement (Singleton et al., 2007, 2010).

Addition of HA to human pulmonary microvascular EC monolayers and measuring transendothelial electrical resistance (TER) revealed that HMW-HA promotes endothelial barrier enhancement while LMW-HA (~2500 Da) induces EC barrier disruption (see Fig. 7.1). This has accompanied by dramatic changes in the actin cytoskeleton. HMW-HA induced a cortical actin "ring" structure while LMW-HA caused stress fiber formation in EC (Singleton et al., 2006).

Mechanistically, we observed that HMW-HA binds to and inhibits the EC barrier disrupting activity of the extracellular serine protease HABP2 (Mambetsariev et al., 2010). In addition, HMW-HA binds to the transmembrane receptor, CD44s (standard form), in specialized plasma membrane domains enriched in the scaffolding protein, caveolin-1, called CEM (Singleton et al., 2006, 2009; Singleton, Dudek, Chiang, & Garcia, 2005). CD44s then transactivates the barrier enhancing $S1P_1$ receptor (Singleton et al., 2006). This results in the serine/threonine kinase, Akt-mediated activation of the Rac1 guanine nucleotide exchange factor, Tiam1, and Rac1-GTP formation leading to cortical actin formation and

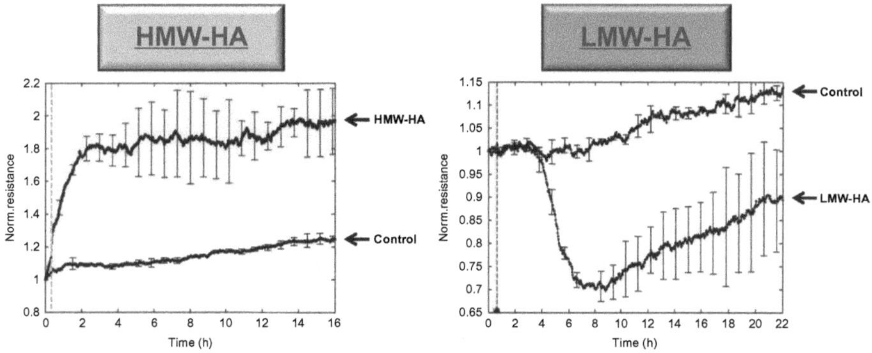

Figure 7.1 *Hyaluronan regulation of endothelial barrier function.* Left: HMW-HA (~1 million Da, 100 nM) induces an increase in human pulmonary microvascular EC barrier function using transendothelial electrical resistance (TER). An increase in resistance (*y*-axis) over time (*x*-axis) indicates increased barrier function. The **upper** line indicates HMW-HA treatment and the **lower** line indicates control (no treatment). Right: LMW-HA (~2500 Da, 100 nM) induces a decrease in human pulmonary microvascular EC barrier function using TER. A decrease in resistance (*y*-axis) over time (*x*-axis) indicates a decrease in barrier function. The **lower** line indicates LMW-HA treatment and the **upper** line indicates control (no treatment).

strengthening of EC–EC contacts (Singleton et al., 2006, 2010). Further, HMW-HA recruits several other actin regulatory proteins to CEM including annexin A2, protein S100-A10, filamin-A, and filamin-B which enhance cortical actin formation and vascular integrity (see Fig. 7.2).

In contrast to HMW-HA, LMW-HA binds to and activates the HA receptor, CD44v10 (variant 10) in CEM (Singleton et al., 2006). CD44v10 then transactivates the barrier disruptive $S1P_3$ receptor (Singleton et al., 2006). In addition, LMW-HA binds to and activates the extracellular serine protease, HABP2, which induces protease-activated receptor activation in EC (Mambetsariev et al., 2010). These events promote RhoA guanine nucleotide exchange factor activation and RhoA-GTP formation which stimulates the serine/threonine kinase, rho kinase (ROCK) (Singleton et al., 2006). This leads to actin stress fiber formation and EC barrier disruption (see Fig. 7.2). In addition, Flynn, Michaud, Canosa, & Madri (2013) have demonstrated that CD44 regulation of the EC barrier involves PECAM-1. Disruption of the endothelium is an important initiating event in tumor angiogenesis and cancer metastasis (Pardue et al., 2008; Reymond et al., 2013; Singleton & Bourguignon, 2002, 2004; Slevin et al., 2007).

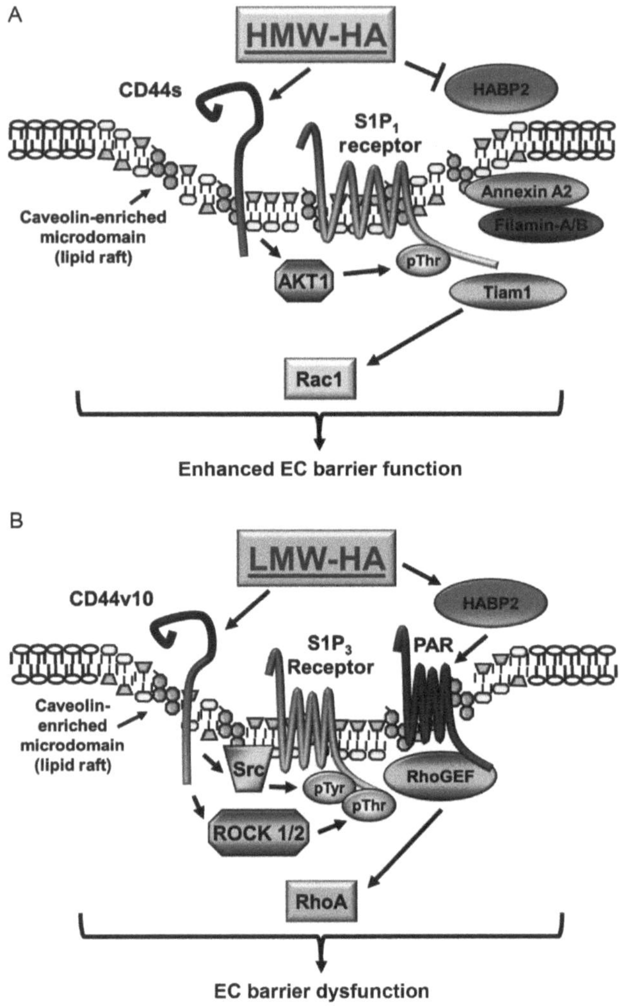

Figure 7.2 *Hyaluronan regulation of normal and impaired vascular integrity*. (A) HMW-HA (~1 million Da) binds to and inhibits the EC barrier disrupting activity of the extracellular serine protease HABP2. In addition, HMW-HA binds to the transmembrane receptor, CD44s (standard form), in specialized plasma membrane domains enriched in the scaffolding protein, caveolin-1, called caveolin-enriched microdomains (CEM). CD44s then transactivates the barrier enhancing $S1P_1$ receptor. This results in Akt-mediated activation of the Rac1 guanine nucleotide exchange factor, Tiam1, and Rac1-GTP formation leading to cortical actin formation and strengthening of EC–EC contacts. Further, HMW-HA recruits several other actin regulatory proteins to CEM including annexin A2, protein S100-A10, filamin-A, and filamin-B, which enhance cortical actin formation and vascular integrity. (B) In contrast to HMW-HA, LMW-HA (~2500 Da) binds to and activates the HA receptor, CD44v10 (variant 10) in CEM. CD44v10 then transactivates the barrier disruptive $S1P_3$ receptor. In addition, LMW-HA binds to and activates the

3. HA REGULATION OF ENDOTHELIAL BARRIER FUNCTION DURING TUMOR ANGIOGENESIS

Angiogenesis is an essential phenotype in a number of physiologic and pathologic processes including growth and development (Flamme, Frolich, & Risau, 1997), wound healing (Arnold & West, 1991), and reproduction (Shimizu, Hoshino, Miyazaki, & Sato, 2012). Inadequate angiogenesis contributes to ulcer formation (Folkman et al., 1991), while excessive angiogenesis contributes to the pathology of arthritis (Semerano, Clavel, Assier, Denys, & Boissier, 2011), psoriasis (Leong, Fearon, & Veale, 2005), and neoplasia (Folkman, 1974). In a series of now classical experiments, Folkman and colleagues demonstrated that solid tumors cannot grow larger than 3–4 mm in diameter unless they induce their own blood supply (Hanahan & Folkman, 1996).

An initial step in the complex process of angiogenesis is the disruption of the barrier of a preexisting blood vessel to allow "budding" of EC to form new capillaries (Otrock, Mahfouz, Makarem, & Shamseddine, 2007; Potente, Gerhardt, & Carmeliet, 2011). We have previously reported that LMW-HA induces EC barrier disruption, which suggests a role for HA fragments in the initiation of tumor angiogenesis (Singleton et al., 2006). Several cancers produce HA fragments in the ~10–15 disaccharide range that have been shown to promote angiogenesis (Lataillade, Albanese, & Uzan, 2010). These HA fragments bind to EC surface receptors such as CD44, RHAMM, and TLR4 (Dang et al., 2013; Gao, Yang, Mo, Liu, & He, 2008; Sironen et al., 2011). Wang, Cao, et al. (2011) reported that CD44 regulates EC proliferation and tubule formation induced by HA fragments in HUVECs. In telomerase-immortalized foreskin microvascular EC, Olofsson, Porsch, and Heldin (2014) demonstrated that inhibiting CD44 or Hyal2 expression prevented tubular network formation in 3D matrices. In addition, CD44 and TLR4 can be physically associated in a signaling complex following exposure to HA and endothelial TLR4 regulates certain responses to HA fragments (Taylor et al., 2004, 2007). Further, Savani et al. (2001)

---

extracellular serine protease, HABP2, which induces protease-activated receptor (PAR) activation in EC. These events promote RhoA guanine nucleotide exchange factor (RhoGEF) activation and RhoA-GTP formation, which stimulates the serine/threonine kinase, rho kinase (ROCK). This leads to actin stress fiber formation and EC barrier disruption. Disruption of the endothelium is an important initiating event in tumor angiogenesis and cancer metastasis. (See the color plate.)

demonstrated that HA fragment-induced HUVEC and H5V (murine EC line) angiogenesis involved the expression of both CD44 and RHAMM. Specifically, CD44 regulated HA-mediated adhesion, proliferation, and tube formation, while RHAMM regulated migration and *in vivo* angiogenesis. Gao et al. (2008) showed that RHAMM is the major receptor involved in HA fragment-induced angiogenesis during wound healing. Further, Matou-Nasri, Gaffney, Kumar, & Slevin (2009) demonstrated that both CD44 and RHAMM are involved in HA fragment-induced BAEC angiogenesis. Additional confirmation of a dual role of CD44 and RHAMM in HA fragment-mediated HUVEC angiogenesis was determined by Park et al. (2012), who demonstrated that HA induced RHAMM–TGFβ receptor complexes through a CD44–PKCδ mechanism. Interestingly, in postmortem human coronary arteries that have endothelial injury and angiogenesis associated with atherosclerotic plaques, there is increased expression of RHAMM and Hyal-1 but not CD44 in the neovessels suggesting differential roles of RHAMM and CD44 in angiogenesis from small versus large blood vessels (Krupinski et al., 2008).

A final step in the process of angiogenesis is the resealing of the endothelial barrier of the newly formed blood vessels from preexisting vessels (Otrock et al., 2007; Potente et al., 2011). We have reported that HMW-HA enhances EC barrier function (Singleton et al., 2006, 2010). The HA in the stroma of several cancers has been implicated in numerous processes including inhibition of apoptosis, stimulation of tumor cell proliferation, migration and invasion, epithelial mesenchymal transition, protection from immune cells, and regulation of intercellular space (Auvinen et al., 2013; Bollyky et al., 2009; Cho et al., 2012; Dickinson & Gerecht, 2010; Sironen et al., 2011; Tammi et al., 2008). Whether this stroma-associated HA also promotes resealing of new capillaries remains to be determined.

4. HA REGULATION OF ENDOTHELIAL BARRIER FUNCTION DURING CANCER METASTASIS

During cancer metastasis, there is a disruption of the endothelial barrier first during tumor cell intravasation from the tumor into a blood vessel and second during tumor cell extravasation out of a blood vessel and into a target tissue (Reymond et al., 2013). For intravasation, the tumor cell must invade through the tumor stroma (Chiang & Massague, 2008; Joyce & Pollard, 2009). HA, such as that found in the HA-rich stroma of tumors, can induce expression of several ECM degrading enzymes on tumor cells including the matrix metalloproteinases MT1-MMP, MMP-2, MMP-7, and MMP-9 as well as EMMPRIN and the cysteine proteases cathepsin D and cathepsin K (Abecassis, Olofsson, Schmid, Zalcman, & Karniguian, 2003; Bourguignon et al., 2004; Chetty et al., 2012; Droller, 2003; Kim et al., 2008; Kosunen et al., 2007; Marrero-Diaz et al., 2009; Mitchell & King, 2014; Murray, Morrin, & McDonnell, 2004; Nalla, Gorantla, Gondi, Lakka, & Rao, 2010; Veeravalli et al., 2010). Once at the blood vessel, the tumor cell can secrete a host of factors to induce disruption of EC junctions and EC retraction including MMP1, ADAM12, and TNF1α to allow for paracellular transendothelial migration (Mierke, 2011). In addition, we have demonstrated that LMW-HA induces EC barrier disruption and can potentially contribute to this process (Singleton et al., 2006).

Within a blood vessel, the tumor cells are subjected to sheer stress and the host immune system (Reymond et al., 2013). Interestingly, Hirose et al. (2012) reported that inhibition of the scavenger receptor, Stabilin-2, in mice increases circulating HA which does not affect primary tumor size but inhibits tumor metastasis. The increased circulating HA was found to be ~40 kDa in size which did not affect tumor cell proliferation, migration, or invasion (Hirose et al., 2012). This size of HA was not tested in our experiments on EC barrier function. These results are intriguing since higher molecular weight HA can promote immune tolerance. While the exact mechanism for this effect is unknown, it illustrates the complex relationship between HA and cancer metastasis.

A final step of tumor cell metastasis is extravasation from a blood vessel, which is a process fundamentally different from intravasation (Reymond et al., 2013). Tumor cells initially adhere to the HA-rich glycocalyx of EC in the vasculature and make initial cell–cell connections through various receptors including E-selectin, N-cadherin, and the HA receptor, CD44 (Draffin, McFarlane, Hill, Johnston, & Waugh, 2004; Kobayashi, Boelte, & Lin, 2007; Mine et al., 2003; Mitchell & King, 2014; Orian-Rousseau, 2010; St Hill, 2011). Then, stable adhesions are formed between tumor cells and EC, which are mediated by integrins, mucins, and CD44 (Fujisaki et al., 1999; Rahn et al., 2005; Wang et al., 2005). Tumor cells secrete various factors to disrupt endothelial cell–cell junctions and form protrusions to migrate between ECs, processes dependent on Rac1, RhoA, ROCK, MLCK, Src, PI3 kinase, and other signaling molecules (Aytes et al., 2013; Chatterjee et al., 2011; Chen, 2008; Khuon et al., 2010; Peinado, Lavotshkin, & Lyden, 2011; Reymond, Riou, & Ridley, 2012). The contribution of tumor cell hyaluronidases could be important in the breakdown

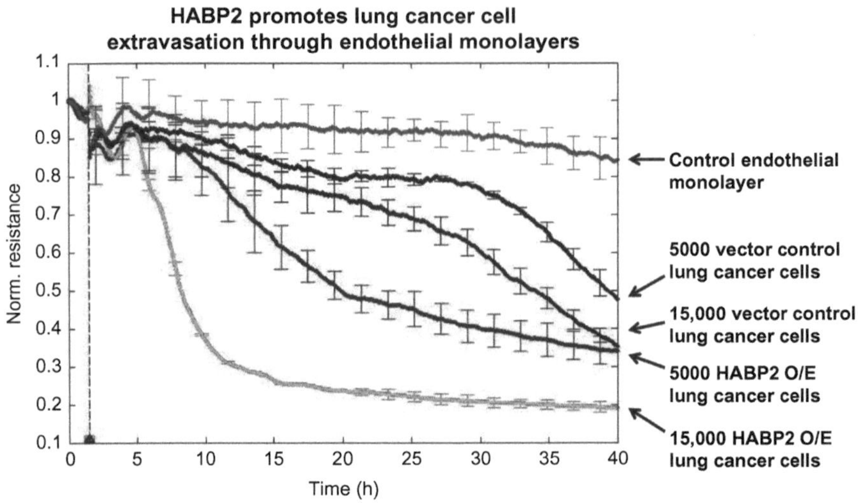

Figure 7.3 Overexpression of the HA-binding extracellular serine protease, HABP2, in human cancer cells promotes extravasation through endothelial monolayers. Human pulmonary microvascular EC were grown to confluency on gold plated electrodes and changes in electrical resistance (*y*-axis) over time (*x*-axis) were measured with addition of no cancer cells (control), 5000 or 15,000 vector control or 5000 or 15,000 HABP2 overexpressing (O/E) SKLU-1human lung cancer cells. As cancer cells extravasate through the EC monolayers, a decrease in resistance is observed which is indicative of EC barrier disruption. HABP2 overexpressing SKLU-1 human lung cancer cells demonstrated a faster and greater magnitude of EC barrier disruption than equal numbers of vector control SKLU-1 cells. (See the color plate.)

of the glycocalyx, which can generate HA fragments and contribute to EC barrier disruption through activation of HA-binding proteins including HABP2 (see Fig. 7.3).

5. POTENTIAL THERAPEUTIC EFFECTS OF HMW-HA IN INHIBITING ENDOTHELIAL BARRIER DISRUPTION DURING CANCER PROGRESSION

Although HMW-HA ($\geq$1 million Da) is produced endogenously and is an integral component of the extracellular matrix, synovial fluid, and vitreous humor, recent attention has been focused on the use of exogenously administered HMW-HA in a variety of diseases including cancer (Benitez et al., 2011; Cantor, 2007; Gaffney, Matou-Nasri, Grau-Olivares, &

Slevin, 2010; Shay, He, Sakurai, & Tseng, 2011). *In vitro*, exogenous administration of HMW-HA inhibits ROS, nitrotyrosine, and inflammatory cytokine production (Bollyky et al., 2009; Miki et al., 2010; Yasuda, 2007). As discussed earlier, HA forms a major part of the vascular glycocalyx (Reitsma, Slaaf, Vink, van Zandvoort, & oude Egbrink, 2007). Loss or disruption of the glycocalyx leads to increased vascular permeability and edema, and may be a contributing factor to tumor angiogenesis and metastasis. Repair of the glycocalyx and restoration of vascular barrier function could potentially be beneficial in the treatment of cancer. Previous studies have shown restoration of the glycocalyx by plasma resuscitation in animal models of hemorrhagic shock (Kozar et al., 2011). Other studies have shown that perfusion of HA can restore the glycocalyx following degradation by hyaluronidase or in response to ischemia/reperfusion injury (Henry & Duling, 1999; Rubio-Gayosso, Platts, & Duling, 2006). In models of pulmonary vascular hyperpermeability, exogenous administration of HMW-HA inhibited vascular barrier disruption (Liu et al., 2008; Singleton et al., 2007, 2010). In addition, HMW-HA and HA conjugates have been successfully utilized to target chemotherapeutics to cancer cells (Karbownik & Nowak, 2013; Misra et al., 2011; Ricciardelli et al., 2013; Wang et al., 2013; Yang et al., 2013). The ability of exogenously administered HMW-HA to restore damaged glycocalyx function and enhance EC barrier integrity make it a novel potential therapeutic strategy for diseases with defects in vascular integrity including cancer progression (Fig. 7.4).

6. CONCLUSIONS

HA is an important regulator of angiogenic and metastatic processes based on its size and the HA-binding proteins present. HMW-HA promotes vascular integrity, while LMW-HA induces EC barrier disruption, angiogenesis, and cancer metastasis. Several HA-binding proteins have been identified to regulate these processes including CD44, RHAMM, TLR4, and hyaluronidases. While, much is known about HA regulation of vascular integrity during cancer progression, several mechanisms have yet to be explained. The role of other HA-binding proteins in EC barrier function during cancer remains poorly defined. Further research into the interactions of HA and tumor cells within blood vessels can lead to novel chemotherapeutic interventions. In addition, the differential roles of HA and HA-binding proteins in angiogenesis from small and large blood

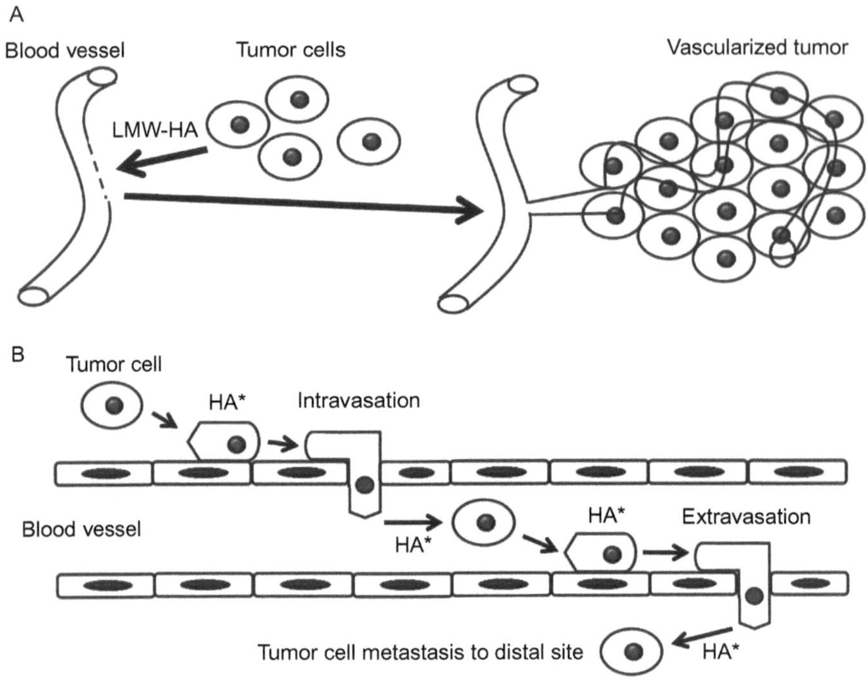

Figure 7.4 The role of hyaluronan regulation of endothelial barrier function in cancer. (A) Tumors that grow larger than 3–4 mm in diameter must undergo angiogenesis. Several types of tumor cells degrade HMW-HA, often found in the tumor stroma, to LMW-HA through hyaluronidase activity and/or ROS. LMW-HA binds to and activates certain HA-binding proteins on EC including CD44v10, HABP2, RHAMM, and TLR4 which, along with other factors, disrupts the EC barrier of a preexisting blood vessel to allow "budding" of EC to form new capillaries and a vascularized tumor. (B) During cancer metastasis, there is a disruption of the endothelial barrier first during tumor cell intravasation from the tumor into a blood vessel and second during tumor cell extravasation out of a blood vessel and into a target tissue. For intravasation, HA fragments along with other factors produced by tumor cells promotes tumor cell invasion through the stroma and disruption of the EC barrier of a blood vessel. Tumor cells then undergo paracellular transendothelial migration into a blood vessel and travel through the bloodstream. For extravasation, tumor cells within blood vessels attach to the HA-rich glycocalyx of the endothelium and subsequently to the surface of ECs in a process regulated by several factors including the HA-binding receptor, CD44. Tumor cells then disrupt EC junctions and undergo paracellular transendothelial migration through the blood vessel wall to distal sites. HA* indicates steps during tumor cell metastasis in which HA and/or HA-binding proteins are involved. (See the color plate.)

vessels represent an intriguing area of research. Thus, HA and HA-binding proteins exhibit a complex regulation of EC barrier disruption, angiogenesis, and cancer metastasis. Understanding the mechanism(s) of these processes can lead to novel therapeutic targets for cancer.

ACKNOWLEDGMENTS

Dr. P. A. S. was supported in part by the American Heart Association National Scientist Development Grant 0730277N, the American Lung Association National Biomedical Research Grant RG-75229-N and NIH NHLBI Grant RO1-HL 095723.

REFERENCES

Abecassis, I., Olofsson, B., Schmid, M., Zalcman, G., & Karniguian, A. (2003). RhoA induces MMP-9 expression at CD44 lamellipodial focal complexes and promotes HMEC-1 cell invasion. *Experimental Cell Research, 291*, 363–376.

Almond, A. (2007). Hyaluronan. *Cellular and Molecular Life Sciences, 64*, 1591–1596.

Arai, E., Nishida, Y., Wasa, J., Urakawa, H., Zhuo, L., Kimata, K., et al. (2011). Inhibition of hyaluronan retention by 4-methylumbelliferone suppresses osteosarcoma cells in vitro and lung metastasis in vivo. *British Journal of Cancer, 105*, 1839–1849.

Arnold, F., & West, D. C. (1991). Angiogenesis in wound healing. *Pharmacology & Therapeutics, 52*, 407–422.

Auvinen, P., Tammi, R., Kosma, V. M., Sironen, R., Soini, Y., Mannermaa, A., et al. (2013). Increased hyaluronan content and stromal cell CD44 associate with HER2 positivity and poor prognosis in human breast cancer. *International Journal of Cancer, 132*, 531–539.

Aytes, A., Mitrofanova, A., Kinkade, C. W., Lefebvre, C., Lei, M., Phelan, V., et al. (2013). ETV4 promotes metastasis in response to activation of PI3-kinase and Ras signaling in a mouse model of advanced prostate cancer. *Proceedings of the National Academy of Sciences of the United States of America, 110*, E3506–E3515.

Benitez, A., Yates, T. J., Lopez, L. E., Cerwinka, W. H., Bakkar, A., & Lokeshwar, V. B. (2011). Targeting hyaluronidase for cancer therapy: Antitumor activity of sulfated hyaluronic acid in prostate cancer cells. *Cancer Research, 71*, 4085–4095.

Bernert, B., Porsch, H., & Heldin, P. (2011). Hyaluronan synthase 2 (HAS2) promotes breast cancer cell invasion by suppression of tissue metalloproteinase inhibitor 1 (TIMP-1). *The Journal of Biological Chemistry, 286*, 42349–42359.

Bollyky, P. L., Falk, B. A., Wu, R. P., Buckner, J. H., Wight, T. N., & Nepom, G. T. (2009). Intact extracellular matrix and the maintenance of immune tolerance: High molecular weight hyaluronan promotes persistence of induced CD4 + CD25 + regulatory T cells. *Journal of Leukocyte Biology, 86*, 567–572.

Bourguignon, L. Y., Singleton, P. A., Diedrich, F., Stern, R., & Gilad, E. (2004). CD44 interaction with Na+-H + exchanger (NHE1) creates acidic microenvironments leading to hyaluronidase-2 and cathepsin B activation and breast tumor cell invasion. *The Journal of Biological Chemistry, 279*, 26991–27007.

Cantor, J. O. (2007). Potential therapeutic applications of hyaluronan in the lung. *International Journal of Chronic Obstructive Pulmonary Disease, 2*, 283–288.

Cantor, J. O., & Nadkarni, P. P. (2006). Hyaluronan: The Jekyll and Hyde molecule. *Inflammation & Allergy Drug Targets, 5*, 257–260.

Chatterjee, M., Sequeira, L., Jenkins-Kabaila, M., Dubyk, C. W., Pathak, S., & van Golen, K. L. (2011). Individual rac GTPases mediate aspects of prostate cancer cell and bone marrow endothelial cell interactions. *Journal of Signal Transduction, 2011*, 541851.

Chen, J. (2008). Is Src the key to understanding metastasis and developing new treatments for colon cancer? *Nature Clinical Practice. Gastroenterology & Hepatology, 5*, 306–307.

Chetty, C., Vanamala, S. K., Gondi, C. S., Dinh, D. H., Gujrati, M., & Rao, J. S. (2012). MMP-9 induces CD44 cleavage and CD44 mediated cell migration in glioblastoma xenograft cells. *Cellular Signalling, 24*, 549–559.

Chiang, A. C., & Massague, J. (2008). Molecular basis of metastasis. *The New England Journal of Medicine, 359*, 2814–2823.

Cho, S. H., Park, Y. S., Kim, H. J., Kim, C. H., Lim, S. W., Huh, J. W., et al. (2012). CD44 enhances the epithelial-mesenchymal transition in association with colon cancer invasion. *International Journal of Oncology*, *41*, 211–218.

Csoka, A. B., Frost, G. I., & Stern, R. (2001). The six hyaluronidase-like genes in the human and mouse genomes. *Matrix Biology*, *20*, 499–508.

Curry, F. R. (2005). Microvascular solute and water transport. *Microcirculation*, *12*, 17–31.

Dang, S., Peng, Y., Ye, L., Wang, Y., Qian, Z., Chen, Y., et al. (2013). Stimulation of TLR4 by LMW-HA induces metastasis in human papillary thyroid carcinoma through CXCR7. *Clinical & Developmental Immunology*, *2013*, 712561.

Dejana, E., Tournier-Lasserve, E., & Weinstein, B. M. (2009). The control of vascular integrity by endothelial cell junctions: Molecular basis and pathological implications. *Developmental Cell*, *16*, 209–221.

Dickinson, L. E., & Gerecht, S. (2010). Micropatterned surfaces to study hyaluronic acid interactions with cancer cells. *Journal of Visualized Experiments*, *46*, 2413.

Draffin, J. E., McFarlane, S., Hill, A., Johnston, P. G., & Waugh, D. J. (2004). CD44 potentiates the adherence of metastatic prostate and breast cancer cells to bone marrow endothelial cells. *Cancer Research*, *64*, 5702–5711.

Droller, M. J. (2003). Expression of cathepsin D in urothelial carcinoma of the urinary bladder: An immunohistochemical study including correlations with extracellular matrix components, CD44, p53, Rb, c-erbB-2 and the proliferation indices. *The Journal of Urology*, *170*, 671–672.

Flamme, I., Frolich, T., & Risau, W. (1997). Molecular mechanisms of vasculogenesis and embryonic angiogenesis. *Journal of Cellular Physiology*, *173*, 206–210.

Flynn, K. M., Michaud, M., Canosa, S., & Madri, J. A. (2013). CD44 regulates vascular endothelial barrier integrity via a PECAM-1 dependent mechanism. *Angiogenesis*, *16*, 689–705.

Folkman, J. (1974). Tumor angiogensis: Role in regulation of tumor growth. *The Symposium/The Society for Developmental Biology*, *30*, 43–52.

Folkman, J., Szabo, S., Stovroff, M., McNeil, P., Li, W., & Shing, Y. (1991). Duodenal ulcer. Discovery of a new mechanism and development of angiogenic therapy that accelerates healing. *Annals of Surgery*, *214*, 414–425, discussion 426–417.

Fraser, J. R., Laurent, T. C., & Laurent, U. B. (1997). Hyaluronan: Its nature, distribution, functions and turnover. *Journal of Internal Medicine*, *242*, 27–33.

Fujisaki, T., Tanaka, Y., Fujii, K., Mine, S., Saito, K., Yamada, S., et al. (1999). CD44 stimulation induces integrin-mediated adhesion of colon cancer cell lines to endothelial cells by up-regulation of integrins and c-Met and activation of integrins. *Cancer Research*, *59*, 4427–4434.

Furlan, S., La Penna, G., Perico, A., & Cesaro, A. (2005). Hyaluronan chain conformation and dynamics. *Carbohydrate Research*, *340*, 959–970.

Futamura, N., Urakawa, H., Arai, E., Kozawa, E., Ishiguro, N., & Nishida, Y. (2013). Hyaluronan synthesis inhibitor supplements the inhibitory effects of zoledronic acid on bone metastasis of lung cancer. *Clinical & Experimental Metastasis*, *30*, 595–606.

Gaffney, J., Matou-Nasri, S., Grau-Olivares, M., & Slevin, M. (2010). Therapeutic applications of hyaluronan. *Molecular BioSystems*, *6*, 437–443.

Gao, F., Yang, C. X., Mo, W., Liu, Y. W., & He, Y. Q. (2008). Hyaluronan oligosaccharides are potential stimulators to angiogenesis via RHAMM mediated signal pathway in wound healing. *Clinical and Investigative Medicine*, *31*, E106–E116.

Girish, K. S., & Kemparaju, K. (2007). The magic glue hyaluronan and its eraser hyaluronidase: A biological overview. *Life Sciences*, *80*, 1921–1943.

Hanahan, D., & Folkman, J. (1996). Patterns and emerging mechanisms of the angiogenic switch during tumorigenesis. *Cell*, *86*, 353–364.

Henry, C. B., & Duling, B. R. (1999). Permeation of the luminal capillary glycocalyx is determined by hyaluronan. *The American Journal of Physiology*, *277*, H508–H514.

Hirose, Y., Saijou, E., Sugano, Y., Takeshita, F., Nishimura, S., Nonaka, H., et al. (2012). Inhibition of Stabilin-2 elevates circulating hyaluronic acid levels and prevents tumor metastasis. *Proceedings of the National Academy of Sciences of the United States of America*, *109*, 4263–4268.

Itano, N., & Kimata, K. (2002). Mammalian hyaluronan synthases. *IUBMB Life*, *54*, 195–199.

Jiang, D., Liang, J., & Noble, P. W. (2007). Hyaluronan in tissue injury and repair. *Annual Review of Cell and Developmental Biology*, *23*, 435–461.

Jiang, D., Liang, J., & Noble, P. W. (2011). Hyaluronan as an immune regulator in human diseases. *Physiological Reviews*, *91*, 221–264.

Joyce, J. A., & Pollard, J. W. (2009). Microenvironmental regulation of metastasis. *Nature Reviews. Cancer*, *9*, 239–252.

Karbownik, M. S., & Nowak, J. Z. (2013). Hyaluronan: Towards novel anti-cancer therapeutics. *Pharmacological Reports*, *65*, 1056–1074.

Khuon, S., Liang, L., Dettman, R. W., Sporn, P. H., Wysolmerski, R. B., & Chew, T. L. (2010). Myosin light chain kinase mediates transcellular intravasation of breast cancer cells through the underlying endothelial cells: A three-dimensional FRET study. *Journal of Cell Science*, *123*, 431–440.

Kim, Y., Lee, Y. S., Choe, J., Lee, H., Kim, Y. M., & Jeoung, D. (2008). CD44-epidermal growth factor receptor interaction mediates hyaluronic acid-promoted cell motility by activating protein kinase C signaling involving Akt, Rac1, Phox, reactive oxygen species, focal adhesion kinase, and MMP-2. *The Journal of Biological Chemistry*, *283*, 22513–22528.

Kobayashi, H., Boelte, K. C., & Lin, P. C. (2007). Endothelial cell adhesion molecules and cancer progression. *Current Medicinal Chemistry*, *14*, 377–386.

Kosunen, A., Pirinen, R., Ropponen, K., Pukkila, M., Kellokoski, J., Virtaniemi, J., et al. (2007). CD44 expression and its relationship with MMP-9, clinicopathological factors and survival in oral squamous cell carcinoma. *Oral Oncology*, *43*, 51–59.

Kozar, R. A., Peng, Z., Zhang, R., Holcomb, J. B., Pati, S., Park, P., et al. (2011). Plasma restoration of endothelial glycocalyx in a rodent model of hemorrhagic shock. *Anesthesia and Analgesia*, *112*, 1289–1295.

Krupinski, J., Ethirajan, P., Font, M. A., Turu, M. M., Gaffney, J., Kumar, P., et al. (2008). Changes in hyaluronan metabolism and RHAMM receptor expression accompany formation of complicated carotid lesions and may be pro-angiogenic mediators of intimal neovessel growth. *Biomarker Insights*, *2*, 361–367.

Lataillade, J. J., Albanese, P., & Uzan, G. (2010). Implication of hyaluronic acid in normal and pathological angiogenesis. Application for cellular engineering. *Annales de Dermatologie et de Vénéréologie*, *137*(Suppl. 1), S15–S22.

Lennon, F. E., & Singleton, P. A. (2011a). Hyaluronan regulation of vascular integrity. *American Journal of Cardiovascular Disease*, *1*, 200–213.

Lennon, F. E., & Singleton, P. A. (2011b). The Role of hyaluronan and hyaluronan-binding proteins in lung pathobiology. *American Journal of Physiology. Lung Cellular and Molecular Physiology*, *301*, L137–L147.

Leong, T. T., Fearon, U., & Veale, D. J. (2005). Angiogenesis in psoriasis and psoriatic arthritis: Clues to disease pathogenesis. *Current Rheumatology Reports*, *7*, 325–329.

Liao, Y. H., Jones, S. A., Forbes, B., Martin, G. P., & Brown, M. B. (2005). Hyaluronan: Pharmaceutical characterization and drug delivery. *Drug Delivery*, *12*, 327–342.

Liu, Y. Y., Lee, C. H., Dedaj, R., Zhao, H., Mrabat, H., Sheidlin, A., et al. (2008). High-molecular-weight hyaluronan—A possible new treatment for sepsis-induced lung injury: A preclinical study in mechanically ventilated rats. *Critical Care*, *12*, R102.

Mambetsariev, N., Mirzapoiazova, T., Mambetsariev, B., Sammani, S., Lennon, F. E., Garcia, J. G., et al. (2010). Hyaluronic acid binding protein 2 is a novel regulator of vascular integrity. *Arteriosclerosis, Thrombosis, and Vascular Biology, 30*, 483–490.

Marrero-Diaz, R., Bravo-Cordero, J. J., Megias, D., Garcia, M. A., Bartolome, R. A., Teixido, J., et al. (2009). Polarized MT1-MMP-CD44 interaction and CD44 cleavage during cell retraction reveal an essential role for MT1-MMP in CD44-mediated invasion. *Cell Motility and the Cytoskeleton, 66*, 48–61.

Matou-Nasri, S., Gaffney, J., Kumar, S., & Slevin, M. (2009). Oligosaccharides of hyaluronan induce angiogenesis through distinct CD44 and RHAMM-mediated signalling pathways involving Cdc2 and gamma-adducin. *International Journal of Oncology, 35*, 761–773.

Mierke, C. T. (2011). Cancer cells regulate biomechanical properties of human microvascular endothelial cells. *The Journal of Biological Chemistry, 286*, 40025–40037.

Miki, Y., Teramura, T., Tomiyama, T., Onodera, Y., Matsuoka, T., Fukuda, K., et al. (2010). Hyaluronan reversed proteoglycan synthesis inhibited by mechanical stress: Possible involvement of antioxidant effect. *Inflammation Research, 59*, 471–477.

Mine, S., Fujisaki, T., Kawahara, C., Tabata, T., Iida, T., Yasuda, M., et al. (2003). Hepatocyte growth factor enhances adhesion of breast cancer cells to endothelial cells in vitro through up-regulation of CD44. *Experimental Cell Research, 288*, 189–197.

Misra, S., Heldin, P., Hascall, V. C., Karamanos, N. K., Skandalis, S. S., Markwald, R. R., et al. (2011). Hyaluronan-CD44 interactions as potential targets for cancer therapy. *The FEBS Journal, 278*, 1429–1443.

Mitchell, M. J., & King, M. R. (2014). Physical Biology in Cancer. 3. The role of cell glycocalyx in vascular transport of circulating tumor cells. *American Journal of Physiology. Cell Physiology, 306*, C89–C97.

Murray, D., Morrin, M., & McDonnell, S. (2004). Increased invasion and expression of MMP-9 in human colorectal cell lines by a CD44-dependent mechanism. *Anticancer Research, 24*, 489–494.

Nalla, A. K., Gorantla, B., Gondi, C. S., Lakka, S. S., & Rao, J. S. (2010). Targeting MMP-9, uPAR, and cathepsin B inhibits invasion, migration and activates apoptosis in prostate cancer cells. *Cancer Gene Therapy, 17*, 599–613.

Nandi, A., Estess, P., & Siegelman, M. H. (2000). Hyaluronan anchoring and regulation on the surface of vascular endothelial cells is mediated through the functionally active form of CD44. *The Journal of Biological Chemistry, 275*, 14939–14948.

Olczyk, P., Komosinska-Vassev, K., Winsz-Szczotka, K., Kuznik-Trocha, K., & Olczyk, K. (2008). Hyaluronan: Structure, metabolism, functions, and role in wound healing. *Postępy Higieny i Medycyny Doświadczalnej (Online), 62*, 651–659.

Olofsson, B., Porsch, H., & Heldin, P. (2014). Knock-down of CD44 regulates endothelial cell differentiation via NFkappaB-mediated chemokine production. *PLoS One, 9*, e90921.

Orian-Rousseau, V. (2010). CD44, a therapeutic target for metastasising tumours. *European Journal of Cancer, 46*, 1271–1277.

Otrock, Z. K., Mahfouz, R. A., Makarem, J. A., & Shamseddine, A. I. (2007). Understanding the biology of angiogenesis: Review of the most important molecular mechanisms. *Blood Cells, Molecules & Diseases, 39*, 212–220.

Pardue, E. L., Ibrahim, S., & Ramamurthi, A. (2008). Role of hyaluronan in angiogenesis and its utility to angiogenic tissue engineering. *Organogenesis, 4*, 203–214.

Park, D., Kim, Y., Kim, H., Kim, K., Lee, Y. S., Choe, J., et al. (2012). Hyaluronic acid promotes angiogenesis by inducing RHAMM-TGFbeta receptor interaction via CD44-PKCdelta. *Molecules and Cells, 33*, 563–574.

Peinado, H., Lavotshkin, S., & Lyden, D. (2011). The secreted factors responsible for pre-metastatic niche formation: Old sayings and new thoughts. *Seminars in Cancer Biology, 21*, 139–146.

Potente, M., Gerhardt, H., & Carmeliet, P. (2011). Basic and therapeutic aspects of angiogenesis. *Cell, 146*, 873–887.
Rahn, J. J., Chow, J. W., Horne, G. J., Mah, B. K., Emerman, J. T., Hoffman, P., et al. (2005). MUC1 mediates transendothelial migration in vitro by ligating endothelial cell ICAM-1. *Clinical & Experimental Metastasis, 22*, 475–483.
Reitsma, S., Slaaf, D. W., Vink, H., van Zandvoort, M. A., & oude Egbrink, M. G. (2007). The endothelial glycocalyx: Composition, functions, and visualization. *Pflugers Archiv: European Journal of Physiology, 454*, 345–359.
Reymond, N., d'Agua, B. B., & Ridley, A. J. (2013). Crossing the endothelial barrier during metastasis. *Nature Reviews. Cancer, 13*, 858–870.
Reymond, N., Riou, P., & Ridley, A. J. (2012). Rho GTPases and cancer cell transendothelial migration. *Methods in Molecular Biology, 827*, 123–142.
Ricciardelli, C., Ween, M. P., Lokman, N. A., Tan, I. A., Pyragius, C. E., & Oehler, M. K. (2013). Chemotherapy-induced hyaluronan production: a novel chemoresistance mechanism in ovarian cancer. *BMC Cancer, 13*, 476.
Rubio-Gayosso, I., Platts, S. H., & Duling, B. R. (2006). Reactive oxygen species mediate modification of glycocalyx during ischemia-reperfusion injury. *American Journal of Physiology. Heart and Circulatory Physiology, 290*, H2247–H2256.
Saito, T., Dai, T., & Asano, R. (2013). The hyaluronan synthesis inhibitor 4-methylumbelliferone exhibits antitumor effects against mesenchymal-like canine mammary tumor cells. *Oncology Letters, 5*, 1068–1074.
Savani, R. C., Cao, G., Pooler, P. M., Zaman, A., Zhou, Z., & DeLisser, H. M. (2001). Differential involvement of the hyaluronan (HA) receptors CD44 and receptor for HA-mediated motility in endothelial cell function and angiogenesis. *The Journal of Biological Chemistry, 276*, 36770–36778.
Scott, J. E., & Heatley, F. (2002). Biological properties of hyaluronan in aqueous solution are controlled and sequestered by reversible tertiary structures, defined by NMR spectroscopy. *Biomacromolecules, 3*, 547–553.
Semerano, L., Clavel, G., Assier, E., Denys, A., & Boissier, M. C. (2011). Blood vessels, a potential therapeutic target in rheumatoid arthritis? *Joint, Bone, Spine: Revue du Rhumatisme, 78*, 118–123.
Shay, E., He, H., Sakurai, S., & Tseng, S. C. (2011). Inhibition of angiogenesis by HC.HA, a complex of hyaluronan and the heavy chain of inter-alpha-inhibitor, purified from human amniotic membrane. *Investigative Ophthalmology & Visual Science, 52*, 2669–2678.
Shimizu, T., Hoshino, Y., Miyazaki, H., & Sato, E. (2012). Angiogenesis and microvasculature in the female reproductive organs: Physiological and pathological implications. *Current Pharmaceutical Design, 18*, 303–309.
Singleton, P. A., & Bourguignon, L. Y. (2002). CD44v10 interaction with Rho-kinase (ROK) activates inositol 1,4,5-triphosphate (IP3) receptor-mediated Ca2+ signaling during hyaluronan (HA)-induced endothelial cell migration. *Cell Motility and the Cytoskeleton, 53*, 293–316.
Singleton, P. A., & Bourguignon, L. Y. (2004). CD44 interaction with ankyrin and IP3 receptor in lipid rafts promotes hyaluronan-mediated Ca2+ signaling leading to nitric oxide production and endothelial cell adhesion and proliferation. *Experimental Cell Research, 295*, 102–118.
Singleton, P. A., Chatchavalvanich, S., Fu, P., Xing, J., Birukova, A. A., Fortune, J. A., et al. (2009). Akt-mediated transactivation of the S1P1 receptor in caveolin-enriched microdomains regulates endothelial barrier enhancement by oxidized phospholipids. *Circulation Research, 104*, 978–986.
Singleton, P. A., Dudek, S. M., Chiang, E. T., & Garcia, J. G. (2005). Regulation of sphingosine 1-phosphate-induced endothelial cytoskeletal rearrangement and barrier

enhancement by S1P1 receptor, PI3 kinase, Tiam1/Rac1, and alpha-actinin. *FASEB Journal, 19*, 1646–1656.

Singleton, P. A., Dudek, S. M., Ma, S. F., & Garcia, J. G. (2006). Transactivation of sphingosine 1-phosphate receptors is essential for vascular barrier regulation. Novel role for hyaluronan and CD44 receptor family. *The Journal of Biological Chemistry, 281*, 34381–34393.

Singleton, P. A., & Lennon, F. E. (2011). Acute lung injury regulation by hyaluronan. *Journal of Allergy & Therapy*, S4-003.

Singleton, P. A., Mirzapoiazova, T., Guo, Y., Sammani, S., Mambetsariev, N., Lennon, F. E., et al. (2010). High-molecular-weight hyaluronan is a novel inhibitor of pulmonary vascular leakiness. *American Journal of Physiology. Lung Cellular and Molecular Physiology, 299*, L639–L651.

Singleton, P. A., Salgia, R., Moreno-Vinasco, L., Moitra, J., Sammani, S., Mirzapoiazova, T., et al. (2007). CD44 regulates hepatocyte growth factor-mediated vascular integrity. Role of c-Met, Tiam1/Rac1, dynamin 2, and cortactin. *The Journal of Biological Chemistry, 282*, 30643–30657.

Sironen, R. K., Tammi, M., Tammi, R., Auvinen, P. K., Anttila, M., & Kosma, V. M. (2011). Hyaluronan in human malignancies. *Experimental Cell Research, 317*, 383–391.

Slevin, M., Krupinski, J., Gaffney, J., Matou, S., West, D., Delisser, H., et al. (2007). Hyaluronan-mediated angiogenesis in vascular disease: Uncovering RHAMM and CD44 receptor signaling pathways. *Matrix Biology: Journal of the International Society for Matrix Biology, 26*, 58–68.

St Hill, C. A. (2011). Interactions between endothelial selectins and cancer cells regulate metastasis. *Frontiers in Bioscience (Landmark Edition), 16*, 3233–3251.

Stern, R. (2003). Devising a pathway for hyaluronan catabolism: Are we there yet? *Glycobiology, 13*, 105R–115R.

Stern, R. (2008). Hyaluronidases in cancer biology. *Seminars in Cancer Biology, 18*, 275–280.

Stern, R., Asari, A. A., & Sugahara, K. N. (2006). Hyaluronan fragments: An information-rich system. *European Journal of Cell Biology, 85*, 699–715.

Stern, R., Kogan, G., Jedrzejas, M. J., & Soltes, L. (2007). The many ways to cleave hyaluronan. *Biotechnology Advances, 25*, 537–557.

Tammi, M. I., Day, A. J., & Turley, E. A. (2002). Hyaluronan and homeostasis: A balancing act. *The Journal of Biological Chemistry, 277*, 4581–4584.

Tammi, R. H., Kultti, A., Kosma, V. M., Pirinen, R., Auvinen, P., & Tammi, M. I. (2008). Hyaluronan in human tumors: Pathobiological and prognostic messages from cell-associated and stromal hyaluronan. *Seminars in Cancer Biology, 18*, 288–295.

Tan, J. X., Wang, X. Y., Li, H. Y., Su, X. L., Wang, L., Ran, L., et al. (2011). HYAL1 over-expression is correlated with the malignant behavior of human breast cancer. *International Journal of Cancer. Journal International du Cancer, 128*, 1303–1315.

Tan, J. X., Wang, X. Y., Su, X. L., Li, H. Y., Shi, Y., Wang, L., et al. (2011). Upregulation of HYAL1 expression in breast cancer promoted tumor cell proliferation, migration, invasion and angiogenesis. *PLoS One, 6*, e22836.

Taylor, K. R., Trowbridge, J. M., Rudisill, J. A., Termeer, C. C., Simon, J. C., & Gallo, R. L. (2004). Hyaluronan fragments stimulate endothelial recognition of injury through TLR4. *The Journal of Biological Chemistry, 279*, 17079–17084.

Taylor, K. R., Yamasaki, K., Radek, K. A., Di Nardo, A., Goodarzi, H., Golenbock, D., et al. (2007). Recognition of hyaluronan released in sterile injury involves a unique receptor complex dependent on Toll-like receptor 4, CD44, and MD-2. *The Journal of Biological Chemistry, 282*, 18265–18275.

Toole, B. P. (2004). Hyaluronan: From extracellular glue to pericellular cue. *Nature Reviews Cancer, 4*, 528–539.

Udabage, L., Brownlee, G. R., Nilsson, S. K., & Brown, T. J. (2005). The over-expression of HAS2, Hyal-2 and CD44 is implicated in the invasiveness of breast cancer. *Experimental Cell Research*, *310*, 205–217.

Vandenbroucke, E., Mehta, D., Minshall, R., & Malik, A. B. (2008). Regulation of endothelial junctional permeability. *Annals of the New York Academy of Sciences*, *1123*, 134–145.

Veeravalli, K. K., Chetty, C., Ponnala, S., Gondi, C. S., Lakka, S. S., Fassett, D., et al. (2010). MMP-9, uPAR and cathepsin B silencing downregulate integrins in human glioma xenograft cells in vitro and in vivo in nude mice. *PLoS One*, *5*, e11583.

Wang, Y. Z., Cao, M. L., Liu, Y. W., He, Y. Q., Yang, C. X., & Gao, F. (2011). CD44 mediates oligosaccharides of hyaluronan-induced proliferation, tube formation and signal transduction in endothelial cells. *Experimental Biology and Medicine (Maywood, N.J.)*, *236*, 84–90.

Wang, A., de la Motte, C., Lauer, M., & Hascall, V. (2011). Hyaluronan matrices in pathobiological processes. *The FEBS Journal*, *278*(9), 1412–1418. http://dx.doi.org/10.1111/j.1742-4658.2011.08069.x. Epub 2011 Mar 25.

Wang, D., Huang, J., Wang, X., Yu, Y., Zhang, H., Chen, Y., et al. (2013). The eradication of breast cancer cells and stem cells by 8-hydroxyquinoline-loaded hyaluronan modified mesoporous silica nanoparticle-supported lipid bilayers containing docetaxel. *Biomaterials*, *34*, 7662–7673.

Wang, H. S., Hung, Y., Su, C. H., Peng, S. T., Guo, Y. J., Lai, M. C., et al. (2005). CD44 cross-linking induces integrin-mediated adhesion and transendothelial migration in breast cancer cell line by up-regulation of LFA-1 (alpha L beta2) and VLA-4 (alpha4beta1). *Experimental Cell Research*, *304*, 116–126.

Weigel, P. H., & DeAngelis, P. L. (2007). Hyaluronan synthases: A decade-plus of novel glycosyltransferases. *The Journal of Biological Chemistry*, *282*, 36777–36781.

Wheeler-Jones, C. P., Farrar, C. E., & Pitsillides, A. A. (2010). Targeting hyaluronan of the endothelial glycocalyx for therapeutic intervention. *Current Opinion in Investigational Drugs*, *11*, 997–1006.

Yang, C., Liu, Y., He, Y., Du, Y., Wang, W., Shi, X., et al. (2013). The use of HA oligosaccharide-loaded nanoparticles to breach the endogenous hyaluronan glycocalyx for breast cancer therapy. *Biomaterials*, *34*, 6829–6838.

Yasuda, T. (2007). Hyaluronan inhibits cytokine production by lipopolysaccharide-stimulated U937 macrophages through down-regulation of NF-kappaB via ICAM-1. *Inflammation Research*, *56*, 246–253.

CHAPTER EIGHT

HAS2 and CD44 in Breast Tumorigenesis

Paraskevi Heldin*,†,1, Kaustuv Basu*, Inna Kozlova*, Helena Porsch*
*Ludwig Institute for Cancer Research, Science for Life Laboratory, Uppsala University, Uppsala, Sweden
†Department of Medical Biochemistry and Microbiology, Uppsala University, Uppsala, Sweden
[1]Corresponding author: e-mail address: evi.heldin@licr.uu.se

Contents

1. Introduction 211
2. Molecular Classification of Breast Cancer 213
3. Role of Stromal HA in Tumor Progression 215
4. Expression of *HAS* Genes and Breast Cancer Malignancy 218
5. HA–CD44 Interactions: A Regulatory Network During TGFβ-Mediated EMT 220
6. Conclusions 223
Acknowledgments 223
References 224

Abstract

Metastatic spread of breast cancer cells, facilitated by the epithelial–mesenchymal transition (EMT) process, is responsible for the majority of breast cancer mortality. Increased levels of hyaluronan due to deregulation of hyaluronan-synthesizing enzymes, like HAS2, and expression of CD44, the key receptor for hyaluronan, are correlated to poor outcome of patients with basal-like breast cancer. TGFβ induces HAS2 and CD44, both of which are required in the course of efficient TGFβ-induced EMT processes by mammary epithelial cells. Elucidation of the molecular mechanisms underlying tumor–stroma interactions in breast cancer including the regulation of HAS2 and CD44 expression may contribute to the development of better strategies to treat breast cancer patients.

1. INTRODUCTION

In solid tumors, the microenvironment around the tumor cells is characterized by abnormal extracellular matrix (ECM). The ECM is a biological complex composed of structural proteins, glycoproteins, proteoglycans, and glycosaminoglycans that forms a bioactive network determining the

Advances in Cancer Research, Volume 123
ISSN 0065-230X
http://dx.doi.org/10.1016/B978-0-12-800092-2.00008-3

mechanical properties of tissues. Moreover, the ECM contains fibroblasts and is infiltrated by immune cells as well as vascular and lymphatic endothelial cells. Following malignant transformation, the composition of ECM components is modulated through the action of growth factors, chemokines, and cytokines, resulting in changes in their organization which subsequently leads to altered phenotype of the cells residing in the tissue (Kadar, Tokes, Kulka, & Robert, 2002; Roskelley & Bissell, 2002).

In the human mammary gland, the ECM undergoes dramatic changes during development, pregnancy, aging, and tumorigenesis. For example, in normal human breast tissue, the glycosaminoglycan hyaluronan (HA) and the protein fibronectin are found at low amounts, whereas a strong accumulation of these ECM molecules is detected in the stroma of breast tumors. In fact, their overproduction has been correlated to severity of the disease. They transmit extracellular signals via binding to their cell surface receptors, such as CD44 and RHAMM for HA, and integrins for fibronectin, thus linking the ECM to signaling events inside cells (Chaturvedi & Hass, 2011; Muschler & Streuli, 2010; Park & Schwarzbauer, 2014).

Glycosaminoglycans are critical components of the ECM which contribute to tissue hydration and architecture (Iozzo, 1998; Taylor & Gallo, 2006). HA is the simplest glycosaminoglycan and is composed of repeating disaccharide units of D-glucuronic acid and *N*-acetyl-D-glucosamine; it possesses unique hygroscopic and viscoelastic properties, as water is trapped inside the structure, resulting in soft and elastic tissues (Laurent & Fraser, 1992). In tumors, HA is aberrantly accumulated due to deregulated functions of HA synthases (HAS1, HAS2, and HAS3), HA-degrading enzymes (HYALs and PH-20), or altered clearance from the tissues via lymphatics (Adamia, Maxwell, & Pilarski, 2005; Heldin et al., 2013; Heldin, Karousou, & Skandalis, 2009; Stern & Jedrzejas, 2006; Tammi et al., 2011; Weigel, Hascall, & Tammi, 1997). The individual physiological roles of HAS1, HAS2, and HAS3 have not yet been elucidated. Growth factors and hormones differentially regulate the three HASes at the transcriptional level (Heldin, Laurent, & Heldin, 1989; Jacobson, Brinck, Briskin, Spicer, & Heldin, 2000; Suzuki, Asplund, Yamashita, Heldin, & Heldin, 1995; Tammi et al., 2011), but the synthetic rate of HA is also modulated due to posttranslational modifications of HAS molecules, such as phosphorylation and ubiquitinylation (Goentzel, Weigel, & Steinberg, 2006; Karousou et al., 2010; Suzuki et al., 1995; Vigetti et al., 2011). Furthermore, different cytoplasmic proteins might specifically interact with each HAS protein, exhibiting accessory or regulatory roles in HA biosynthesis

(Brinck & Heldin, 1999). Among the HYALs, HYAL1 and HYAL2 are ubiquitously distributed and cooperate to degrade high-molecular-mass HA. Their functional roles in tumor progression and suppression are complex and the underlying molecular mechanisms are not well understood (Benitez et al., 2011; Jacobson, Rahmanian, Rubin, & Heldin, 2002; Simpson & Lokeshwar, 2008; Stern, 2008).

HA modulates cellular behavior in a size-dependent manner through binding to its major receptors, CD44 and RHAMM, whereby cell migration, differentiation, and proliferation are promoted (Heldin et al., 2013; Ponta, Sherman, & Herrlich, 2003; Toole, 2004; Turley, Noble, & Bourguignon, 2002). CD44 exhibits wide structural diversity due to alternative splicing of the pre-mRNA transcript and posttranslational modifications, resulting in the standard form of CD44 (CD44s) and CD44 variant forms (CD44v1–10) (Goodison, Urquidi, & Tarin, 1999). HA molecules of high-molecular mass are present in healthy tissues, whereas fragmented HA is found during inflammation and cancer. Interestingly, accumulation of HA of a polydisperse character in the lungs of irradiated rats promotes the production of collagen I and III by fibroblasts, and thus fibrosis (Li et al., 2000). How high amounts of HA contribute to tumorigenesis is not understood. In order to understand the role of HA in breast cancer progression, it is essential to unravel the molecular mechanisms responsible for its synthesis and degradation, as well as the regulation of the expression of the proteins that HA interacts with. This review will focus on the emerging role of HA in the specialized stroma accompanying breast cancers. In particular, the molecular mechanisms underlying the increase in HAS2 and CD44 expression within aggressive breast cancer will be highlighted.

2. MOLECULAR CLASSIFICATION OF BREAST CANCER

The human mammary gland consists of ducts giving rise to a branching ductal-alveolar tree lined by an inner layer of secretory epithelial (luminal) cells and surrounded by a basal layer of myoepithelial cells. Based on gene expression profiling, breast tumors can be classified into five subtypes: Luminal A (expressing estrogen and progesterone receptors; ER^{+}/PR^{+}), luminal B (expressing ER but not PR, possibly expressing HER2, having higher proliferative capacity than luminal A), HER2-enriched (lacking ER), basal-like (also designated as triple-negative, i.e., $ER^{-}/PR^{-}/HER2^{-}$; overexpressing epidermal growth factor receptor (EGFR) and keratin 5/6), and the normal-like subclass (overexpressing genes of adipose and other nonepithelial cells). A new subclass has also been identified and designated as Claudin Low (characterized by the low gene expression of tight junction proteins claudin 3, 4, 7, and E-cadherin) (Charafe-Jauffret et al., 2006; Malhotra, Zhao, Band, & Band, 2010; Nakshatri, Srour, & Badve, 2009; Neve et al., 2006; Sorlie et al., 2001) (Fig. 8.1). However, breast cancers can display overlapping markers and it may be difficult to classify them as certain subtypes. Both the prognosis and treatment of patients are guided by stage, hormone receptor status, and HER2 (ErbB2) status. Additional markers associated with the abnormal behavior of specific cell types of the breast are required.

In addition to luminal and myoepithelial differentiated cells, there are stem cells with high self-renewal capacity and potential for multilineage differentiation whereby they can reconstitute the entire mammary gland. In 2003, Al-Hajj, Wicha, Benito-Hernandez, Morrison, and Clarke (2003) demonstrated that breast cancer stem cells (CSCs) characterized by being $CD44^{+}/CD24^{low-}$ give rise to tumors. More recently, aldehyde dehydrogenase (ALDH) was demonstrated to be expressed both by human normal

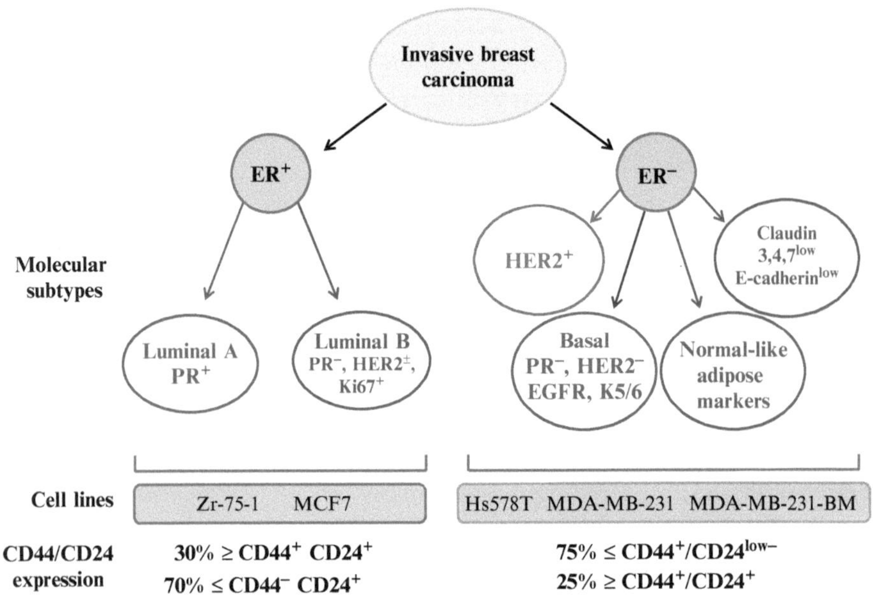

Figure 8.1 Intrinsic subtype classification of breast cancer, cell lines, and their link to CSCs. Molecular classification of breast cancer based on microarray analysis of patients (Malhotra et al., 2010). Expression of CD44 and CD24 in cell lines derived from luminal and basal-like breast cancer. (See the color plate.)

and malignant breast stem cells. $ALDH^+$ breast cancers are associated with basal-like and $HER2^+$ subtypes, whereas $CD44^+/CD24^{low-}$ breast cancers are associated with basal-like and luminal B cancers (Nakshatri et al., 2009; Tsang et al., 2012). The $CD44^+/CD24^{low-}$ signature is expressed at less than 30% in luminal (Zr-75-1, MCF7) and higher than 75% in basal-like (Hs578T, MDA-MB-231, MDA-MB-231-BM) breast cancer cell lines (Fig. 8.1) (Marotta et al., 2011).

The epithelial–mesenchymal transition (EMT) process not only plays a critical role during embryonal development but also promotes the development of more invasive tumors characterized by a mesenchymal phenotype. The process is transient and reversible; under specific conditions, the mesenchymal cells can reacquire the epithelial phenotype, so-called mesenchymal–epithelial transition (MET) (Moustakas & Heldin, 2014; Thiery, Acloque, Huang, & Nieto, 2009). New breast CSC findings demonstrate that breast CSCs display a transitory and reversible phenotype, similar to EMT state (express $CD44^+/CD24^{low-}$, are dormant and localized at the tumor invasive edge) and MET state ($ALDH^+$, capable of growing and are located more centrally) (Liu et al., 2014). The molecular mechanisms underlying the two states of breast CSCs are not known. It will be of interest to elucidate if HA–CD44 interactions are required for fine tuning the transition between these two states.

3. ROLE OF STROMAL HA IN TUMOR PROGRESSION

As carcinomas evolve from epithelial cells, likely due to accumulated genetic mutations, and recruited diverse cells alter the composition of the surrounding matrix, tumor cells, and their microenvironment coevolve because of reciprocal interactions (Bissell & Radisky, 2001; Bissell, Rizki, & Mian, 2003; Polyak, Haviv, & Campbell, 2009). Notably, the cross-talk between epithelial cells and stromal resident cells, including cancer-associated fibroblasts, infiltrating immune cells, and endothelial cells, is cell-type and tissue specific (Bhowmick & Moses, 2005). Alterations of fibroblasts and other nonmalignant cells in the stroma can promote epithelial cancer through the release of growth factors, chemokines, cytokines, and metalloproteinases (MMPs), in a paracrine manner; stimuli include fibroblast growth factor (FGF), transforming growth factor β (TGFβ), platelet-derived growth factor (PDGF-BB), hepatocyte growth factor (HGF), insulin-like growth factor, Wnt, the cytokines IL-6 and IL-8, and the chemokine CXCL7, as well as MMP-2 and MMP-9 (Korkaya, Liu, & Wicha, 2011).

The infiltration of inflammatory cells into tumor stroma causes release of growth factors and cytokines including FGF, TGFβ, PDGF-BB, HGF that induce the expression of HASes (in particular, *HAS2* mRNA) and thereby HA production (Angelucci et al., 2012; Asplund, Versnel, Laurent, & Heldin, 1993; Heldin et al., 1989; Jacobson et al., 2000; Tyan et al., 2011) (Fig. 8.2). Also, tumor-promoting agents, such as phorbol 12-myristate 13-acetate, powerfully induce HA production (Suzuki et al., 1995).

In addition to stimulating HA synthesis, growth factors, such as PDGF-BB and TGFβ, regulate the synthesis of HA-binding proteins, including versican and CD44 (Evanko et al., 2001). Similar to HA and CD44, the accumulation of versican has been associated with breast cancer invasiveness and bone metastasis. HA–versican complexes interact with CD44 expressed by leukocytes, thereby promoting the infiltration of inflammatory cells and thus sustain inflammation in reactive stroma. Furthermore, these complexes play an important role in the formation of the pericellular matrix surrounding tumor cells and activated stromal cells. The formation of pericellular matrices is regulated in response to growth factors and they provide a viscoelastic extracellular environment that supports cell proliferation and

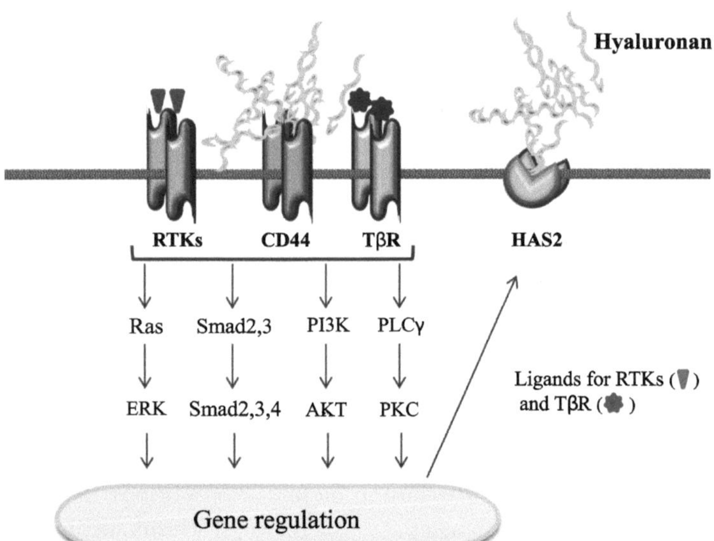

Figure 8.2 Signaling pathways leading to induction of HAS2 expression. Growth factors in an autocrine or paracrine manner activate their cognate receptor which leads to the induction of the *HAS2* gene and HA synthesis. Then, HA-activated CD44 in cooperation with growth factor receptors modulates cellular fate. (See the color plate.)

migration (Heldin & Pertoft, 1993; Li & Heldin, 2001; Wight, 2002). Such matrices might also serve in tumor–stromal interactions by functioning as reservoirs for growth factors under physiological conditions and releasing them in conjunction with breast cancer development.

In addition to mediating ECM signals independently, cell adhesion receptors such as CD44 cooperate with receptor tyrosine kinases, thereby affecting cellular fate (Orian-Rousseau & Ponta, 2008) (Fig. 8.2). For example, HA-stimulated CD44s positively cooperates with ErbB2 (Bourguignon et al., 1997). The CD44v3 and CD44v6 isoforms bind growth factors, including FGF, heparin binding-epidermal growth factor, HGF, and vascular endothelial growth factor, resulting in the activation of each one of their cognate receptors (Jones, Tussey, Athanasou, & Jackson, 2000; Tremmel et al., 2009; Yu, Woessner, McNeish, & Stamenkovic, 2002). The effects of HA on cellular behavior can be concentration dependent. At high HA concentrations, CD44 modulates the response of human dermal fibroblasts to PDGF-BB by suppressing the activity of its receptor and cell migration (Li, Heldin, & Heldin, 2006), whereas low amounts of HA support PDGF-BB-mediated cell proliferation (Li, Asteriou, Bernert, Heldin, & Heldin, 2007).

Rudolph Virchow introduced the term "myxoma" in 1858, describing a soft tissue tumor that structurally resembles umbilical cord (Willems, Wiweger, van Roggen, & Hogendoorn, 2010). Myxoid areas/changes in the ECM are found in various nonneoplastic and neoplastic lesions. Structural molecules, such as collagen, fibronectin, tenascin C, proteoglycans, and glycosaminoglycans, are present in the myxoid ECM. HA is the most commonly changed constituent of the myxoid regions and contributes to the edematous phenotype of myxoid ECM (Willems et al., 2010). Notably, breast cancer stromal myxoid areas with high HA content, which stain positively with Alcian Blue, are associated with positive lymph nodes and high tumor grade (Vignal, Meslet, Romeo, & Feuilhade, 2002; Wernicke et al., 2003). As the tumor progresses, an aberrant accumulation of HA is detected in the surrounding matrix (Auvinen et al., 2000; Boregowda et al., 2006; Corte et al., 2010). These observations further reinforce the importance of HA in breast cancer invasion and metastasis.

Naoki Itano and his colleagues established a *HAS2* conditional knockout mouse strain to address the importance of stromal HA in mammary tumors. The analysis revealed that high amounts of HA in the stroma led to recruitment of tumor-associated macrophages (TAMs) and subsequent tumor neovascularization, in cooperation with versican (Kobayashi et al., 2010;

Koyama et al., 2007, 2008). Thus, better understanding of the molecular mechanisms regulating HA amounts and its interactions in tumor stroma may improve cancer treatment.

4. EXPRESSION OF *HAS* GENES AND BREAST CANCER MALIGNANCY

Much of our knowledge of the molecular mechanisms modulating HA production during breast tumorigenesis is based on *in vivo* experiments and *in vitro* studies performed using breast cancer cell lines. The cell lines provide a convenient experimental resource and sustain their phenotype for a long time in culture and thus provide a suitable system to study gene expression without an impact on their stromal environment (Lacroix & Leclercq, 2004). A few of them, namely, the luminal cancer cells Zr-75-1 and MCF7, as well as the basal-like cancer cells Hs578T, MDA-MB-231 and a clone that was selected for its ability to metastasize to bones, designated MDA-MB-231-BM, have been used to study the expression of CD44, HASes, and HYALs (Bernert, Porsch, & Heldin, 2011; Heldin, de la Torre, Ytterberg, & Bergh, 1996; Li, Li, Brown, & Heldin, 2007; Udabage, Brownlee, Nilsson, & Brown, 2005) (Table 8.1). The expression pattern of clinically critical molecules such as ER, PR, HER2, and EGFR expression has also been investigated (Kao et al., 2009; Lacroix & Leclercq, 2004; Subik et al., 2010) (Table 8.1).

HAS1 mRNA is not detectable or expressed at very low levels in the breast cancer cells studied (Table 8.1); however, *HAS1* has been identified among the 0.9% genes that are required for the growth of $CD44^{+}/CD24^{low-}$ stem cell-like breast cancer cells and its suppression favored reduction in Stat3 activation (Marotta et al., 2011). Malignant transformation of the rat 3Y1 fibroblast cell line with *v-src* and/or *v-fos* induced the expression of both *HAS1* and *HAS2*, and ectopic expression of *HAS1* and *HAS2* genes promoted the growth of subcutaneous tumors (Itano et al., 2004).

Compared to HAS1, HAS2 mRNA is abundantly expressed in the hormone-negative breast cancer cell lines that exhibit aggressive and invasive mesenchymal-like traits (Table 8.1). Interestingly, the HA-synthesizing capacity and expression level of HAS2 is powerfully increased in bone-metastatic cells, (MDA-MB-231-BM), compared to the parental MDA-MB-231 cells. HAS2 expression regulates the invasive phenotype of MDA-MB-231-BM cells by suppressing the expression of tissue metalloproteinase inhibitor 1 (TIMP1). To gain insights about the

Table 8.1 Molecular profiles and expression of HASes and HYALs in different breast cancer cell lines

Cell line	Zr-75-1	MCF7	Hs578T	MDA-MB-231	MDA-MB-231-BM
Subtype[a]	L	L	B	B	B
ER[a]	+	+	–	–	–
PR[a]	+	+	–	–	–
HER2[a]	–	–	–	–	–
EGFR[a]	+	+	+	+	+
CD44[a,b]	–	±	++++	++++	++++
HAS1[b]	±	+	+	+	+
HAS2[b]	±	±	++	++	++++
HAS3[b]	++	++	++	++	++
HYAL1[b]	+	+	+	+	+
HYAL2[b]	+++	+++	+++	+++	++++

[a]Information from ATCC and references: Heldin et al. (1996), Kao et al. (2009), Lacroix and Leclercq (2004), Olofsson et al. (2014), Subik et al. (2010).
[b]Bernert et al. (2011), Li, Asteriou, et al. (2007), Li, Li, et al. (2007), Udabage, Brownlee, Nilsson, et al. (2005), Udabage, Brownlee, Waltham, et al. (2005).
Features of cell lines derived from breast. L, luminal; B, basal.

mechanisms of HAS2 expression and HA production in the invasive phenotype of bone-seeking cells, we examined the activation status of signaling pathways. The analysis revealed that HAS2 expression is required for EGF-mediated activation of Akt, which is controlled by silencing of FAK and/or addition of an inhibitor of PI3K (LY294002). These data suggest that FAK acts downstream of the activated EGF receptor to activate the PI3K/Akt signaling pathway and HAS2 expression is interconnected with signaling and cell invasion (Bernert et al., 2011).

The molecular mechanisms behind the growth and self-renewal of breast CSCs in distant organs are still poorly understood. Importance of HAS2-mediated HA for the growth and self-renewal of the $CD44^{+}/CD24^{low-}$ bone-metastasizing breast CSCs illustrates the clinical significance of HAS2 in breast cancer progression. HA polymerized by HAS2 in bone marrow-metastasizing CSCs couples with CD44 on TAMs to secrete PDGF-BB that in turn activates other stromal cells to create a niche to nourish the metastasizing cancer cells (Okuda et al., 2011). HAS2-synthesized HA has also been shown to stimulate migration and invasion of other aggressive breast cancer cells (Li, Li et al., 2007; Udabage, Brownlee, Waltham, et al., 2005). Thus, HAS2-synthesized HA does confer an invasive and prometastatic phenotype to basal-like breast cancer cells and creates a suitable milieu promoting their growth through interaction with CD44 on bone marrow stromal cells. This underscores the significance of HAS2 as a characteristic of bone-metastasizing breast cancer cells.

Recent studies implicate HAS2 expression in inflammation. For example, HAS2 elevation causes the wrinkled skin phenotype of Shar-Pei dogs and is associated with the periodic fever syndrome (Olsson et al., 2011), as well as autoinflammatory diseases like glaucomatous neurodegeneration (Basu et al., 2012). The intimate relation between a high level of HAS2 and inflammation may further contribute to its protumorigenic effect. Importantly, HAS2 expression was altered in 28% of radiation transformed breast cells and the expression was elevated in grade 3 tumors (Unger et al., 2010). Furthermore, HAS2 is reported to be among 0.3% of highly expressed genes that confer increased invasiveness of pancreatic cancer cells, as also associated with tumor–stromal interactions both in pancreatic and breast cancer (Sato, Maehara, & Goggins, 2004). In contrast, the expression levels of HAS3, HYAL1, and HYAL2 differ only slightly among the cell lines studied. As shown in Table 8.1, a marked difference in the expression of CD44 was detected between cells derived from luminal cancer and those from basal-like cancer.

The expression levels of each HAS isoform appear to be strictly regulated in breast cancer subpopulations, but their significance for the transformation-mediated HA accumulation in breast tumors remains to be elucidated. However, HA synthesized by breast carcinomas has been related to relapses of human breast cancer patients (Auvinen et al., 2013). Thus, the abundant expression of HAS2 in the hormone-negative breast cancer cell lines (Table 8.1) might have a significant clinical relevance in controlling basal-like breast cancer distant metastasis.

5. HA–CD44 INTERACTIONS: A REGULATORY NETWORK DURING TGFβ-MEDIATED EMT

Changes in the expression and/or function of adhesion molecules, such as CD44, on the cell surface of epithelial cells, alter cell–cell and cell–microenvironment interactions. Such modulations include EMT, during which the epithelial cancer cells lose contacts with each other due to a progressive loss of adherence junction proteins, such as E-cadherin, and a concomitant gain of the mesenchymal marker *N*-cadherin, and acquire a mesenchymal migratory phenotype (Moustakas & Heldin, 2014; Thiery et al., 2009). Several growth factors and cytokines, such as TGFβ, promote EMT by inducing certain transcription factors, including Snail2, Twist, Zeb1 and 2, noncoding RNAs, and alternative splicing (De Craene & Berx, 2013).

CD44 undergoes alternative splicing in conjunction with EMT. A switch in the expression of CD44v isoforms to the CD44s isoform occurs by inhibition of splicing activity, thus promoting EMT as well as breast cancer progression by suppression of E-cadherin via activated PI3K/Akt pathway (Brown et al., 2011). The production of CD44v isoforms by alternative splicing is regulated by a group of splicing factors including epithelial splicing regulatory protein 1 and 2 (ESRP1 and 2) that are downregulated by the EMT-induced transcription factors Snail1, Zeb1, and Zeb2, respectively (Reinke, Xu, & Cheng, 2012). Notably, TGFβ downregulates ESRP genes (Horiguchi et al., 2012), suggesting a correlation between TGFβ stimulation and CD44 splicing. The EMT process is also controlled by microRNAs that target EMT-stimulatory proteins (De Craene & Berx, 2013). In breast cancer, complete EMT is associated with poor prognosis of patients with carcinosarcomas (Sarrio et al., 2008); breast cancer cells derived from patients with a basal-like phenotype abundantly express CD44s that exhibits high HA-binding capacity (Bernert et al., 2011; Heldin et al., 1996). Despite extensive research, the function of CD44 in breast cancer tumorigenesis is unclear. CD44 can function as a molecular switch regulating cell growth, adhesion, or metastasis, depending on its interactions with high- or low-molecular-mass HA and subsequent recruitment of specific cytoplasmic signaling mediators (Lopez et al., 2005; Louderbough & Schroeder, 2011).

Binding of HA leads to clustering of CD44, thereby triggering downstream signaling on its own and/or in cooperation with growth factor receptors, such as the receptors for PDGF-BB, TGFβ, and ErbB2 (also designated HER2), as well as by interacting with key regulatory cytoplasmic proteins, such as ERM-related proteins, merlin, and IQGAP1 (Ghatak, Misra, & Toole, 2005; Li et al., 2006; Skandalis, Kozlova, Engstrom, Hellman, & Heldin, 2010; Toole & Slomiany, 2008). Previous studies demonstrated that HA-activated CD44 plays an important role in breast cancer metastasis by promoting the invasion and adhesion of tumor cells to bone marrow endothelial cells; the underlying mechanisms are, however, poorly understood (Hill et al., 2006). As shown in Table 8.1, basal-like cells that are highly metastatic exhibit high expression of CD44s. It has been demonstrated that although TGFβ signaling is the major regulator of EMT and maintains the stem cell phenotype (Scheel et al., 2011), CD44s, but not the variant isoforms, plays an important role in TGFβ-mediated mesenchymal phenotype associated with vimentin expression and high levels of Smad2 in nuclei (Mima et al., 2012; Ouhtit et al., 2013). Nevertheless, recent studies in our laboratory revealed that signals triggered via the binding of HAS2-synthesized HA on bone-metastasizing breast cancer cells, to CD44 expressed by microvascular endothelial cells, promotes the adhesiveness between breast cancer cells and endothelial cells and thus cancer cell intravasation (Olofsson, Porsch, & Heldin, 2014). Furthermore, the high HA content in the glycocalyx most likely becomes fragmented by the action of HYAL2; it is possible that such fragments can bind to CD44 on endothelial cells and promote tubulogenesis by inducing the expression of the cytokines CXCL9 and CXCL12, as well as their receptors (Olofsson et al., 2014). These results indicate that high expression of CD44s by breast cancer cells and stromal microvascular cells promotes the aggressive phenotype.

TGFβ stimulates HA production in immortal NMuMG breast epithelial cultures by induction of HAS2 expression via activation of the Smad and p38 MAPK pathways (Porsch et al., 2013). Most intriguingly, TGFβ-dependent HAS2 expression, independently of its HA-producing activity and the CD44 expression of the cells, was shown to have a regulatory role in TGFβ-induced EMT. Suppression of HAS2 resulted in a significant decrease of the EMT markers fibronectin, Snail1, and Zeb1 in NMuMG cells, as well as inhibition of TGFβ-induced cell migration, whereas CD44 knockdown did not. This finding implies a novel functional role of HAS2 independent of its HA-synthesizing capacity. Notably, TIMP-1, the expression of which is reversely correlated to HAS2 expression (Bernert et al., 2011), has a modulatory function during EMT, independently of its matrix metalloproteinase inhibition functions (Jung et al., 2012). Other studies have also demonstrated the efficiency of HAS2 overexpression in mediating an invasive malignant phenotype via suppression of E-cadherin and concomitant nuclear staining of β-catenin, as well as stimulation of the PI3K/Akt pathway (Koyama et al., 2007; Zoltan-Jones, Huang, Ghatak, & Toole, 2003). A better understanding of the molecular mechanisms underlying the functions of HAS2 and CD44 in TGFβ-mediated EMT phenotype of breast cancer cells is required. Figure 8.3 depicts some of the signaling pathways through which HAS2 and CD44 affect breast cancer progression.

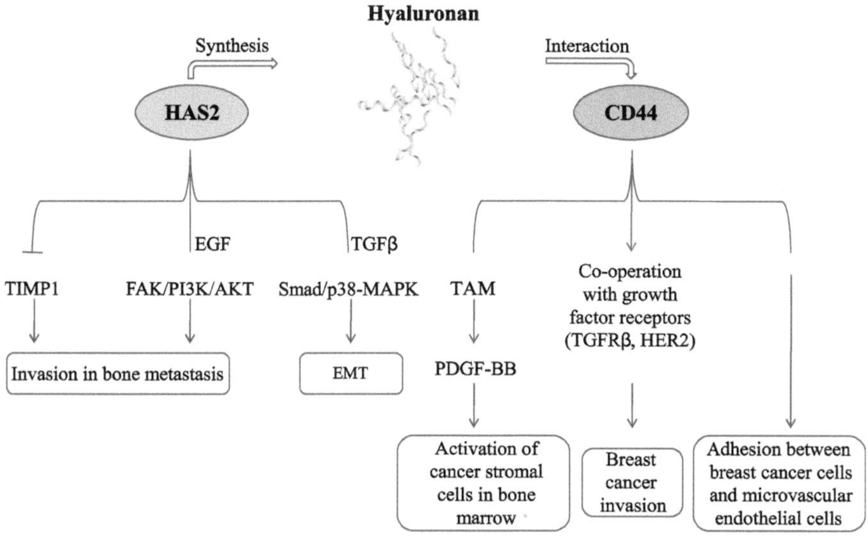

Figure 8.3 Importance of HAS2 and CD44 in breast cancer tumorigenesis. Illustration of signaling pathways mediating HA-activated CD44 function during breast cancer progression. (See the color plate.)

6. CONCLUSIONS

Breast cancer is the most common cancer in females worldwide and metastasis is thought to be the primary cause of treatment failure of the disease. During breast cancer metastasis, there is an intimate interplay between cells and ECM. For instance, HA–CD44 signaling most likely plays a role in basal-like breast cancer progression by endowing these subtypes with a highly hydrated environment facilitating cellular invasiveness. The different regulatory mechanisms modulating the expression of *HAS1*, *HAS2*, and *HAS3* genes in luminal and basal cell-like breast cancer cells lead to different physiological responses; a better understanding of the underlying molecular mechanisms is warranted. Furthermore, the importance of HAS2 in TGFβ-induced breast cancer EMT needs to be explored further.

ACKNOWLEDGMENTS

We thank Carl-Henrik Heldin for his critical and insightful comments on this review. We also thank past and present members of our group for their contributions to our studies. We acknowledge funding by the Ludwig Institute for Cancer Research, the Swedish Cancer Society, and Uppsala University.

REFERENCES

Adamia, S., Maxwell, C. A., & Pilarski, L. M. (2005). HA and HA synthases: Potential therapeutic targets in cancer. *Current Drug Targets: Cardiovascular & Haematological Disorders*, *5*(1), 3–14.

Al-Hajj, M., Wicha, M. S., Benito-Hernandez, A., Morrison, S. J., & Clarke, M. F. (2003). Prospective identification of tumorigenic breast cancer cells. *Proceedings of the National Academy of Sciences of the United States of America*, *100*(7), 3983–3988.

Angelucci, C., Maulucci, G., Lama, G., Proietti, G., Colabianchi, A., Papi, M., et al. (2012). Epithelial-stromal interactions in human breast cancer: Effects on adhesion, plasma membrane fluidity and migration speed and directness. *PLoS One*, 7(12), e50804. http://dx.doi.org/10.1371/journal.pone.0050804.

Asplund, T., Versnel, M. A., Laurent, T. C., & Heldin, P. (1993). Human mesothelioma cells produce factors that stimulate the production of HA by mesothelial cells and fibroblasts. *Cancer Research*, *53*, 388–392.

Auvinen, P., Tammi, R., Kosma, V. M., Sironen, R., Soini, Y., Mannermaa, A., et al. (2013). Increased HA content and stromal cell CD44 associate with HER2 positivity and poor prognosis in human breast cancer. *International Journal of Cancer*, *132*(3), 531–539.

Auvinen, P., Tammi, R., Parkkinen, J., Tammi, M., Agren, U., Johansson, R., et al. (2000). HA in peritumoral stroma and malignant cells associates with breast cancer spreading and predicts survival. *The American Journal of Pathology*, *156*(2), 529–536.

Basu, K., Sen, A., Ray, K., Ghosh, I., Datta, K., & Mukhopadhyay, A. (2012). Genetic association and gene-gene interaction of HAS2, HABP1 and HYAL3 implicate HA metabolic genes in glaucomatous neurodegeneration. *Disease Markers*, *33*(3), 145–154.

Benitez, A., Yates, T. J., Lopez, L. E., Cerwinka, W. H., Bakkar, A., & Lokeshwar, V. B. (2011). Targeting hyaluronidase for cancer therapy: Antitumor activity of sulfated hyaluronic acid in prostate cancer cells. *Cancer Research*, *71*(12), 4085–4095.

Bernert, B., Porsch, H., & Heldin, P. (2011). HA synthase 2 (HAS2) promotes breast cancer cell invasion by suppression of tissue metalloproteinase inhibitor 1 (TIMP-1). *The Journal of Biological Chemistry*, *286*(49), 42349–42359.

Bhowmick, N. A., & Moses, H. L. (2005). Tumor-stroma interactions. *Current Opinion in Genetics & Development*, *15*(1), 97–101.

Bissell, M. J., & Radisky, D. (2001). Putting tumours in context. *Nature Reviews Cancer*, *1*(1), 46–54.

Bissell, M. J., Rizki, A., & Mian, I. S. (2003). Tissue architecture: The ultimate regulator of breast epithelial function. *Current Opinion in Cell Biology*, *15*(6), 753–762.

Boregowda, R. K., Appaiah, H. N., Siddaiah, M., Kumarswamy, S. B., Sunila, S., Thimmaiah, K. N., et al. (2006). Expression of HA in human tumor progression. *Journal of Carcinogenesis*, *5*, 2.

Bourguignon, L. Y., Zhu, H., Chu, A., Iida, N., Zhang, L., & Hung, M. C. (1997). Interaction between the adhesion receptor, CD44, and the oncogene product, p185HER2, promotes human ovarian tumor cell activation. *The Journal of Biological Chemistry*, *272*(44), 27913–27918.

Brinck, J., & Heldin, P. (1999). Expression of recombinant HA synthase (HAS) isoforms in CHO cells reduces cell migration and cell surface CD44. *Experimental Cell Research*, *252*, 342–351.

Brown, R. L., Reinke, L. M., Damerow, M. S., Perez, D., Chodosh, L. A., Yang, J., et al. (2011). CD44 splice isoform switching in human and mouse epithelium is essential for epithelial-mesenchymal transition and breast cancer progression. *The Journal of Clinical Investigation*, *121*(3), 1064–1074.

Charafe-Jauffret, E., Ginestier, C., Monville, F., Finetti, P., Adelaide, J., Cervera, N., et al. (2006). Gene expression profiling of breast cell lines identifies potential new basal markers. *Oncogene*, *25*(15), 2273–2284.

Chaturvedi, S., & Hass, R. (2011). Extracellular signals in young and aging breast epithelial cells and possible connections to age-associated breast cancer development. *Mechanisms of Ageing and Development, 132*(5), 213–219.

Corte, M. D., Gonzalez, L. O., Junquera, S., Bongera, M., Allende, M. T., & Vizoso, F. J. (2010). Analysis of the expression of HA in intraductal and invasive carcinomas of the breast. *Journal of Cancer Research and Clinical Oncology, 136*(5), 745–750.

De Craene, B., & Berx, G. (2013). Regulatory networks defining EMT during cancer initiation and progression. *Nature Reviews Cancer, 13*(2), 97–110.

Evanko, S. P., Johnson, P. Y., Braun, K. R., Underhill, C. B., Dudhia, J., & Wight, T. N. (2001). Platelet-derived growth factor stimulates the formation of versican-HA aggregates and pericellular matrix expansion in arterial smooth muscle cells. *Archives of Biochemistry and Biophysics, 394*(1), 29–38.

Ghatak, S., Misra, S., & Toole, B. P. (2005). HA constitutively regulates ErbB2 phosphorylation and signaling complex formation in carcinoma cells. *The Journal of Biological Chemistry, 280*(10), 8875–8883.

Goentzel, B. J., Weigel, P. H., & Steinberg, R. A. (2006). Recombinant human HA synthase 3 is phosphorylated in mammalian cells. *The Biochemical Journal, 396*(2), 347–354.

Goodison, S., Urquidi, V., & Tarin, D. (1999). CD44 cell adhesion molecules. *Molecular Pathology, 52*(4), 189–196.

Heldin, P., Basu, K., Olofsson, B., Porsch, H., Kozlova, I., & Kahata, K. (2013). Deregulation of HA synthesis, degradation and binding promotes breast cancer. *Journal of Biochemistry, 154*(5), 395–408.

Heldin, P., de la Torre, M., Ytterberg, D., & Bergh, J. (1996). Differential synthesis and binding of HA by human breast cancer cell lines: Relationship to hormone receptor status. *Oncology Reports, 3*, 1011–1016.

Heldin, P., Karousou, E., & Skandalis, S. (2009). *Growth factor regulation of HA deposition in malignancies*. R. Stern (Ed.), (1st ed.). USA: Elsevier Press.

Heldin, P., Laurent, T. C., & Heldin, C.-H. (1989). Effect of growth factors on HA synthesis in cultured human fibroblasts. *The Biochemical Journal, 258*, 919–922.

Heldin, P., & Pertoft, H. (1993). Synthesis and assembly of the HA-containing coats around normal human mesothelial cells. *Experimental Cell Research, 208*, 422–429.

Hill, A., McFarlane, S., Mulligan, K., Gillespie, H., Draffin, J. E., Trimble, A., et al. (2006). Cortactin underpins CD44-promoted invasion and adhesion of breast cancer cells to bone marrow endothelial cells. *Oncogene, 25*(45), 6079–6091.

Horiguchi, K., Sakamoto, K., Koinuma, D., Semba, K., Inoue, A., Inoue, S., et al. (2012). TGF-beta drives epithelial-mesenchymal transition through deltaEF1-mediated downregulation of ESRP. *Oncogene, 31*(26), 3190–3201.

Iozzo, R. V. (1998). Matrix proteoglycans: From molecular design to cellular function. *Annual Review of Biochemistry, 67*, 609–652.

Itano, N., Sawai, T., Atsumi, F., Miyaishi, O., Taniguchi, S., Kannagi, R., et al. (2004). Selective expression and functional characteristics of three mammalian HA synthases in oncogenic malignant transformation. *The Journal of Biological Chemistry, 279*(18), 18679–18687.

Jacobson, A., Brinck, J., Briskin, M. J., Spicer, A. P., & Heldin, P. (2000). Expression of human HA synthases in response to external stimuli. *The Biochemical Journal, 348*, 29–35.

Jacobson, A., Rahmanian, M., Rubin, K., & Heldin, P. (2002). Expression of HA synthase 2 or hyaluronidase 1 differentially affect the growth rate of transplantable colon carcinoma cell tumors. *International Journal of Cancer, 102*, 212–219.

Jones, M., Tussey, L., Athanasou, N., & Jackson, D. G. (2000). Heparan sulfate proteoglycan isoforms of the CD44 HA receptor induced in human inflammatory macrophages can function as paracrine regulators of fibroblast growth factor action. *The Journal of Biological Chemistry, 275*(11), 7964–7974.

Jung, Y. S., Liu, X. W., Chirco, R., Warner, R. B., Fridman, R., & Kim, H. R. (2012). TIMP-1 induces an EMT-like phenotypic conversion in MDCK cells independent of its MMP-inhibitory domain. *PLoS One*, 7(6), e38773.

Kadar, A., Tokes, A. M., Kulka, J., & Robert, L. (2002). Extracellular matrix components in breast carcinomas. *Seminars in Cancer Biology*, *12*(3), 243–257.

Kao, J., Salari, K., Bocanegra, M., Choi, Y. L., Girard, L., Gandhi, J., et al. (2009). Molecular profiling of breast cancer cell lines defines relevant tumor models and provides a resource for cancer gene discovery. *PLoS One*, *4*(7), e6146.

Karousou, E., Kamiryo, M., Skandalis, S. S., Ruusala, A., Asteriou, T., Passi, A., et al. (2010). The activity of HA synthase 2 is regulated by dimerization and ubiquitination. *The Journal of Biological Chemistry*, *285*(31), 23647–23654.

Kobayashi, N., Miyoshi, S., Mikami, T., Koyama, H., Kitazawa, M., Takeoka, M., et al. (2010). HA deficiency in tumor stroma impairs macrophage trafficking and tumor neovascularization. *Cancer Research*, *70*(18), 7073–7083.

Korkaya, H., Liu, S., & Wicha, M. S. (2011). Breast cancer stem cells, cytokine networks, and the tumor microenvironment. *The Journal of Clinical Investigation*, *121*(10), 3804–3809.

Koyama, H., Hibi, T., Isogai, Z., Yoneda, M., Fujimori, M., Amano, J., et al. (2007). Hyperproduction of HA in neu-induced mammary tumor accelerates angiogenesis through stromal cell recruitment: Possible involvement of versican/PG-M. *The American Journal of Pathology*, *170*(3), 1086–1099.

Koyama, H., Kobayashi, N., Harada, M., Takeoka, M., Kawai, Y., Sano, K., et al. (2008). Significance of tumor-associated stroma in promotion of intratumoral lymphangiogenesis: Pivotal role of a HA-rich tumor microenvironment. *The American Journal of Pathology*, *172*(1), 179–193.

Lacroix, M., & Leclercq, G. (2004). Relevance of breast cancer cell lines as models for breast tumours: An update. *Breast Cancer Research and Treatment*, *83*(3), 249–289.

Laurent, T. C., & Fraser, J. R. E. (1992). Hyaluronan. *FASEB Journal*, *6*, 2397–2404.

Li, L., Asteriou, T., Bernert, B., Heldin, C. H., & Heldin, P. (2007). Growth factor regulation of HA synthesis and degradation in human dermal fibroblasts: Importance of HA for the mitogenic response of PDGF-BB. *The Biochemical Journal*, *404*(2), 327–336.

Li, Y., & Heldin, P. (2001). HA production increases the malignant properties of mesothelioma cells. *British Journal of Cancer*, *85*, 600–607.

Li, L., Heldin, C. H., & Heldin, P. (2006). Inhibition of platelet-derived growth factor-BB-induced receptor activation and fibroblast migration by HA activation of CD44. *The Journal of Biological Chemistry*, *281*(36), 26512–26519.

Li, Y., Li, L., Brown, T. J., & Heldin, P. (2007). Silencing of HA synthase 2 suppresses the malignant phenotype of invasive breast cancer cells. *International Journal of Cancer*, *120*(12), 2557–2567.

Li, Y., Rahmanian, M., Widstrom, C., Lepperdinger, G., Frost, G. I., & Heldin, P. (2000). Irradiation-induced expression of HA (HA) synthase 2 and hyaluronidase 2 genes in rat lung tissue accompanies active turnover of HA and induction of types I and III collagen gene expression. *American Journal of Respiratory Cell and Molecular Biology*, *23*(3), 411–418.

Liu, S., Cong, Y., Wang, D., Sun, Y., Deng, L., Liu, Y., et al. (2014). Breast cancer stem cells transition between epithelial and mesenchymal states reflective of their normal counterparts. *Stem Cell Reports*, *2*(1), 78–91.

Lopez, J. I., Camenisch, T. D., Stevens, M. V., Sands, B. J., McDonald, J., & Schroeder, J. A. (2005). CD44 attenuates metastatic invasion during breast cancer progression. *Cancer Research*, *65*(15), 6755–6763.

Louderbough, J. M., & Schroeder, J. A. (2011). Understanding the dual nature of CD44 in breast cancer progression. *Molecular Cancer Research*, *9*(12), 1573–1586.

Malhotra, G. K., Zhao, X., Band, H., & Band, V. (2010). Histological, molecular and functional subtypes of breast cancers. *Cancer Biology & Therapy*, *10*(10), 955–960.

Marotta, L. L., Almendro, V., Marusyk, A., Shipitsin, M., Schemme, J., Walker, S. R., et al. (2011). The JAK2/STAT3 signaling pathway is required for growth of CD44(+)CD24(-) stem cell-like breast cancer cells in human tumors. *The Journal of Clinical Investigation, 121*(7), 2723–2735.

Mima, K., Okabe, H., Ishimoto, T., Hayashi, H., Nakagawa, S., Kuroki, H., et al. (2012). CD44s regulates the TGF-beta-mediated mesenchymal phenotype and is associated with poor prognosis in patients with hepatocellular carcinoma. *Cancer Research, 72*(13), 3414–3423.

Moustakas, A., & Heldin, P. (2014). TGFbeta and matrix-regulated epithelial to mesenchymal transition. *Biochim. Biophys. Acta,* http://dx.doi.org/10.1016/j.bbagen.2014.02.004.

Muschler, J., & Streuli, C. H. (2010). Cell-matrix interactions in mammary gland development and breast cancer. *Cold Spring Harb Perspect Biol,* http://dx.doi.org/10.1101/cshperspect.a003202.

Nakshatri, H., Srour, E. F., & Badve, S. (2009). Breast cancer stem cells and intrinsic subtypes: Controversies rage on. *Current Stem Cell Research & Therapy, 4*(1), 50–60.

Neve, R. M., Chin, K., Fridlyand, J., Yeh, J., Baehner, F. L., Fevr, T., et al. (2006). A collection of breast cancer cell lines for the study of functionally distinct cancer subtypes. *Cancer Cell, 10*(6), 515–527.

Okuda, H., Kobayashi, A., Xia, B., Watabe, M., Pai, S. K., Hirota, S., et al. (2011). HA synthase HAS2 promotes tumor progression in bone by stimulating the interaction of breast cancer stem-like cells with macrophages and stromal cells. *Cancer Research, 72*(2), 537–547.

Olofsson, B., Porsch, H., & Heldin, P. (2014). Knock-down of CD44 regulates endothelial cell differentiation via NFkappaB-mediated chemokine production. *PLoS One, 9*(3), e90921.

Olsson, M., Meadows, J. R., Truve, K., Rosengren Pielberg, G., Puppo, F., Mauceli, E., et al. (2011). A novel unstable duplication upstream of HAS2 predisposes to a breed-defining skin phenotype and a periodic fever syndrome in Chinese Shar-Pei dogs. *PLoS Genetics,* 7(3), e1001332.

Orian-Rousseau, V., & Ponta, H. (2008). Adhesion proteins meet receptors: A common theme? *Advances in Cancer Research, 101*, 63–92.

Ouhtit, A., Madani, S., Gupta, I., Shanmuganathan, S., Abdraboh, M. E., Al-Riyami, H., et al. (2013). TGF-beta2: A novel target of CD44-promoted breast cancer invasion. *Journal of Cancer, 4*(7), 566–572.

Park, J., & Schwarzbauer, J. E. (2014). Mammary epithelial cell interactions with fibronectin stimulate epithelial-mesenchymal transition. *Oncogene, 33*(13), 1649–1657.

Polyak, K., Haviv, I., & Campbell, I. G. (2009). Co-evolution of tumor cells and their microenvironment. *Trends in Genetics, 25*(1), 30–38.

Ponta, H., Sherman, L., & Herrlich, P. A. (2003). CD44: From adhesion molecules to signalling regulators. *Nature Reviews Molecular Cell Biology, 4*(1), 33–45.

Porsch, H., Bernert, B., Mehic, M., Theocharis, A. D., Heldin, C. H., & Heldin, P. (2013). Efficient TGFbeta-induced epithelial-mesenchymal transition depends on HA synthase HAS2. *Oncogene, 32*(37), 4355–4365.

Reinke, L. M., Xu, Y., & Cheng, C. (2012). Snail represses the splicing regulator epithelial splicing regulatory protein 1 to promote epithelial-mesenchymal transition. *The Journal of Biological Chemistry, 287*(43), 36435–36442.

Roskelley, C. D., & Bissell, M. J. (2002). The dominance of the microenvironment in breast and ovarian cancer. *Seminars in Cancer Biology, 12*(2), 97–104.

Sarrio, D., Rodriguez-Pinilla, S. M., Hardisson, D., Cano, A., Moreno-Bueno, G., & Palacios, J. (2008). Epithelial-mesenchymal transition in breast cancer relates to the basal-like phenotype. *Cancer Research, 68*(4), 989–997.

Sato, N., Maehara, N., & Goggins, M. (2004). Gene expression profiling of tumor-stromal interactions between pancreatic cancer cells and stromal fibroblasts. *Cancer Research, 64*(19), 6950–6956.

Scheel, C., Eaton, E. N., Li, S. H., Chaffer, C. L., Reinhardt, F., Kah, K. J., et al. (2011). Paracrine and autocrine signals induce and maintain mesenchymal and stem cell states in the breast. *Cell, 145*(6), 926–940.

Simpson, M. A., & Lokeshwar, V. B. (2008). HA and hyaluronidase in genitourinary tumors. *Frontiers in Bioscience, 13*, 5664–5680.

Skandalis, S. S., Kozlova, I., Engstrom, U., Hellman, U., & Heldin, P. (2010). Proteomic identification of CD44 interacting proteins. *IUBMB Life, 62*(11), 833–840.

Sorlie, T., Perou, C. M., Tibshirani, R., Aas, T., Geisler, S., Johnsen, H., et al. (2001). Gene expression patterns of breast carcinomas distinguish tumor subclasses with clinical implications. *Proceedings of the National Academy of Sciences of the United States of America, 98*(19), 10869–10874.

Stern, R. (2008). Hyaluronidases in cancer biology. *Seminars in Cancer Biology, 18*(4), 275–280.

Stern, R., & Jedrzejas, M. J. (2006). Hyaluronidases: Their genomics, structures, and mechanisms of action. *Chemical Reviews, 106*(3), 818–839.

Subik, K., Lee, J. F., Baxter, L., Strzepek, T., Costello, D., Crowley, P., et al. (2010). The expression patterns of ER, PR, HER2, CK5/6, EGFR, Ki-67 and AR by immunohistochemical analysis in breast cancer cell lines. *Breast Cancer (Auckland), 4*, 35–41.

Suzuki, M., Asplund, T., Yamashita, H., Heldin, C.-H., & Heldin, P. (1995). Stimulation of HA biosynthesis by platelet-derived growth factor-BB and transforming growth factor-β1 involves activation of protein kinase C. *The Biochemical Journal, 307*, 817–821.

Tammi, R. H., Passi, A. G., Rilla, K., Karousou, E., Vigetti, D., Makkonen, K., et al. (2011). Transcriptional and post-translational regulation of HA synthesis. *The FEBS Journal, 278*, 1419–1428.

Taylor, K. R., & Gallo, R. L. (2006). Glycosaminoglycans and their proteoglycans: Host-associated molecular patterns for initiation and modulation of inflammation. *FASEB Journal, 20*(1), 9–22.

Thiery, J. P., Acloque, H., Huang, R. Y., & Nieto, M. A. (2009). Epithelial-mesenchymal transitions in development and disease. *Cell, 139*(5), 871–890.

Toole, B. P. (2004). HA: From extracellular glue to pericellular cue. *Nature Reviews Cancer, 4*(7), 528–539.

Toole, B. P., & Slomiany, M. G. (2008). HA: A constitutive regulator of chemoresistance and malignancy in cancer cells. *Seminars in Cancer Biology, 18*(4), 244–250.

Tremmel, M., Matzke, A., Albrecht, I., Laib, A. M., Olaku, V., Ballmer-Hofer, K., et al. (2009). A CD44v6 peptide reveals a role of CD44 in VEGFR-2 signaling and angiogenesis. *Blood, 114*(25), 5236–5244.

Tsang, J. Y., Huang, Y. H., Luo, M. H., Ni, Y. B., Chan, S. K., Lui, P. C., et al. (2012). Cancer stem cell markers are associated with adverse biomarker profiles and molecular subtypes of breast cancer. *Breast Cancer Research and Treatment, 136*(2), 407–417.

Turley, E. A., Noble, P. W., & Bourguignon, L. Y. (2002). Signaling properties of HA receptors. *The Journal of Biological Chemistry, 277*(7), 4589–4592.

Tyan, S. W., Kuo, W. H., Huang, C. K., Pan, C. C., Shew, J. Y., Chang, K. J., et al. (2011). Breast cancer cells induce cancer-associated fibroblasts to secrete hepatocyte growth factor to enhance breast tumorigenesis. *PLoS One, 6*(1), e15313.

Udabage, L., Brownlee, G. R., Nilsson, S. K., & Brown, T. J. (2005). The over-expression of HAS2, Hyal-2 and CD44 is implicated in the invasiveness of breast cancer. *Experimental Cell Research, 310*(1), 205–217.

Udabage, L., Brownlee, G. R., Waltham, M., Blick, T., Walker, E. C., Heldin, P., et al. (2005). Antisense-mediated suppression of HA synthase 2 inhibits the tumorigenesis and progression of breast cancer. *Cancer Research, 65*(14), 6139–6150.

Unger, K., Wienberg, J., Riches, A., Hieber, L., Walch, A., Brown, A., et al. (2010). Novel gene rearrangements in transformed breast cells identified by high-resolution breakpoint analysis of chromosomal aberrations. *Endocrine-Related Cancer, 17*(1), 87–98.

Vigetti, D., Clerici, M., Deleonibus, S., Karousou, E., Viola, M., Moretto, P., et al. (2011). HA synthesis is inhibited by adenosine monophosphate-activated protein kinase through the regulation of HAS2 activity in human aortic smooth muscle cells. *Journal of Biological Chemistry, 286*(10), 7917–7924.

Vignal, P., Meslet, M. R., Romeo, J. M., & Feuilhade, F. (2002). Sonographic morphology of infiltrating breast carcinoma: Relationship with the shape of the HA extracellular matrix. *Journal of Ultrasound in Medicine, 21*(5), 532–538.

Weigel, P. H., Hascall, V. C., & Tammi, M. (1997). HA synthases. *The Journal of Biological Chemistry, 272*(22), 13997–14000.

Wernicke, M., Pineiro, L. C., Caramutti, D., Dorn, V. G., Raffo, M. M., Guixa, H. G., et al. (2003). Breast cancer stromal myxoid changes are associated with tumor invasion and metastasis: A central role for HA. *Modern Pathology, 16*(2), 99–107.

Wight, T. N. (2002). Versican: A versatile extracellular matrix proteoglycan in cell biology. *Current Opinion in Cell Biology, 14*(5), 617–623.

Willems, S. M., Wiweger, M., van Roggen, J. F. G., & Hogendoorn, P. C. W. (2010). Running GAGs: Myxoid matrix in tumor pathology revisited. *Virchows Archiv, 456*(2), 181–192.

Yu, W. H., Woessner, J. F., Jr., McNeish, J. D., & Stamenkovic, I. (2002). CD44 anchors the assembly of matrilysin/MMP-7 with heparin-binding epidermal growth factor precursor and ErbB4 and regulates female reproductive organ remodeling. *Genes & Development, 16*(3), 307–323.

Zoltan-Jones, A., Huang, L., Ghatak, S., & Toole, B. P. (2003). Elevated HA production induces mesenchymal and transformed properties in epithelial cells. *The Journal of Biological Chemistry, 278*(46), 45801–45810.

CHAPTER NINE

CD44 is a Multidomain Signaling Platform that Integrates Extracellular Matrix Cues with Growth Factor and Cytokine Signals

Véronique Orian-Rousseau*[,1], Jonathan Sleeman*[,†]
*Karlsruhe Institute of Technology, Institute of Toxicology and Genetics, Karlsruhe, Germany
†Centre for Biomedicine and Medical Technology Mannheim (CBTM), Universitätsmedizin Mannheim, University of Heidelberg, Mannheim, Germany
[1]Corresponding author: e-mail address: veronique.orian-rousseau@kit.edu

Contents

1. Introduction 232
2. Ligation of ECM Components by CD44 234
   2.1 Hyaluronic acid 234
   2.2 Other ECM components that are CD44 ligands 237
3. CD44: Coreceptor for Cell-Surface Receptors 238
   3.1 The role of CD44 in regulating HGF-induced Met activation 238
   3.2 CD44-regulated activation of other RTKs: VEGRF, PDGFRβ, FGFR, and EGFR 241
   3.3 Involvement of CD44 in the signaling of TGF family members 242
   3.4 CD44 and GPCRs 243
   3.5 Other receptors 244
4. CD44 as a Multidomain Signal Integration Platform 244
   4.1 HA cross-regulation of the CD44 coreceptor function 245
   4.2 Integration of other ECM signals with the activation of RTKs 246
   4.3 Signal integration through cytoplasmic partners 247
5. Concluding Remarks 248
References 249

Abstract

The reception and integration of the plethora of signals a cell receives from its microenvironment is decisive in determining cell behavior. Perturbation of extracellular cues, or an inappropriate response to or integration of these signals lies at the root of many diseases such as cancer. The transmembrane protein CD44 contributes to the reception of a broad variety of microenvironmental components, including extracellular matrix constituents such as hyaluronic acid, as well as growth factors and cytokines. In this

Advances in Cancer Research, Volume 123
ISSN 0065-230X
http://dx.doi.org/10.1016/B978-0-12-800092-2.00009-5

chapter, we review the range of extracellular cues that are recognized by CD44, and show how CD44 serves to integrate this information at several levels through the mechanisms by which it contributes to transduction of these various microenvironmental signals.

1. INTRODUCTION

The behavior and properties of a cell are tightly controlled by the microenvironment of the cell. Thus, signals from the microenvironment control a wide range of cellular processes, including proliferation, survival, differentiation, identity, motility, secretion, morphology, metabolism, and size (reviewed in Kholodenko, 2003). Important targets for these signals include signal transduction components, the cytoskeleton, junctions with other cells, metabolic pathways, the epigenome, gene transcription, and alternative splicing. External signals that regulate cell behavior and properties come in many shapes, forms, and spatial conformations, including signaling molecules such as growth factors and cytokines, extracellular matrix (ECM) components, contacts with other cells, mechanical stress, the redox status of the surrounding milieu, and the local oxygen tension. All of these concomitant microenvironmental signals have to be interpreted and integrated by the cell, so that an appropriate coordinated response ensues. Perturbation of microenvironmental cues, or inappropriate reception and integration of these external signals lies at the root of many diseases. For example, it is increasingly recognized that changes in the extracellular environment can drive tumor progression and metastasis, and can act dominantly over genetic aberrations in the tumor cells (reviewed in Ruckert, Grutzmann, & Pilarsky, 2012; Soto & Sonnenschein, 2011).

Our understanding of the integration of microenvironmental signals by cells is still at an early stage, but seems to occur at many levels (reviewed in Pawson & Scott, 2010). Cell-surface receptors for particular ligands can cross-regulate the activity of receptors for other ligands, whether through direct complex formation within the plasma membrane, regulation of ligand accessibility or affinity, or intracellularly through the complex of proteins that interact with the cytoplasmic portion of the receptors concerned (reviewed in Dudanova & Klein, 2013; Groves & Kuriyan, 2010). For example, integrins bind to a number of ECM components and can act synergistically with growth factor receptors through common components of their respective receptor-associated intracellular protein complexes (reviewed in Kholodenko, Hancock, & Kolch, 2010). Downstream, posttranslation modification and interactions between signal transduction components that transmit signals from the cell surface to the nucleus also play an important role in integrating extracellular signals (Deribe, Pawson, & Dikic, 2010; Dibble & Manning, 2013; Irvine, 2012; McNeill & Woodgett, 2010; Sweatt, 2001). At the genomic level, activation of specific transcription factors by particular external cues and their subsequent combined effects on transcription also contributes to signal integration (Panne, 2008). Clearly, much remains to be learned about how cells integrate signals in their response to extracellular cues. In turn, this will allow us to understand how inappropriate signal reception and integration can drive diseases such as cancer, allowing more effective intervention strategies to be developed.

CD44 is a family of transmembrane glycoproteins that is emerging as a multidomain platform, which integrates a variety of extracellular information, thereby regulating cell–cell and cell–matrix adhesion, and controlling cell proliferation, differentiation, migration, and survival. CD44 proteins have a common structure consisting of three major domains (reviewed in Orian-Rousseau, 2010). The extracellular portion or ectodomain (ECD) is composed of an N-terminal globular domain and a stem membrane-proximal region. In addition, there is a transmembrane domain and a cytoplasmic or intracellular domain (ICD). All CD44 isoforms are encoded by one single gene present on chromosome 11 in humans and on chromosome 2 in mice. The *CD44* gene is composed of 20 exons that are called constant and variable exons. Exons 1–5 and 16–20 are constant exons that encode the N-terminal and C-terminal regions of CD44, respectively. Exon 19 is spliced out in most CD44 isoforms as its inclusion generates a short five amino acid cytoplasmic tail (Goldstein et al., 1989). Ten variant exons (v1–v10), encoded by exons 6–15, can be alternatively spliced and thereby included in various combinations in the stem region, giving rise to many CD44 variant isoforms. In human cells, exon v1 (exon 6) encodes a stop codon and is not expressed (Screaton, Bell, Bell, & Jackson, 1993). The heterogeneity of the CD44 family of proteins is not only created by alternative splicing but also by extensive N- and O-linked glycosylation.

The smallest CD44 isoform, namely CD44s, does not contain any variant exon in its ECD and is virtually ubiquitously expressed. In contrast, the expression of CD44 variant isoforms is restricted to specific tissues (reviewed in Goodison, Urquidi, & Tarin, 1999). A CD44 epithelial isoform (CD44E) has been described that includes the exons v8–v10 in addition to the constant sequences. CD44v6 is mainly expressed in proliferative tissues, including the skin (Seiter et al., 1996) and the crypts of the intestine (Wielenga et al., 1993). Alternative splicing is tightly controlled in the crypt-villus axis, as CD44s is expressed in transit amplifying cells along the villi, whereas CD44v4–v10 isoforms are found on $lgr5^{+}$ intestinal stem cells (Zeilstra et al., 2014). The largest human CD44 isoform containing exons v2–v10 in the stem region is expressed in keratinocytes (Bloor, Jelvagharan, White, & Odell, 2001). In many cancers, the expression of CD44s and also CD44 variant isoforms correlates with advanced stages of carcinogenesis. As an example, expression of CD44v6 isoforms correlates with aggressive stages of colorectal cancers (Wielenga et al., 1993).

CD44 participates in the reception of a variety of microenvironmental signals, including the presence and size of hyaluronic acid (HA) in the ECM, as well a number of growth factors and cytokines, by coordinating with cell-surface receptors such as receptor tyrosine kinases (RTKs) and G protein-coupled receptors (GPCRs). These activities are regulated by its alternative splicing and glycosylation status. An emerging picture is that CD44 serves to integrate the response to these and other extracellular cues through a variety of molecular mechanisms. In this review, we survey the role of CD44 in the reception of signals from the external cellular environment and describe our current state of knowledge about how it integrates these signals and coordinates the cellular response.

2. LIGATION OF ECM COMPONENTS BY CD44

2.1. Hyaluronic acid

CD44 is the main receptor for hyaluronan (reviewed in Underhill, 1992). The binding of HA to CD44 regulates many cellular processes including inflammation, wound healing, neutrophil migration and endothelial cell activation, tumor growth, and metastasis (reviewed in Noble & Jiang, 2006). The N-terminal ECD of CD44 contains the binding sites for HA. HA binding is mediated in part by the "link domain," a stretch of 90 amino acids (residues 32–123) that shows high homology to the HA-binding cartilage link protein. Site-directed mutagenesis has defined additional HA-binding sites outside the link domain (Peach, Hollenbaugh, Stamenkovic, & Aruffo, 1993). Inspection of the NMR and X-ray crystallographic structures for the HA-binding domain of CD44 reveals that sequences flanking the link domain form a lobular extension, creating a larger HA-binding domain (Teriete et al., 2004).

The HA-binding ability of CD44 is regulated by several factors (Fig. 9.1A). A threshold density of CD44 proteins is required on the cell surface to enable HA binding and determines in part whether cells bind HA in a constitutive or inducible manner (English, Lesley, & Hyman, 1998; Lesley et al., 2002). Consistently, certain CD44v isoforms have a much higher HA-binding capacity than CD44s, which is associated with clustering of the larger CD44v isoforms as compared to CD44s (Sleeman et al., 1996). This oligomerization of the CD44v isoforms was lost upon treatment with tunicamycin, indicating that N-linked glycosylation might regulate CD44-mediated HA binding. Indeed, both the N- and O-linked glycosylation present on the CD44 ECD have been reported to positively or negatively modulate HA binding to CD44. For example, N-linked glycosylation has been reported to be necessary for CD44-mediated HA binding (Bartolazzi, Nocks, Aruffo, Spring, & Stamenkovic, 1996), while in contrast, other studies have shown that removal of N-linked glycosylation induced HA binding (Katoh, Zheng, Oritani, Shimozato, & Kincade, 1995; Lesley, English, Perschl, Gregoroff, & Hyman, 1995), suggesting that the cellular context may determine how glycosylation regulates CD44 binding to HA. Furthermore, the spatial organization of HA molecules in the ECM regulates the binding of CD44 to HA (Kaindl et al., 2012).

Intracellularly, CD44 can bind to molecules such as ERM (ezrin–radixin–moesin) family members and ankyrin that link it to the cytoskeleton (reviewed in Ponta, Sherman, & Herrlich, 2003). CD44 also interacts with signaling molecules such as Tiam (Bourguignon, Zhu, Shao, & Chen, 2000), Vav2 (Bourguignon et al., 2001), RhoGEF/RhoA-activated ROK (Bourguignon, Zhu, Shao, Zhu, & Chen, 1999), and cdc42-IQGAP1 (Bourguignon, Gilad, Rothman, & Peyrollier, 2005). The cytoplasmic domain of CD44 has been suggested to be required for HA binding, since a truncated version of CD44 which contained only the six membrane-proximal amino acids exhibited decreased binding to soluble HA (Lesley et al., 1992). However, the requirement of the CD44 cytoplasmic domain for HA binding is still under debate, as other studies suggest that CD44 cytoplasmic and transmembrane domain sequences are not involved in regulating HA binding by CD44 (Lesley, English, Charles, & Hyman, 2000). Another study showed that phosphorylation of the CD44 ICD on Ser (325) is not required for efficient HA binding, but is essential for CD44-mediated cell motility on HA substrates (Peck & Isacke, 1996, 1998). Nevertheless,

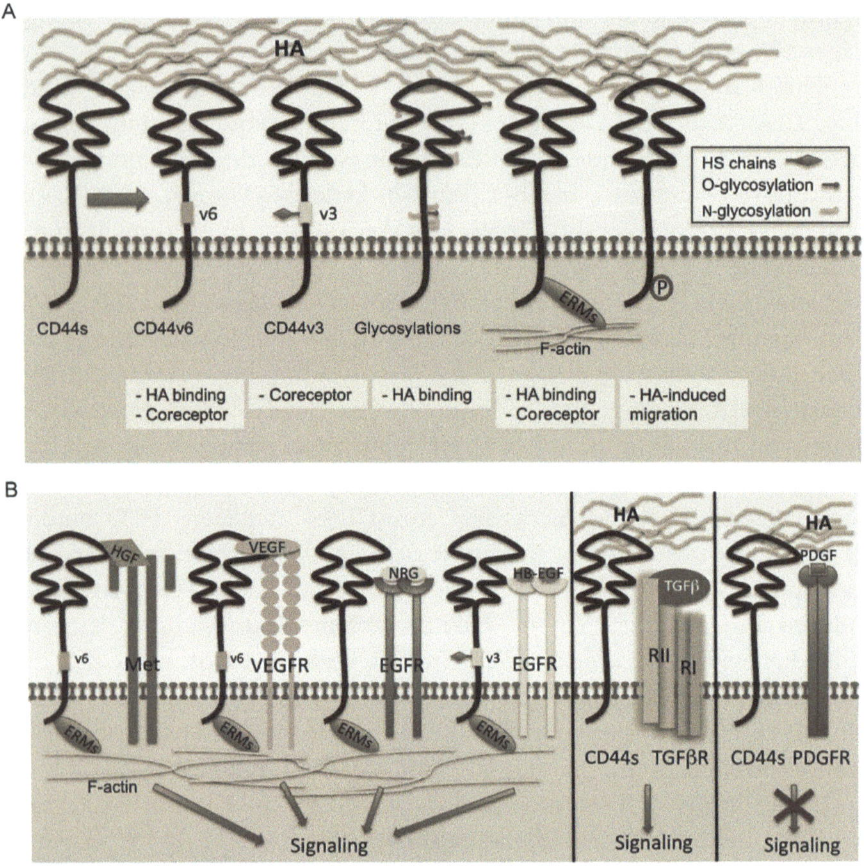

Figure 9.1 (A) Characteristics of CD44 proteins that regulate either their HA binding or coreceptor function. The inclusion of variant exons favors clustering of CD44 molecules and thereby increases binding to HA (Sleeman et al., 1996). Inclusion of the variant exon v6 allows it to function as a coreceptor for Met and VEGFR-2 (reviewed in Orian-Rousseau, 2010). CD44v3 is the only CD44 isoform that can be heparan sulfated (Greenfield et al., 1999), which allows it to function as a coreceptor for several RTKs. N- and O-linked glycosylation can modulate binding of CD44 to HA both positively and negatively. Phosphorylation of ERM proteins has been shown to be required for the coreceptor function of CD44v6 for Met and VEGFR-2, and for HA binding (Brown et al., 2005). Phosphorylation of the CD44 ICD is also important for cell motility on HA surfaces (Peck & Isacke, 1996, 1998). (B) Coreceptor function of CD44 isoforms for RTKs. CD44v6 isoforms act as a coreceptor for Met, VEGFR-2 (reviewed in Orian-Rousseau, 2010), and neuregulin (NRG)-induced ErbB2–ErbB3 heterodimerization is enhanced by CD44s (Sherman, Rizvi, Karyala, & Ratner, 2000). Pro-HB-EGF maturation is dependent on heparan sulfation of CD44v3 (Yu, Woessner, McNeish, & Stamenkovic, 2002). HA induces a complex between CD44s and members of the TGFRβ family of RTKs and promotes signaling (Bourguignon, Singleton, Zhu, & Zhou, 2002). In contrast, HA inhibits activation of PDGFR (Li, Heldin, & Heldin, 2006). (See the color plate.)

CD44-mediated HA binding can be regulated by activated phosphorylated members of the ERM family that link CD44 with the actin cytoskeleton (Brown et al., 2005).

2.2. Other ECM components that are CD44 ligands

Posttranslational modification regulates the ability of CD44 to bind to a number of ECM components. For example, chondroitin sulfate modification of CD44 on the amino acid sequence encoded by exon 5, as well as heparan sulfate (HS) modification of Ser-Gly sequences on exon v3 allows CD44 to interact with collagen, fibronectin, and laminin (Faassen et al., 1992; Jalkanen & Jalkanen, 1992; Naor, Sionov, & IshShalom, 1997). The degree of sulfation of CD44 sugar side chains also regulates its ability to bind to fibrin (Alves & Konstantopoulos, 2012), while sialylation inhibits the interaction between CD44 and HA (Jackson, 2009).

Serglycin is a peptide proteoglycan secreted by hematopoietic cells and endothelial cells that is modified with heparan or chondroitin sulfate (reviewed in Korpetinou et al., 2014). Secreted serglycin is incorporated into the ECM, and can interact with several proteins, including fibronectin, collagen, and CD44 (reviewed in Korpetinou et al., 2014). A number of observations suggest that serglycin plays a role in tumor progression and metastasis (reviewed in Korpetinou et al., 2014). Recent studies have implicated CD44–serglycin interactions in the growth of malignant melanomas, in part through the regulation of angiogenesis (Purushothaman & Toole, 2014).

Osteopontin (OPN) is a highly negatively charged secreted protein that functions as both an extracellular structural protein and a cytokine (reviewed in Anborgh, Mutrie, Tuck, & Chambers, 2010). It is secreted in the bone and other tissues by fibroblasts, endothelial cells, activated T cells, and macrophages. OPN has a number of functions, including biomineralization and bone remodeling, as well as the regulation of immune responses through its chemotactic and antiapoptotic properties and the regulation of T cells (reviewed in Wai & Kuo, 2008). Together with integrins, CD44 acts as a cell-surface receptor for OPN (Katagiri et al., 1999; Weber, Ashkar, Glimcher, & Cantor, 1996). The interactions between OPN, integrins, and CD44 are involved in adhesion, migration, homing, survival and proliferation of tumor, and stromal cells (reviewed in Wai & Kuo, 2008). Compared to HA, OPN binding to CD44 has different effects on chemotaxis and aggregation (reviewed in Weber et al., 1996). However, OPN also plays a key role in cancer progression and metastasis. A comparison of downstream targets of Ras and Rho family members revealed that OPN is a major regulator of Ras-induced tumorigenesis. Binding of OPN to CD44 induces foci formation, invasion, and tumorigenesis of H-Ras-V12-transformed cells (Teramoto et al., 2005). These events are mediated through Rac signaling. OPN binding to CD44 also increases clonogenicity of colon cancer cells through c-jun-jnk signaling (Rao et al., 2013).

3. CD44: CORECEPTOR FOR CELL-SURFACE RECEPTORS

The observation that HS-modified CD44 isoforms are able to bind to growth factors (Bennett et al., 1995) gave the first indication that CD44 might contribute to growth factor signaling through RTKs. The two serine–glycine motifs that specify CD44 HS modification are exclusively present in CD44v3 (Greenfield et al., 1999). HS-modified CD44 can bind HB-EGF (heparin-binding epidermal growth factor), FGF-2 (fibroblast growth factor-2), VEGF (vascular endothelial growth factor), HGF (hepatocyte growth factor) but interestingly, not Amphiregulin (Bennett et al., 1995; Jones, Tussey, Athanasou, & Jackson, 2000). Subsequently, various CD44 isoforms have been demonstrated to regulate the signaling of RTKs (Fig. 9.1B) such as Met, VEGFR-2 (vascular endothelial growth factor receptor-2), PDGFR (platelet-derived growth factor receptor), and EGFR (epidermal growth factor receptor). However, this activity can be both HS-dependent and independent.

3.1. The role of CD44 in regulating HGF-induced Met activation

HGF was found to bind to CD44v3–v10-expressing Namalwa cells in a heparitinase-sensitive manner (van der Voort et al., 1999), in contrast to cells expressing CD44s, or CD44v8–v10 transfectants. Importantly, CD44v3–v10 isoforms were shown to promote HGF-induced phosphorylation of Met and downstream targets such as ERK. Furthermore, heparitinase treatment decreased HGF-induced ERK signaling in these CD44v3–v10-expressing Namalwa cells. Other studies also suggest that HGF-induced Met signaling can be dependent on HS modification, as mutation of the three amino acids essential for HS binding in the NK1 domain of HGF rendered this molecule a potent inhibitor that could bind to Met but could not promote its activation (Cecchi et al., 2012).

Despite these observations, binding of HGF to HS proteoglycans is not absolutely required for its activity. Indeed, a mutant form of HGF with a greatly decreased (more than 50-fold) affinity for heparin could still bind to Met, retained full mitogenic and motogenic effects on cultured cells, and increased proliferation in the murine liver *in vivo*, suggesting that this mutant form of HGF is a more potent inducer of Met (Hartmann et al., 1998). Accordingly, other mechanisms have been uncovered that promote HGF-induced Met signaling in an HS-independent manner. Among these, non-HS-modified CD44 isoforms have been found to play a key role in regulating Met signaling.

HGF-mediated activation of the Met receptor in several contexts is exclusively dependent on CD44v6-containing isoforms (reviewed in Orian-Rousseau, 2010; Fig. 9.1B) a non-heparan-sulfated form of CD44 (Greenfield et al., 1999). Indeed, in Bsp73 AS rat pancreatic carcinoma cells transfected with various CD44 isoforms, CD44 isoforms bearing exon v6 were found to be necessary and sufficient for HGF-mediated activation of Met and downstream signaling to ERK (Orian-Rousseau, Chen, Sleeman, Herrlich, & Ponta, 2002). In these cells and in HT29 cells, heparinase treatment had no impact on HGF-induced ERK activation, demonstrating that HS is not required for Met activation in these contexts. The dependency of Met activation by HGF on CD44v6 coexpression has also been demonstrated in a variety of other cells, including primary keratinocytes (Orian-Rousseau et al., 2002), primary liver cells (Olaku et al., 2011), and primary endothelial cells (Tremmel et al., 2009), as well as several different cancer cell lines (reviewed in Orian-Rousseau, 2010). CD44v6, HGF, and Met form a ternary complex in Bsp73ASs6 cells that express CD44v6 and CD44s (Orian-Rousseau et al., 2002). A complex between CD44 and Met has also been identified in HT29 cells in response to HGF and to NK1, a short version of HGF consisting of the N-terminal domain of HGF linked to the first kringle domain (Cecchi et al., 2012).

Signaling from the Met receptor depends on the binding of the CD44 ICD to cytoplasmic proteins such as ERM family members that link CD44 to the cytoskeleton (Orian-Rousseau et al., 2002, 2007) Accordingly, a CD44v4–v7 mutant in which the cytoplasmic domain was deleted supported ligand-induced autophosphorylation of Met, but not downstream signaling (Orian-Rousseau et al., 2002, 2007). The CD44 ICD contains binding sites for various cytoplasmic proteins including ankyrin (Lokeshwar & Bourguignon, 1991), IQGAP (Bourguignon et al., 2005) and ERMs (Legg & Isacke, 1998), all of which are involved in signal transduction. Transfection of HEK293 cells with a CD44 cytoplasmic ICD bearing ERM-binding sites blocked Met signaling, possibly by competing with the endogenous CD44 for key cytoplasmic components. Transfection of a mutant CD44 ICD mutated in the ERM-binding sites had no such effect, indicating that ERM proteins are critical cytoplasmic mediators in Met signaling. Consistently, knockdown of ezrin completely inhibited HGF-induced ERK phosphorylation (Orian-Rousseau et al., 2007). Furthermore, a dominant negative version of ezrin that is unable to bind to actin did not support HGF-induced ERK activation, demonstrating that the linking of CD44v6 to the cytoskeleton through ERMs is essential for Met signaling. This notion is substantiated by the observation that a CD44 fusion protein containing the CD44v6 ECD and transmembrane domain linked to ezrin was able to promote Met signaling, while a similar fusion protein lacking the ezrin actin-binding domain was unable to fulfill this function (Orian-Rousseau et al., 2007).

Taken together, these data suggest that CD44v6-containing isoforms exert a dual function during HGF-induced Met signaling (reviewed in Orian-Rousseau, 2010). On one hand, the ECD of CD44 controls Met activation and on the other hand, the CD44 ICD binds to ERMs and to the cytoskeleton, providing an intracellular platform for Met signaling.

The coreceptor function of CD44v6 for Met is also involved in Met internalization. It is now well accepted that the internalization process is not only a means to terminate signaling, but that specific signaling pathways are activated by RTKs as they traffic through various internal compartments (reviewed in Wiley & Burke, 2001). Like other RTKs, Met is internalized via clathrin-mediated endocytosis and continues to signal from internal compartments (reviewed in Lefebvre et al., 2012). Met-induced STAT3 signaling, for example, is dependent on the perinuclear localization of Met (Kermorgant & Parker, 2008). CD44v6 and Met cointernalize into early endosomes, and the internalization of Met is CD44v6-dependent (Hasenauer et al., 2013). Again, the link between CD44v6 and the cytoskeleton through ezrin is essential for this process.

Mouse genetics has demonstrated the physiological relevance of the collaboration between CD44v6 and Met. Like HGF and Gab1 (a docking protein essential for Met signaling), Met is haploinsufficient in the context of the *cd44−/−* mice (Matzke et al., 2007). The *cd44−/−met+/−* mice (as well as *cd44−/−;hgf+/−*; and *cd44−/−;gab1+/−* mice) die at birth from a breathing defect that results from impaired synaptic transmission in the respiratory rhythm-generating network and alterations in the phrenic nerve. The collaboration between CD44 and Met is also instrumental in the context of liver regeneration, because CD44v6 specific antibodies inhibited liver regeneration after partial hepatectomy (Olaku et al., 2011), a process that is strictly Met dependent (Borowiak et al., 2004).

CD44v6-containing isoforms have long been implicated in the metastatic process (Gunthert et al., 1991). The interaction between CD44v6 and Met is likely pivotal in this process as both Met and CD44v6 have been reported to be required for metastasis of colorectal cancer spheres injected orthotopically into the mouse caecum (Todaro et al., 2014). Consistently, mutational analysis identified three amino acids in the v6 exon of CD44v6 that are required for Met activation (Matzke, Herrlich, Ponta, & Orian-Rousseau, 2005), and peptides encompassing these three amino acids and composed of 5–14 amino acids were able to block Met activation, as well as angiogenesis and consequently tumor growth of human pancreatic tumor cells implanted into orthotopic sites (Tremmel et al., 2009). The ability of these peptides to also inhibit VEGFR-2 activation may additional contribute to these antitumor effects.

3.2. CD44-regulated activation of other RTKs: VEGRF, PDGFRβ, FGFR, and EGFR

A key molecular regulator of angiogenesis (the new growth of blood vessels) is ligand-induced activation of VEGFR-2 by the VEGF family of growth factors (reviewed in Chung & Ferrara, 2011). CD44v6 acts as a coreceptor for VEGFR-2 in a manner that is similar to that described for Met (Tremmel et al., 2009; Fig. 9.1B). The function of CD44v6 is again twofold. On one hand, the CD44 ECD promotes VEGF165- and VEGF121-induced activation of VEGFR-2, as demonstrated with several types of primary endothelial cells, including HUVECs, HCMECs, and HAOECs (Tremmel et al., 2009). On the other hand, the CD44 ICD is involved in downstream signaling, an event that requires the binding of ERMs to the cytoskeleton. CD44v6 and VEGFR-2 are constitutively associated as shown by coimmunoprecipitation experiments. This contrasts to Met, which needs its HGF ligand to interact with CD44v6 (Orian-Rousseau et al., 2002). Several angiogenic processes, including migration, sprouting, and tubule formation could be blocked by a CD44v6 peptide (Tremmel et al., 2009). Development of blood vessels from grafted endothelial cell spheroids *in vivo* was also impaired by treatment with the CD44v6 peptide. Moreover, tumor angiogenesis was shown to be dependent on CD44v6, as growth of vascularized tumors formed by L3.6pl human pancreatic carcinoma cells in the pancreas were dramatically reduced upon treatment of the animals with the CD44v6 peptide due to the inhibition of angiogenesis (Tremmel et al., 2009).

CD44 isoforms are also involved in the regulation of PDGFRβ and FGFR activation. CD44 and PDGFRβ interact with each other, in a manner that suppresses PDGF-induced signaling (Li et al., 2006). Collaboration between CD44v3/CD44v6 and FGFR is involved in the interactions between epithelial cells in the AER (apical ectodermal ridge) and the underlying mesenchymal cells that are required for vertebrate limb development (Sherman, Wainwright, Ponta, & Herrlich, 1998). In this context, CD44v6 and v3-specific antibodies blocked limb outgrowth when applied to the AER, and the same antibodies inhibited presentation of FGFs to limb bud mesenchymal cells *in vitro*. FGF4 and FGF8 were able to bind to CD44v3–v10 isoforms but not to CD44s or CD44v8–v10, as evidenced by FACS analysis. In addition, binding of FGFs to CD44v3–v10 was inhibited by treatment with heparitinase, indicating that the HS moieties present on the v3 exon are required for binding. However, ligand-induced activation of the FGFR was not addressed in this study.

Collaboration between CD44 and members of the EGFR family of RTKs has been demonstrated in various contexts. In Schwann cells, the CD44s isoform was found to associate with ErbB2 and ErbB3 in response to neuregulin (Sherman et al., 2000). CD44s also enhanced neuregulin-induced phosphorylation of ErbB2 and ErbB2–ErbB3 heterodimerization. CD44v3 on the other hand was shown to recruit matrix metalloproteinase 7 (MMP-7, matrilysin) and pro-HB-EGF to the surface of both tumor and nontransformed cells, an event that facilitated the processing of HB-EGF and ErbB4 activation (Yu et al., 2002). Consistently, ErbB4 phosphorylation was observed in Namalwa cells transfected with HS-modified CD44v3 and to a much lesser extent in v6–v10 or v7–v10 transfectants (Yu et al., 2002). These data indicate that different CD44 isoforms regulate activation of EGFR family members in a ligand-specific manner.

Together, these observations indicate that the unique ligand-binding properties of specific CD44 isoforms underlie their ability to support ligand-induced RTK activation, as well as the specific requirement of a given RTK for particular CD44 isoforms in its signaling (e.g., CD44v6 for Met and VEGFR-2, CD44v3 for FGFR). This notion begins to explain at the functional level why the CD44 family is such a diverse group of proteins.

3.3. Involvement of CD44 in the signaling of TGF family members

TGFβ-RI is a membrane serine/threonine kinase receptor that binds with high affinity to CD44v3 (Bourguignon et al., 2002). CD44-dependent activation of TGFβ has been shown during the migration of fibroblasts in the context of injury/wound healing (Acharya et al., 2008). Fibroblasts from CD44 deficient mice exhibited a defect in TGFβ activation and decreased TGFβ signaling, as evidenced by reduced Smad2 nuclear translocation in wounded CD44-deficient fibroblasts. This role of CD44 in TGFβ activation was shown to be MMP-dependent. In human MDA-MB-231 metastatic breast tumor cells, complexes were detected between CD44 and TGFβ-RI and to a lesser extent TGFβ-RII, consistent with a role for CD44 in TGFβ signaling within aggressive tumor cells as well (Bourguignon et al., 2002). Other studies have reported that CD44 serves as a docking platform for MMP-9, thereby inducing maturation of the latent form of TGFβ. Mature TGFβ in turn induces tumor vascularization (Yu & Stamenkovic, 1999).

The activation of Smad1 induced by BMP-7 signaling, another member of the TGFβ superfamily, has also been shown to be dependent on CD44 (Peterson et al., 2004). A yeast-two-hybrid screen and subsequent validation revealed that the ICD of CD44 binds to Smad1, a downstream target of BMPs that transduces signals from BMP receptors to the nucleus. Expression of a CD44 mutant truncated in the cytoplasmic domain inhibited Smad1 nuclear translocation upon BMP-7 induction, indicating that the ability of Smad1 to transduce BMP-7 signaling to the nucleus is dependent on the interaction between CD44 and Smad1.

3.4. CD44 and GPCRs

Trafficking of human CD34+ stem/progenitor cells is regulated by chemokine receptors present at their surface. Expression of the cognate chemokines in tissues such as the bone marrow and spleen guides the homing of these progenitor cells (reviewed in Burger & Kipps, 2006). CXCL12 (SDF-1α) is a major chemoattractant for HPCs/HSCs (hematopoietic progenitor cells/ hematopoietic stem cells) in these tissues, which acts by binding to and activating the GPCR CXCR4 that is expressed on HPCs/HSCs (reviewed in Burger & Kipps, 2006). A first indication that CD44 might contribute to CXCL12 signaling came from the observation that CD44 antibodies block the chemotaxis of HSCs toward CXCL12 (Avigdor et al., 2004). More recently, we found that CD44 and CXCR4 physically interact in the presence of CXCL12 (Fuchs et al., 2013), suggesting that CD44 may be required for CXCL12-induced activation of CXCR4 (Fig. 9.2). This notion was further substantiated by the observation that knockdown of CD44 abrogated CXCL12-induced CXCR4 signaling (Fuchs et al., 2013).

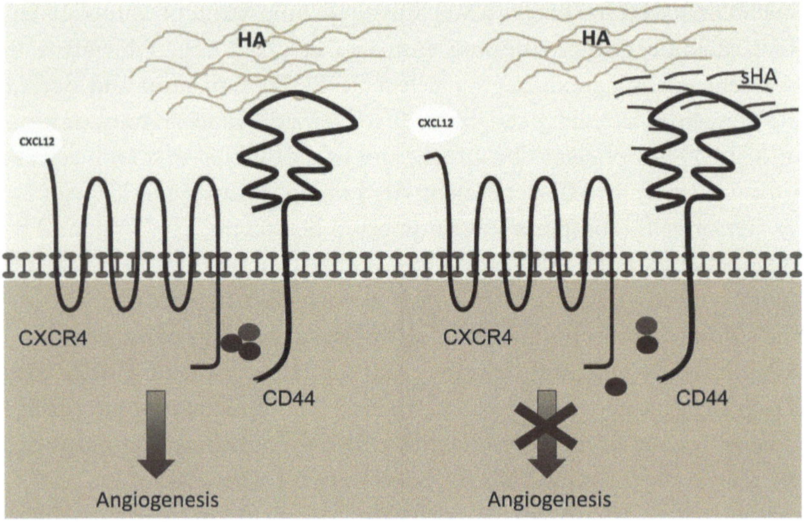

Figure 9.2 Coreceptor function of CD44 for GPCRs. CXCL12-induced CXCR4 signaling is augmented by HA and inhibited by small HA fragments (Fuchs et al., 2013). CD44/HA-dependent CXCL12-induced CXCR4 activation is required for angiogenesis. (See the color plate.)

3.5. Other receptors

One of the earliest proteins identified as interacting with CD44 was the chondroitin-sulfated form of CD74 (also known as the MHC class II-associated invariant chain) (Naujokas, Morin, Anderson, Peterson, & Miller, 1993). CD74 is a homotrimeric membrane protein that acts as a receptor for macrophage migration inhibitory factor (MIF) (Leng et al., 2003). MIF plays a role in innate immunity and subsequent adaptive responses (reviewed in Calandra & Roger, 2003). Accordingly, MIF has a wide range of inflammatory and proproliferative effects on a number of cell types, and has also been implicated in tumorigenesis (reviewed in Candido & Hagemann, 2013). While CD74 alone binds to MIF, MIF-induced signaling requires the interaction between CD44 and CD74 and is dependent on the cytoplasmic domain of CD44 (Shi et al., 2006). This axis has recently been shown to form an autocrine loop that promotes the growth of renal cell carcinoma (Du et al., 2013).

4. CD44 AS A MULTIDOMAIN SIGNAL INTEGRATION PLATFORM

An increasing number of publications indicate that CD44 acts at the cell surface to coordinate and integrate signal transduction in response to a variety of extracellular stimuli. In many instances, the interaction between CD44 and HA has been shown to cross-regulate its function as a coreceptor for a variety of growth factors and cytokines. Similar observations have also been made with other ECM components interacting with CD44. Signal integration occurs both extracellularly through the regulation of ligand-induced receptor activation, as well as intracellularly through modulation of signal transduction components.

4.1. HA cross-regulation of the CD44 coreceptor function

The binding of HA to CD44 can both positively and negatively regulate the ability of CD44 to promote growth factor and cytokine-mediated signaling via the cognate receptor, depending on the context. Furthermore, several studies indicate that the size of HA bound by CD44 can be decisive in determining whether positive or negative regulation of signaling occurs.

HA positively regulates BMP-7 signaling via Smad proteins in a CD44-dependent manner (Peterson et al., 2004). Depletion of HA by means of hyaluronidase prevented BMP-7-induced translocation of Smad1 to the nucleus, and HA-CD44 binding increased BMP-7-induced phosphorylation of Smad1. This crosstalk between HA and CD44 is necessary for the response of chondrocytes to damage or matrix disruption (Peterson et al., 2004). Similarly, HA induced the interaction between CD44 and TGFβ-RI in MDA-MB-231 cells and promoted the activation of TGFβ-RI in a CD44-dependent manner (Bourguignon et al., 2002; Fig. 9.1B). Although HA-induced activation of TGFβ-RI was dependent on CD44 in this context, TGFβ-dependent induction of TGFβ-RI did not rely on CD44. Moreover, TGFβ promotes interaction between CD44 and EGFR during the early development of progressive tissue fibrosis, leading to activation of the MAPK/ERK pathway and consequently proliferation of fibroblasts (Meran et al., 2011). HA was found to facilitate this process, providing an intriguing insight into how CD44 integrates ECM cues with multiple RTK-activating signaling molecules (Meran et al., 2011).

In contrast to these findings, binding of CD44 to HA negatively regulates ligand-induced PDGFRβ activation (Li et al., 2006; Fig. 9.1B). HA was found to inhibit PDGF-BB-induced migration of primary human dermal fibroblasts as well as PDGFRβ activation, and antibodies that block the binding of CD44 to HA could neutralize these effects. Mechanistically, there is evidence to suggest that the interaction between CD44 and HA might recruit phosphatases to the CD44–PDGFRβ complex, thereby suppressing ligand-induced activation of PDGFRβ (Li et al., 2006). It has been speculated that this signal integration might be physiologically relevant during wound healing, as the synthesis of HA would inhibit the activity of PDGF-BB, preventing the infiltration of fibroblasts into the provisional matrix before matrix remodeling occurs.

The interaction between CD44 and HA plays a key role in regulating CXCL12-dependent homing of HSCs, as evidenced by the observation that treatment of NOD/SCID mice with either anti-CD44 antibodies that block HA binding, soluble HA, or hyaluronidase, inhibits the homing of HPCs to the bone marrow and to the spleen (Avigdor et al., 2004). The HA–CD44 interaction regulates CXCL12-induced activation of CXCR4 in a manner that is strongly dependent on the size of HA (Fuchs et al., 2013). High-molecular-weight HA (hHA) augments CXCL12-induced signaling in a CD44 dependent manner, and increases CXCL12-induced migration of cancer cells and endothelial cells, as well as sprouting and *in vivo* angiogenesis (Fuchs et al., 2013). In contrast, small HA oligosaccharides (sHA) of 6–10 disaccharides in length block CXCL12-induced CXCR4 activation. The ability of CD44 to form a complex with CXCR4 in the presence of CXCL12 was also suppressed by sHA, suggesting that sHA competes with hHA for CD44 binding and thereby inhibits the function of hHA and CD44 in CXCR4 activation. Consistent with these findings, hHA stimulated CXCL12-induced recruitment of bone marrow cells to the myocardium following myocardial infarction (Purcell, Elser, Mu, Margulies, & Burdick, 2012). Along the same lines, hHA was shown to induce phosphorylation of ErbB2 in a CD44-dependent manner, whereas sHA decreased it (Ghatak, Misra, & Toole, 2005). A complex between CD44, ErbB2, PI3K, ezrin, and chaperone proteins was found in these studies to assemble in lipid rafts, enabling signaling through ErbB2.

4.2. Integration of other ECM signals with the activation of RTKs

OPN has recently been shown to modulate CD44-dependent activation of Met in hepatocellular carcinoma (HCC). Overexpression of LSF (late SV40 factor) is detected in more than 90% of HCC cases. LSF controls expression of OPN, which in turn activates Met through interacting with CD44 (Yoo et al., 2011). OPN induces phosphorylation of Met in the absence of its authentic ligand HGF, an effect that can be blocked by antibodies against

CD44, suggesting that OPN can substitute for HGF in the interaction between CD44 and Met, leading to Met autophosphorylation and activation.

4.3. Signal integration through cytoplasmic partners

As outlined above, HA binding by CD44 can act extracellularly to regulate growth factor and cytokine signaling by modulating the interaction of the ligand with its cognate receptor. However, other observations indicate that signal integration also occurs intracellularly. For example, HA promotes CD44v3–Vav2 interaction with grb2–p185Her2 complexes, thereby inducing signaling through Rac1 and Ras during ovarian tumor cell migration and growth (Bourguignon et al., 2001). Similarly, HA was found to stimulate complex formation between N-WASP and CD44 (Bourguignon, Peyrollier, Gilad, & Brightman, 2007). Enhanced association between CD44 and N-WASP in turn induced ErbB2 kinase activity, which resulted in β-catenin tyrosine phosphorylation and translocation to the nucleus, and subsequent TCF/LEF-induced gene transcription. As a consequence, HA fostered migration of the ovarian tumor cells (Bourguignon et al., 2007). In a similar vein, the interaction between HA and CD44 was found to promote the formation of CD44–ErbB2 complexes in melanoma cells. In turn, this stimulated ErbB2-dependent regulation of PKC and Akt signaling, resulting in increased motility of the cells (Kim et al., 2008). Further cross-regulation is illustrated by HA/CD44-mediated activation of TGFβ-RI, where activated TGFβ-RI in turn phosphorylates CD44v3, resulting in enhanced association of CD44 with ankyrin (Bourguignon et al., 2002).

Several extracellular cues are dependent on the interaction of ERM proteins with the CD44 ICD for their signaling, and these interactions serve to integrate some of these signals. The ERM-related protein merlin (Trofatter et al., 1993) is a tumor suppressor that is encoded by the neurofibromatosis-2 gene (NF2) and interacts with the CD44 ICD (Stamenkovic & Yu, 2010). Increased expression of merlin suppresses schwannoma cell growth (Bai et al., 2007; Morrison et al., 2001), but only at high cell density (Bai et al., 2007; Morrison et al., 2001) and requires the interaction between merlin and CD44 (Bai et al., 2007). At low cell density, merlin is phosphorylated and associates together with ezrin and moesin on the CD44 ICD, while at high cell density only merlin associates with the CD44 ICD and is dephosphorylated (Morrison et al., 2001). Similarly, HA binding to CD44 suppresses the growth of the schwannoma cells and induces merlin dephosphorylation (Morrison et al., 2001). Enhanced levels of merlin expression also inhibit CD44–HA interactions in schwannoma cells (Bai et al., 2007; Morrison et al., 2001). Together these observations suggest that CD44 transduces both contact inhibition cues and HA-mediated signaling through the differential interaction of merlin and ERM proteins with its ICD, and that in turn these interactions can regulate CD44-mediated HA binding.

5. CONCLUDING REMARKS

The picture that has emerged in recent years is that CD44 acts as a multidomain, transmembrane platform that serves to integrate a wide variety of extracellular signals. In many cases, this function of CD44 occurs by regulating ligand-dependent activation of RTKs and GPCRs in a manner that is controlled by the presence and size of HA. This integration appears to work at a number of levels, including the control of receptor–ligand interactions, and cross-regulation of intracellular signaling molecules. A variety of other mechanisms exquisitely control CD44-mediated signal integration, including differential glycosylation and alternative splicing of CD44.

Despite these advances in knowledge, a number of open questions remain to be answered. The ability of HA to regulate the coreceptor function of CD44 has not been investigated for all RTKs and GPCRs that interact with CD44. Furthermore, the ability of other CD44 ECM ligands such as serglycin, collagen, fibronectin, and laminin to modulate the coreceptor function of CD44 remains unclear. Moreover, CD44 may utilize other mechanisms to integrate signaling. For example, it has recently been shown that CXCR2 and CXCR4 both act as cell-surface receptors for MIF (Bernhagen et al., 2007). Our recent results demonstrating that CD44 and HA tightly regulate CXCR4 signaling (Fuchs et al., 2013) could thereby extend to MIF signaling, a possible new axis that deserves to be investigated. Finally, both HGF-induced Met activation and the MIF/CD74 axis have a survival function in B-cells, and MIF recruits Met to the CD74/CD44 complex in these cells (Gordin et al., 2010). Given the central role of CD44 as a coreceptor in Met/HGF signaling, and its regulation by HA, these observations clearly suggest that CD44 is likely to coordinate multiple cell-surface signaling complexes and to combine these with ECM signals. It will be exciting to investigate these possibilities in order to understand and appreciate the full potential of CD44 as a signal-integrating platform.

REFERENCES

Acharya, P. S., Majumdar, S., Jacob, M., Hayden, J., Mrass, P., Weninger, W., et al. (2008). Fibroblast migration is mediated by CD44-dependent TGF beta activation. *Journal of Cell Science*, *121*, 1393–1402.

Alves, C. S., & Konstantopoulos, K. (2012). PDGF suppresses the sulfation of CD44v and potentiates CD44v-Mediated binding of colon carcinoma cells to fibrin under flow. *PLoS One*, *7*, e41472.

Anborgh, P. H., Mutrie, J. C., Tuck, A. B., & Chambers, A. F. (2010). Role of the metastasis-promoting protein osteopontin in the tumour microenvironment. *Journal of Cellular and Molecular Medicine*, *14*, 2037–2044.

Avigdor, A., Goichberg, P., Shivtiel, S., Dar, A., Peled, A., Samira, S., et al. (2004). CD44 and hyaluronic acid cooperate with SDF-1 in the trafficking of human CD34+ stem/progenitor cells to bone marrow. *Blood*, *103*, 2981–2989.

Bai, Y., Liu, Y. J., Wang, H., Xu, Y., Stamenkovic, I., & Yu, Q. (2007). Inhibition of the hyaluronan-CD44 interaction by merlin contributes to the tumor-suppressor activity of merlin. *Oncogene*, *26*, 836–850.

Bartolazzi, A., Nocks, A., Aruffo, A., Spring, F., & Stamenkovic, I. (1996). Glycosylation of CD44 is implicated in CD44-mediated cell adhesion to hyaluronan. *The Journal of Cell Biology*, *132*, 1199–1208.

Bennett, K. L., Jackson, D. G., Simon, J. C., Tanczos, E., Peach, R., Modrell, B., et al. (1995). CD44 isoforms containing exon V3 are responsible for the presentation of heparin-binding growth factor. *The Journal of Cell Biology*, *128*, 687–698.

Bernhagen, J., Krohn, R., Lue, H., Gregory, J. L., Zernecke, A., Koenen, R. R., et al. (2007). MIF is a noncognate ligand of CXC chemokine receptors in inflammatory and atherogenic cell recruitment. *Nature Medicine*, *13*, 587–596.

Bloor, B. K., Jelvagharan, M., White, K. N., & Odell, E. W. (2001). Characterization of CD44 splicing patterns in normal keratinocytes, dysplastic and squamous carcinoma cell lines. *International Journal of Oncology*, *18*, 1053–1059.

Borowiak, M., Garratt, A. N., Wustefeld, T., Strehle, M., Trautwein, C., & Birchmeier, C. (2004). Met provides essential signals for liver regeneration. *Proceedings of the National Academy of Sciences of the United States of America*, *101*, 10608–10613.

Bourguignon, L. Y., Gilad, E., Rothman, K., & Peyrollier, K. (2005). Hyaluronan-CD44 interaction with IQGAP1 promotes Cdc42 and ERK signaling, leading to actin binding, Elk-1/estrogen receptor transcriptional activation, and ovarian cancer progression. *The Journal of Biological Chemistry*, *280*, 11961–11972.

Bourguignon, L. Y., Peyrollier, K., Gilad, E., & Brightman, A. (2007). Hyaluronan-CD44 interaction with neural Wiskott-Aldrich syndrome protein (N-WASP) promotes actin polymerization and ErbB2 activation leading to beta-catenin nuclear translocation, transcriptional up-regulation, and cell migration in ovarian tumor cells. *The Journal of Biological Chemistry*, *282*, 1265–1280.

Bourguignon, L. Y., Singleton, P. A., Zhu, H., & Zhou, B. (2002). Hyaluronan promotes signaling interaction between CD44 and the transforming growth factor beta receptor I in metastatic breast tumor cells. *The Journal of Biological Chemistry*, *277*, 39703–39712.

Bourguignon, L. Y., Zhu, H., Shao, L., & Chen, Y. W. (2000). CD44 interaction with tiam1 promotes Rac1 signaling and hyaluronic acid-mediated breast tumor cell migration. *The Journal of Biological Chemistry*, *275*, 1829–1838.

Bourguignon, L. Y., Zhu, H., Shao, L., Zhu, D., & Chen, Y. W. (1999). Rho-kinase (ROK) promotes CD44v(3,8-10)-ankyrin interaction and tumor cell migration in metastatic breast cancer cells. *Cell Motility and the Cytoskeleton*, *43*, 269–287.

Bourguignon, L. Y., Zhu, H., Zhou, B., Diedrich, F., Singleton, P. A., & Hung, M. C. (2001). Hyaluronan promotes CD44v3-Vav2 interaction with Grb2-p185(HER2) and induces Rac1 and Ras signaling during ovarian tumor cell migration and growth. *The Journal of Biological Chemistry, 276*, 48679–48692.

Brown, K. L., Birkenhead, D., Lai, J. C., Li, L., Li, R., & Johnson, P. (2005). Regulation of hyaluronan binding by F-actin and colocalization of CD44 and phosphorylated ezrin/radixin/moesin (ERM) proteins in myeloid cells. *Experimental Cell Research, 303*, 400–414.

Burger, J. A., & Kipps, T. J. (2006). CXCR4: A key receptor in the crosstalk between tumor cells and their microenvironment. *Blood, 107*, 1761–1767.

Calandra, T., & Roger, T. (2003). Macrophage migration inhibitory factor: A regulator of innate immunity. *Nature Reviews Immunology, 3*, 791–800.

Candido, J., & Hagemann, T. (2013). Cancer-related inflammation. *Journal of Clinical Immunology, 33*(Suppl. 1), S79–S84.

Cecchi, F., Pajalunga, D., Fowler, C. A., Uren, A., Rabe, D. C., Peruzzi, B., et al. (2012). Targeted disruption of heparan sulfate interaction with hepatocyte and vascular endothelial growth factors blocks normal and oncogenic signaling. *Cancer Cell, 22*, 250–262.

Chung, A. S., & Ferrara, N. (2011). Developmental and pathological angiogenesis. *Annual Review of Cell and Developmental Biology, 27*, 563–584.

Deribe, Y. L., Pawson, T., & Dikic, I. (2010). Post-translational modifications in signal integration. *Nature Structural & Molecular Biology, 17*, 666–672.

Dibble, C. C., & Manning, B. D. (2013). Signal integration by mTORC1 coordinates nutrient input with biosynthetic output. *Nature Cell Biology, 15*, 555–564.

Du, W., Wright, B. M., Li, X., Finke, J., Rini, B. I., Zhou, M., et al. (2013). Tumor-derived macrophage migration inhibitory factor promotes an autocrine loop that enhances renal cell carcinoma. *Oncogene, 32*, 1469–1474.

Dudanova, I., & Klein, R. (2013). Integration of guidance cues: Parallel signaling and crosstalk. *Trends in Neurosciences, 36*, 295–304.

English, N. M., Lesley, J. F., & Hyman, R. (1998). Site-specific de-N-glycosylation of CD44 can activate hyaluronan binding, and CD44 activation states show distinct threshold densities for hyaluronan binding. *Cancer Research, 58*, 3736–3742.

Faassen, A. E., Schrager, J. A., Klein, D. J., Oegema, T. R., Couchman, J. R., & Mccarthy, J. B. (1992). A cell-surface chondroitin sulfate proteoglycan, immunologically related to Cd44, is involved in Type-I collagen-mediated melanoma cell motility and invasion. *Journal of Cell Biology, 116*, 521–531.

Fuchs, K., Hippe, A., Schmaus, A., Homey, B., Sleeman, J. P., & Orian-Rousseau, V. (2013). Opposing effects of high- and low-molecular weight hyaluronan on CXCL12-induced CXCR4 signaling depend on CD44. *Cell Death Disease, 4*, e819.

Ghatak, S., Misra, S., & Toole, B. P. (2005). Hyaluronan constitutively regulates ErbB2 phosphorylation and signaling complex formation in carcinoma cells. *The Journal of Biological Chemistry, 280*, 8875–8883.

Goldstein, L. A., Zhou, D. F., Picker, L. J., Minty, C. N., Bargatze, R. F., Ding, J. F., et al. (1989). A human lymphocyte homing receptor, the hermes antigen, is related to cartilage proteoglycan core and link proteins. *Cell, 56*, 1063–1072.

Goodison, S., Urquidi, V., & Tarin, D. (1999). CD44 cell adhesion molecules. *Molecular Pathology, 52*, 189–196.

Gordin, M., Tesio, M., Cohen, S., Gore, Y., Lantner, F., Leng, L., et al. (2010). c-Met and its ligand hepatocyte growth factor/scatter factor regulate mature B cell survival in a pathway induced by CD74. *The Journal of Immunology, 185*, 2020–2031.

Greenfield, B., Wang, W. C., Marquardt, H., Piepkorn, M., Wolff, E. A., Aruffo, A., et al. (1999). Characterization of the heparan sulfate and chondroitin sulfate assembly sites in CD44. *The Journal of Biological Chemistry, 274*, 2511–2517.

Groves, J. T., & Kuriyan, J. (2010). Molecular mechanisms in signal transduction at the membrane. *Nature Structural & Molecular Biology, 17*, 659–665.

Gunthert, U., Hofmann, M., Rudy, W., Reber, S., Zoller, M., Haussmann, I., et al. (1991). A new variant of glycoprotein CD44 confers metastatic potential to rat carcinoma cells. *Cell, 65*, 13–24.

Hartmann, G., Prospero, T., Brinkmann, V., Ozcelik, C., Winter, G., Hepple, J., et al. (1998). Engineered mutants of HGF/SF with reduced binding to heparan sulphate proteoglycans, decreased clearance and enhanced activity in vivo. *Current Biology, 8*, 125–134.

Hasenauer, S., Malinger, D., Koschut, D., Pace, G., Matzke, A., von Au, A., et al. (2013). Internalization of Met requires the co-receptor CD44v6 and its link to ERM proteins. *PLoS One, 8*, e62357.

Irvine, K. D. (2012). Integration of intercellular signaling through the Hippo pathway. *Seminars in Cell & Developmental Biology, 23*, 812–817.

Jackson, D. G. (2009). Immunological functions of hyaluronan and its receptors in the lymphatics. *Immunological Reviews, 230*, 216–231.

Jalkanen, S., & Jalkanen, M. (1992). Lymphocyte Cd44 binds the cooh-terminal heparin-binding domain of fibronectin. *Journal of Cell Biology, 116*, 817–825.

Jones, M., Tussey, L., Athanasou, N., & Jackson, D. G. (2000). Heparan sulfate proteoglycan isoforms of the CD44 hyaluronan receptor induced in human inflammatory macrophages can function as paracrine regulators of fibroblast growth factor action. *The Journal of Biological Chemistry, 275*, 7964–7974.

Kaindl, T., Rieger, H., Kaschel, L. M., Engel, U., Schmaus, A., Sleeman, J., et al. (2012). Spatio-temporal patterns of pancreatic cancer cells expressing CD44 isoforms on supported membranes displaying hyaluronic acid oligomers arrays. *PLoS One, 7*, e42991.

Katagiri, Y. U., Sleeman, J., Fujii, H., Herrlich, P., Hotta, H., Tanaka, K., et al. (1999). CD44 variants but not CD44s cooperate with beta1-containing integrins to permit cells to bind to osteopontin independently of arginine-glycine-aspartic acid, thereby stimulating cell motility and chemotaxis. *Cancer Research, 59*, 219–226.

Katoh, S., Zheng, Z., Oritani, K., Shimozato, T., & Kincade, P. W. (1995). Glycosylation of CD44 negatively regulates its recognition of hyaluronan. *The Journal of Experimental Medicine, 182*, 419–429.

Kermorgant, S., & Parker, P. J. (2008). Receptor trafficking controls weak signal delivery: A strategy used by c-Met for STAT3 nuclear accumulation. *The Journal of Cell Biology, 182*, 855–863.

Kholodenko, B. N. (2003). Four-dimensional organization of protein kinase signaling cascades: The roles of diffusion, endocytosis and molecular motors. *The Journal of Experimental Biology, 206*, 2073–2082.

Kholodenko, B. N., Hancock, J. F., & Kolch, W. (2010). Signalling ballet in space and time. *Nature Reviews Molecular Cell Biology, 11*, 414–426.

Kim, Y., Lee, Y. S., Choe, J., Lee, H., Kim, Y. M., & Jeoung, D. (2008). CD44-epidermal growth factor receptor interaction mediates hyaluronic acid-promoted cell motility by activating protein kinase C signaling involving Akt, Rac1, Phox, reactive oxygen species, focal adhesion kinase, and MMP-2. *The Journal of Biological Chemistry, 283*, 22513–22528.

Korpetinou, A., Skandalis, S. S., Labropoulou, V. T., Smirlaki, G., Noulas, A., Karamanos, N. K., et al. (2014). Serglycin: At the crossroad of inflammation and malignancy. *Frontiers in Oncology, 3*, 327.

Lefebvre, J., Ancot, F., Leroy, C., Muharram, G., Lemiere, A., & Tulasne, D. (2012). Met degradation: More than one stone to shoot a receptor down. *FASEB Journal, 26*, 1387–1399.

Legg, J. W., & Isacke, C. M. (1998). Identification and functional analysis of the ezrin-binding site in the hyaluronan receptor, CD44. *Current Biology, 8*, 705–708.

Leng, L., Metz, C. N., Fang, Y., Xu, J., Donnelly, S., Baugh, J., et al. (2003). MIF signal transduction initiated by binding to CD74. *The Journal of Experimental Medicine, 197*, 1467–1476.

Lesley, J., English, N., Charles, C., & Hyman, R. (2000). The role of the CD44 cytoplasmic and transmembrane domains in constitutive and inducible hyaluronan binding. *European Journal of Immunology, 30*, 245–253.

Lesley, J., English, N. M., Gal, I., Mikecz, K., Day, A. J., & Hyman, R. (2002). Hyaluronan binding properties of a CD44 chimera containing the link module of TSG-6. *The Journal of Biological Chemistry, 277*, 26600–26608.

Lesley, J., English, N., Perschl, A., Gregoroff, J., & Hyman, R. (1995). Variant cell lines selected for alterations in the function of the hyaluronan receptor CD44 show differences in glycosylation. *The Journal of Experimental Medicine, 182*, 431–437.

Lesley, J., He, Q., Miyake, K., Hamann, A., Hyman, R., & Kincade, P. W. (1992). Requirements for hyaluronic acid binding by CD44: A role for the cytoplasmic domain and activation by antibody. *The Journal of Experimental Medicine, 175*, 257–266.

Li, L., Heldin, C. H., & Heldin, P. (2006). Inhibition of platelet-derived growth factor-BB-induced receptor activation and fibroblast migration by hyaluronan activation of CD44. *The Journal of Biological Chemistry, 281*, 26512–26519.

Lokeshwar, V. B., & Bourguignon, L. Y. (1991). Post-translational protein modification and expression of ankyrin-binding site(s) in GP85 (Pgp-1/CD44) and its biosynthetic precursors during T-lymphoma membrane biosynthesis. *The Journal of Biological Chemistry, 266*, 17983–17989.

Matzke, A., Herrlich, P., Ponta, H., & Orian-Rousseau, V. (2005). A five-amino-acid peptide blocks Met- and Ron-dependent cell migration. *Cancer Research, 65*, 6105–6110.

Matzke, A., Sargsyan, V., Holtmann, B., Aramuni, G., Asan, E., Sendtner, M., et al. (2007). Haploinsufficiency of c-Met in cd44-/- mice identifies a collaboration of CD44 and c-Met in vivo. *Molecular and Cellular Biology, 27*, 8797–8806.

McNeill, H., & Woodgett, J. R. (2010). When pathways collide: Collaboration and connivance among signalling proteins in development. *Nature Reviews Molecular Cell Biology, 11*, 404–413.

Meran, S., Luo, D. D., Simpson, R., Martin, J., Wells, A., Steadman, R., et al. (2011). Hyaluronan facilitates transforming growth factor-beta1-dependent proliferation via CD44 and epidermal growth factor receptor interaction. *The Journal of Biological Chemistry, 286*, 17618–17630.

Morrison, H., Sherman, L. S., Legg, J., Banine, F., Isacke, C., Haipek, C. A., et al. (2001). The NF2 tumor suppressor gene product, merlin, mediates contact inhibition of growth through interactions with CD44. *Genes & Development, 15*, 968–980.

Naor, D., Sionov, R. V., & IshShalom, D. (1997). CD44: Structure, function, and association with the malignant process. *Advances in Cancer Research, 71*, 241–319.

Naujokas, M. F., Morin, M., Anderson, M. S., Peterson, M., & Miller, J. (1993). The chondroitin sulfate form of invariant chain can enhance stimulation of T cell responses through interaction with CD44. *Cell, 74*, 257–268.

Noble, P. W., & Jiang, D. (2006). Matrix regulation of lung injury, inflammation, and repair: The role of innate immunity. *Proceedings of the American Thoracic Society, 3*, 401–404.

Olaku, V., Matzke, A., Mitchell, C., Hasenauer, S., Sakkaravarthi, A., Pace, G., et al. (2011). c-Met recruits ICAM-1 as a coreceptor to compensate for the loss of CD44 in Cd44 null mice. *Molecular Biology of the Cell, 22*, 2777–2786.

Orian-Rousseau, V. (2010). CD44, a therapeutic target for metastasising tumours. *European Journal of Cancer, 46*, 1271–1277.

Orian-Rousseau, V., Chen, L., Sleeman, J. P., Herrlich, P., & Ponta, H. (2002). CD44 is required for two consecutive steps in HGF/c-Met signaling. *Genes & Development, 16*, 3074–3086.

Orian-Rousseau, V., Morrison, H., Matzke, A., Kastilan, T., Pace, G., Herrlich, P., et al. (2007). Hepatocyte growth factor-induced Ras activation requires ERM proteins linked to both CD44v6 and F-actin. *Molecular Biology of the Cell, 18*, 76–83.

Panne, D. (2008). The enhanceosome. *Current Opinion in Structural Biology, 18*, 236–242.

Pawson, C. T., & Scott, J. D. (2010). Signal integration through blending, bolstering and bifurcating of intracellular information. *Nature Structural & Molecular Biology, 17*, 653–658.

Peach, R. J., Hollenbaugh, D., Stamenkovic, I., & Aruffo, A. (1993). Identification of hyaluronic acid binding sites in the extracellular domain of CD44. *The Journal of Cell Biology, 122*, 257–264.

Peck, D., & Isacke, C. M. (1996). CD44 phosphorylation regulates melanoma cell and fibroblast migration on, but not attachment to, a hyaluronan substratum. *Current Biology, 6*, 884–890.

Peck, D., & Isacke, C. M. (1998). Hyaluronan-dependent cell migration can be blocked by a CD44 cytoplasmic domain peptide containing a phosphoserine at position 325. *Journal of Cell Science, 111*(Pt. 11), 1595–1601.

Peterson, R. S., Andhare, R. A., Rousche, K. T., Knudson, W., Wang, W., Grossfield, J. B., et al. (2004). CD44 modulates Smad1 activation in the BMP-7 signaling pathway. *The Journal of Cell Biology, 166*, 1081–1091.

Ponta, H., Sherman, L., & Herrlich, P. A. (2003). CD44: From adhesion molecules to signalling regulators. *Nature Reviews Molecular Cell Biology, 4*, 33–45.

Purcell, B. P., Elser, J. A., Mu, A., Margulies, K. B., & Burdick, J. A. (2012). Synergistic effects of SDF-1alpha chemokine and hyaluronic acid release from degradable hydrogels on directing bone marrow derived cell homing to the myocardium. *Biomaterials, 33*, 7849–7857.

Purushothaman, A., & Toole, B. P. (2014). Serglycin proteoglycan is required for multiple myeloma cell adhesion, in vivo growth and vascularization. *The Journal of Biological Chemistry, 289*, 5499–5509.

Rao, G., Wang, H., Li, B., Huang, L., Xue, D., Wang, X., et al. (2013). Reciprocal interactions between tumor-associated macrophages and CD44-positive cancer cells via osteopontin/CD44 promote tumorigenicity in colorectal cancer. *Clinical Cancer Research, 19*, 785–797.

Ruckert, F., Grutzmann, R., & Pilarsky, C. (2012). Feedback within the inter-cellular communication and tumorigenesis in carcinomas. *PLoS One*, 7, e36719.

Screaton, G. R., Bell, M. V., Bell, J. I., & Jackson, D. G. (1993). The identification of a new alternative exon with highly restricted tissue expression in transcripts encoding the mouse Pgp-1 (CD44) homing receptor. Comparison of all 10 variable exons between mouse, human, and rat. *The Journal of Biological Chemistry, 268*, 12235–12238.

Seiter, S., Tilgen, W., Herrmann, K., Schadendorf, D., Patzelt, E., Moller, P., et al. (1996). Expression of CD44 splice variants in human skin and epidermal tumours. *Virchows Archiv, 428*, 141–149.

Sherman, L. S., Rizvi, T. A., Karyala, S., & Ratner, N. (2000). CD44 enhances neuregulin signaling by Schwann cells. *The Journal of Cell Biology, 150*, 1071–1084.

Sherman, L., Wainwright, D., Ponta, H., & Herrlich, P. (1998). *Genes & Development, 12*, 1058–1071.

Shi, X., Leng, L., Wang, T., Wang, W., Du, X., Li, J., et al. (2006). CD44 is the signaling component of the macrophage migration inhibitory factor-CD74 receptor complex. *Immunity, 25*, 595–606.

Sleeman, J., Rudy, W., Hofmann, M., Moll, J., Herrlich, P., & Ponta, H. (1996). Regulated clustering of variant CD44 proteins increases their hyaluronate binding capacity. *The Journal of Cell Biology, 135*, 1139–1150.

Soto, A. M., & Sonnenschein, C. (2011). The tissue organization field theory of cancer: A testable replacement for the somatic mutation theory. *BioEssays, 33*, 332–340.

Stamenkovic, I., & Yu, Q. (2010). Merlin, a "magic" linker between extracellular cues and intracellular signaling pathways that regulate cell motility, proliferation, and survival. *Current Protein & Peptide Science, 11*, 471–484.

Sweatt, J. D. (2001). The neuronal MAP kinase cascade: A biochemical signal integration system subserving synaptic plasticity and memory. *Journal of Neurochemistry, 76*, 1–10.

Teramoto, H., Castellone, M. D., Malek, R. L., Letwin, N., Frank, B., Gutkind, J. S., et al. (2005). Autocrine activation of an osteopontin-CD44-Rac pathway enhances invasion and transformation by H-RasV12. *Oncogene, 24*, 489–501.

Teriete, P., Banerji, S., Noble, M., Blundell, C. D., Wright, A. J., Pickford, A. R., et al. (2004). Structure of the regulatory hyaluronan binding domain in the inflammatory leukocyte homing receptor CD44. *Molecular Cell, 13*, 483–496.

Todaro, M., Gaggianesi, M., Catalano, V., Benfante, A., Iovino, F., Biffoni, M., et al. (2014). CD44v6 is a marker of constitutive and reprogrammed cancer stem cells driving colon cancer metastasis. *Cell Stem Cell, 14*, 342–356.

Tremmel, M., Matzke, A., Albrecht, I., Laib, A. M., Olaku, V., Ballmer-Hofer, K., et al. (2009). A CD44v6 peptide reveals a role of CD44 in VEGFR-2 signaling and angiogenesis. *Blood, 114*, 5236–5244.

Trofatter, J. A., MacCollin, M. M., Rutter, J. L., Murrell, J. R., Duyao, M. P., Parry, D. M., et al. (1993). A novel moesin-, ezrin-, radixin-like gene is a candidate for the neurofibromatosis 2 tumor suppressor. *Cell, 72*, 791–800.

Underhill, C. (1992). CD44: The hyaluronan receptor. *Journal of Cell Science, 103*(Pt. 2), 293–298.

van der Voort, R., Taher, T. E., Wielenga, V. J., Spaargaren, M., Prevo, R., Smit, L., et al. (1999). Heparan sulfate-modified CD44 promotes hepatocyte growth factor/scatter factor-induced signal transduction through the receptor tyrosine kinase c-Met. *The Journal of Biological Chemistry, 274*, 6499–6506.

Wai, P. Y., & Kuo, P. C. (2008). Osteopontin: Regulation in tumor metastasis. *Cancer Metastasis Reviews, 27*, 103–118.

Weber, G. F., Ashkar, S., Glimcher, M. J., & Cantor, H. (1996). Receptor-ligand interaction between CD44 and osteopontin (Eta-1). *Science, 271*, 509–512.

Wielenga, V. J., Heider, K. H., Offerhaus, G. J., Adolf, G. R., van den Berg, F. M., Ponta, H., et al. (1993). Expression of CD44 variant proteins in human colorectal cancer is related to tumor progression. *Cancer Research, 53*, 4754–4756.

Wiley, H. S., & Burke, P. M. (2001). Regulation of receptor tyrosine kinase signaling by endocytic trafficking. *Traffic, 2*, 12–18.

Yoo, B. K., Gredler, R., Chen, D., Santhekadur, P. K., Fisher, P. B., & Sarkar, D. (2011). c-Met activation through a novel pathway involving osteopontin mediates oncogenesis by the transcription factor LSF. *Journal of Hepatology, 55*, 1317–1324.

Yu, Q., & Stamenkovic, I. (1999). Localization of matrix metalloproteinase 9 to the cell surface provides a mechanism for CD44-mediated tumor invasion. *Genes & Development, 13*, 35–48.

Yu, W. H., Woessner, J. F., Jr., McNeish, J. D., & Stamenkovic, I. (2002). CD44 anchors the assembly of matrilysin/MMP-7 with heparin-binding epidermal growth factor precursor and ErbB4 and regulates female reproductive organ remodeling. *Genes & Development, 16*, 307–323.

Zeilstra, J., Joosten, S. P., van Andel, H., Tolg, C., Berns, A., Snoek, M., et al. (2014). Stem cell CD44v isoforms promote intestinal cancer formation in Apc(min) mice downstream of Wnt signaling. *Oncogene, 33*, 665–670.

CHAPTER TEN

Hyaluronan–CD44 Interaction Promotes Oncogenic Signaling, microRNA Functions, Chemoresistance, and Radiation Resistance in Cancer Stem Cells Leading to Tumor Progression

Lilly Y.W. Bourguignon*[,1], **Marisa Shiina***, **Jian-Jian Li**[†]
*Department of Medicine, University of California at San Francisco & Endocrine Unit (111N), VA Medical Center, San Francisco, California, USA
[†]Department of Radiation Oncology, University of California Davis, Sacramento, California, USA
[1]Corresponding author: e-mail address: lilly.bourguignon@ucsf.edu

Contents

1. Introduction 256
   1.1 Hyaluronan in tumor progression 256
   1.2 CD44 in tumor progression 256
2. Regulation of Tumor Progression by HA/CD44 258
   2.1 Ankyrin function 258
   2.2 RhoGTPase signaling 259
   2.3 Matrix metalloproteinases activities 260
3. Activation of CSCs by HA/CD44 261
   3.1 CD44v3, a newly identified CSC marker 263
   3.2 Nanog/Oct4/Sox2 & miR-302-regulated CSC functions 265
   3.3 Nanog & miR-21-regulated CSC functions 267
   3.4 ErbB2, p53, and CD44 signaling in CSCs 268
4. Conclusion 269
Acknowledgments 269
References 270

Abstract

Hyaluronan (HA), a major component of the extracellular matrix (ECM), is enriched in many types of tumors. There is good evidence linking high levels of HA production in human carcinomas to an aggressive phenotype and tumor metastasis. HA is generally bound to CD44 isoforms (so-called CD44s and CD44v3) which are ubiquitous, abundant, and functionally important cell surface receptors. This chapter describes the evidence

Advances in Cancer Research, Volume 123
ISSN 0065-230X
http://dx.doi.org/10.1016/B978-0-12-800092-2.00010-1

for HA/CD44v3-mediated activation of the cytoskeleton (e.g., ankyrin and GTPases) and matrix metalloproteinase (MMP) signaling during tumor progression.

A special focus is placed on the role of HA–CD44v3 interaction in cancer stem cells (CSCs). Matrix HA is known to be present in CSC niches. Since CD44v3 serves as a CSC marker, it provides an important physical linkage between matrix HA and various transcription factors that regulate tumor cell functions through distinct signaling pathways. CSCs are known to be chemoresistant and/or radiation resistant and to cause cancer relapse. The purpose of this chapter is to review the most current research on the cellular and molecular biology of CSCs. The emphasis will be placed on both CSC niche and matrix HA-induced microRNA signaling plus various CSC functions (e.g., self-renewal, differentiation, and chemoresistance) during cancer progression. Understanding the regulation of CSCs is critically important for designing CSC-specific therapeutic targets to prevent cancer development and progression.

1. INTRODUCTION

1.1. Hyaluronan in tumor progression

Myriad studies attempt to identify specific adhesion molecule(s) expressed in solid tumor cancer (e.g., breast cancer (BC), ovarian cancer, and head and neck squamous cell carcinoma (HNSCC)) that correlate with tumor cell invasive behavior(s). Among such candidate molecules is matrix hyaluronan (HA), a nonsulfated, unbranched glycosaminoglycan consisting of repeating disaccharide units, D-glucuronic acid and *N*-acetyl-D-glucosamine (Lee & Spicer, 2000) which is found in the extracellular matrix (ECM) of most mammalian tissues (Laurent & Fraser, 1992; Toole, 2001). In cancer patients, HA concentrations are usually higher in malignant tumors than in corresponding benign or normal tissues, and in some tumor types, the level of HA is predictive of malignancy. HA is often overexpressed at tumor cell attachment sites and appears to play an important role in promoting tumor cell-specific behaviors (Lee & Spicer, 2000; Toole, 2001; Turley, Nobel, & Bourguignon, 2002). The biosynthesis of HA is regulated by three mammalian HA synthase isozymes: HA synthase1 (HAS1), HA synthase 2 (HAS2), and HA synthase 3 (HAS3) (Itano & Kimata, 1996; Spicer & Nguyen, 1999; Weige, Hascall, & Tammi, 1997; Yamada et al., 1998).

1.2. CD44 in tumor progression

CD44 denotes a family of cell surface glycoproteins which are expressed in a variety of cells and tissues including tumor cells and carcinoma tissues (Auvinen, Tammi, Tammi, Johansson, & Kosma, 2005; Bourguignon,

2001; Bourguignon, Gilad, Brightman, Diedrich, & Singleton, 2006; Bourguignon, Gunja-Smith, et al., 1998; Bourguignon, Peyrollier, Gilad, & Brightman, 2007; Bourguignon, Singleton, Diedrich, Stern, & Gilad, 2004; Bourguignon, Singleton, Zhu, & Diedrich, 2003; Bourguignon, Singleton, Zhu, & Zhou, 2002; Bourguignon et al., 1997; Bourguignon, Zhu, Shao, & Chen, 2000; Bourguignon, Zhu, Shao, & Chen, 2001; Bourguignon, Zhu, Shao, Zhu, & Chen, 1999; Bourguignon, Zhu, Zhou, et al., 2001; Bourguignon, Zhu, & Zhu, 1998; Gunthert et al., 1991; Iida & Bourguignon, 1995, 1997; Turley et al., 2002; Zhu & Bourguignon, 1998). Nucleotide sequence analyses reveal that many CD44 isoforms (derived by alternative splicing mechanisms) are variants of the standard form, CD44s isoforms (Screaton et al., 1992). The presence of high levels of CD44 variant (e.g., CD44v3) isoforms is emerging as an important metastatic tumor marker in a number of cancers (Auvinen et al., 2005; Bourguignon, Gunja-Smith, et al., 1998; Bourguignon, Zhu, Shao, et al., 2001; Bourguignon, Zhu, Zhou, et al., 2001; Bourguignon, Zhu, et al., 1998; Bourguignon et al., 1997, 1999, 2000, 2002, 2003; Franzmann, Weed, Civantos, Goodwin, & Bourguignon, 2001; Iida & Bourguignon, 1997; Iida & Bourguignon, 1995; Gunthert et al., 1991; Turley et al., 2002; Zhu & Bourguignon, 1998). Recent studies have shown that CD44 is also expressed in tumor stem cells which have the unique ability to initiate tumor cell-specific properties (Al-Hajj, Wicha, Benito-Hernandez, Morrison, & Clarke, 2003). In fact, CD44 is proposed to be one of the important surface markers for cancer stem cells (CSCs) (Al-Hajj et al., 2003). All CD44 isoforms contain a HA-binding site in their extracellular domain and thereby serve as a major cell surface receptor for HA (Underhill, 1992). Both CD44 isoforms and HA are overexpressed at sites of tumor attachment (Toole & Hascall, 2002). As the histologic grade of each tumor progresses, the percentage of lesions expressing an associated CD44 isoform increases. In particular, the CD44 isoforms are expressed preferentially on highly malignant cancer tissue samples such as BC (Fig. 10.1). In fact, there is a direct correlation between CD44 isoform expression and increased histologic grade of the malignancy in head and neck cancer (Wang & Bourguignon, 2011; Wang, Wong, de Heer, Xia, & Bourguignon, 2009). Several studies indicate that tumor expression of the CD44 isoform may be used as an accurate predictor of overall survival (e.g., nodal status, tumor size, and grade) (Wang & Bourguignon, 2011; Wang, Wong, et al., 2009). It is likely that some of these CD44 isoforms on epithelial cells may act as surface modulators to facilitate unwanted growth factor receptor–growth factor interactions

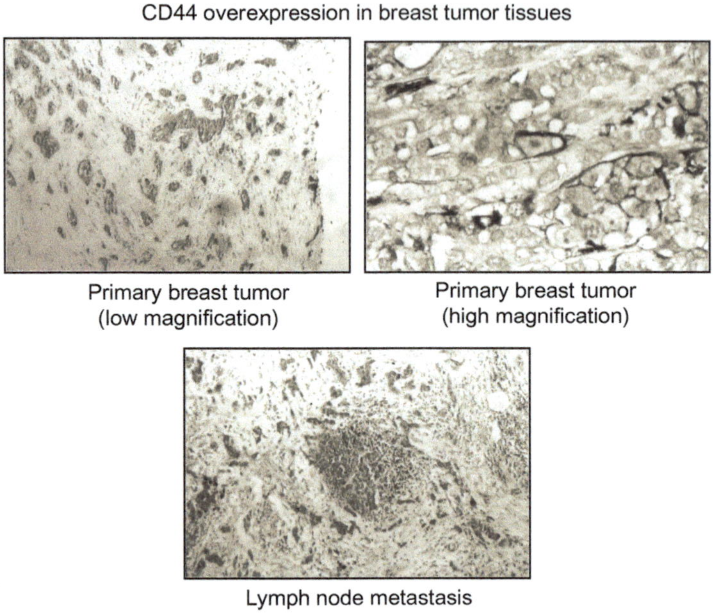

Figure 10.1 Immunoperoxidase staining of CD44 in human primary breast tumor and lymph node metastasis samples. (See the color plate.)

(Bourguignon et al., 2006; Wang & Bourguignon, 2006) and subsequent tumor formation. It is also possible that these CD44 isoforms may interact with ECM materials (e.g., HA) such that epithelial cells undergo abnormal mitogenesis and migration. This could be one of the critical steps in tumor invasion and metastasis.

2. REGULATION OF TUMOR PROGRESSION BY HA/CD44

The invasive phenotype of tumor cells is linked to cytoskeletal function. Dissection of the transmembrane pathways controlling these cellular processes should aid in understanding the regulatory mechanisms underlying tumor invasion and metastases.

2.1. Ankyrin function

CD44 interacts with a number of membrane-associated cytoskeletal proteins, such as ankyrin, which are expressed in a variety of biological systems, including epithelial cells and tissues (Bourguignon, 2001; Bourguignon, Gunja-Smith, et al., 1998; Bourguignon, Zhu, et al., 1998; Turley et al.,

2002; Zhu & Bourguignon, 1998). At the present time, three ankyrin genes have been identified as follows: ankyrin 1 (Ank 1 or ankyrin R), ankyrin 2 (Ank2 or ankyrin B), and ankyrin 3 (Ank3 or ankyrin G), all of which code for ankyrin monomers composed of highly conserved domains and one variable domain (Bennett & Baines, 2001; Lambert et al., 1990). The conserved domain contains a membrane-binding site (~89–95 kDa), also called the ankyrin repeat domain (ARD) in the N-terminal region, and a spectrin-binding domain (~62 kDa). The striking feature shared by all ankyrin N-terminal domain is a repeating 33-amino acid motif present in 24 contiguous copies (Lambert et al., 1990; Lux, John, & Bennett, 1990). The 24 ankyrin repeats within the ARD are known to form binding sites for several distinct membrane protein families, including CD44 (Zhu & Bourguignon, 1998). Overexpression of ARD in cells transfected with ARDcDNA not only interferes with CD44–ankyrin binding but also impairs HA-dependent and CD44-specific biological activities (Lokeshwar, Fregien, & Bourguignon, 1994; Zhu & Bourguignon, 1998). CD44–ankyrin interaction causes cytoskeleton activation and results in several important HA-mediated functions such as cell adhesion, proliferation, and migration (Bourguignon et al., 2007). These observations support the notion that HA-induced CD44–ankyrin interaction is very important for most HA-mediated CD44 functions and tumor-specific behaviors.

2.2. RhoGTPase signaling

RhoGTPases (e.g., RhoA, Rac1, and Cdc42) are members of the Rho subclass of the Ras superfamily (Hall, 1998) and are known to cycle between an active GTP-bound state and an inactive GDP-bound state to transmit diverse signals from cell surface receptors to intracellular targets. They function as molecular switches that, in response to external stimuli, regulate key signaling pathways and control a variety of cellular activities (Fig. 10.2) including cytoskeleton reorganization, tumor cell migration, and invasion (Hall, 1998; Li & Lim, 2003). The high incidence of RhoGTPase overexpression in human cancers suggests that these proteins are involved in both cancer onset and progression (Li & Lim, 2003). In fact, overexpression of certain RhoGTPases in human tumors often correlates with a poor prognosis (Fritz, Just, & Kaina, 1999; Suwa et al., 1998). Coordinated RhoGTPase signaling is considered to be a possible mechanism underlying cell proliferation, migration, and invasion, an obvious prerequisite for metastasis (Li & Lim, 2003). HA/CD44-mediated tumor cell-specific

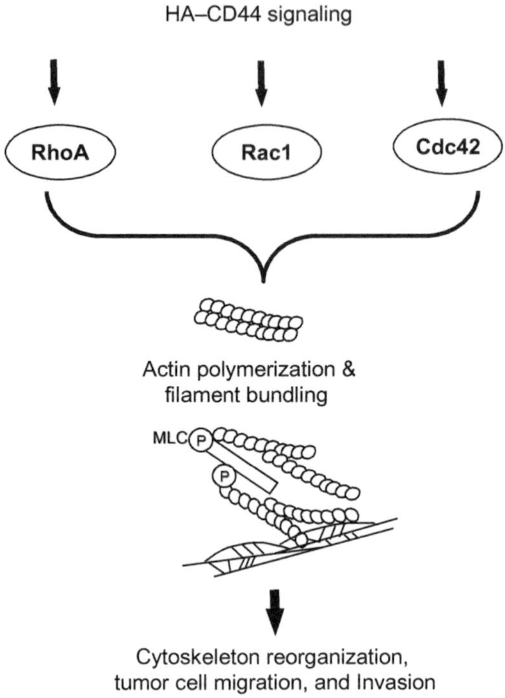

Figure 10.2 A proposed model for HA/CD44-mediated RhoGTPase signaling and cytoskeleton-mediated functions.

phenotypes are closely linked to the small GTP-binding proteins such as RhoA, Rac1, and Cdc42. Activation of RhoGTPases has been shown to produce specific structural changes in actin assembly, cytoskeleton reorganization, transcriptional activation, tumor cell growth, survival, migration, and invasion (Bourguignon, 2008).

2.3. Matrix metalloproteinases activities

It has been suggested that a number of different matrix metalloproteinases (MMPs) (including the 72 kDa gelatinase (gelatinase A, type IV collagenase, MMP-2), the 92 kDa gelatinase (gelatinase B, type V collagenase, MMP-9), the stromelysins (MMP-3, MMP-11), and the interstitial fibroblast-type collagenase (MMP-1)) are thought to play an important role in degrading ECM materials during tumor invasion and metastases (Chen, 1996; Gum et al., 1996). A membrane-type matrix metalloproteinase (MT-MMP, a transmembrane MMP) together with TIMP-2, a tissue inhibitor of metalloproteinase-2 (MMP-2), has been reported to be involved in the activation of MMP-2 on the cell surface (Polette et al., 1996). Brooks et al. (1996) determined that MMP-2 is localized in a proteolytically active form on the surface of invasive cells based on its ability to bind directly to integrin $\alpha v\beta 3$. CD44 isoform (e.g., CD44v3) also found to be closely associated with MMP-9 (gelatinase B) in the plasma membrane of breast tumor cells (Met-1 cell line) (Bourguignon, Gunja-Smith, et al., 1998; Bourguignon, Zhu, et al., 1998). Furthermore, CD44v3-associated MMP-9 appears to be present in a proteolytically active form and preferentially localized at the "invadopodia" structure of the Met-1 cells (Bourguignon, Gunja-Smith, et al., 1998; Bourguignon, Zhu, et al., 1998). Previous findings showed that certain protease(s) is(are) localized on "invadopodia" of human malignant melanoma cells (Chen, 1996). The close interaction between CD44v3 and the active form of MMP-9 in the "invadopodia" structure of Met-1 tumor cells may be required for the degradation of ECM for tumor cell invasion and metastasis. Therefore, we believe that CD44v3 plays an important role in linking ankyrin to the membrane-associated actomyosin contractile system required for upregulating metastatic tumor cell behavior (e.g., invadopodia formation and matrix degradation) and promoting angiogenesis during BC progression.

3. ACTIVATION OF CSCS BY HA/CD44

The inexorable progression and therapy-resistant phenotype of cancer cells are driven by uncontrolled growth regulation and adaptive signaling pathways. Despite a trend toward overall improvement in early detection and therapy outcomes, the mortality of solid tumors such as breast cancer (BC) and head and neck cancer (HNSCCs) remains unacceptably high, especially for patients with recurrent or metastatic tumors.

Multidrug resistance is a challenging clinical problem in the treatment of a variety of cancers (Harnett, Kirsten, & Tattersall, 1986). The most commonly used antitumor agents in the treatment of disseminated breast or ovarian cancers include doxorubicin and paclitaxel (taxol) (Gewirtz, 1999). The ability of doxorubicin to bind DNA and/or produce free radicals is thought to be a possible mechanism to induce cytotoxic effects on tumor cells (Gewirtz, 1999). Paclitaxel is known to bind a subunit of the tubulin heterodimers that form cellular microtubules (Bhalla, 2003). The paclitaxel–cytoskeleton interaction accelerates tubulin polymerization and inhibits the depolymerization of microtubules resulting in an inactivation of the mitotic checkpoint and tumor cell killing (Bhalla, 2003). Resistance to chemotherapeutic drugs used in breast and ovarian cancer treatments (e.g., doxorubicin and paclitaxel) is often associated with overexpression of the multidrug resistance gene 1 (MDR1 or P-glycoprotein (P-gp)). P-gp, the product of the MDR1 (ABCB1) gene, is a transmembrane ATP-dependent transporter molecule known to play a critical role in drug fluxes and chemotherapeutic resistance in a variety of cancers (Baker & El-Osta, 2004; Higgins, 1992). The MDR1 gene product functions as a drug efflux pump actively reducing intracellular drug concentrations in resistant tumor cells (Baker & El-Osta, 2004; Higgins, 1992). Interestingly, recent studies indicate that both HA and CD44 are also involved in chemotherapeutic drug resistance in many cancer types (Bourguignon, Peyrollier, Xia, & Gilad, 2008; Bourguignon et al., 2002; Wang & Bourguignon, 2011). Specifically, HA binding promotes MDR1 expression, cytoskeletal protein (ankyrin)-induced drug fluxes, and chemoresistance in tumor cells (Bourguignon et al., 2008). Furthermore, HA/CD44-mediated ErbB2 signaling and PI3 kinase/AKT-related survival pathways are also involved in chemotherapeutic drug resistance in tumor cells (Misra, Ghatak, Zoltan-Jones, & Toole, 2003).

Accumulating evidence suggests that tumors contain CSCs which contribute at least a part in tumor aggressiveness and resistance to anticancer chemo- and radiotherapy. Accumulating evidence has indicated that most tumors contain a small population of cells which persistently initiate tumor growth and promote tumor progression. These "CSCs" (also called tumor-initiating cells (TICs)) share several hallmarks of normal stem cells (Al-Hajj et al., 2003; Bonnet & Dick, 1997). CSCs are known to undergo self-renewal, maintain quiescence, display multipotency, and express survival protein/antiapoptosis proteins (Al-Hajj et al., 2003; Bonnet & Dick, 1997; Bourguignon, Earle, Wong, Spevak, & Krueger, 2012; Bourguignon, Wong, Earle, & Chen, 2012). Another well-known property of CSCs is their ability to expand the stem cell population by undergoing cell proliferation/survival and/or clone formation and differentiation (Al-Hajj et al., 2003; Bonnet & Dick, 1997; Bourguignon, Earle, et al., 2012; Bourguignon, Wong, et al., 2012). These CSCs have been shown both experimentally and clinically to be resistant to radiation and chemotherapy potentially resulting in residual disease that leads to recurrence. A number of studies have identified specific molecules expressed in CSCs that correlate with both stem cell properties and tumor cell behaviors. CD44 is expressed in both normal stem cells and CSCs and serves as an important stem cell marker (Al-Hajj et al., 2003). A number of studies now indicate that tumor cells with high levels of CD44 expression appear to exhibit CSC properties in many cancers (Bourguignon, Earle, et al., 2012; Bourguignon, Wong, et al., 2012).

3.1. CD44v3, a newly identified CSC marker

All stem cells are thought to exist in specialized microenvironments known as niches (Astachov, Vago, Aviv, & Nevo, 2011; Haylock & Nilsson, 2006). Components present in the niches can regulate stem cell behavior through direct binding to stem cell surface receptors or via indirect activation of paracrine signaling (Astachov et al., 2011; Haylock & Nilsson, 2006). ECM components including HA are known to be present in one of the stem cell niches (Astachov et al., 2011; Haylock & Nilsson, 2006). HA–CD44 interaction enhances sphere formation (cell–cell adhesion) (Fig. 10.3A), self-renewal/growth, and clonal formation in highly tumorigenic CD44v3^{high}ALDH1^{high} tumor cell population (Bourguignon, Wong, et al.,, 2012). These observations support the contention that HA signaling is directly involved in the regulation of CSC-like properties in CD44v3^{high}ALDH1^{high} cells.

CD44 is commonly expressed in various isoforms generated by alternative mRNA splicing of variant exons inserted into an extracellular membrane-proximal site (Screaton et al., 1992). CD44 is expressed in both

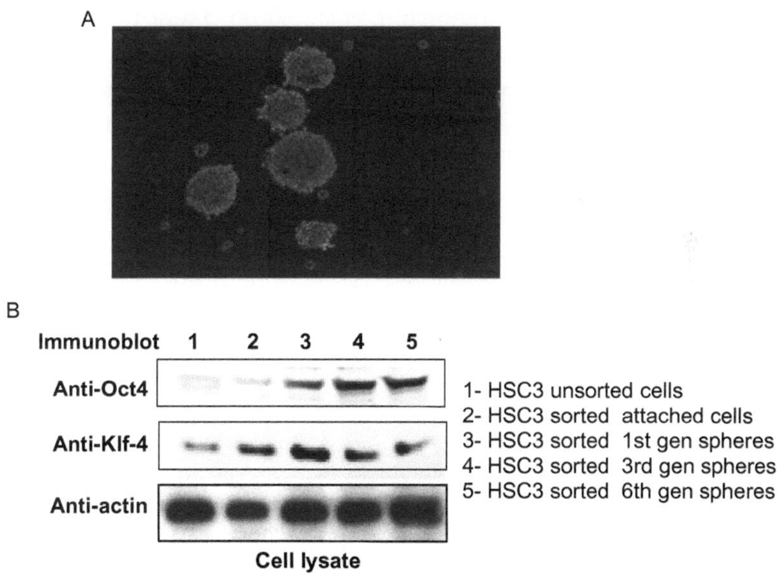

Figure 10.3 HA/CD44-mediated sphere formation (A) and stem cell marker expression in sphere cells with multiple generations of passages (B).

normal and CSCs and serves as an important stem cell marker (Al-Hajj et al., 2003). It is now known that neither CD44s nor CD44v6 distinguishes normal from benign or malignant epithelia of head and neck (Mack & Gires, 2008). Consequently, the identification of other specific CD44 isoform-related CSC markers is needed. In this regard, the role of CD44v3 isoform in head and neck cancer progression has been highlighted in studies that have identified an association of the v3-containing isoform with head and neck cancer growth, migration, and MMP expression (Wang & Bourguignon, 2011; Wang, Peyrollier, & Bourguignon, 2007; Wang, Wreesmann, & Bourguignon, 2007). Transfection of the CD44v3 isoform into a non-expressing head and neck cancer cell line results in significantly increased tumor cell migration (Wang, Peyrollier, et al., 2007; Wang, Wreesmann, et al., 2007). In addition, treatment of a CD44v3 isoform-expressing head and neck cancer cell line with anti-CD44v3 isoform antibody decreased *in vitro* proliferation and increased cisplatin sensitivity (Wang, Peyrollier, et al., 2007; Wang, Wreesmann, et al., 2007). Using the same anti-CD44v3 isoform antibody, immunohistochemical tissue analysis revealed that the CD44v3 isoform is preferentially expressed in metastatic lymph nodes. Furthermore, CD44v3 isoform expression in primary tumors is significantly associated with advanced T status and positive lymph nodes (Wang & Bourguignon, 2011; Wang, Peyrollier, et al., 2007; Wang, Wong, et al., 2009; Wang, Wreesmann, et al., 2007). These findings suggest that the CD44v3 isoform is closely associated with head and neck cancer development and progression and may serve as a new CSC marker.

Recently, an elevated level of aldehyde dehydrogenenase 1 (ALDH1) activity was also shown to identify CSCs in head and neck cancer (Bourguignon, Wong, et al., 2012). The subpopulation of $CD44v3^{high}ALDH1^{high}$ cells expressing stem cell markers such as Oct4 and Klf4 (isolated from head and neck tumor cells) appears to undergo cell growth and self-renewing after multiple passages from spheres (demonstrated by serial passage of sphere-forming ability and long-term tumor cell growth) (Fig. 10.3B). Importantly, these $CD44v3^{high}ALDH1^{high}$ cells exhibit CSC-like phenotypes and are highly tumorigenic for generating tumors starting with as few as 50 cells. In breast cancer, a $CD44^{high}CD24^{low}$ tumor population has also been shown to be highly tumorigenic (Bourguignon, Earle, et al., 2012; Bourguignon, Wong, et al., 2012).

One of the characteristics of CSCs is overexpression of stem cell-specific transcription factors such as Oct4, Sox2, and Nanog (Boyer et al., 2005;

Kashyap et al., 2009). Previous studies indicate that Oct4, Sox2, and Nanog often function in combinatorial complexes to regulate the expression of gene loci which are involved in self-renewal, proliferation, and differentiation (Boyer et al., 2005; Kashyap et al., 2009). These three molecules are also closely associated with higher histological grades and poorer clinical survival in head and neck cancer (Bourguignon, Earle, et al., 2012; Dong, Wilhelm, & Koopman, 2004; Wang, Wong, et al., 2009). One of the characteristics of CSCs is overexpression of stem cell-specific transcription factors such as Oct4, Sox2, and Nanog (Boyer et al., 2005). These three molecules are also closely associated with higher histological grades and poorer clinical survival in head and neck cancer (Bourguignon, Earle, et al., 2012; Dong et al., 2004; Wang, Wong, et al., 2009). Moreover, a high level of Oct4, Sox2, and Nanog expression is detected in CD44v3^{high}ALDH1^{high} cell-induced mouse tumors and in human head and neck cancer patient specimens (Bourguignon, Wong, et al., 2012). These findings indicate that CD44v3^{high}ALDH1^{high} cells are likely to be generating tumors through the stem cell processes of self-renewal, growth, and differentiation. These cells can persist in tumors as a distinct population, possibly causing chemotherapy resistance and tumor recurrence at a later time. The mechanism by which the regulatory network consisting of Oct4, Sox2, and Nanog contributes to the establishment of CSC-specific phenotypes in human cancer is described below.

3.2. Nanog/Oct4/Sox2 & miR-302-regulated CSC functions

Previous genetic studies using mouse models revealed that transcription factors such as Oct4, Sox2, and Nanog have distinct roles but could use similar signaling pathways to maintain "stemness" functions during development (Boyer et al., 2005; Kashyap et al., 2009). Some studies have also indicated that Oct4, Sox2, and Nanog cooccupy the promoter of miR-302 which has been shown to target genes required for development and oncogenesis (Card et al., 2008; Liu et al., 2011). At the transcriptional level, Oct4, Sox2, and Nanog form a positive autoregulatory loop that is important for the maintenance of the undifferentiated state. At the post-translational level, miR-302 family is emerging as a key player in the control of cell proliferation and cell fate determination during differentiation (Card et al., 2008; Liu et al., 2011). Recently, new evidence indicates that the HA–CD44v3 interaction activates Oct4, Sox2, and Nanog complex formation in CD44v3-expressing CSCs. Furthermore, our results indicate that the miR-302 cluster is controlled by a promoter containing Oct4/Sox2/Nanog-binding sites in these cells. The production of miR-302 is Oct4/Sox2/Nanog dependent in CSCs. Overexpression of miR-302a and miR-302b in both mouse tumors induced by CD44v3-expressing CSC cells can be detected in human head and neck cancer patient samples (Bourguignon, Wong, et al., 2012). These findings clearly established the existence of a close association between miR-302 clusters (e.g., miR-302a and miR-302b) and head and neck cancer development.

Inhibitors of the apoptosis family of proteins (IAPs, e.g., cIAP-1, cIAP-2, and XIAP) are also frequently overexpressed by CSCs. Importantly, upregulation of IAPs increases cell survival by the binding of IAPs to caspases and suppressing apoptosis (Hunter, LaCasse, & Korneluk, 2007). A recent study indicated that HA enhances antiapoptotic effects and decreases the ability of cisplatin to induce apoptosis and cell death of these $CD44v3^{high}ALDH1^{high}$ cells leading to the enhancement of chemoresistance (Bourguignon, Wong, et al., 2012). These findings suggest that HA-mediated cell survival contributes to both a decrease in apoptosis/cell death and an increase in cisplatin resistance in these CSC-like $CD44v3^{high}ALDH1^{high}$ cells. Furthermore, downregulation of miR-302a or miR-302b by transfecting $CD44v3^{high}ALDH1^{high}$ cells with either anti-miR-302a or miR-302b inhibitor effectively reduces HA-mediated IAP protein (e.g., cIAP-1, cIAP-2, and XIAP) expression in $CD44v3^{high}ALDH1^{high}$ cells. Consequently, the level of cisplatin-induced apoptosis and cell death of these cells is also significantly elevated. Taken together, we believe that the miR-302 cluster (e.g., miR-302a and miR-302b) plays a critical role in regulating HA/CD44-mediated chemotherapy resistance in these tumorigenic and CSC-like $CD44v3^{high}ALDH1^{high}$ cell populations from head and neck cancer. Several years ago, a small molecule IAP protein inhibitor (bivalent SM-164) was shown to concurrently target cIAP-1/cIAP-2 for degradation and antagonize XIAP leading to an enhancement of tumor cell apoptosis and tumor regression (Lu et al., 2008). Recently, we reported that chemosensitivity in CSC-like $CD44v3^{high}ALDH1^{high}$ cells increased with the newly developed IAP protein inhibitor, SM164 (Bourguignon, Wong, et al., 2012). These findings suggest that the use of a combination of IAP protein inhibitor (SM164) and cisplatin may significantly improve the efficacy of chemotherapeutic drug treatment by targeting CSCs in head and neck cancer.

3.3. Nanog & miR-21-regulated CSC functions

Cisplatin is one of the most common antihead and neck cancer chemotherapy drugs used today. Various mechanisms of cisplatin resistance in head and neck cancer have been proposed with potential molecular effectors of resistance, including p53, glutathione *S*-transferase, and ErbB2 (Bradford et al., 2003; Kato et al., 2000; Shiga et al., 2000). The ability of this drug to induce tumor cell death is also counteracted by the presence of antiapoptotic regulators leading to chemoresistance (Wang & Bourguignon, 2011). Activation of several other HA–CD44-mediated oncogenic signaling pathways (e.g., intracellular Ca^{2+} mobilization (Wang & Bourguignon, 2006; Bourguignon et al., 2006), epidermal growth factor receptor-mediated ERK signaling (Bourguignon et al., 2006; Wang & Bourguignon, 2006), and topoisomerase activation (Wang, Peyrollier, et al., 2007; Wang, Wreesmann, et al., 2007)) leads to multidrug resistance in a number of tumor cells. However, the mechanism by which HA/CD44 activates oncogenesis and drug resistance in HNSCC has not been fully elucidated.

The functional significance of miR-21 has been elucidated in several recent studies following the discovery of its specific targets (Asangani et al., 2008), making miR-21 one of the most-studied miRNAs due to its involvement in cancer progression. Importantly, miR-21 plays a role in the inhibition of tumor suppressor proteins such as program cell death 4 (PDCD4), via a conserved site within the 3′-UTR (3′-untranslated region) of the mRNA (Asangani et al., 2008; Bourguignon, Spevak, Wong, Xia, & Gilad, 2009). Downregulation of PDCD4 expression by miR-21 leads to tumor cell growth, survival, chemoresistance, invasion, and metastasis (Bourguignon et al., 2009). Thus, miR-21 is currently considered to be an oncogenic miRNA. In previous studies, a new matrix HA/CD44-mediated Nanog/Stat-3 signaling mechanism was reported to be involved in miR-21 production and chemotherapy chemoresistance in breast tumor cells and HNSCC cells (Bourguignon, Earle, et al., 2012). Specifically, these results indicated that HA/CD44 activates Nanog–Stat-3 signaling which, in turn, stimulates miR-21 expression and function. These events lead to the reduction of the tumor suppressor protein, PDCD4, upregulation of survival proteins, IAP family (e.g., cIAP-1, cIAP-2 and XIAP), and cisplatin chemoresistance in HNSCC cells (Bourguignon, Earle, et al., 2012). Inhibition of either Nanog/Stat-3 signaling or silencing miR-21 expression/function results in PDCD4 upregulation and causes a reduction of survival protein expression and an enhancement of chemosensitivity to cisplatin (Bourguignon, Earle, et al., 2012). Thus, these findings strongly support the contention that Nanog, Stat-3, and miR-21 form a functional signaling axis that regulates tumor cell survival and cisplatin chemoresistance in tumor cells.

3.4. ErbB2, p53, and CD44 signaling in CSCs

The p53 tumor suppressor is a key mediator of cellular responses to various stresses. A previous study showed that p53 inhibits CD44 expression via binding to a noncanonical p53-binding sequence in the CD44 promoter (Godar et al., 2008). This interaction enables an untransformed cell to respond to stress-induced, p53-dependent cytostatic and apoptotic signals that would otherwise be blocked by the actions of CD44. In the absence of p53 function, the resulting derepressed CD44 expression is essential for the growth and tumor-initiating ability of highly tumorigenic mammary epithelial cells. Thus, CD44 appears to function as a key tumor-promoting agent in transformed tumor cells lacking p53 function.

CSCs from breast cancer cells are able to form colonies or new tumors (Al-Hajj et al., 2003; Bonnet & Dick, 1997). Specific cell surface markers such as $CD44^{high}/CD24^{-/low}$ in breast CSCs (Al-Hajj et al., 2003) are shown to have increased expression of proinvasive genes required for metastasis such as IL-1α, IL-6, IL-8, and urokinase plasminogen activator (Sheridan et al., 2006). Additionally, CSCs appear to exhibit notable chemo- and radioresistance. For example, in tumor cells, under genotoxic stress conditions (such as ionizing radiation), activation of prosurvival pathways and inhibition of proapoptotic pathways are responsible for tumor radioresistance. The molecules involved in these pathways are potential targets to enhance tumor radiosensitivity. Clinical data suggest that breast cancer patients with tumors overexpressing ErbB2/HER2/neu, a member of ErbB family of receptor tyrosine kinases, show a poor prognosis compared to ErbB2/HER2/neu-negative tumors, and the ErbB2/HER2 gene enhancement is correlated with the time to relapse of the disease (Slamon et al., 1987). The poorer prognosis of women with ErbB2/HER2/Neu-amplified tumors is entirely abrogated if they are treated with anti-ErbB2/HER2 therapy. Upon irradiation, cells overexpressing ErbB2/HER2 are able to activate NFκB, a key transcription factor in stress response that regulates a prosurvival network (Guo et al., 2004). Importantly, the NFκB-binding motif in the ErbB2/HER2 gene promoter region is found to be responsible for radiation-induced HER2 expression (Cao et al., 2009), and a recent study indicated that HER2-positive CSCs are present in "ErbB2/HER2-negative" breast cancer cells and the HER2-positive CSCs are increased in the recurrent/metastatic tumors (Duru et al., 2012), indicating that ErbB2/HER2 is a dynamic feature of CSCs in breast cancer. Defining the central role of ErbB2/HER2-positive CSCs may offer effective targets for breast cancer treatment. These findings suggest that the existence of different subpopulations of CSCs with varied sets of mutations and genomic alterations may be likely since heterogeneous tumors consist of unstable genomes and different responses to anticancer therapy. During chemo- and/or radiotherapy, the most resistant CSCs would be selected and continue to sustain the tumor under treatments. These findings shed light onto a new conceptual paradigm of how CSCs or TICs contribute to the radiation response. Identification of CSC-associated radioresistance needs to be further evaluated in clinical studies.

4. CONCLUSION

HA/CD44-mediated tumor cell-specific phenotype is closely linked to cytoskeletal functions which involve membrane-associated cytoskeletal proteins (e.g., ankyrin and GTPases) and MMPs. Activation of ankyrin, GTPases, and MMPs produces specific structural changes in actin assembly, cytoskeleton reorganization, transcriptional activation, tumor cell growth, survival, migration, and invasion. These events generate coordinated "cross talk" among multiple signaling pathways required for oncogenesis. Accumulating evidence also indicates that Oct4/Sox2/Nanog signaling and microRNA (e.g., miR-302 and miR-21) expression occur in HA/CD44-activated CSCs. Furthermore, CD44-expressing CSCs require downregulation of p53 and upregulation of ErbB2. All these events contribute to CSC self-renewal, clonal formation, antiapoptosis/survival, radiation resistance, and chemoresistance leading to tumor progression. A current model illustrating HA-dependent and CD44-specific signaling pathways is described in Fig. 10.4.

ACKNOWLEDGMENTS

We gratefully acknowledge the assistance of Dr. Gerard J. Bourguignon in the preparation and review of this manuscript. This work was supported by Veterans Affairs (VA) Merit Review Awards (RR & D-1I01 RX000601 and BLR & D-5I01 BX000628), United States Public Health Grants (R01 CA66163). L. Y. W. B is a VA Senior Research Career Scientist.

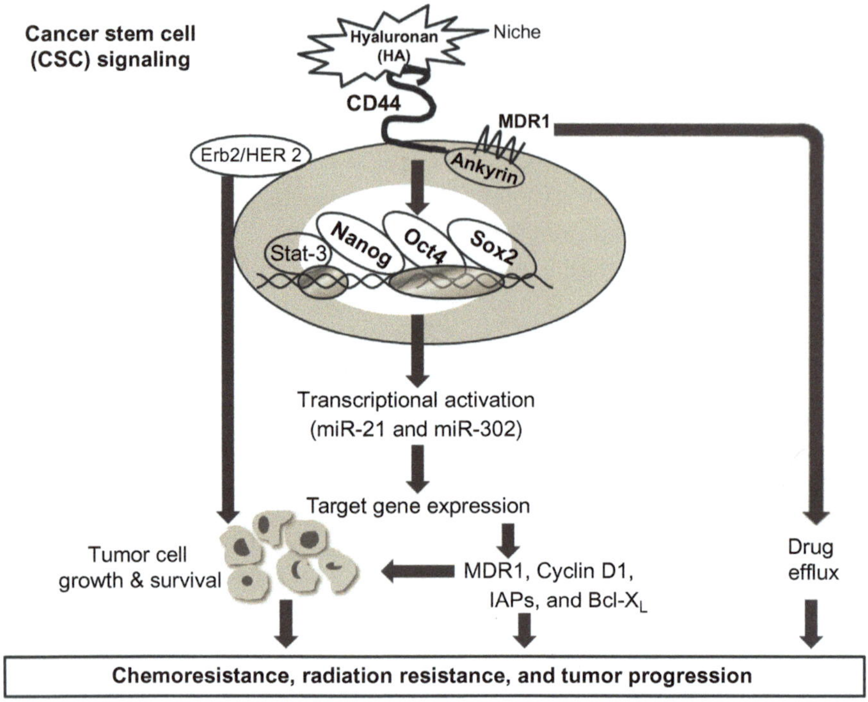

Figure 10.4 A proposed model for cancer stem cell signaling and tumor progression.

REFERENCES

Al-Hajj, M., Wicha, M. S., Benito-Hernandez, A., Morrison, S. J., & Clarke, M. F. (2003). Prospective identification of tumorigenic breast cancer cells. *Proceedings of the National Academy of Sciences of the United States of America, 100*, 3983–3988.

Asangani, I. A., Rasheed, S. A., Nikolova, D. A., Leupold, J. H., Colburn, N. H., Post, S., et al. (2008). MicroRNA-21 (miR-21) post-transcriptionally downregulates tumor suppressor Pdcd4 and stimulates invasion, intravasation and metastasis in colorectal cancer. *Oncogene, 27*, 2128–2136.

Astachov, L., Vago, R., Aviv, M., & Nevo, Z. (2011). Hyaluronan and mesenchymal stem cells: From germ layer to cartilage and bone. *Frontiers in Bioscience (Landmark Edition), 16*, 261–276.

Auvinen, P., Tammi, R., Tammi, M., Johansson, R., & Kosma, V. M. (2005). Expression of CD44s, CD44v3 and CD44v6 in benign and malignant breast lesions: Correlation and colocalization with hyaluronan. *Histopathology, 47*, 420–428.

Baker, E. K., & El-Osta, A. (2004). MDR1, chemotherapy and chromatin remodeling. *Cancer Biology & Therapy, 3*, 819–824.

Bennett, V., & Baines, A. J. (2001). Spectrin and ankyrin-based pathways: Metazoan inventions for integrating cells into tissues. *Physiological Reviews, 81*, 1353–1392.

Bhalla, K. N. (2003). Microtubule-targeted anticancer agents and apoptosis. *Oncogene, 22*, 9075–9086.

Bonnet, D., & Dick, J. E. (1997). Human acute myeloid leukemia is organized as a hierarchy that originates from a primitive hematopoietic cell. *Nature Medicine, 3*, 730–737.

Bourguignon, L. Y. (2001). CD44-mediated oncogenic signaling and cytoskeleton activation during mammary tumor progression. *Journal of Mammary Gland Biology and Neoplasia, 6*, 287–297.

Bourguignon, L. Y. (2008). Hyaluronan-mediated CD44 activation of RhoGTPase signaling and cytoskeleton function promotes tumor progression. *Seminars in Cancer Biology, 18*, 251–259.

Bourguignon, L. Y., Earle, C., Wong, G., Spevak, C. C., & Krueger, K. (2012). Stem cell marker (Nanog) and Stat-3 signaling promote MicroRNA-21 expression and chemoresistance in hyaluronan/CD44-activated head and neck squamous cell carcinoma cells. *Oncogene, 31*, 149–160.

Bourguignon, L. Y., Gilad, E., Brightman, A., Diedrich, F., & Singleton, P. (2006). Hyaluronan-CD44 interaction with leukemia-associated RhoGEF and epidermal growth factor receptor promotes Rho/Ras co-activation, phospholipase C epsilon-Ca2+ signaling, and cytoskeleton modification in head and neck squamous cell carcinoma cells. *The Journal of Biological Chemistry, 281*, 14026–14040.

Bourguignon, L. Y., Gunja-Smith, Z., Iida, N., Zhu, H. B., Young, L. J., Muller, W. J., et al. (1998). CD44v(3,8-10) is involved in cytoskeleton-mediated tumor cell migration and matrix metalloproteinase (MMP-9) association in metastatic breast cancer cells. *Journal of Cellular Physiology, 176*, 206–215.

Bourguignon, L. Y., Peyrollier, K., Gilad, E., & Brightman, A. (2007). Hyaluronan-CD44 interaction with neural Wiskott-Aldrich syndrome protein (N-WASP) promotes actin polymerization and ErbB2 activation leading to beta-catenin nuclear translocation, transcriptional up-regulation, and cell migration in ovarian tumor cells. *The Journal of Biological Chemistry, 282*, 1265–1280.

Bourguignon, L. Y., Peyrollier, K., Xia, W., & Gilad, E. (2008). Hyaluronan-CD44 interaction activates stem cell marker Nanog, Stat-3-mediated MDR1 gene expression, and ankyrin-regulated multidrug efflux in breast and ovarian tumor cells. *The Journal of Biological Chemistry, 283*, 17635–17651.

Bourguignon, L. Y., Singleton, P. A., Diedrich, F., Stern, R., & Gilad, E. (2004). CD44 interaction with Na+-H+ exchanger (NHE1) creates acidic microenvironments leading to hyaluronidase-2 and cathepsin B activation and breast tumor cell invasion. *The Journal of Biological Chemistry, 279*, 26991–27007.

Bourguignon, L. Y., Singleton, P. A., Zhu, H., & Diedrich, F. (2003). Hyaluronan-mediated CD44 interaction with RhoGEF and Rho kinase promotes Grb2-associated binder-1 phosphorylation and phosphatidylinositol 3-kinase signaling leading to cytokine (macrophage-colony stimulating factor) production and breast tumor progression. *The Journal of Biological Chemistry, 278*, 29420–29434.

Bourguignon, L. Y., Singleton, P. A., Zhu, H., & Zhou, B. (2002). Hyaluronan promotes signaling interaction between CD44 and the transforming growth factor beta receptor I in metastatic breast tumor cells. *The Journal of Biological Chemistry, 277*, 39703–39710.

Bourguignon, L. Y., Spevak, C. C., Wong, G., Xia, W., & Gilad, E. (2009). Hyaluronan-CD44 interaction with protein kinase C(epsilon) promotes oncogenic signaling by the stem cell marker Nanog and the production of microRNA-21, leading to down-regulation of the tumor suppressor protein PDCD4, anti-apoptosis, and chemotherapy resistance in breast tumor cells. *The Journal of Biological Chemistry, 284*, 26533–26546.

Bourguignon, L. Y., Wong, G., Earle, C., & Chen, L. (2012). Hyaluronan-CD44v3 interaction with Oct4-Sox2-Nanog promotes miR-302 expression leading to self-renewal, clonal formation, and cisplatin resistance in cancer stem cells from head and neck squamous cell carcinoma. *The Journal of Biological Chemistry, 287*, 32800–32824.

Bourguignon, L. Y., Zhu, H., Chu, A., Iida, N., Zhang, L., & Hung, M. C. (1997). Interaction between the adhesion receptor, CD44, and the oncogene product, p185HER2, promotes human ovarian tumor cell activation. *The Journal of Biological Chemistry, 272*, 27913–27918.

Bourguignon, L. Y., Zhu, H., Shao, L., & Chen, Y. W. (2000). CD44 interaction with tiam1 promotes Rac1 signaling and hyaluronic acid-mediated breast tumor cell migration. *The Journal of Biological Chemistry, 275*, 1829–1838.

Bourguignon, L. Y., Zhu, H., Shao, L., & Chen, Y. W. (2001). CD44 interaction with c-Src kinase promotes cortactin-mediated cytoskeleton function and hyaluronic acid-dependent ovarian tumor cell migration. *The Journal of Biological Chemistry, 276*, 7327–7336.

Bourguignon, L. Y., Zhu, H., Shao, L., Zhu, D., & Chen, Y. W. (1999). Rho-kinase (ROK) promotes CD44v(3,8-10)-ankyrin interaction and tumor cell migration in metastatic breast cancer cells. *Cell Motility and the Cytoskeleton, 43*, 269–287.

Bourguignon, L. Y., Zhu, H., Zhou, B., Diedrich, F., Singleton, P. A., & Hung, M. C. (2001). Hyaluronan promotes CD44v3-Vav2 interaction with Grb2-p185(HER2) and induces Rac1 and Ras signaling during ovarian tumor cell migration and growth. *The Journal of Biological Chemistry, 276*, 48679–48692.

Bourguignon, L. Y., Zhu, D., & Zhu, H. (1998). CD44 isoform-cytoskeleton interaction in oncogenic signaling and tumor progression. *Frontiers in Bioscience, 3*, d637–d649.

Boyer, L. A., Lee, T. I., Cole, M. F., Johnstone, S. E., Levine, S. S., Zucker, J. P., et al. (2005). Core transcriptional regulatory circuitry in human embryonic stem cells. *Cell, 122*, 947–956.

Bradford, C. R., Zhu, S., Ogawa, H., Ogawa, T., Ubell, M., Narayan, A., et al. (2003). P53 mutation correlates with cisplatin sensitivity in head and neck squamous cell carcinoma lines. *Head & Neck, 25*, 654–661.

Brooks, P. C., Strömblad, S., Sanders, L. C., von Schalscha, T. L., Aimes, R. T., Stetler-Stevenson, W. G., et al. (1996). Localization of matrix metalloproteinase MMP-2 to the surface of invasive cells by interaction with integrin alpha v beta 3. *Cell, 85*, 683–693.

Cao, N., Li, S., Wang, Z., Ahmed, K. M., Degnan, M. E., Fan, M., et al. (2009). NF-kappaB-mediated HER2 overexpression in radiation-adaptive resistance. *Radiation Research, 171*, 9–21.

Card, D. A., Hebbar, P. B., Li, L., Trotter, K. W., Komatsu, Y., Mishina, Y., et al. (2008). Oct4/Sox2-regulated miR-302 targets cyclin D1 in human embryonic stem cells. *Molecular and Cellular Biology, 28*, 6426–6438.

Chen, W. T. (1996). Proteases associated with invadopodia, and their role in degradation of extracellular matrix. *Enzyme & Protein, 49*, 59–71.

Dong, C., Wilhelm, D., & Koopman, P. (2004). Sox genes and cancer. *Genome Research, 105*, 442–447.

Duru, N., Fan, M., Candas, D., Menaa, C., Liu, H. C., Nantajit, D., et al. (2012). HER2-associated radioresistance of breast cancer stem cells isolated from HER2-negative breast cancer cells. *Clinical Cancer Research, 18*, 6634–6647.

Franzmann, E. J., Weed, D. T., Civantos, F. J., Goodwin, W. J., & Bourguignon, L. Y. (2001). A novel CD44 v3 isoform is involved in head and neck squamous cell carcinoma progression. *Otolaryngology and Head and Neck Surgery, 124*, 426–432.

Fritz, G., Just, I., & Kaina, B. (1999). RhoGTPase over-expressed in human tumors. *International Journal of Cancer, 81*, 682–687.

Gewirtz, D. A. (1999). A critical evaluation of the mechanisms of action proposed for the antitumor effects of the anthracycline antibiotics adriamycin and daunorubicin. *Biochemical Pharmacology, 57*, 727–741.

Godar, S., Ince, T. A., Bell, G. W., Feldser, D., Donaher, J. L., Bergh, J., et al. (2008). Growth-inhibitory and tumor- suppressive functions of p53 depend on its repression of CD44 expression. *Cell, 134*, 62–73.

Gum, R., Lengyel, E., Juarez, J., Chen, J. H., Sato, H., Seiki, M., et al. (1996). Stimulation of 92-kDa gelatinase B promoter activity by ras is mitogen-activated protein kinase kinase 1-independent and requires multiple transcription factor binding sites including closely spaced PEA3/ets and AP-1 sequences. *The Journal of Biological Chemistry*, *271*, 10672–10680.

Gunthert, U., Hoffman, M., Rudy, W., Reber, S., Zoller, M., Haußmann, I., et al. (1991). A new variant of glycoprotein CD44 confers metastatic potential to rat carcinoma cells. *Cell*, *65*, 13–24.

Guo, G., Wang, T., Gao, Q., Tamae, D., Wong, P., Chen, T., et al. (2004). Expression of ErbB2 enhances radiation-induced NF-kappaB activation. *Oncogene*, *23*, 535–545.

Hall, A. (1998). Rho GTPase and the actin cytoskeleton. *Science*, *279*, 509–514.

Harnett, P. R., Kirsten, F., & Tattersall, M. H. (1986). Drug resistance in clinical practice: Patterns of treatment failure in advanced breast and ovarian cancer. *Journal of Clinical Oncology*, *4*, 952–957.

Haylock, D. N., & Nilsson, S. K. (2006). The role of hyaluronic acid in hemopoietic stem cell biology. *Regenerative Medicine*, *1*, 437–445.

Higgins, C. F. (1992). ABC transporters: From microorganisms to man. *Annual Review of Cell Biology*, *8*, 67–113.

Hunter, A. M., LaCasse, E. C., & Korneluk, R. G. (2007). The inhibitors of apoptosis (IAPs) as cancer targets. *Apoptosis*, *12*, 1543–1568.

Iida, N., & Bourguignon, L. Y. (1995). New CD44 splice variants associated with human breast cancers. *Journal of Cellular Physiology*, *162*, 127–133.

Iida, N., & Bourguignon, L. Y. (1997). Coexpression of CD44 variant (v10/ex14) and CD44S in human mammary epithelial cells promotes tumorigenesis. *Journal of Cellular Physiology*, *171*, 152–160.

Itano, N., & Kimata, K. (1996). Expression cloning and molecular characterization of HAS protein, a eukaryotic hyaluronan synthase. *The Journal of Biological Chemistry*, *271*, 9875–9878.

Kashyap, V., Rezende, N. C., Scotland, K. B., Shaffer, S. M., Persson, J. L., Gudas, L. J., et al. (2009). Regulation of stem cell pluripotency and differentiation involves a mutual regulatory circuit of the NANOG, OCT4, and SOX2 pluripotency transcription factors with polycomb repressive complexes and stem cell microRNAs. *Stem Cells and Development*, *18*, 1093–1108.

Kato, T., Duffey, D. C., Ondrey, F. G., Dong, G., Chen, Z., Cook, J. A., et al. (2000). Cisplatin and radiation sensitivity in human head and neck squamous carcinomas are independently modulated by glutathione and transcription factor NF-kappaB. *Head & Neck*, *22*, 748–759.

Lambert, S., Yu, H., Prchal, J. T., Lawler, J., Ruff, P., Speicher, D., et al. (1990). cDNA sequence for human erythrocyte ankyrin. *Proceedings of the National Academy of Sciences of the United States of America*, *87*, 1730–1734.

Laurent, T. C., & Fraser, J. R. E. (1992). Hyaluronan. *FASEB Journal*, *6*, 2397–2404.

Lee, J. Y., & Spicer, A. P. (2000). Hyaluronan: A multifunctional, megadalton, stealth molecule. *Current Opinion in Cell Biology*, *12*, 581–586.

Li, X., & Lim, B. (2003). RhoGTPases and their role in cancer. *Oncology Research*, *13*, 323–331.

Liu, H., Deng, S., Zhao, Z., Zhang, H., Xiao, J., Song, W., et al. (2011). Oct4 regulates the miR-302 cluster in P19 mouse embryonic carcinoma cells. *Molecular Biology Reports*, *38*, 2155–2160.

Lokeshwar, V. B., Fregien, N., & Bourguignon, L. Y. (1994). Ankyrin-binding domain of CD44(GP85) is required for the expression of hyaluronic acid-mediated adhesion function. *The Journal of Cell Biology*, *126*, 1099–1109.

Lu, J., Bai, L., Sun, H., Nikolovska-Coleska, Z., McEachern, D., Qiu, S., et al. (2008). SM-164: A novel, bivalent Smac mimetic that induces apoptosis and tumor regression by concurrent removal of the blockade of cIAP-1/2 and XIAP. *Cancer Research*, *68*, 9384–9393.

Lux, S. E., John, K. M., & Bennett, V. (1990). Analysis of cDNA for human erythrocyte ankyrin indicates a repeated structure with homology to tissue-differentiation and cell-cycle control proteins. *Nature*, *344*, 36–42.

Mack, B., & Gires, O. (2008). CD44s and CD44v6 expression in head and neck epithelia. *PLoS One*, *3*, e3360.

Misra, S., Ghatak, S., Zoltan-Jones, A., & Toole, B. P. (2003). Regulation of multidrug resistance in cancer cells by hyaluronan. *The Journal of Biological Chemistry*, *278*, 25285–25288.

Polette, M., Nawrocki, B., Gilles, C., Sato, H., Seiki, M., Tournier, J. M., et al. (1996). MT-MMP expression and localisation in human lung and breast cancers. *Virchows Archiv*, *428*, 29–35.

Screaton, G. R., Bell, M. V., Jackson, D. G., Cornelis, F. B., Gerth, U., & Bell, J. I. (1992). Genomic structure of DNA coding the lymphocyte homing receptor CD44 reveals 12 alternatively spliced exons. *Proceedings of the National Academy of Sciences of the United States of America*, *89*, 12160–12164.

Sheridan, C., Kishimoto, H., Fuchs, R. K., Mehrotra, S., Bhat-Nakshatri, P., Turner, C. H., et al. (2006). CD44+/CD24- breast cancer cells exhibit enhanced invasive properties: An early step necessary for metastasis. *Breast Cancer Research*, *8*, R59.

Shiga, H., Rasmussen, A. A., Johnston, P. G., Langmacher, M., Baylor, A., Lee, M., et al. (2000). Prognostic value of c-erbB2 and other markers in patients treated with chemotherapy for recurrent head and neck cancer. *Head & Neck*, *22*, 599–608.

Slamon, D. J., Clark, G. M., Wong, S. G., Levin, W. J., Ullrich, A., & McGuire, W. L. (1987). Human breast cancer: Correlation of relapse and survival with amplification of the HER-2/neu oncogene. *Science*, *235*, 177–182.

Spicer, A. P., & Nguyen, T. K. (1999). Mammalian hyaluronan synthases: Investigation of functional relationships in vivo. *Biochemical Society Transactions*, *27*, 109–115.

Suwa, H., Ohshio, G., Imamura, T., Watanabe, G., Arii, S., Imamura, M., et al. (1998). Overexpression of the rho C gene correlates with progression of ductal adenocarcinoma of the pancreas. *British Journal of Cancer*, *77*, 147–152.

Toole, B. P. (2001). Hyaluronan in morphogenesis. *Seminars in Cell & Developmental Biology*, *12*, 79–87.

Toole, B. P., & Hascall, V. C. (2002). Hyaluronan and tumor growth. *The American Journal of Pathology*, *161*, 745–747.

Turley, E. A., Nobel, P. W., & Bourguignon, L. Y. W. (2002). Signaling properties of hyaluronan receptors. *The Journal of Biological Chemistry*, *277*, 4589–4592.

Underhill, C. (1992). CD44: The hyaluronan receptor. *Journal of Cell Science*, *103*, 293–298.

Wang, S. J., & Bourguignon, L. Y. (2006). Hyaluronan and the interaction between CD44 and epidermal growth factor receptor in oncogenic signaling and chemotherapy resistance in head and neck cancer. *Archives of Otolaryngology—Head & Neck Surgery*, *132*, 771–778.

Wang, S. J., & Bourguignon, L. Y. (2011). Role of hyaluronan-mediated CD44 signaling in head and neck squamous cell carcinoma progression and chemoresistance. *The American Journal of Pathology*, *178*, 956–963.

Wang, S. J., Peyrollier, K., & Bourguignon, L. Y. (2007). The influence of hyaluronan-CD44 interaction on topoisomerase II activity and etoposide cytotoxicity in head and neck cancer. *Archives of Otolaryngology—Head & Neck Surgery*, *133*, 281–288.

Wang, S. J., Wong, G., de Heer, A. M., Xia, W., & Bourguignon, L. Y. (2009). CD44 variant isoforms in head and neck squamous cell carcinoma progression. *Laryngoscope*, *119*, 1518–1530.

Wang, S. J., Wreesmann, V. B., & Bourguignon, L. Y. (2007). Association of CD44 V3-containing isoforms with tumor cell growth, migration, matrix metalloproteinase expression, and lymph node metastasis in head and neck cancer. *Head & Neck*, *29*, 550–558.

Weige, P. H., Hascall, V. C., & Tammi, M. (1997). Hyaluronan synthases. *The Journal of Biological Chemistry*, *272*, 13997–14000.

Yamada, Y., Itano, N., Zako, M., Yoshida, M., Lenas, P., Niimi, A., et al. (1998). The gene structure and promoter sequence of mouse hyaluronan synthase 1. *The Biochemical Journal*, *330*, 1223–1227.

Zhu, D., & Bourguignon, L. Y. (1998). The ankyrin-binding domain of CD44s is involved in regulating hyaluronic acid-mediated functions and prostate tumor cell transformation. *Cell Motility and the Cytoskeleton*, *39*, 209–222.

CHAPTER ELEVEN

Advances and Advantages of Nanomedicine in the Pharmacological Targeting of Hyaluronan-CD44 Interactions and Signaling in Cancer

Spyros S. Skandalis*, Chrisostomi Gialeli*,†, Achilleas D. Theocharis*, Nikos K. Karamanos*,†,1

*Biochemistry, Biochemical Analysis and Matrix Pathobiology Research Group, Laboratory of Biochemistry, Department of Chemistry, University of Patras, Patras, Greece

†Foundation for Research and Technology, Institute of Chemical Engineering Sciences (FORTH/ICE-HT), Patras, Greece

[1]Corresponding author: e-mail address: n.k.karamanos@upatras.gr

Contents

1. Introduction 278
2. Importance of Targeting Hyaluronan–CD44 in Tumors 279
3. Therapeutic Interventions/Strategies That Target Hyaluronan and/or CD44 to Perturb Hyaluronan-CD44 Interactions in Tumors 284
4. Advances in Nanomedicine Related with Hyaluronan-CD44 Targeting 295
   - 4.1 Targeted drug delivery 295
   - 4.2 Advantages of nanomedicine 295
   - 4.3 Nanoparticle properties 296
   - 4.4 Tissue-specific deletion of CD44 signaling 297
   - 4.5 Targeting chemotherapeutics to CD44 with hyaluronan conjugates as drug carriers 299
   - 4.6 Application of polysaccharides (hyaluronan) in nanomedicine 300
5. Concluding Remarks/Conclusions 304

Acknowledgments 304

References 305

Abstract

Extensive experimental evidence in cell and animal tumor models show that hyaluronan–CD44 interactions are crucial in both malignancy and resistance to cancer therapy. Because of the intimate relationship between the hyaluronan–CD44 system and tumor cell survival and growth, it is an increasingly investigated area for applications to anticancer chemotherapeutics. Interference with the hyaluronan–CD44 interaction

Advances in Cancer Research, Volume 123
ISSN 0065-230X
http://dx.doi.org/10.1016/B978-0-12-800092-2.00011-3

by targeting drugs to CD44, targeting drugs to the hyaluronan matrix, or interfering with hyaluronan matrix/tumor cell-associated CD44 interactions is a viable strategy for cancer treatment. Many of these methods can decrease tumor burden in animal models but have yet to show significant clinical utility. Recent advances in nanomedicine have offered new valuable tools for cancer detection, prevention, and treatment. The enhanced permeability and retention effect has served as key rationale for using nanoparticles to treat solid tumors. However, the targeted and uniform delivery of these particles to all regions of tumors in sufficient quantities requires optimization. An ideal nanocarrier should be equipped with selective ligands that are highly or exclusively expressed on target cells and thus endow the carriers with specific targeting capabilities. In this review, we describe how the hyaluronan–CD44 system may provide such an alternative in tumors expressing specific CD44 variants.

1. INTRODUCTION

Extracellular matrix (ECM) was initially considered an inert, space-filling material that provided only mechanical strength to tissues and organs. Currently, this simplified description of ECM has been substantially changed since numerous studies support that ECM is a dynamic structure that interacts with cells and generates signals through feedback loops to control the behavior of cells. Matrix macromolecules are bioactive and modulate cellular events such as adhesion, migration, proliferation/survival, and differentiation (Daley, Peters, & Larsen, 2008). In tumors, the microenvironment (involving ECM macromolecules, ECM-associated growth factors and cytokines, as well as stromal cells) not only responds to and supports carcinogenesis but also actively contributes to tumor progression and metastasis. Cell migration, invasion, metastasis, and angiogenesis, all of significance in cancer progression, are dependent on the ECM properties of the tumor itself and the surrounding microenvironment, involving alterations in structure and/or gene expression as well as activity modifications that take place during this process (Hascall & Karamanos, 2011; Iozzo & Karamanos, 2010; Nagase & Karamanos, 2011).

Lately, the active role of ECM in both physiological and pathological conditions has attracted considerable attention in cancer treatment research. Several of the currently available drugs can modulate ECM metabolism and assembly beyond their primary targets. Their effects on the ECM, however, are often neither potent nor specific enough. Moreover, several drugs may exert deleterious effects on the ECM. Therefore, novel drugs influencing the ECM in a more direct way are still desired. In recent years, the number of drug development projects aiming to primarily target the ECM has increased (Huxley-Jones, Foord, & Barnes, 2008; Jarvelainen, Sainio, Koulu, Wight, & Penttinen, 2009). In these projects, cell- and gene-based strategies and particularly gene and antisense therapies as well as RNA interference (RNAi) and recombinant technologies are used. Although these approaches offer fascinating new ways to treat various human diseases, they have not yet been reached clinical practice. The multiplicity and functional complexity of the ECM molecules and the diversity of diseases related to the ECM offer significant challenges for future targeting strategies. However, the molecular complexity of diseases suggests that agents targeting not only a single ECM molecule but concomitantly several aspects of the ECM will also have an important position in future drug development. As a consequence, pharmacological targeting of the tumor ECM unveils a promising tumor cell-directed therapy (Gialeli, Theocharis, & Karamanos, 2011; Karamanos & Tzanakakis, 2012; Misra et al., 2011; Theocharis, Skandalis, Tzanakakis, & Karamanos, 2010). In this chapter, potential future pharmacotherapeutics specifically targeting ECM and particularly the perturbed hyaluronan–CD44 system in tumors are presented and discussed.

2. IMPORTANCE OF TARGETING HYALURONAN–CD44 IN TUMORS

Hyaluronan is a very large, linear, negatively charged heteropolysaccharide, a glycosaminoglycan, that is composed of repeating disaccharides of glucuronic acid and *N*-acetylglucosamine. Although simple in structure, hyaluronan is produced by three hyaluronan synthases (HAS1/HAS2/HAS3), which are integral plasma membrane proteins whose active sites are located at the intracellular face of the membrane and are subject to several posttranslational modifications (such as ubiquitination, O-GlcNAcylation) that control their activity and stability (Karousou et al., 2010; Vigetti, Viola, Karousou, De Luca, & Passi, 2014; Weigel & DeAngelis, 2007). Newly synthesized hyaluronan is extruded as it is elongated, and then targeted to the cell surface or to pericellular and ECMs. On the other hand, hyaluronan is degraded by hyaluronidases (HYAL-1/HYAL-2/HYAL-3/HYAL-4/PH-20), which are endoglycosidases as they degrade the β-*N*-acetyl-D-glucosaminidic linkages in the hyaluronan polymer producing fragments of various sizes with diverse biological functions. Hyaluronan is distributed ubiquitously in vertebrate tissues but is especially concentrated in regions of cell division and invasion (Skandalis, Theocharis,

Papageorgakopoulou, & Zagris, 2003; Toole, 2001). In adult tissues such as synovial fluid, vitreous body, cartilage, and dermis, it clearly plays a structural role that depends on its unique hydrodynamic properties and its interactions with other ECM components. On the other hand, hyaluronan has an instructive, cell signaling role during dynamic cell processes such as morphogenesis, inflammation, wound repair, and cancer, wherein hyaluronan–receptor interactions are activated and collaborate in driving numerous signaling pathways (Noulas, Skandalis, Feretis, Theocharis, & Karamanos, 2004; Theocharis et al., 2008; Toole, 2004; Turley, Noble, & Bourguignon, 2002). In addition to signal transduction, hyaluronan–receptor interactions participate in at least two other important physiological processes, i.e., endocytosis of hyaluronan and assembly of pericellular matrices (Evanko, Tammi, Tammi, & Wight, 2007; Knudson, Chow, & Knudson, 2002).

CD44 is the major cell surface receptor for hyaluronan. CD44 is a single-chain, single-pass, transmembrane glycoprotein encoded by a single gene, but due to alternative splicing, multiple forms of CD44 are generated that further are modified by N- and O-linked glycosylations. The smallest CD44 isoform that lacks variant exons, the standard CD44 (CD44s), is abundantly expressed by both normal and cancer cells, whereas the variant CD44 (CD44v) isoforms that contain a variable number of exon insertions (v1–v10) at the proximal plasma membrane external region are expressed mostly by cancer cells. In addition, the CD44 ectodomain can be decorated with chondroitin sulfate and/or heparan sulfate enabling CD44 to bind growth factors, including fibroblast growth factor, vascular endothelial growth factor, or hepatocyte growth factor (Orian-Rousseau, 2010; Ponta, Sherman, & Herrlich, 2003). CD44 proteins exist in three states with respect to hyaluronan binding: nonbinding, nonbinding unless activated by physiological stimuli, or constitutively binding (Lesley, English, Charles, & Hyman, 2000; Lesley, Hascall, Tammi, & Hyman, 2000; Lesley & Hyman, 1992). Hyaluronan induces signaling when it binds to constitutively activated CD44 variants (He, Lesley, Hyman, Ishihara, & Kincade, 1992; Sleeman et al., 1996). CD44 can also react with other molecules including collagen, fibronectin, osteopontin, growth factors, and matrix metalloproteinases, but functional roles of such interactions are less well known (Ponta et al., 2003).

In several types of cancer cells, binding of hyaluronan to CD44 results in direct or indirect interaction of CD44 with signaling receptors, such as ErbB2, EGFR, PDGFRβ, and TGF-beta receptor type I, and influences the activity of these receptors (Li, Heldin, & Heldin, 2006; Misra, Toole, & Ghatak, 2006; Toole, 2004; Toole & Slomiany, 2008). It can also lead to interaction with and altered activity of nonreceptor kinases of the Src family or Ras family GTPases (Bourguignon, 2008; Bourguignon, Zhu, Shao, & Chen, 2001). Complex formation with adaptor proteins such as Vav2, Grb2, and Gab-1 mediates interaction of CD44 with upstream effectors, e.g., RhoA, Rac1, and Ras, that drive intracellular signaling pathways (Bourguignon, 2008). Thus, hyaluronan–CD44 binding influences the activity of a variety of downstream signaling pathways, especially the MAP kinase and PI3 kinase/Akt pathways, and consequently promotes tumor cell proliferation, survival, motility, invasiveness, and chemoresistance (Ponta et al., 2003; Toole, 2004; Turley et al., 2002). In addition, binding of hyaluronan to CD44 stimulates multidrug and metabolic transporters of crucial importance in therapy resistance (Bourguignon, Singleton, Diedrich, Stern, & Gilad, 2004; Colone et al., 2008; Miletti-Gonzalez et al., 2005; Slomiany, Dai, et al., 2009; Slomiany, Grass, et al., 2009) and presentation of proteases that facilitate invasion (Ponta et al., 2003). In addition to its action as a coreceptor or coactivator of membrane-associated signaling molecules, the multiple cellular functions of CD44 rely also on its cytoplasmic tail association with partner proteins that regulate cell migration, growth, survival, and differentiation (Bourguignon, 2008; Kozlova, Ruusala, Voytyuk, Skandalis, & Heldin, 2012; Skandalis, Kozlova, Engstrom, Hellman, & Heldin, 2010) (Fig. 11.1).

CD44 is endogenously expressed in low levels on various cell types of normal tissues (Cichy & Pure, 2003; Mackay et al., 1994) and requires activation before binding to hyaluronan (Lesley, Hascall, et al., 2000; Lesley et al., 1992; Lesley, Howes, Perschl, & Hyman, 1994; Lesley & Hyman, 1992; Sneath & Mangham, 1998). The CD44 structure on normal cells is distinct from that on cancer cells because under various physiological and pathological conditions, the local environmental pressure (stromal factors) influences alternate splicing and posttranslational modification of CD44 gene and protein, respectively (Naor, Nedvetzki, Golan, Melnik, & Faitelson, 2002; van Weering, Baas, & Bos, 1993). Pathological conditions that stimulate alternate splicing and posttranslational modifications produce diversified CD44 molecules with enhanced hyaluronan binding that leads to increased tumorigenicity (Bourguignon, Singleton, Zhu, & Zhou, 2002; Bourguignon, Zhu, Shao, & Chen, 2000; Bourguignon et al., 2001; Bourguignon, Zhu, Shao, Zhu, & Chen, 1999; Naor et al., 2002). This diversification allows the production of specific targeting agents that would be useful for both diagnosis and therapy. Glycosylation is required for spliced variant formation of CD44, which has high affinity to bind hyaluronan on certain cell types, whereas glycosylation rich in sialic acid decreases hyaluronan binding (Katoh, Zheng, Oritani, Shimozato, & Kincade, 1995; Rochman et al.,

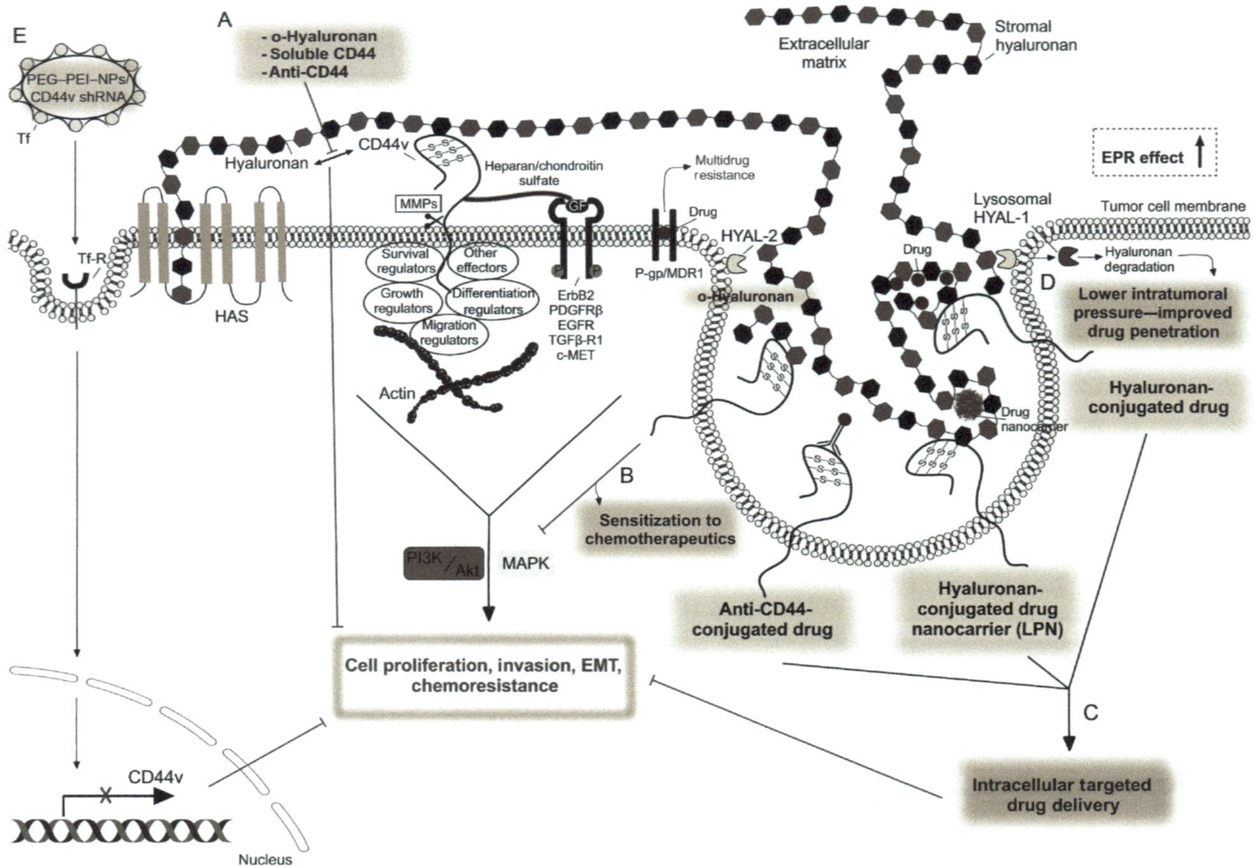

E
PEG–PEI–NPs/
CD44v shRNA
Tf
Tf-R
A
- o-Hyaluronan
- Soluble CD44
- Anti-CD44
Hyaluronan
CD44v
HAS
MMPs
Heparan/chondroitin
sulfate
GF
Survival
regulators
Other
effectors
Growth
regulators
Differentiation
regulators
Migration
regulators
Actin
ErbB2
PDGFRβ
EGFR
TGFβ-R1
c-MET
Multidrug
resistance
Drug
P-gp/MDR1
HYAL-2
o-Hyaluronan
Extracellular
matrix
Stromal
hyaluronan
EPR effect
Lysosomal
HYAL-1
Tumor cell membrane
Drug
Hyaluronan
degradation
D
Lower intratumoral
pressure—improved
drug penetration
Hyaluronan-
conjugated drug
Drug
nanocarrier
B
Sensitization to
chemotherapeutics
PI3K / Akt
MAPK
Anti-CD44-
conjugated drug
Hyaluronan-
conjugated drug
nanocarrier (LPN)
Cell proliferation, invasion, EMT,
chemoresistance
C
Intracellular targeted
drug delivery
CD44v
Nucleus

2000; Skelton, Zeng, Nocks, & Stamenkovic, 1998). For example, circulating lymphocytes express CD44 but do not bind hyaluronan until CD44 is deglycosylated upon lymphocyte activation (Hathcock, Hirano, Murakami, & Hodes, 1993), and to internalize CD44 it must be acylated (Thankamony & Knudson, 2006). CD44 variant interaction with hyaluronan is known for its role in the metastatic cascade as this interaction regulates the ability of malignant cells to activate receptor tyrosine kinases, and to stimulate migration, invasion of basement membranes/ECMs, and migration to distant sites (Heldin et al., 2008; Herold-Mende et al., 1996; Marhaba & Zoller, 2004; Misra, Ghatak, & Toole, 2005; Misra et al., 2008, 2006). Hyaluronan induces intracellular signaling when it binds to "constitutively activated" CD44 variants during cell dynamic processes, but does not do so under conditions of adult tissue homeostasis, which generally involve the standard form of CD44 (CD44s). Overexpression of variant, high molecular weight (HMW) isoforms CD44v4-v7 and CD44v6-v9 in various cancers (Ayhan, Tok, Bildirici, & Ayhan, 2001; Foekens et al., 1999; Gunthert et al., 1991; Ishida, 2000) as well as downregulation of the standard CD44s isoform in colon cancer is postulated to result in increased tumorigenicity (Choi, Takahashi, Eto, Yoon, & Tanabe, 2000), emphasizing the potential importance of CD44 splice variants in cancer (Misra et al., 2011).

A puzzling aspect of many studies of hyaluronan-induced oncogenic signaling is their basis in experiments in which exogenous hyaluronan is added to cultured tumor cells. Although these studies have resulted in apparently solid data indicative of strong effects on the pathways in question, they are difficult to reconcile with the long history of safe utilization of hyaluronan in numerous reconstructive or regenerative capacities in human patients. For example, hyaluronan is employed widely in eye and knee surgeries

Figure 11.1 Schematic representation of potential targeting of hyaluronan–CD44 interactions and signaling in tumors. Hyaluronan may be retained in the pericellular milieu by hyaluronan synthases (HAS) or CD44, and endocytosed by CD44. Multivalent interactions of hyaluronan with CD44 promote the activation of multiple signaling receptors (ErbB2, EGFR, PDGFRβ, c-MET, TGFβR1) and induce oncogenic signaling pathways (e.g., the MAP kinase and PI3 kinase/Akt cell proliferation and survival pathways) as well as interaction with various transporters, such as P-gp. The approaches that utilize the hyaluronan–CD44 system include (A) interference with hyaluronan–CD44 interactions (by o-hyaluronan, soluble CD44, anti-CD44 antibodies); (B) hyaluronan-induced sensitization to chemotherapeutics; (C) intracellular-targeted drug delivery with anti-CD44 antibody- or hyaluronan-conjugated drugs/drug carriers; (D) hyaluronidase-mediated degradation of hyaluronan matrix; and (E) tumor-specific delivery of CD44v shRNA with plasmid–Tf–PEG–PEI nanoparticles. (See the color plate.)

and in prevention of adhesions (Prestwich & Kuo, 2008). Hyaluronan-based hydrogels are also being developed for a variety of purposes including drug delivery, encapsulation of progenitor cells, and tissue engineering (Allison & Grande-Allen, 2006; Prestwich, 2008). Such studies imply that the oncogenic effects of hyaluronan only occur in the context of the tumor microenvironment and that stromal hyaluronan, as well as tumor cell-produced hyaluronan, plays an important role in tumorigenesis, a conclusion supported by correlative studies of numerous human tumor types, including laryngeal, pancreatic, and gastric cancers (Papadas et al., 2002; Skandalis, Stylianou, Vynios, Papageorgakopoulou, & Theocharis, 2007; Tammi et al., 2008; Theocharis, Tsara, Papageorgacopoulou, Karavias, & Theocharis, 2000; Theocharis, Vynios, Papageorgakopoulou, Skandalis, & Theocharis, 2003; Vynios et al., 2008). Strong supporting evidence for the tumor-promoting effects of hyaluronan comes from studies in which tumor hyaluronan levels and interactions with receptors were manipulated *in vivo*.

Although the published literature on hyaluronan–CD44 interactions in cancer is riddled with paradoxes and apparent contradictions (Toole, 2009), these interactions offer an important target for translation into the clinic. A frequently expressed concern is the widespread expression and functions of hyaluronan and CD44 in normal physiology. However, two observations provide promise that therapeutic interventions can be developed that target oncogenic events with some degree of specificity or differential sensitivity. First is the finding that many of the interactions described above involve variants of CD44 that are amplified greatly in many tumor types in comparison to normal processes (Naor, Wallach-Dayan, Zahalka, & Sionov, 2008; Skotheim & Nees, 2007). Second is the nature of "activation" of hyaluronan–CD44 interactions in malignant tumor cells. Although these processes may have overlapping features with immune and inflammatory pathways, there are also clear differences that may be possible to exploit and may ultimately have therapeutic value but these have not yet reached the clinic. Some important approaches that appear to show promise in approaching this important objective as well as recent advances in nanomedicine in this field are described below.

3. THERAPEUTIC INTERVENTIONS/STRATEGIES THAT TARGET HYALURONAN AND/OR CD44 TO PERTURB HYALURONAN-CD44 INTERACTIONS IN TUMORS

Interference with the hyaluronan–CD44 interaction by targeting drugs to CD44, targeting drugs to the hyaluronan matrix, or interfering with hyaluronan matrix/tumor cell-associated CD44 interactions is a viable strategy for cancer treatments (Akima et al., 1996; Misra et al., 2011; Platt & Szoka, 2008; Sy, Guo, & Stamenkovic, 1992; Toole, 2009) (Fig. 11.1). Hyaluronan-degrading enzymes, i.e., hyaluronidases, have long been added to anticancer regimens. Evidence for the anticancer effects of hyaluronidase comes from experimental model systems (Table 11.1). Tumors previously resistant to chemotherapy became sensitive when hyaluronidase was added (Baumgartner, 1998; Klocker, Sabitzer, Raunik, Wieser, & Schumer, 1998). The enzyme may decrease intratumoral pressure, permitting drugs to penetrate the malignancy (Fig. 11.1D). However, data have also accumulated that hyaluronidase correlates with tumor progression. These apparent contradictable findings can be reconciled when considering the double-edged sword role of hyaluronidase. Delicate balance determines whether hyaluronidase enhances tumor progression and metastases by destroying the matrix resistance to cell migration or retards them by tuning and targeting the enzymatic activity to the hyaluronan-depended tumor migration phase. The inconsistencies that abound suggest that additional work must be performed in order to uncover the wide range properties of hyaluronidase in tumors and their potential utilization as anticancer chemotherapeutic agents.

Disrupting association of CD44-expressing tumor cells with the ECMs can reduce the tumor burden by causing antigrowth signaling or by physically interfering with tumor expansion into surrounding tissue. Several studies have shown that soluble CD44, either expressed via stable transfection or added as immunoglobulin fusion constructs, efficiently decreased tumor growth (Table 11.1) (Fig. 11.1A). In these studies, the therapeutic effects of the soluble CD44 were correlated with hyaluronan binding since administration of mutant CD44 that could not bind hyaluronan had no antitumor benefits.

In another line of research, treatment of tumors with anti-CD44 antibodies showed that metastatic progression could be efficiently blocked, but such treatment had no effect on the primary tumor growth (Fig. 11.1A). Therefore, an important parameter that should be considered when treating with anti-CD44 antibodies is the timing of treatment, since these antibodies were most efficacious when given either with tumor implantation or after tumor implantation but before metastasis (Table 11.1). For example, intravenous administration of anti-CD44 antibody in a human melanoma mouse model effectively prevented metastasis but not the growth of the primary tumor (Guo et al., 1994). Lymphoma cell dissemination was efficiently blocked by a bispecific antibody that recognizes both CD44 and a cell surface marker of the malignant

Table 11.1 Therapeutic interventions/strategies targeting the hyaluronan-CD44 system

Acting moiety/strategy	Cell line/tumor model	Effect	Reference
Hyaluronidases			
Overexpression of HYAL-1	Colon carcinoma	Suppression of tumorigenicity	Jacobson, Rahmanian, Rubin, and Heldin (2002)
Treatment with testicular hyaluronidase (PH-20)	Murine model for T-cell lymphoma	Prevention of lymph node invasion	Zahalka, Okon, Gosslar, Holzmann, and Naor (1995)
	Mouse mammary tumor cells (EMT-6)	Reversion of drug resistance	St Croix, Man, and Kerbel (1998)
	Human breast cancer cell lines (MCF-7, ZR-75-1, and T-47-D) and MXT mammary tumor model	Enhancement of the anticancer effects of adriamycin	Beckenlehner et al. (1992)
	Human breast carcinoma grown in SCID mice	Reduction of tumor growth	Shuster, Frost, Csoka, Formby, and Stern (2002)
Soluble CD44			
Overexpression of soluble CD44	Mammary carcinoma cells and primary malignant melanoma tumors	Decreased hyaluronan binding *in vitro* and reduced tumor progression *in vivo*	Ahrens et al. (2001) and Peterson, Yu, Stamenkovic, and Toole (2000)
Administration of CD44-immunoglobulin fusion proteins	Murine models of lymphoma and melanoma tumors	Inhibition of tumor invasion. Suppression of cancer progression	Bartolazzi, Peach, Aruffo, and Stamenkovic (1994) and Sy et al. (1992)

Anti-CD44 antibodies			
Administration before metastasis formation	Rat pancreatic adenocarcinoma model	Substantial reduction of colonization	Seiter et al. (1993)
Treatment soon after tumor cell implantation	Murine T-cell lymphoma model	Prevention of lymph node invasion	Zahalka et al. (1995)
Intravenous administration	Human melanoma mouse model	Prevention of metastasis but not growth of primary tumor	Guo et al. (1994)
Bispecific antibody (anti-CD44 and anti-B lymphocyte surface marker)	Murine cell lymphoma	Blockage of lymphoma cell dissemination	Avin, Haimovich, and Hollander (2004)
In vitro treatments	Human glioma cell lines	Suppression of invasion via decreased adhesion	Merzak, Koocheckpour, and Pilkington (1994) and Radotra and McCormick (1997)
	Gliosarcoma and glioblastoma multiform brain tumors	CD44-specific, dose-dependent cell detachment	Breyer et al. (2000)
	Human acute myeloid leukemia (AML)	Promotion of differentiation and apoptosis	Charrad et al. (1999) and Song et al. (2004)
Exogenous hyaluronan			
o-Hyaluronan	Lung, mammary, and colon carcinoma cell lines	Suppression of PI3 kinase/Akt cell survival pathway. Inhibition of anchorage-independent growth	Ghatak, Misra, and Toole (2002)
	T-lymphoma cells	Induction of apoptosis	Alaniz et al. (2006)

Continued

Table 11.1 Therapeutic interventions/strategies targeting the hyaluronan-CD44 system—cont'd

Acting moiety/strategy	Cell line/tumor model	Effect	Reference
Injection of o-hyaluronan	Murine melanoma, osteosarcoma, lung carcinoma model	Inhibition of tumor growth	Ghatak et al. (2002) and Zeng, Toole, Kinney, Kuo, and Stamenkovic (1998)
Intratumoral injection of octasaccharide hyaluronan	Murine osteosarcoma model	Inhibition of primary tumor growth. Decreased lung metastasis	Hosono et al. (2007)
Injection of o-hyaluronan near the tumor site	Rat spinal cord glioma model	Suppression of tumor growth by downregulating Akt-related antiapoptotic signaling	Gilg et al. (2008)
Daily intratumoral injection of HMW hyaluronan	Murine xenograft breast cancer model	Complete tumor regression in 25% of cases	Herrera-Gayol and Jothy (2002)
Hyaluronan-induced chemotherapy			
o-Hyaluronan	Drug-resistant human mammary carcinoma cells	Suppression of MAP kinase and PI3 kinase pathways. Sensitization of cells to several chemotherapy agents	Misra, Ghatak, Zoltan-Jones, and Toole (2003)
	Drug-resistant lymphoma cells	Inhibition of drug efflux	Cordo Russo et al. (2008)
Vaccines			
Murine cytotoxic T-cells transduced with fusion construct of a single-chain anti-CD44v6 antibody and a T-cell receptor signal transduction domain	Pancreatic carcinoma cells	Recognition and lysis of CD44v6-expressing cancer cells	Hekele et al. (1996)

Daily intratumoral injection of cytotoxic T-cells with fusion construct of a single-chain anti-CD44v7/8 antibody and a T-cell receptor signal transduction domain	Murine cervical carcinoma model	Growth inhibition	Dall et al. (2005)
Intraperitoneal injection of anti-CD44v6 anticomplement receptor CR3 bispecific antibody	Rat adenocarcinoma model	Increased mortality	Somasundaram, Arch, Matzku, and Zoller (1996)
cDNA vaccination of CD44 variant (CD44v)	Experimental mammary tumor of BALB/c mice	Reduction of tumor mass and metastatic potential	Wallach-Dayan, Rubinstein, Hand, Breuer, and Naor (2008)
Gene therapy			
Stable transfection of plasmids encoding antisense CD44 cDNA	Human colon adenocarcinoma cells	Reduced ability to bind hyaluronan *in vitro*	Harada et al. (2001)
	Intrasplenical injection in mice	No metastatic liver colonies and fewer peritoneal nodules	
Stable transfection of plasmids encoding siRNA for CD44v6	Human metastatic colorectal carcinoma cells		Reeder, Gotley, Walsh, Fawcett, and Antalis (1998)
		No metastatic ability when injected in the spleen of nude mice	

Continued

Table 11.1 Therapeutic interventions/strategies targeting the hyaluronan-CD44 system—cont'd

Acting moiety/strategy	**Cell line/tumor model**	**Effect**	**Reference**
Stable transfection of plasmids with siRNA against the transmembrane domain of CD44	Human metastatic colon carcinoma cells	Decreased adhesion and survival *in vitro*	Subramaniam, Vincent, Gilakjan, and Jothy (2007)
		Formation of smaller tumors when subcutaneously injected in mice	
Transfection with plasmid encoding the CD44v9/v10—prodrug-converting enzyme	Prostate adenocarcinoma cells	Sensitization to drugs	Hayes et al. (2002) and Hayes, Dougherty, Davis, and Dougherty (2004)
Hyaluronan/CD44-based drug nanocarriers			
Anti-CD44 antibody conjugates			
BIWA 4 (*Bivatuzumab*) against CD44v6 conjugated with the radioisotope rhenium-186	Refractory head and neck cancers	Stabilized disease	Borjesson et al. (2003), Colnot et al. (2003), Tijink et al. (2006), and Verel et al. (2002)
Tissue-specific deletion of CD44			
Tf–PEG–PEI–NPs/CD44v6 shRNA	Apc Min/+ mice	Inhibition of distant colon tumor growth	Misra, Hascall, De Giovanni, Markwald, and Ghatak (2009) and Misra et al. (2011)
Hyaluronan conjugates			
Hyaluronan–paclitaxel	Breast, colon, and ovarian carcinoma cell lines	Increased cytotoxic delivery	Auzenne et al. (2007)
Hyaluronan–paclitaxel injection locally to the tumor	Human ovarian carcinoma xenograft mouse model	Increased mean survival time	Auzenne et al. (2007)

Intraperitoneal chemotherapy with hyaluronan–paclitaxel hydrogel	Human ovarian carcinoma xenograft mouse model	Improved regional drug delivery	Bajaj, Kim, Mohammed, and Yeo (2012)
Hyaluronan–sodium butyrate injection intratumorally or subcutaneously	Murine Lewis lung carcinoma and melanoma model	Reduced primary tumor growth and lung metastasis	Coradini, Pellizzaro, et al. (2004) and Coradini, Zorzet, et al. (2004)
Hyaluronan–mitomycin c	Murine model of Lewis lung adenocarcinoma	Antimetastatic effects	Akima et al. (1996)
Hyaluronan–doxorubicin	Human breast cancer (HBL-100), ovarian cancer (SKOV-3), and colon cancer (HCT-116) cell lines	Increased cytotoxic delivery	Luo, Bernshaw, Lu, Kopecek, and Prestwich (2002)
Hyaluronan–doxorubicin	Human breast cancer cell line (MCF-7)	Increased cytotoxic delivery	Wu, Liu, Wang, and Huang (2012)
Hyaluronan–PEG–camptothecin	Human breast cancer cell line (MDA-MB-231), squamous cell carcinoma cell line (SCC7)	Increased cytotoxic delivery	Choi et al. (2011)
Hyaluronan–carborane	Human bladder carcinoma, colorectal adenocarcinoma, and ovarian adenocarcinoma cell lines	High accumulation	Di Meo et al. (2008) and Meo et al. (2007)

Continued

Table 11.1 Therapeutic interventions/strategies targeting the hyaluronan-CD44 system—cont'd

Acting moiety/strategy	Cell line/tumor model	Effect	Reference
Hyaluronan–LNPs	Murine models of intraperitoneal ascites tumors, solid colon, pancreatic carcinoma tumors, metastatic melanoma	Increased mean survival time	Peer and Margalit (2004a, 2004b)
Hyaluronan–LNPs/P-gp siRNA	Human lung adenocarcinoma cell line (A549)	Specific delivery combined with downregulation of multidrug resistance extrusion pump (P-gp)	Landesman-Milo et al. (2013)
Hyaluronan–PEI for siRNA delivery	B16F1 murine melanoma cells	Effective gene silencing effect *in vitro* and *in vivo*	Han et al. (2009) and Jiang et al. (2005)
		More efficient gene silencing in cells overexpressing LYVE-1	Jiang et al. (2008)

B lymphocyte (Avin et al., 2004). Therapeutic efficacy of anti-CD44 antibodies lies on their ability to modulate tumor cell adhesion, differentiation, and antigrowth signaling pathways (Table 11.1). Importantly, treatment of tumors with anti-CD44 antibodies should target the relevant CD44 variant on the tumor cells, rather than targeting constant epitopes that are shared by both CD44 variants and standard CD44. The exclusive targeting of the tumor-expressing CD44 variant(s) should rescue all normal cells that express the other CD44 isoforms, including a substantial number of cells that express only the standard CD44. In other words, limiting the therapeutic targeting to the CD44 variant(s) of the tumor cells should markedly reduce the potential of a side-effect damage, which would ultimately be generated in case that the standard CD44 is targeted. Therefore, to cope with such a challenge, one must first identify the CD44 variant(s) of the tumor, then show that this variant(s) is essential for the tumor growth and/or metastasis, and finally design the appropriate approach that would selectively silence or kill the relevant tumor.

HMW hyaluronan and, especially, hyaluronan-oligomers (o-hyaluronan) have been shown to be—in a cancer cell surrounding milieu-dependent manner—effective *in vivo* therapeutic agents for the treatment of cancer (Table 11.1) (Fig. 11.1A). For example, in lung, mammary, and colon carcinoma cell lines, antigrowth effects were seen with exogenous added o-hyaluronan but not with HMW hyaluronan, chitin, or chondroitin sulfate (Ghatak et al., 2002). In another study, daily intratumoral injection of HMW hyaluronan in a murine xenograft breast cancer model resulted in complete tumor regression in 25% of animals studied (Herrera-Gayol & Jothy, 2002). The consequences of introducing free hyaluronan of any size into the body at concentrations high enough to exhibit antitumor effects may result in side effects due to its interactions with endogenous CD44 on other cell types, such as leukocytes within the epithelium of the skin (Mackay et al., 1994; Tijink et al., 2006).

Regulation of the signaling pathways evoked by hyaluronan–CD44 interaction may cause cells to become more sensitive to codelivered chemotherapeutic agents. This is evident from studies in which administration of o-hyaluronan in drug-resistant human mammary carcinoma cells caused the suppression of MAP kinase and phosphoinositide 3-kinase pathways, which sensitized the cells to several chemotherapeutics, such as doxorubicin, taxol, vincristine, methotrexate, and BCNU (Cordo Russo et al., 2008; Misra et al., 2003) (Table 11.1) (Fig. 11.1B). However, there are studies which suggest that the administration of hyaluronan may also be antagonistic in certain hyaluronan–drug combinations. For example, in T-cells, o-hyaluronan of approximately 18 disaccharides upregulated P-glycoprotein (P-gp), resulting in increased efflux of dexamethasone (Tsujimura, Saito, Kohno, & Tanaka, 2006). Overall, these data suggest that sensitization may be specific to the drug, hyaluronan size, or cell type (Table 11.1).

Apart from anti-CD44 antibodies against the relevant tumor antigen (e.g., CD44 variant), a cDNA vaccine against CD44 variant(s) might also advance cancer treatment efficacy since DNA vaccinations have some advantages over passive immunization by antibody. DNA vaccinations provide prolonged antigen exposure that continuously stimulates the immune system and offer a unique opportunity to enhance the immunogenicity of the antigen in question by modifying the potential immunizing cDNA sequence. Most cancer vaccines target tumors through stimulation of cytotoxic T-lymphocytes that recognize tumor-specific antigens. Cytotoxic T-cells can be designed to recognize CD44 by genetically engineering a response into the T-cells. This modification gives them the ability to lyse CD44 variant-expressing cells. Wallach-Dayan et al. generated cDNA vaccines against either CD44 variant or standard CD44 in experimental mammary tumor of BALB/c mice and showed that vaccination with the former cDNA, but not with standard CD44, generated resistance to the tumor growth (Wallach-Dayan et al., 2008). Such immunotherapy methods successfully initiated a response from the host immune system directly against the tumor (Table 11.1). However, their clinical usage may be limited since the host immune response may vary from patient to patient, and therefore, this treatment must be optimized on the patient's clinical profile.

Data from several studies support that gene therapies which aim at reducing the concentration of cancer cell surface CD44 may prevent metastatic spread of primary tumors (Table 11.1). For example, human metastatic colorectal carcinoma cells stably transfected with siRNA plasmids for CD44v6 could efficiently proliferate or bind to hyaluronan *in vitro*. However, their metastatic ability was drastically reduced since when they were injected into the spleen of nude mice, these mice developed no metastatic liver tumors (Reeder et al., 1998). Although repeated intratumoral injection of plasmid DNA encoding siRNA against CD44 (variants or domains) provided some suppression of tumor growth, their clinical usage under this form is questionable unless robust delivery systems, such as specific nanoformulations, are developed.

Anti-CD44 antibody conjugates (with radiolabels or cytotoxic drugs) successfully target to tumor-expressing CD44 (Table 11.1) (Fig. 11.1C).

Humanized monoclonal antibodies (such as BIWA 4; *Bivatuzumab*) against CD44v6 stabilized disease in refractory head and neck cancers when coupled to the radioisotope rhenium-186 (Borjesson et al., 2003; Colnot et al., 2003; Tijink et al., 2006; Verel et al., 2002). However, conjugation of Bivatuzumab to the cytotoxic drug mertansine resulted in dose-limiting toxicities associated with skin-related disorders (Sauter et al., 2007). Moreover, this antibody–drug conjugate caused unfavorable off-target effects in patients with CD44v6-positive metastatic breast cancer (Rupp et al., 2007). To avoid accumulation of the conjugate in nontarget organs, efficient and robust gene carriers that could deliver the genes specifically to tumor sites should be developed. Strategies developed for novel and promising therapeutic approaches are discussed in detail below.

4. ADVANCES IN NANOMEDICINE RELATED WITH HYALURONAN-CD44 TARGETING

4.1. Targeted drug delivery

Targeted delivery has been developed to increase drug distribution in tumor tissues, reduce side effects in normal tissues, and improve therapeutic index. Drug targeting can be classified into two types: passive and active targeting. Passive targeting, depending on the enhanced permeability and retention (EPR) effect, means the accumulation of macromolecular agents and nanocarriers in solid tumor tissues, which have unique pathophysiological characteristics, including extensive angiogenesis, hypervasculature, defective vascular architecture, and an impaired lymphatic drainage/recovery system (Fang, Nakamura, & Maeda, 2011). The EPR effect has now become a major concept in anticancer drug delivery designing (Matsumura & Maeda, 1986; Torchilin, 2011). Active targeting process proceeds after the passive accumulation of nanocarriers in tumors, to achieve the further selective targeting delivery of drugs to tumor cells, through specific interactions, such as antigen–antibody and ligand–receptor binding (Danhier, Feron, & Preat, 2010).

4.2. Advantages of nanomedicine

Nanomedicine for cancer therapy is advantageous over conventional medicine because it has the potential to enable the preferential delivery of drugs to tumors owing to the EPR effect and can deliver more than one therapeutic agent for combination therapy (Jain & Stylianopoulos, 2010). Nanotechnology has played a great role in the delivery of drugs due to the unique nanosize (1–100 nm) and large surface-to-volume ratio. First, the water solubility of drugs can be significantly improved through conjugating it to nanomaterials or encapsulating it inside nanocarriers. Moreover, nanocarriers can improve the stability of drugs by protecting them from the surrounding environment, converting their pharmacokinetic properties, and lead to improved therapeutic efficacy due to controlled drug-release kinetics. When intravenously injected into the body, nanoformulations can protect drugs from metabolizing enzymes until they are released and avoid them from being rapidly cleared through the renal system due to their increased size compared to the free drug, thereby often resulting in a prolonged blood circulation time with an increased chance of accumulation in the target tissue. Nanoformulations can avoid side effects, thus might render the applied drugs more suitable for clinical utilization. In addition, nanocarriers can leak out of fenestrated tumor vascular walls and accumulate in the tumor tissue that frequently lacks effective lymphatic drainage by means of EPR-effect-mediated passive targeting. Other tissues with leaky endothelial walls, including those of the liver, spleen, and bone marrow, usually contribute to the uptake of nanocarriers. The macrophages residing in these tissues have been responsible for the major loss of injected nanocarriers (Li & Huang, 2008). Modification of the surface of nanocarriers can reduce the uptake of the reticuloendothelial system (RES) and prolong circulation time. Other advantages of nanomedicine include specific binding of drugs to targets in cancer cells or the tumor microenvironment, simultaneous visualization of tumors using innovative imaging techniques, and superior dose scheduling for improved patient compliance. Furthermore, many widely used conventional chemotherapeutics, such as taxanes, include synthetic solvents (for example, castor oil and polysorbate 80) that directly contribute to adverse effects (Jain & Stylianopoulos, 2010). Therefore, the adverse effects and systemic toxicity of drugs can be reduced and therapeutic efficacy can be improved significantly due to the application of nanotechnology in drug delivery.

4.3. Nanoparticle properties

In the treatment and diagnosis of cancer, nanoparticles (NPs) are considered promising carriers of drugs and imaging agents due to their ability to accumulate in solid tumors. A well-known explanation for this property is the EPR effect, which allows for the extravasation of NPs and their accumulation in tumors. NPs require intracellular delivery, and cellular internalization would depend on size, configuration, and charge (Jain & Stylianopoulos, 2010). Cationic NPs preferentially target tumor endothelial cells compared with their neutral or anionic counterparts. On the other hand, neutral NPs diffuse faster and distribute more homogeneously inside the tumor interstitial space than cationic and anionic NPs. With respect to shape, NPs with linear, semi-flexible configurations diffuse more efficiently in the interstitial matrix than do comparable sized, rigid spherical particles.

For the efficient use of the EPR effect, NPs should be able to circulate in the body, avoiding clearance by the RES (Kao & Juliano, 1981). In general, NP clearance from the circulation starts with the opsonization of NPs, followed by receptor-mediated phagocytosis (Patel & Moghimi, 1998). Since hydrophobic and/or charged NPs are prone to opsonization, NPs are usually coated with electrically neutral hydrophilic polymers. This protective surface coating, also known as stealth coating, can extend the half-life of NPs significantly (Moghimi, Hunter, & Murray, 2001). For higher circulation times and stabilization, NPs containing polyethylene glycol (PEG) or modified PEG, a nonionic hydrophilic polymer, are the most common approach to prevent destruction by serum proteins and phagocytosis by Kuppfer cells or hepatocytes (Jain & Stylianopoulos, 2010). PEG conjugation (PEGylation) has been widely used to protect NPs, such as liposomes (Klibanov, Maruyama, Torchilin, & Huang, 1990), polymeric NPs (Gref et al., 1994), and micelles (Lee et al., 2011), from premature clearance during circulation. The PEG that covers NPs forms a hydrated layer, which allows the NPs to evade opsonization and subsequently phagocytosis.

Over 20 NP therapeutics have been approved by the FDA for clinical use (Davis, Chen, & Shin, 2008; Zhang et al., 2008). NP formulations for the treatment of solid tumors include liposomes (such as PEGylated liposomal doxorubicin and liposomal daunorubicin), albumin-bound paclitaxel, polymeric particles (such as methoxy-PEG-poly[D,L-lactide]taxol), and many more formulations that are in preclinical and/or clinical trials. However, these agents are still associated with adverse effects, although less toxic than conventional therapies.

4.4. Tissue-specific deletion of CD44 signaling

The development of new strategies for cancer treatment and cure is a growing field of research, particularly with gene therapeutic approaches. Synthesis of interfering RNAs is an important method for transiently blocking protein expression that causes the disease. RNAi is a potent and specific gene silencing event that can exploit an innate biological pathway and disrupt the flow of genetic information from RNA by degrading a target mRNA that has a complementary sequence to the siRNA. The inhibition of expression of the gene of interest occurs in a sequence-specific manner without induction of the interferon response. Thus, the siRNA or vectors that transcribe shRNA can be used to prevent the production of disease-causing proteins, rather than preventing their damage or neutralizing them once they are produced. Compared to the transient phenotypic changes induced by siRNAs, the technique of using shRNA in an expression vector is an alternative strategy to stably suppress selected gene expression, which suggests that the use of shRNA expression vectors holds potential promise for therapeutic approaches for silencing disease-causing genes (Paul, Good, Winer, & Engelke, 2002).

Delivering of shRNA in cancer cells is achieved by either using a viral vector or a nonviral vector. Systemic targeting by viral vectors toward the desired tissue is difficult because the host immune responses activate viral clearance. Systemic administration of a large amount of adenovirus (e.g., into the liver) can be a serious health hazard that even caused the death of one patient (Raper et al., 2003). However, there has been a considerable interest in developing nonviral vectors for gene therapy. In this regard, nonviral vectors, such as PEGylated positively charged polyethyleneimine (PEI) complexes, can be used safely to avoid the nonspecific interactions with nontarget cells and blood components (Kursa et al., 2003). Nonviral vectors were once limited for their low gene transfer efficiency. Importantly, the incorporation of various ligands, such as peptides, growth factors, and proteins, or antibodies for targets highly expressed on cancer cells, circumvented this obstacle (Bellocq, Pun, Jensen, & Davis, 2003).

Nonviral vectors can acquire high gene transfer efficiency due to EPR effect of ligand-coated vectors around the receptors of tumor cells thus increasing chances for high probability of interaction with the cells. This concept has been successfully tested by Bellocq et al. and Misra et al. who prepared nonviral vector NPs with plasmids packed inside an outer PEG–PEI layer coated with transferrin (Tf), an iron transporting protein that binds with Tf receptor (Tf-R), which is ubiquitously expressed at low levels in most normal human tissues but at several fold higher in tumor cells, with high affinity (Bellocq et al., 2003; Misra et al., 2009). Association of Tf with the NPs significantly enhances transfection efficiency of shRNA generator plasmids by promoting the internalization of NPs in dividing and nondividing cells through receptor-mediated endocytosis. Finally, the uptake of NPs carrying multiple functional domains (surface shielding particles Tf–PEG–PEI, shRNA generator plasmids, tissue-specific promoter-driven Cre recombinase, and conditionally silenced plasmid) can overcome the intracellular barriers for successful delivery of the shRNA gene. Misra et al. proposed a cell-specific shRNA delivery approach by targeting the hyaluronan/CD44v6-induced signaling pathway, which successfully inhibited distant colon tumor growth in Apc Min/+ mice (Misra et al., 2009, 2011). The proposed approach minimizes the uptake of the drug by normal cells and enhances the influx and retention of the drug in cancer cells. This delivery system includes (a) an apoptosis-inducing agent that perturbs hyaluronan–CD44 interaction (CD44v6shRNA), (b) a targeting moiety-penetration enhancer (Tf), and (c) a carrier (PEG and PEI) (Table 11.1) (Fig. 11.1E). The tissue-specific shRNA delivery was made possible by the use of Cre recombinase produced in response to a colon tissue-specific promoter, which deletes the interruption between the U6 promoter and the CD44v6shRNA oligonucleotide. This approach has successfully demonstrated that the CD44v6shRNA is localized into the colon tumor cells by an end point assay of CD44v6 expression, and by perturbation of hyaluronan–CD44v6 interaction as reflected in the reduction in number of tumors (Misra et al., 2009).

4.5. Targeting chemotherapeutics to CD44 with hyaluronan conjugates as drug carriers

Hyaluronan has several chemical groups to which drugs can be conjugated (Pouyani & Prestwich, 1994). The carboxylate on the glucuronic acid, the *N*-acetylglucosamine hydroxyl, and the reducing end have all been successfully utilized in conjugation reactions with drugs. Drug conjugation on hyaluronan creates a macromolecular prodrug; the conjugated drug becomes active upon release from the hyaluronan chain. Upon binding to tumor-expressing CD44, hyaluronan can be endocytosed carrying conjugated cargo, such as drugs and drug carriers, into the tumor cell (Luo & Prestwich, 1999; Luo, Ziebell, & Prestwich, 2000). Cellular uptake increases therapeutic efficiency of hyaluronan-conjugated drugs (Coradini, Pellizzaro, et al., 2004; Eliaz, Nir, & Szoka, 2004). A number of cytotoxic anticancer drugs had increased CD44-specific internalization and cell cytotoxicity when conjugated to hyaluronan (Table 11.1) (Fig. 11.1C). In particular, hyaluronan–paclitaxel increased cytotoxic delivery to CD44 overexpressing breast, colon, and ovarian carcinoma cell lines. Furthermore, hyaluronan–paclitaxel conjugates injected locally to the tumor increased the mean survival time in a human ovarian carcinoma xenograft mouse model (Auzenne et al., 2007). In another study, sodium butyrate–hyaluronan conjugates injected intratumorally or subcutaneously reduced primary tumor growth and lung metastasis in a murine Lewis lung carcinoma model and a murine melanoma model (Coradini, Pellizzaro, et al., 2004; Coradini, Zorzet, et al., 2004). Studies performed by Akima et al. showed potent antimetastatic effects of hyaluronan–mitomycin c in a murine model of Lewis lung adenocarcinoma (Akima et al., 1996). Other cytotoxic anticancer drugs that had increased CD44-specific internalization when conjugated to hyaluronan include doxorubicin and epirubicin (Akima et al., 1996; Luo et al., 2002).

Nanoconjugates composed of drug molecules encapsulated in quantum dots (QDs) attract enormous attention due to their promising bioimaging and biomedical applications. Carborane pharmacophores in combination with QDs or other anticancer drugs may be viable for efficient cancer diagnosis and chemotherapy. Carboranes, used in boron neutron capture therapy, must be localized to the tumor at high enough concentrations to deliver a therapeutic dose of thermal neutrons. To improve the uptake of carboranes, they were conjugated to hyaluronan. These conjugates accumulated to a high concentration within human bladder carcinoma, colorectal adenocarcinoma, and ovarian adenocarcinoma cell lines due to CD44-specific uptake mechanism (Table 11.1) (Di Meo et al., 2008; Meo et al., 2007). This promising therapeutic approach together with the development of specific nanoformulations might significantly improve targeted anticancer treatment.

4.6. Application of polysaccharides (hyaluronan) in nanomedicine

Although PEG allows NPs to evade opsonization and subsequent phagocytosis, the PEG-protective surface also interferes with the desired interactions between NPs and targeted cells (Du, Chandaroy, & Hui, 1997). For example, PEGylated liposomal doxorubicin showed greater plasma area under the curve but lower accumulation within the tumor than those of the non-PEGylated counterpart, indicating the paradoxical effect of PEGylation (Hong et al., 1999). A review article described that the PEGylated multifunctional envelope-type nanodevice was less effective in delivering genes into cells than the non-PEGylated device (Hatakeyama, Akita, & Harashima, 2011). Furthermore, Lehtinen et al. demonstrated, using a computational model, that some targeting moieties could lose the functionality due to the steric hindrance effect of the PEG layer (Lehtinen et al., 2012).

The increased stability by PEGylation can also hinder the endosomal escape of NPs, a critical step for effective intracellular delivery of nucleic acid therapeutics (Mishra, Webster, & Davis, 2004; Remaut, Lucas, Braeckmans, Demeester, & De Smedt, 2007). These challenges have prompted a search for new strategies to protect NP surface. Examples of recent efforts include the use of different polymers of synthetic or natural origin, biomimetic coating, and conditional removal of PEG effect (Amoozgar & Yeo, 2012). NPs decorated with polysaccharides showed more efficient cellular uptake than uncoated NPs, due to their specific interactions with various receptors on target cells (Jiang et al., 2008; Li et al., 2012; Rivkin et al., 2010; Yuan et al., 2009). For these reasons, polysaccharides (including hyaluronan) have gained increasing interest in the field of nanomedicine as an alternative surface modification strategy (Doh & Yeo, 2012).

NPs incorporating polysaccharides are produced in various methods, as thoroughly reviewed elsewhere (Lemarchand, Gref, & Couvreur, 2004; Liu, Jiao, Wang, Zhou, & Zhang, 2008). Polysaccharides can form surface layers on core NPs via electrostatic interactions or by anchoring on the NP surface during the formation of NPs. Alternatively, hydrophilic polysaccharides are grafted with hydrophobic low-molecular-weight molecules such as cholesterol or bile acids to allow for the formation of self-assembled NPs, which carry hydrophobic drugs in the core. NPs can be made of conjugates of polysaccharides and synthetic polymers, created by various conjugation methods. For example, linear hydrophobic polymers are grafted to a polysaccharide as side chains, creating a branched copolymer. Alternatively, a linear hydrophobic polymer can be conjugated to a polysaccharide terminus to create a linear diblock copolymer. Depending on the configuration of the copolymer, polysaccharides form a loop- or a brush-style surface layer.

Hyaluronan has been incorporated into a wide variety of nanomedicine constructs. Hyaluronan has been used for the delivery of several anticancer drugs, such as paclitaxel (Bajaj et al., 2012; Li et al., 2012; Luo & Prestwich, 1999; Rivkin et al., 2010), doxorubicin (Eliaz & Szoka, 2001; Luo et al., 2002; Wu et al., 2012), and camptothecin (Choi et al., 2011) (Table 11.1) (Fig. 11.1C). In addition, hyaluronan is used along with cationic liposomes and polymers to reduce cytotoxicity and their interaction with serum proteins (Fischer, Li, Ahlemeyer, Krieglstein, & Kissel, 2003). For example, hyaluronan conjugated with PEI (hyaluronan–PEI) was used for the delivery of siRNA and has shown effective gene silencing effect *in vitro* (Han et al., 2009) and *in vivo* (Jiang et al., 2005; Park et al., 2011). The presence of hyaluronan prevented particle aggregation in blood and death related to pulmonary embolism (Park et al., 2011). The most likely mechanism by which the siRNA–hyaluronan–PEI complex enters cells is via receptor-mediated endocytosis. Several studies have shown that the siRNA–hyaluronan–PEI complex, unlike siRNA–PEI, tends to accumulate in the tissues with hyaluronan receptors. For example, more efficient gene silencing was observed in B16F1 murine melanoma cells overexpressing LYVE-1 than in HEK-293 human kidney cells without HA receptors indicating that LYVE-1 acted as a receptor for hyaluronan-conjugated NPs (Jiang et al., 2008). In another study, tail vein injection of a pDNA–hyaluronan–poly-L-lysine complex mainly accumulated in the sinusoidal endothelial cells of the liver and expressed a reporter gene there via HARE-mediated endocytosis (Takei et al., 2004).

Hyaluronan has been also shown to contribute to gene transfection by increasing extracellular stability. Xu et al. created a ternary gene complex consisting of pDNA, disulfide cross-linked PEI, and hyaluronan coating (DPH complex) that was superior to a binary complex of pDNA and cationic polymers (i.e., cross-linked PEI) in the stability under the treatment with DNase and/or heparin and the transfection efficiency in the serum-containing medium (Xu, Quick, & Yeo, 2009). These authors suggested that hyaluronan not only protects the complex from unwanted interactions with serum proteins but also helps the complex dissociate in a timely manner once internalized by cells. Moreover, it was suggested that the intracellular decomplexation of the hyaluronan-containing complex facilitates the access of transcription factors to pDNA and subsequent gene expression processes (Ito et al., 2006; Xu et al., 2009).

In a recent study, Landesman-Milo et al. proposed a noncationic lipid transfection strategy based on hyaluronan that was grafted on the surface of lipid-based nanoparticles (LNPs), which were devised from neutral phospholipids and cholesterol that self-assembled (Landesman-Milo et al., 2013). Hyaluronan–LNPs were originally developed as vehicles (nanocarriers) for delivering small-molecule chemotherapeutics to target cells *in vivo* (Peer et al., 2007; Peer & Margalit, 2004a, 2004b; Yerushalmi, Arad, & Margalit, 1994) (Fig. 11.1C). In their study, Landesman-Milo et al. observed that hyaluronan–LNPs bound and internalized specifically into cancer cells via receptor (CD44/RHAMM)-mediated endocytosis. Moreover, loaded with siRNAs against the multidrug resistance extrusion pump, P-gp, hyaluronan–LNPs efficiently and specifically reduced mRNA and P-gp protein levels compared with control, noncoated particles, and with hyaluronan–LNPs loaded with control, nontargeted siRNAs (Table 11.1).

Main advantages of polysaccharides are well-established biocompatibility, availability, and abundant functional groups amenable to chemical modifications. Hyaluronan, a naturally occurring GAG, is a nontoxic and nonimmunogenic compound, already approved for use in eye surgery, joint therapy, wound healing, as well as aesthetic medicine. However, some adverse effect cases have been reported in nearly all applications (Alijotas-Reig, 2009; Alijotas-Reig & Garcia-Gimenez, 2008; Bernauer et al., 2006; Castillo-Vico, Checa-Vizcaino, Paya-Panades, Rueda-Garcia, & Carreras-Collado, 2007; Zardawi, 2003; Zardawi & Chan, 2001). One of the main reasons is the high degree of variability with respect to hyaluronan molecular weight, a property that critically influences its biological activities. The molecular weight of hyaluronan has a remarkable impact on its effects on cells of the immune system. HMW hyaluronan is synthesized and accumulated as an ECM component by cell types such as fibroblasts and keratinocytes. There is growing evidence that under certain conditions, inflammation and tissue injury, turnover of HMW hyaluronan is strongly enhanced (Jiang, Liang, & Noble, 2007). Dendritic cells (DC) and T-cells regulate the production and degradation of hyaluronan by expression of three HAS genes and four genes with hyaluronidase activity (Mummert et al., 2002). Depending on their size, the ensuing hyaluronan fragments have a variety of effects on cells of the immune system and can act as immunomodulators by activating antigen-presenting cells, such as DC and macrophages (Hodge-Dufour et al., 1997; McKee et al., 1997, 1996). Although contaminants in hyaluronan preparations can complicate matters, studies show that these fragments are potent modulators of immune cells and highly purified hyaluronan oligosaccharide preparations potently induce DC maturation (Mummert, 2005; Termeer, Sleeman, & Simon, 2003; Yang, Yan, Chen, Hansson, & Kiessling, 2002). In fact, there is clear evidence from different groups that hyaluronan oligosaccharides induce direct signaling differently from HMW hyaluronan. Oligosaccharides smaller than hexasaccharide length cannot compete with HMW hyaluronan for CD44 binding, and it remains unclear to what extent CD44 is involved in the induction of signaling through hyaluronan oligosaccharides. In contrast, hyaluronan oligosaccharides can directly activate DC and macrophages through the Toll-like receptor (TLR) 4 and TLR2 (Jiang et al., 2005; Termeer et al., 2002). Therefore, a subtle difference in hyaluronan monomers' molecular weight can result in significant differences in hyaluronan-based nanomedicines. Recently, Mizrahy et al. addressed this issue by presenting a small library of lipid NPs distinguished only by the MW of their surface-anchored hyaluronan (ranging from 6.4 to 1500 kDa) (Mizrahy et al., 2011). The authors observed that all NPs, regardless of their surface-anchored hyaluronan MW, revealed no apparent effect on cellular proliferation. In addition, no macrophage activation or complement activation were detected. The affinity toward CD44, however, was shown to be controlled solely by adjusting the MW of the NP surface hyaluronan, from weak binding for LMW hyaluronan to binding with high affinity for HMW hyaluronan. Therefore, these observations may have important implications for the development of future drug delivery systems as LMW hyaluronan bearing NPs may safely replace PEG when passive delivery is required, since complement activation can be bypassed. Moreover, HMW hyaluronan NPs were suggested to be used for active targeting of CD44-overexpressing tumors and aberrantly activated leukocytes in inflammation.

5. CONCLUDING REMARKS/CONCLUSIONS

Reorganization of tumor ECM is directly related to tumor initiation, growth, and metastasis. Multivalent interactions of matrix hyaluronan with tumor cell-associated CD44, especially CD44 variants, collaborate in driving numerous tumor-promoting signaling pathways and transporter activities. It is widely accepted that hyaluronan–CD44 interactions are crucial in both malignancy and resistance to therapy. As critically reviewed in this chapter, many investigators have shown increased efficacy in cell and animal tumor models by conjugating either drugs to hyaluronan or CD44 antibody or by incorporating drugs or shRNAs into vehicles such as liposomes, hydrogels, and NPs that have been decorated with hyaluronan or antibodies against CD44. These studies highlight the potential significance of nanomedicine, a rapidly growing research field, in the pharmacological targeting of hyaluronan–CD44 interactions and signaling in cancer, and suggest that these interactions are an important target for translation of research preclinical data to the clinic.

ACKNOWLEDGMENTS

This research has been co-financed by the European Union (European Social Fund—ESF) and Greek National Funds through the Operational Program "Education and Lifelong Learning" of the National Strategic Reference Framework (NSRF) Research Funding Program: Thales. Investing in knowledge society through the European Social Fund.

REFERENCES

Ahrens, T., Sleeman, J. P., Schempp, C. M., Howells, N., Hofmann, M., Ponta, H., et al. (2001). Soluble CD44 inhibits melanoma tumor growth by blocking cell surface CD44 binding to hyaluronic acid. *Oncogene*, *20*(26), 3399–3408. http://dx.doi.org/10.1038/sj.onc.1204435.

Akima, K., Ito, H., Iwata, Y., Matsuo, K., Watari, N., Yanagi, M., et al. (1996). Evaluation of antitumor activities of hyaluronate binding antitumor drugs: Synthesis, characterization and antitumor activity. *Journal of Drug Targeting*, *4*(1), 1–8. http://dx.doi.org/10.3109/10611869609046255.

Alaniz, L., Garcia, M. G., Gallo-Rodriguez, C., Agusti, R., Sterin-Speziale, N., Hajos, S. E., et al. (2006). Hyaluronan oligosaccharides induce cell death through PI3-K/Akt pathway independently of NF-kappaB transcription factor. *Glycobiology*, *16*(5), 359–367. http://dx.doi.org/10.1093/glycob/cwj085.

Alijotas-Reig, J. (2009). Recurrent urticarial vasculitis related to nonanimal hyaluronic acid skin filler injection. *Dermatologic Surgery*, *35*(Suppl. 1), 395–397. http://dx.doi.org/10.1111/j.1524-4725.2008.01049.x, discussion 397–398.

Alijotas-Reig, J., & Garcia-Gimenez, V. (2008). Delayed immune-mediated adverse effects related to hyaluronic acid and acrylic hydrogel dermal fillers: Clinical findings, long-term follow-up and review of the literature. *Journal of the European Academy of Dermatology and Venereology*, *22*(2), 150–161. http://dx.doi.org/10.1111/j.1468-3083.2007.02354.x.

Allison, D. D., & Grande-Allen, K. J. (2006). Review. Hyaluronan: A powerful tissue engineering tool. *Tissue Engineering*, *12*(8), 2131–2140. http://dx.doi.org/10.1089/ten.2006.12.2131.

Amoozgar, Z., & Yeo, Y. (2012). Recent advances in stealth coating of nanoparticle drug delivery systems. *Wiley Interdisciplinary Reviews. Nanomedicine and Nanobiotechnology*, *4*(2), 219–233. http://dx.doi.org/10.1002/wnan.1157.

Auzenne, E., Ghosh, S. C., Khodadadian, M., Rivera, B., Farquhar, D., Price, R. E., et al. (2007). Hyaluronic acid-paclitaxel: Antitumor efficacy against CD44(+) human ovarian carcinoma xenografts. *Neoplasia*, *9*(6), 479–486.

Avin, E., Haimovich, J., & Hollander, N. (2004). Anti-idiotype × anti-CD44 bispecific antibodies inhibit invasion of lymphoid organs by B cell lymphoma. *The Journal of Immunology*, *173*(7), 4736–4743.

Ayhan, A., Tok, E. C., Bildirici, I., & Ayhan, A. (2001). Overexpression of CD44 variant 6 in human endometrial cancer and its prognostic significance. *Gynecologic Oncology*, *80*(3), 355–358. http://dx.doi.org/10.1006/gyno.2000.6014.

Bajaj, G., Kim, M. R., Mohammed, S. I., & Yeo, Y. (2012). Hyaluronic acid-based hydrogel for regional delivery of paclitaxel to intraperitoneal tumors. *Journal of Controlled Release*, *158*(3), 386–392. http://dx.doi.org/10.1016/j.jconrel.2011.12.001.

Bartolazzi, A., Peach, R., Aruffo, A., & Stamenkovic, I. (1994). Interaction between CD44 and hyaluronate is directly implicated in the regulation of tumor development. *The Journal of Experimental Medicine*, *180*(1), 53–66.

Baumgartner, G. (1998). The impact of extracellular matrix on chemoresistance of solid tumors—Experimental and clinical results of hyaluronidase as additive to cytostatic chemotherapy. *Cancer Letters*, *131*(1), 1–2.

Beckenlehner, K., Bannke, S., Spruss, T., Bernhardt, G., Schonenberg, H., & Schiess, W. (1992). Hyaluronidase enhances the activity of adriamycin in breast cancer models in vitro and in vivo. *Journal of Cancer Research and Clinical Oncology*, *118*(8), 591–596.

Bellocq, N. C., Pun, S. H., Jensen, G. S., & Davis, M. E. (2003). Transferrin-containing, cyclodextrin polymer-based particles for tumor-targeted gene delivery. *Bioconjugate Chemistry*, *14*(6), 1122–1132. http://dx.doi.org/10.1021/bc034125f.

Bernauer, W., Thiel, M. A., Kurrer, M., Heiligenhaus, A., Rentsch, K. M., Schmitt, A., et al. (2006). Corneal calcification following intensified treatment with sodium hyaluronate artificial tears. *The British Journal of Ophthalmology, 90*(3), 285–288. http://dx.doi.org/10.1136/bjo.2005.082792.

Borjesson, P. K., Postema, E. J., Roos, J. C., Colnot, D. R., Marres, H. A., van Schie, M. H., et al. (2003). Phase I therapy study with (186)Re-labeled humanized monoclonal antibody BIWA 4 (bivatuzumab) in patients with head and neck squamous cell carcinoma. *Clinical Cancer Research, 9*(10 Pt. 2), 3961S–3972S.

Bourguignon, L. Y. (2008). Hyaluronan-mediated CD44 activation of RhoGTPase signaling and cytoskeleton function promotes tumor progression. *Seminars in Cancer Biology, 18*(4), 251–259. http://dx.doi.org/10.1016/j.semcancer.2008.03.007.

Bourguignon, L. Y., Singleton, P. A., Diedrich, F., Stern, R., & Gilad, E. (2004). CD44 interaction with Na+ – H+ exchanger (NHE1) creates acidic microenvironments leading to hyaluronidase-2 and cathepsin B activation and breast tumor cell invasion. *The Journal of Biological Chemistry, 279*(26), 26991–27007. http://dx.doi.org/10.1074/jbc.M311838200.

Bourguignon, L. Y., Singleton, P. A., Zhu, H., & Zhou, B. (2002). Hyaluronan promotes signaling interaction between CD44 and the transforming growth factor beta receptor I in metastatic breast tumor cells. *The Journal of Biological Chemistry, 277*(42), 39703–39712. http://dx.doi.org/10.1074/jbc.M204320200.

Bourguignon, L. Y., Zhu, H., Shao, L., & Chen, Y. W. (2000). CD44 interaction with tiam1 promotes Rac1 signaling and hyaluronic acid-mediated breast tumor cell migration. *The Journal of Biological Chemistry, 275*(3), 1829–1838.

Bourguignon, L. Y., Zhu, H., Shao, L., & Chen, Y. W. (2001). CD44 interaction with c-Src kinase promotes cortactin-mediated cytoskeleton function and hyaluronic acid-dependent ovarian tumor cell migration. *The Journal of Biological Chemistry, 276*(10), 7327–7336. http://dx.doi.org/10.1074/jbc.M006498200.

Bourguignon, L. Y., Zhu, H., Shao, L., Zhu, D., & Chen, Y. W. (1999). Rho-kinase (ROK) promotes CD44v(3,8-10)-ankyrin interaction and tumor cell migration in metastatic breast cancer cells. *Cell Motility and the Cytoskeleton, 43*(4), 269–287. http://dx.doi.org/10.1002/(SICI)1097-0169(1999)43:4<269::AID-CM1>3.0.CO;2-5.

Breyer, R., Hussein, S., Radu, D. L., Putz, K. M., Gunia, S., Hecker, H., et al. (2000). Disruption of intracerebral progression of C6 rat glioblastoma by in vivo treatment with anti-CD44 monoclonal antibody. *Journal of Neurosurgery, 92*(1), 140–149. http://dx.doi.org/10.3171/jns.2000.92.1.0140.

Castillo-Vico, M. T., Checa-Vizcaino, M. A., Paya-Panades, A., Rueda-Garcia, C., & Carreras-Collado, R. (2007). Periurethral granuloma following injection with dextranomer/hyaluronic acid copolymer for stress urinary incontinence. *International Urogynecology Journal and Pelvic Floor Dysfunction, 18*(1), 95–97. http://dx.doi.org/10.1007/s00192-005-0049-7.

Charrad, R. S., Li, Y., Delpech, B., Balitrand, N., Clay, D., Jasmin, C., et al. (1999). Ligation of the CD44 adhesion molecule reverses blockage of differentiation in human acute myeloid leukemia. *Nature Medicine, 5*(6), 669–676. http://dx.doi.org/10.1038/9518.

Choi, S. H., Takahashi, K., Eto, H., Yoon, S. S., & Tanabe, K. K. (2000). CD44s expression in human colon carcinomas influences growth of liver metastases. *International Journal of Cancer, 85*(4), 523–526.

Choi, K. Y., Yoon, H. Y., Kim, J. H., Bae, S. M., Park, R. W., Kang, Y. M., et al. (2011). Smart nanocarrier based on PEGylated hyaluronic acid for cancer therapy. *ACS Nano, 5*(11), 8591–8599. http://dx.doi.org/10.1021/nn202070n.

Cichy, J., & Pure, E. (2003). The liberation of CD44. *The Journal of Cell Biology, 161*(5), 839–843. http://dx.doi.org/10.1083/jcb.200302098.

Colnot, D. R., Roos, J. C., de Bree, R., Wilhelm, A. J., Kummer, J. A., Hanft, G., et al. (2003). Safety, biodistribution, pharmacokinetics, and immunogenicity of 99mTc-labeled humanized monoclonal antibody BIWA 4 (bivatuzumab) in patients with squamous cell carcinoma of the head and neck. *Cancer Immunology, Immunotherapy*, *52*(9), 576–582. http://dx.doi.org/10.1007/s00262-003-0396-5.

Colone, M., Calcabrini, A., Toccacieli, L., Bozzuto, G., Stringaro, A., Gentile, M., et al. (2008). The multidrug transporter P-glycoprotein: A mediator of melanoma invasion? *The Journal of Investigative Dermatology*, *128*(4), 957–971. http://dx.doi.org/10.1038/sj.jid.5701082.

Coradini, D., Pellizzaro, C., Abolafio, G., Bosco, M., Scarlata, I., Cantoni, S., et al. (2004). Hyaluronic-acid butyric esters as promising antineoplastic agents in human lung carcinoma: A preclinical study. *Investigational New Drugs*, *22*(3), 207–217. http://dx.doi.org/10.1023/B:DRUG.0000026247.72656.8a.

Coradini, D., Zorzet, S., Rossin, R., Scarlata, I., Pellizzaro, C., Turrin, C., et al. (2004). Inhibition of hepatocellular carcinomas in vitro and hepatic metastases in vivo in mice by the histone deacetylase inhibitor HA-But. *Clinical Cancer Research*, *10*(14), 4822–4830. http://dx.doi.org/10.1158/1078-0432.CCR-04-0349.

Cordo Russo, R. I., Garcia, M. G., Alaniz, L., Blanco, G., Alvarez, E., & Hajos, S. E. (2008). Hyaluronan oligosaccharides sensitize lymphoma resistant cell lines to vincristine by modulating P-glycoprotein activity and PI3K/Akt pathway. *International Journal of Cancer*, *122*(5), 1012–1018. http://dx.doi.org/10.1002/ijc.23122.

Daley, W. P., Peters, S. B., & Larsen, M. (2008). Extracellular matrix dynamics in development and regenerative medicine. *Journal of Cell Science*, *121*(Pt. 3), 255–264. http://dx.doi.org/10.1242/jcs.006064.

Dall, P., Herrmann, I., Durst, B., Stoff-Khalili, M. A., Bauerschmitz, G., Hanstein, B., et al. (2005). In vivo cervical cancer growth inhibition by genetically engineered cytotoxic T cells. *Cancer Immunology, Immunotherapy*, *54*(1), 51–60. http://dx.doi.org/10.1007/s00262-004-0569-x.

Danhier, F., Feron, O., & Preat, V. (2010). To exploit the tumor microenvironment: Passive and active tumor targeting of nanocarriers for anti-cancer drug delivery. *Journal of Controlled Release*, *148*(2), 135–146. http://dx.doi.org/10.1016/j.jconrel.2010.08.027.

Davis, M. E., Chen, Z. G., & Shin, D. M. (2008). Nanoparticle therapeutics: An emerging treatment modality for cancer. *Nature Reviews Drug Discovery*, 7(9), 771–782. http://dx.doi.org/10.1038/nrd2614.

Di Meo, C., Panza, L., Campo, F., Capitani, D., Mannina, L., Banzato, A., et al. (2008). Novel types of carborane-carrier hyaluronan derivatives via "click chemistry" *Macromolecular Bioscience*, *8*(7), 670–681. http://dx.doi.org/10.1002/mabi.200700304.

Doh, K. O., & Yeo, Y. (2012). Application of polysaccharides for surface modification of nanomedicines. *Therapeutic Delivery*, *3*(12), 1447–1456. http://dx.doi.org/10.4155/tde.12.105.

Du, H., Chandaroy, P., & Hui, S. W. (1997). Grafted poly-(ethylene glycol) on lipid surfaces inhibits protein adsorption and cell adhesion. *Biochimica et Biophysica Acta*, *1326*(2), 236–248.

Eliaz, R. E., Nir, S., & Szoka, F. C., Jr. (2004). Interactions of hyaluronan-targeted liposomes with cultured cells: Modeling of binding and endocytosis. *Methods in Enzymology*, *387*, 16–33. http://dx.doi.org/10.1016/S0076-6879(04)87002-2.

Eliaz, R. E., & Szoka, F. C., Jr. (2001). Liposome-encapsulated doxorubicin targeted to CD44: A strategy to kill CD44-overexpressing tumor cells. *Cancer Research*, *61*(6), 2592–2601.

Evanko, S. P., Tammi, M. I., Tammi, R. H., & Wight, T. N. (2007). Hyaluronan-dependent pericellular matrix. *Advanced Drug Delivery Reviews*, *59*(13), 1351–1365. http://dx.doi.org/10.1016/j.addr.2007.08.008.

Fang, J., Nakamura, H., & Maeda, H. (2011). The EPR effect: Unique features of tumor blood vessels for drug delivery, factors involved, and limitations and augmentation of the effect. *Advanced Drug Delivery Reviews*, *63*(3), 136–151. http://dx.doi.org/10.1016/j.addr.2010.04.009.

Fischer, D., Li, Y., Ahlemeyer, B., Krieglstein, J., & Kissel, T. (2003). In vitro cytotoxicity testing of polycations: Influence of polymer structure on cell viability and hemolysis. *Biomaterials*, *24*(7), 1121–1131.

Foekens, J. A., Dall, P., Klijn, J. G., Skroch-Angel, P., Claassen, C. J., Look, M. P., et al. (1999). Prognostic value of CD44 variant expression in primary breast cancer. *International Journal of Cancer*, *84*(3), 209–215.

Ghatak, S., Misra, S., & Toole, B. P. (2002). Hyaluronan oligosaccharides inhibit anchorage-independent growth of tumor cells by suppressing the phosphoinositide 3-kinase/Akt cell survival pathway. *The Journal of Biological Chemistry*, *277*(41), 38013–38020. http://dx.doi.org/10.1074/jbc.M202404200.

Gialeli, C., Theocharis, A. D., & Karamanos, N. K. (2011). Roles of matrix metalloproteinases in cancer progression and their pharmacological targeting. *The FEBS Journal*, *278*(1), 16–27. http://dx.doi.org/10.1111/j.1742-4658.2010.07919.x.

Gilg, A. G., Tye, S. L., Tolliver, L. B., Wheeler, W. G., Visconti, R. P., Duncan, J. D., et al. (2008). Targeting hyaluronan interactions in malignant gliomas and their drug-resistant multipotent progenitors. *Clinical Cancer Research*, *14*(6), 1804–1813. http://dx.doi.org/10.1158/1078-0432.CCR-07-1228.

Gref, R., Minamitake, Y., Peracchia, M. T., Trubetskoy, V., Torchilin, V., & Langer, R. (1994). Biodegradable long-circulating polymeric nanospheres. *Science*, *263*(5153), 1600–1603.

Gunthert, U., Hofmann, M., Rudy, W., Reber, S., Zoller, M., Haussmann, I., et al. (1991). A new variant of glycoprotein CD44 confers metastatic potential to rat carcinoma cells. *Cell*, *65*(1), 13–24.

Guo, Y., Ma, J., Wang, J., Che, X., Narula, J., Bigby, M., et al. (1994). Inhibition of human melanoma growth and metastasis in vivo by anti-CD44 monoclonal antibody. *Cancer Research*, *54*(6), 1561–1565.

Han, S. E., Kang, H., Shim, G. Y., Kim, S. J., Choi, H. G., Kim, J., et al. (2009). Cationic derivatives of biocompatible hyaluronic acids for delivery of siRNA and antisense oligonucleotides. *Journal of Drug Targeting*, *17*(2), 123–132. http://dx.doi.org/10.1080/10611860802472461.

Harada, N., Mizoi, T., Kinouchi, M., Hoshi, K., Ishii, S., Shiiba, K., et al. (2001). Introduction of antisense CD44S CDNA down-regulates expression of overall CD44 isoforms and inhibits tumor growth and metastasis in highly metastatic colon carcinoma cells. *International Journal of Cancer*, *91*(1), 67–75.

Hascall, V., & Karamanos, N. (2011). Regulatory roles of hyaluronan in health and disease. *The FEBS Journal*, *278*(9), 1411. http://dx.doi.org/10.1111/j.1742-4658.2011.08068.x.

Hatakeyama, H., Akita, H., & Harashima, H. (2011). A multifunctional envelope type nano device (MEND) for gene delivery to tumours based on the EPR effect: A strategy for overcoming the PEG dilemma. *Advanced Drug Delivery Reviews*, *63*(3), 152–160. http://dx.doi.org/10.1016/j.addr.2010.09.001.

Hathcock, K. S., Hirano, H., Murakami, S., & Hodes, R. J. (1993). CD44 expression on activated B cells. Differential capacity for CD44-dependent binding to hyaluronic acid. *The Journal of Immunology*, *151*(12), 6712–6722.

Hayes, G. M., Carpenito, C., Davis, P. D., Dougherty, S. T., Dirks, J. F., & Dougherty, G. J. (2002). Alternative splicing as a novel of means of regulating the expression of therapeutic genes. *Cancer Gene Therapy*, *9*(2), 133–141. http://dx.doi.org/10.1038/sj.cgt.7700427.

Hayes, G. M., Dougherty, S. T., Davis, P. D., & Dougherty, G. J. (2004). Molecular mechanisms regulating the tumor-targeting potential of splice-activated gene expression. *Cancer Gene Therapy*, *11*(12), 797–807. http://dx.doi.org/10.1038/sj.cgt.7700759.

He, Q., Lesley, J., Hyman, R., Ishihara, K., & Kincade, P. W. (1992). Molecular isoforms of murine CD44 and evidence that the membrane proximal domain is not critical for hyaluronate recognition. *The Journal of Cell Biology*, *119*(6), 1711–1719.

Hekele, A., Dall, P., Moritz, D., Wels, W., Groner, B., Herrlich, P., et al. (1996). Growth retardation of tumors by adoptive transfer of cytotoxic T lymphocytes reprogrammed by CD44v6-specific scFv:zeta-chimera. *International Journal of Cancer*, *68*(2), 232–238. http://dx.doi.org/10.1002/(SICI)1097-0215(19961009)68:2<232::AID-IJC16>3.0.CO;2-C.

Heldin, P., Karousou, E., Bernert, B., Porsch, H., Nishitsuka, K., & Skandalis, S. S. (2008). Importance of hyaluronan-CD44 interactions in inflammation and tumorigenesis. *Connective Tissue Research*, *49*(3), 215–218. http://dx.doi.org/10.1080/03008200802143323.

Herold-Mende, C., Seiter, S., Born, A. I., Patzelt, E., Schupp, M., Zoller, J., et al. (1996). Expression of CD44 splice variants in squamous epithelia and squamous cell carcinomas of the head and neck. *The Journal of Pathology*, *179*(1), 66–73. http://dx.doi.org/10.1002/(SICI)1096-9896(199605)179:1<66::AID-PATH544>3.0.CO;2-5.

Herrera-Gayol, A., & Jothy, S. (2002). Effect of hyaluronan on xenotransplanted breast cancer. *Experimental and Molecular Pathology*, *72*(3), 179–185. http://dx.doi.org/10.1006/exmp.2002.2437.

Hodge-Dufour, J., Noble, P. W., Horton, M. R., Bao, C., Wysoka, M., Burdick, M. D., et al. (1997). Induction of IL-12 and chemokines by hyaluronan requires adhesion-dependent priming of resident but not elicited macrophages. *The Journal of Immunology*, *159*(5), 2492–2500.

Hong, R. L., Huang, C. J., Tseng, Y. L., Pang, V. F., Chen, S. T., Liu, J. J., et al. (1999). Direct comparison of liposomal doxorubicin with or without polyethylene glycol coating in C-26 tumor-bearing mice: Is surface coating with polyethylene glycol beneficial? *Clinical Cancer Research*, *5*(11), 3645–3652.

Hosono, K., Nishida, Y., Knudson, W., Knudson, C. B., Naruse, T., Suzuki, Y., et al. (2007). Hyaluronan oligosaccharides inhibit tumorigenicity of osteosarcoma cell lines MG-63 and LM-8 in vitro and in vivo via perturbation of hyaluronan-rich pericellular matrix of the cells. *The American Journal of Pathology*, *171*(1), 274–286. http://dx.doi.org/10.2353/ajpath.2007.060828.

Huxley-Jones, J., Foord, S. M., & Barnes, M. R. (2008). Drug discovery in the extracellular matrix. *Drug Discovery Today*, *13*(15–16), 685–694. http://dx.doi.org/10.1016/j.drudis.2008.05.005.

Iozzo, R. V., & Karamanos, N. (2010). Proteoglycans in health and disease: Emerging concepts and future directions. *The FEBS Journal*, *277*(19), 3863. http://dx.doi.org/10.1111/j.1742-4658.2010.07796.x.

Ishida, T. (2000). Immunohistochemical expression of the CD44 variant 6 in colorectal adenocarcinoma. *Surgery Today*, *30*(1), 28–32.

Ito, T., Iida-Tanaka, N., Niidome, T., Kawano, T., Kubo, K., Yoshikawa, K., et al. (2006). Hyaluronic acid and its derivative as a multi-functional gene expression enhancer: Protection from non-specific interactions, adhesion to targeted cells, and transcriptional activation. *Journal of Controlled Release*, *112*(3), 382–388. http://dx.doi.org/10.1016/j.jconrel.2006.03.013.

Jacobson, A., Rahmanian, M., Rubin, K., & Heldin, P. (2002). Expression of hyaluronan synthase 2 or hyaluronidase 1 differentially affect the growth rate of transplantable colon carcinoma cell tumors. *International Journal of Cancer*, *102*(3), 212–219. http://dx.doi.org/10.1002/ijc.10683.

Jain, R. K., & Stylianopoulos, T. (2010). Delivering nanomedicine to solid tumors. *Nature Reviews. Clinical Oncology*, *7*(11), 653–664. http://dx.doi.org/10.1038/nrclinonc.2010.139.

Jarvelainen, H., Sainio, A., Koulu, M., Wight, T. N., & Penttinen, R. (2009). Extracellular matrix molecules: Potential targets in pharmacotherapy. *Pharmacological Reviews*, *61*(2), 198–223. http://dx.doi.org/10.1124/pr.109.001289.

Jiang, D., Liang, J., Fan, J., Yu, S., Chen, S., Luo, Y., et al. (2005). Regulation of lung injury and repair by Toll-like receptors and hyaluronan. *Nature Medicine*, *11*(11), 1173–1179. http://dx.doi.org/10.1038/nm1315.

Jiang, D., Liang, J., & Noble, P. W. (2007). Hyaluronan in tissue injury and repair. *Annual Review of Cell and Developmental Biology*, *23*, 435–461. http://dx.doi.org/10.1146/annurev.cellbio.23.090506.123337.

Jiang, G., Park, K., Kim, J., Kim, K. S., Oh, E. J., Kang, H., et al. (2008). Hyaluronic acid-polyethyleneimine conjugate for target specific intracellular delivery of siRNA. *Biopolymers*, *89*(7), 635–642. http://dx.doi.org/10.1002/bip.20978.

Kao, Y. J., & Juliano, R. L. (1981). Interactions of liposomes with the reticuloendothelial system. Effects of reticuloendothelial blockade on the clearance of large unilamellar vesicles. *Biochimica et Biophysica Acta*, *677*(3–4), 453–461.

Karamanos, N. K., & Tzanakakis, G. N. (2012). Glycosaminoglycans: From "cellular glue" to novel therapeutical agents. *Current Opinion in Pharmacology*, *12*(2), 220–222. http://dx.doi.org/10.1016/j.coph.2011.12.003.

Karousou, E., Kamiryo, M., Skandalis, S. S., Ruusala, A., Asteriou, T., Passi, A., et al. (2010). The activity of hyaluronan synthase 2 is regulated by dimerization and ubiquitination. *The Journal of Biological Chemistry*, *285*(31), 23647–23654. http://dx.doi.org/10.1074/jbc.M110.127050.

Katoh, S., Zheng, Z., Oritani, K., Shimozato, T., & Kincade, P. W. (1995). Glycosylation of CD44 negatively regulates its recognition of hyaluronan. *The Journal of Experimental Medicine*, *182*(2), 419–429.

Klibanov, A. L., Maruyama, K., Torchilin, V. P., & Huang, L. (1990). Amphipathic polyethyleneglycols effectively prolong the circulation time of liposomes. *FEBS Letters*, *268*(1), 235–237.

Klocker, J., Sabitzer, H., Raunik, W., Wieser, S., & Schumer, J. (1998). Hyaluronidase as additive to induction chemotherapy in advanced squamous cell carcinoma of the head and neck. *Cancer Letters*, *131*(1), 113–115.

Knudson, W., Chow, G., & Knudson, C. B. (2002). CD44-mediated uptake and degradation of hyaluronan. *Matrix Biology*, *21*(1), 15–23.

Kozlova, I., Ruusala, A., Voytyuk, O., Skandalis, S. S., & Heldin, P. (2012). IQGAP1 regulates hyaluronan-mediated fibroblast motility and proliferation. *Cellular Signalling*, *24*(9), 1856–1862. http://dx.doi.org/10.1016/j.cellsig.2012.05.013.

Kursa, M., Walker, G. F., Roessler, V., Ogris, M., Roedl, W., Kircheis, R., et al. (2003). Novel shielded transferrin-polyethylene glycol-polyethylenimine/DNA complexes for systemic tumor-targeted gene transfer. *Bioconjugate Chemistry*, *14*(1), 222–231. http://dx.doi.org/10.1021/bc0256087.

Landesman-Milo, D., Goldsmith, M., Leviatan Ben-Arye, S., Witenberg, B., Brown, E., Leibovitch, S., et al. (2013). Hyaluronan grafted lipid-based nanoparticles as RNAi carriers for cancer cells. *Cancer Letters*, *334*(2), 221–227. http://dx.doi.org/10.1016/j.canlet.2012.08.024.

Lee, S. W., Yun, M. H., Jeong, S. W., In, C. H., Kim, J. Y., Seo, M. H., et al. (2011). Development of docetaxel-loaded intravenous formulation, Nanoxel-PM using polymer-based delivery system. *Journal of Controlled Release*, *155*(2), 262–271. http://dx.doi.org/10.1016/j.jconrel.2011.06.012.

Lehtinen, J., Magarkar, A., Stepniewski, M., Hakola, S., Bergman, M., Rog, T., et al. (2012). Analysis of cause of failure of new targeting peptide in PEGylated liposome: Molecular modeling as rational design tool for nanomedicine. *European Journal of Pharmaceutical Sciences*, *46*(3), 121–130. http://dx.doi.org/10.1016/j.ejps.2012.02.009.

Lemarchand, C., Gref, R., & Couvreur, P. (2004). Polysaccharide-decorated nanoparticles. *European Journal of Pharmaceutics and Biopharmaceutics*, *58*(2), 327–341. http://dx.doi.org/10.1016/j.ejpb.2004.02.016.

Lesley, J., English, N., Charles, C., & Hyman, R. (2000). The role of the CD44 cytoplasmic and transmembrane domains in constitutive and inducible hyaluronan binding. *European Journal of Immunology*, *30*(1), 245–253. http://dx.doi.org/10.1002/1521-4141(200001)30:1<245::AID-IMMU245>3.0.CO;2-X.

Lesley, J., Hascall, V. C., Tammi, M., & Hyman, R. (2000). Hyaluronan binding by cell surface CD44. *The Journal of Biological Chemistry*, *275*(35), 26967–26975. http://dx.doi.org/10.1074/jbc.M002527200.

Lesley, J., He, Q., Miyake, K., Hamann, A., Hyman, R., & Kincade, P. W. (1992). Requirements for hyaluronic acid binding by CD44: A role for the cytoplasmic domain and activation by antibody. *The Journal of Experimental Medicine*, *175*(1), 257–266.

Lesley, J., Howes, N., Perschl, A., & Hyman, R. (1994). Hyaluronan binding function of CD44 is transiently activated on T cells during an in vivo immune response. *The Journal of Experimental Medicine*, *180*(1), 383–387.

Lesley, J., & Hyman, R. (1992). CD44 can be activated to function as an hyaluronic acid receptor in normal murine T cells. *European Journal of Immunology*, *22*(10), 2719–2723. http://dx.doi.org/10.1002/eji.1830221036.

Li, L., Heldin, C. H., & Heldin, P. (2006). Inhibition of platelet-derived growth factor-BB-induced receptor activation and fibroblast migration by hyaluronan activation of CD44. *The Journal of Biological Chemistry*, *281*(36), 26512–26519. http://dx.doi.org/10.1074/jbc.M605607200.

Li, S. D., & Huang, L. (2008). Pharmacokinetics and biodistribution of nanoparticles. *Molecular Pharmaceutics*, *5*(4), 496–504. http://dx.doi.org/10.1021/mp800049w.

Li, J., Huo, M., Wang, J., Zhou, J., Mohammad, J. M., Zhang, Y., et al. (2012). Redox-sensitive micelles self-assembled from amphiphilic hyaluronic acid-deoxycholic acid conjugates for targeted intracellular delivery of paclitaxel. *Biomaterials*, *33*(7), 2310–2320. http://dx.doi.org/10.1016/j.biomaterials.2011.11.022.

Liu, Z., Jiao, Y., Wang, Y., Zhou, C., & Zhang, Z. (2008). Polysaccharides-based nanoparticles as drug delivery systems. *Advanced Drug Delivery Reviews*, *60*(15), 1650–1662. http://dx.doi.org/10.1016/j.addr.2008.09.001.

Luo, Y., Bernshaw, N. J., Lu, Z. R., Kopecek, J., & Prestwich, G. D. (2002). Targeted delivery of doxorubicin by HPMA copolymer-hyaluronan bioconjugates. *Pharmaceutical Research*, *19*(4), 396–402.

Luo, Y., & Prestwich, G. D. (1999). Synthesis and selective cytotoxicity of a hyaluronic acid-antitumor bioconjugate. *Bioconjugate Chemistry*, *10*(5), 755–763.

Luo, Y., Ziebell, M. R., & Prestwich, G. D. (2000). A hyaluronic acid-taxol antitumor bioconjugate targeted to cancer cells. *Biomacromolecules*, *1*(2), 208–218.

Mackay, C. R., Terpe, H. J., Stauder, R., Marston, W. L., Stark, H., & Gunthert, U. (1994). Expression and modulation of CD44 variant isoforms in humans. *The Journal of Cell Biology*, *124*(1–2), 71–82.

Marhaba, R., & Zoller, M. (2004). CD44 in cancer progression: Adhesion, migration and growth regulation. *Journal of Molecular Histology*, *35*(3), 211–231.

Matsumura, Y., & Maeda, H. (1986). A new concept for macromolecular therapeutics in cancer chemotherapy: Mechanism of tumoritropic accumulation of proteins and the antitumor agent smancs. *Cancer Research*, *46*(12 Pt. 1), 6387–6392.

McKee, C. M., Lowenstein, C. J., Horton, M. R., Wu, J., Bao, C., Chin, B. Y., et al. (1997). Hyaluronan fragments induce nitric-oxide synthase in murine macrophages through a nuclear factor kappaB-dependent mechanism. *The Journal of Biological Chemistry*, *272*(12), 8013–8018.

McKee, C. M., Penno, M. B., Cowman, M., Burdick, M. D., Strieter, R. M., Bao, C., et al. (1996). Hyaluronan (HA) fragments induce chemokine gene expression in alveolar macrophages. The role of HA size and CD44. *The Journal of Clinical Investigation*, *98*(10), 2403–2413. http://dx.doi.org/10.1172/JCI119054.

Meo, C. D., Panza, L., Capitani, D., Mannina, L., Banzato, A., Rondina, M., et al. (2007). Hyaluronan as carrier of carboranes for tumor targeting in boron neutron capture therapy. *Biomacromolecules*, *8*(2), 552–559. http://dx.doi.org/10.1021/bm0607426.

Merzak, A., Koocheckpour, S., & Pilkington, G. J. (1994). CD44 mediates human glioma cell adhesion and invasion in vitro. *Cancer Research*, *54*(15), 3988–3992.

Miletti-Gonzalez, K. E., Chen, S., Muthukumaran, N., Saglimbeni, G. N., Wu, X., Yang, J., et al. (2005). The CD44 receptor interacts with P-glycoprotein to promote cell migration and invasion in cancer. *Cancer Research*, *65*(15), 6660–6667. http://dx.doi.org/10.1158/0008-5472.CAN-04-3478.

Mishra, S., Webster, P., & Davis, M. E. (2004). PEGylation significantly affects cellular uptake and intracellular trafficking of non-viral gene delivery particles. *European Journal of Cell Biology*, *83*(3), 97–111. http://dx.doi.org/10.1078/0171-9335-00363.

Misra, S., Ghatak, S., & Toole, B. P. (2005). Regulation of MDR1 expression and drug resistance by a positive feedback loop involving hyaluronan, phosphoinositide 3-kinase, and ErbB2. *The Journal of Biological Chemistry*, *280*(21), 20310–20315. http://dx.doi.org/10.1074/jbc.M500737200.

Misra, S., Ghatak, S., Zoltan-Jones, A., & Toole, B. P. (2003). Regulation of multidrug resistance in cancer cells by hyaluronan. *The Journal of Biological Chemistry*, *278*(28), 25285–25288. http://dx.doi.org/10.1074/jbc.C300173200.

Misra, S., Hascall, V. C., De Giovanni, C., Markwald, R. R., & Ghatak, S. (2009). Delivery of CD44 shRNA/nanoparticles within cancer cells: Perturbation of hyaluronan/CD44v6 interactions and reduction in adenoma growth in Apc Min/+ MICE. *The Journal of Biological Chemistry*, *284*(18), 12432–12446. http://dx.doi.org/10.1074/jbc.M806772200.

Misra, S., Heldin, P., Hascall, V. C., Karamanos, N. K., Skandalis, S. S., Markwald, R. R., et al. (2011). Hyaluronan-CD44 interactions as potential targets for cancer therapy. *The FEBS Journal*, *278*(9), 1429–1443. http://dx.doi.org/10.1111/j.1742-4658.2011.08071.x.

Misra, S., Obeid, L. M., Hannun, Y. A., Minamisawa, S., Berger, F. G., Markwald, R. R., et al. (2008). Hyaluronan constitutively regulates activation of COX-2-mediated cell survival activity in intestinal epithelial and colon carcinoma cells. *The Journal of Biological Chemistry*, *283*(21), 14335–14344. http://dx.doi.org/10.1074/jbc.M703811200.

Misra, S., Toole, B. P., & Ghatak, S. (2006). Hyaluronan constitutively regulates activation of multiple receptor tyrosine kinases in epithelial and carcinoma cells. *The Journal of Biological Chemistry*, *281*(46), 34936–34941. http://dx.doi.org/10.1074/jbc.C600138200.

Mizrahy, S., Raz, S. R., Hasgaard, M., Liu, H., Soffer-Tsur, N., Cohen, K., et al. (2011). Hyaluronan-coated nanoparticles: The influence of the molecular weight on CD44-hyaluronan interactions and on the immune response. *Journal of Controlled Release*, *156*(2), 231–238. http://dx.doi.org/10.1016/j.jconrel.2011.06.031.

Moghimi, S. M., Hunter, A. C., & Murray, J. C. (2001). Long-circulating and target-specific nanoparticles: Theory to practice. *Pharmacological Reviews*, *53*(2), 283–318.

Mummert, M. E. (2005). Immunologic roles of hyaluronan. *Immunologic Research*, *31*(3), 189–206. http://dx.doi.org/10.1385/IR:31:3:189.

Mummert, M. E., Mummert, D., Edelbaum, D., Hui, F., Matsue, H., & Takashima, A. (2002). Synthesis and surface expression of hyaluronan by dendritic cells and its potential role in antigen presentation. *The Journal of Immunology*, *169*(8), 4322–4331.

Nagase, H., & Karamanos, N. (2011). Metalloproteinases in health and disease: Challenges and the future prospects. *The FEBS Journal, 278*(1), 1.

Naor, D., Nedvetzki, S., Golan, I., Melnik, L., & Faitelson, Y. (2002). CD44 in cancer. *Critical Reviews in Clinical Laboratory Sciences, 39*(6), 527–579. http://dx.doi.org/10.1080/10408360290795574.

Naor, D., Wallach-Dayan, S. B., Zahalka, M. A., & Sionov, R. V. (2008). Involvement of CD44, a molecule with a thousand faces, in cancer dissemination. *Seminars in Cancer Biology, 18*(4), 260–267. http://dx.doi.org/10.1016/j.semcancer.2008.03.015.

Noulas, A. V., Skandalis, S. S., Feretis, E., Theocharis, D. A., & Karamanos, N. K. (2004). Variations in content and structure of glycosaminoglycans of the vitreous gel from different mammalian species. *Biomedical Chromatography, 18*(7), 457–461. http://dx.doi.org/10.1002/bmc.336.

Orian-Rousseau, V. (2010). CD44, a therapeutic target for metastasising tumours. *European Journal of Cancer, 46*(7), 1271–1277. http://dx.doi.org/10.1016/j.ejca.2010.02.024.

Papadas, T. A., Stylianou, M., Mastronikolis, N. S., Papageorgakopoulou, N., Skandalis, S., Goumas, P., et al. (2002). Alterations in the content and composition of glycosaminoglycans in human laryngeal carcinoma. *Acta Oto-Laryngologica, 122*(3), 330–337.

Park, K., Hong, S. W., Hur, W., Lee, M. Y., Yang, J. A., Kim, S. W., et al. (2011). Target specific systemic delivery of TGF-beta siRNA/(PEI-SS)-g-HA complex for the treatment of liver cirrhosis. *Biomaterials, 32*(21), 4951–4958. http://dx.doi.org/10.1016/j.biomaterials.2011.03.044.

Patel, H. M., & Moghimi, S. M. (1998). Serum-mediated recognition of liposomes by phagocytic cells of the reticuloendothelial system—The concept of tissue specificity. *Advanced Drug Delivery Reviews, 32*(1–2), 45–60.

Paul, C. P., Good, P. D., Winer, I., & Engelke, D. R. (2002). Effective expression of small interfering RNA in human cells. *Nature Biotechnology, 20*(5), 505–508. http://dx.doi.org/10.1038/nbt0502-505.

Peer, D., Karp, J. M., Hong, S., Farokhzad, O. C., Margalit, R., & Langer, R. (2007). Nanocarriers as an emerging platform for cancer therapy. *Nature Nanotechnology, 2*(12), 751–760. http://dx.doi.org/10.1038/nnano.2007.387.

Peer, D., & Margalit, R. (2004a). Loading mitomycin C inside long circulating hyaluronan targeted nano-liposomes increases its antitumor activity in three mice tumor models. *International Journal of Cancer, 108*(5), 780–789. http://dx.doi.org/10.1002/ijc.11615.

Peer, D., & Margalit, R. (2004b). Tumor-targeted hyaluronan nanoliposomes increase the antitumor activity of liposomal doxorubicin in syngeneic and human xenograft mouse tumor models. *Neoplasia, 6*(4), 343–353. http://dx.doi.org/10.1593/neo.03460.

Peterson, R. M., Yu, Q., Stamenkovic, I., & Toole, B. P. (2000). Perturbation ofhyaluronan interactions by soluble CD44 inhibits growth of murine mammary carcinoma cells in ascites. *The American Journal of Pathology, 156*(6), 2159–2167. http://dx.doi.org/10.1016/S0002-9440(10)65086-9.

Platt, V. M., & Szoka, F. C., Jr. (2008). Anticancer therapeutics: Targeting macromolecules and nanocarriers to hyaluronan or CD44, a hyaluronan receptor. *Molecular Pharmaceutics, 5*(4), 474–486. http://dx.doi.org/10.1021/mp800024g.

Ponta, H., Sherman, L., & Herrlich, P. A. (2003). CD44: From adhesion molecules to signalling regulators. *Nature Reviews. Molecular Cell Biology, 4*(1), 33–45. http://dx.doi.org/10.1038/nrm1004.

Pouyani, T., & Prestwich, G. D. (1994). Functionalized derivatives of hyaluronic acid oligosaccharides: Drug carriers and novel biomaterials. *Bioconjugate Chemistry, 5*(4), 339–347.

Prestwich, G. D. (2008). Engineering a clinically-useful matrix for cell therapy. *Organogenesis, 4*(1), 42–47.

Prestwich, G. D., & Kuo, J. W. (2008). Chemically-modified HA for therapy and regenerative medicine. *Current Pharmaceutical Biotechnology, 9*(4), 242–245.

Radotra, B., & McCormick, D. (1997). Glioma invasion in vitro is mediated by CD44-hyaluronan interactions. *The Journal of Pathology, 181*(4), 434–438. http://dx.doi.org/10.1002/(SICI)1096-9896(199704)181:4<434::AID-PATH797>3.0.CO;2-S.

Raper, S. E., Chirmule, N., Lee, F. S., Wivel, N. A., Bagg, A., Gao, G. P., et al. (2003). Fatal systemic inflammatory response syndrome in a ornithine transcarbamylase deficient patient following adenoviral gene transfer. *Molecular Genetics and Metabolism, 80*(1–2), 148–158.

Reeder, J. A., Gotley, D. C., Walsh, M. D., Fawcett, J., & Antalis, T. M. (1998). Expression of antisense CD44 variant 6 inhibits colorectal tumor metastasis and tumor growth in a wound environment. *Cancer Research, 58*(16), 3719–3726.

Remaut, K., Lucas, B., Braeckmans, K., Demeester, J., & De Smedt, S. C. (2007). Pegylation of liposomes favours the endosomal degradation of the delivered phosphodiester oligonucleotides. *Journal of Controlled Release, 117*(2), 256–266. http://dx.doi.org/10.1016/j.jconrel.2006.10.029.

Rivkin, I., Cohen, K., Koffler, J., Melikhov, D., Peer, D., & Margalit, R. (2010). Paclitaxel-clusters coated with hyaluronan as selective tumor-targeted nanovectors. *Biomaterials, 31*(27), 7106–7114. http://dx.doi.org/10.1016/j.biomaterials.2010.05.067.

Rochman, M., Moll, J., Herrlich, P., Wallach, S. B., Nedvetzki, S., Sionov, R. V., et al. (2000). The CD44 receptor of lymphoma cells: Structure-function relationships and mechanism of activation. *Cell Adhesion and Communication*, 7(4), 331–347.

Rupp, U., Schoendorf-Holland, E., Eichbaum, M., Schuetz, F., Lauschner, I., Schmidt, P., et al. (2007). Safety and pharmacokinetics of bivatuzumab mertansine in patients with CD44v6-positive metastatic breast cancer: Final results of a phase I study. *Anti-Cancer Drugs, 18*(4), 477–485. http://dx.doi.org/10.1097/CAD.0b013e32801403f4.

Sauter, A., Kloft, C., Gronau, S., Bogeschdorfer, F., Erhardt, T., Golze, W., et al. (2007). Pharmacokinetics, immunogenicity and safety of bivatuzumab mertansine, a novel CD44v6-targeting immunoconjugate, in patients with squamous cell carcinoma of the head and neck. *International Journal of Oncology, 30*(4), 927–935.

Seiter, S., Arch, R., Reber, S., Komitowski, D., Hofmann, M., Ponta, H., et al. (1993). Prevention of tumor metastasis formation by anti-variant CD44. *The Journal of Experimental Medicine*, 177(2), 443–455.

Shuster, S., Frost, G. I., Csoka, A. B., Formby, B., & Stern, R. (2002). Hyaluronidase reduces human breast cancer xenografts in SCID mice. *International Journal of Cancer, 102*(2), 192–197. http://dx.doi.org/10.1002/ijc.10668.

Skandalis, S. S., Kozlova, I., Engstrom, U., Hellman, U., & Heldin, P. (2010). Proteomic identification of CD44 interacting proteins. *IUBMB Life, 62*(11), 833–840. http://dx.doi.org/10.1002/iub.392.

Skandalis, S. S., Stylianou, M., Vynios, D. H., Papageorgakopoulou, N., & Theocharis, D. A. (2007). The structural and compositional changes of glycosaminoglycans are closely associated with tissue type in human laryngeal cancer. *Biochimie, 89*(12), 1573–1580. http://dx.doi.org/10.1016/j.biochi.2007.07.006.

Skandalis, S. S., Theocharis, A. D., Papageorgakopoulou, N., & Zagris, N. (2003). Glycosaminoglycans in early chick embryo. *International Journal of Developmental Biology, 47*(4), 311–314.

Skelton, T. P., Zeng, C., Nocks, A., & Stamenkovic, I. (1998). Glycosylation provides both stimulatory and inhibitory effects on cell surface and soluble CD44 binding to hyaluronan. *The Journal of Cell Biology, 140*(2), 431–446.

Skotheim, R. I., & Nees, M. (2007). Alternative splicing in cancer: Noise, functional, or systematic? *International Journal of Biochemistry and Cell Biology, 39*(7–8), 1432–1449. http://dx.doi.org/10.1016/j.biocel.2007.02.016.

Sleeman, J., Rudy, W., Hofmann, M., Moll, J., Herrlich, P., & Ponta, H. (1996). Regulated clustering of variant CD44 proteins increases their hyaluronate binding capacity. *The Journal of Cell Biology*, *135*(4), 1139–1150.

Slomiany, M. G., Dai, L., Bomar, P. A., Knackstedt, T. J., Kranc, D. A., Tolliver, L., et al. (2009). Abrogating drug resistance in malignant peripheral nerve sheath tumors by disrupting hyaluronan-CD44 interactions with small hyaluronan oligosaccharides. *Cancer Research*, *69*(12), 4992–4998. http://dx.doi.org/10.1158/0008-5472.CAN-09-0143.

Slomiany, M. G., Grass, G. D., Robertson, A. D., Yang, X. Y., Maria, B. L., Beeson, C., et al. (2009). Hyaluronan, CD44, and emmprin regulate lactate efflux and membrane localization of monocarboxylate transporters in human breast carcinoma cells. *Cancer Research*, *69*(4), 1293–1301. http://dx.doi.org/10.1158/0008-5472.CAN-08-2491.

Sneath, R. J., & Mangham, D. C. (1998). The normal structure and function of CD44 and its role in neoplasia. *Molecular Pathology*, *51*(4), 191–200.

Somasundaram, C., Arch, R., Matzku, S., & Zoller, M. (1996). Development of a bispecific F(ab′)2 conjugate against the complement receptor CR3 of macrophages and a variant CD44 antigen of rat pancreatic adenocarcinoma for redirecting macrophage-mediated tumor cytotoxicity. *Cancer Immunology, Immunotherapy*, *42*(6), 343–350.

Song, G., Liao, X., Zhou, L., Wu, L., Feng, Y., & Han, Z. C. (2004). HI44a, an anti-CD44 monoclonal antibody, induces differentiation and apoptosis of human acute myeloid leukemia cells. *Leukemia Research*, *28*(10), 1089–1096. http://dx.doi.org/10.1016/j.leukres.2004.02.005.

St Croix, B., Man, S., & Kerbel, R. S. (1998). Reversal of intrinsic and acquired forms of drug resistance by hyaluronidase treatment of solid tumors. *Cancer Letters*, *131*(1), 35–44.

Subramaniam, V., Vincent, I. R., Gilakjan, M., & Jothy, S. (2007). Suppression of human colon cancer tumors in nude mice by siRNA CD44 gene therapy. *Experimental and Molecular Pathology*, *83*(3), 332–340. http://dx.doi.org/10.1016/j.yexmp.2007.08.013.

Sy, M. S., Guo, Y. J., & Stamenkovic, I. (1992). Inhibition of tumor growth in vivo with a soluble CD44-immunoglobulin fusion protein. *The Journal of Experimental Medicine*, *176*(2), 623–627.

Takei, Y., Maruyama, A., Ferdous, A., Nishimura, Y., Kawano, S., Ikejima, K., et al. (2004). Targeted gene delivery to sinusoidal endothelial cells: DNA nanoassociate bearing hyaluronan-glycocalyx. *The FASEB Journal*, *18*(6), 699–701. http://dx.doi.org/10.1096/fj.03-0494fje.

Tammi, R. H., Kultti, A., Kosma, V. M., Pirinen, R., Auvinen, P., & Tammi, M. I. (2008). Hyaluronan in human tumors: Pathobiological and prognostic messages from cell-associated and stromal hyaluronan. *Seminars in Cancer Biology*, *18*(4), 288–295. http://dx.doi.org/10.1016/j.semcancer.2008.03.005.

Termeer, C., Benedix, F., Sleeman, J., Fieber, C., Voith, U., Ahrens, T., et al. (2002). Oligosaccharides of hyaluronan activate dendritic cells via toll-like receptor 4. *The Journal of Experimental Medicine*, *195*(1), 99–111.

Termeer, C., Sleeman, J. P., & Simon, J. C. (2003). Hyaluronan—Magic glue for the regulation of the immune response? *Trends in Immunology*, *24*(3), 112–114.

Thankamony, S. P., & Knudson, W. (2006). Acylation of CD44 and its association with lipid rafts are required for receptor and hyaluronan endocytosis. *The Journal of Biological Chemistry*, *281*(45), 34601–34609. http://dx.doi.org/10.1074/jbc.M601530200.

Theocharis, D. A., Skandalis, S. S., Noulas, A. V., Papageorgakopoulou, N., Theocharis, A. D., & Karamanos, N. K. (2008). Hyaluronan and chondroitin sulfate proteoglycans in the supramolecular organization of the mammalian vitreous body. *Connective Tissue Research*, *49*(3), 124–128. http://dx.doi.org/10.1080/03008200802148496.

Theocharis, A. D., Skandalis, S. S., Tzanakakis, G. N., & Karamanos, N. K. (2010). Proteoglycans in health and disease: Novel roles for proteoglycans in malignancy and their pharmacological targeting. *The FEBS Journal*, *277*(19), 3904–3923.

Theocharis, A. D., Tsara, M. E., Papageorgacopoulou, N., Karavias, D. D., & Theocharis, D. A. (2000). Pancreatic carcinoma is characterized by elevated content of hyaluronan and chondroitin sulfate with altered disaccharide composition. *Biochimica et Biophysica Acta*, *1502*(2), 201–206.

Theocharis, A. D., Vynios, D. H., Papageorgakopoulou, N., Skandalis, S. S., & Theocharis, D. A. (2003). Altered content composition and structure of glycosaminoglycans and proteoglycans in gastric carcinoma. *International Journal of Biochemistry and Cell Biology*, *35*(3), 376–390.

Tijink, B. M., Buter, J., de Bree, R., Giaccone, G., Lang, M. S., Staab, A., et al. (2006). A phase I dose escalation study with anti-CD44v6 bivatuzumab mertansine in patients with incurable squamous cell carcinoma of the head and neck or esophagus. *Clinical Cancer Research*, *12*(20 Pt. 1), 6064–6072. http://dx.doi.org/10.1158/1078-0432.CCR-06-0910.

Toole, B. P. (2001). Hyaluronan in morphogenesis. *Seminars in Cell and Developmental Biology*, *12*(2), 79–87. http://dx.doi.org/10.1006/scdb.2000.0244.

Toole, B. P. (2004). Hyaluronan: From extracellular glue to pericellular cue. *Nature Reviews Cancer*, *4*(7), 528–539. http://dx.doi.org/10.1038/nrc1391.

Toole, B. P. (2009). Hyaluronan-CD44 interactions in cancer: Paradoxes and possibilities. *Clinical Cancer Research*, *15*(24), 7462–7468. http://dx.doi.org/10.1158/1078-0432.CCR-09-0479.

Toole, B. P., & Slomiany, M. G. (2008). Hyaluronan, CD44 and emmprin: Partners in cancer cell chemoresistance. *Drug Resistance Updates*, *11*(3), 110–121. http://dx.doi.org/10.1016/j.drup.2008.04.002.

Torchilin, V. (2011). Tumor delivery of macromolecular drugs based on the EPR effect. *Advanced Drug Delivery Reviews*, *63*(3), 131–135. http://dx.doi.org/10.1016/j.addr.2010.03.011.

Tsujimura, S., Saito, K., Kohno, K., & Tanaka, Y. (2006). Fragmented hyaluronan induces transcriptional up-regulation of the multidrug resistance-1 gene in CD4+ T cells. *The Journal of Biological Chemistry*, *281*(49), 38089–38097. http://dx.doi.org/10.1074/jbc.M601030200.

Turley, E. A., Noble, P. W., & Bourguignon, L. Y. (2002). Signaling properties of hyaluronan receptors. *The Journal of Biological Chemistry*, *277*(7), 4589–4592. http://dx.doi.org/10.1074/jbc.R100038200.

van Weering, D. H., Baas, P. D., & Bos, J. L. (1993). A PCR-based method for the analysis of human CD44 splice products. *PCR Methods and Applications*, *3*(2), 100–106.

Verel, I., Heider, K. H., Siegmund, M., Ostermann, E., Patzelt, E., Sproll, M., et al. (2002). Tumor targeting properties of monoclonal antibodies with different affinity for target antigen CD44V6 in nude mice bearing head-and-neck cancer xenografts. *International Journal of Cancer*, *99*(3), 396–402. http://dx.doi.org/10.1002/ijc.10369.

Vigetti, D., Viola, M., Karousou, E., De Luca, G., & Passi, A. (2014). Metabolic control of hyaluronan synthases. *Matrix Biology*, *35*, 8–13. http://dx.doi.org/10.1016/j.matbio.2013.10.002.

Vynios, D. H., Theocharis, D. A., Papageorgakopoulou, N., Papadas, T. A., Mastronikolis, N. S., Goumas, P. D., et al. (2008). Biochemical changes of extracellular proteoglycans in squamous cell laryngeal carcinoma. *Connective Tissue Research*, *49*(3), 239–243. http://dx.doi.org/10.1080/03008200802147662.

Wallach-Dayan, S. B., Rubinstein, A. M., Hand, C., Breuer, R., & Naor, D. (2008). DNA vaccination with CD44 variant isoform reduces mammary tumor local growth and lung metastasis. *Molecular Cancer Therapeutics*, *7*(6), 1615–1623. http://dx.doi.org/10.1158/1535-7163.MCT-07-2383.

Weigel, P. H., & DeAngelis, P. L. (2007). Hyaluronan synthases: A decade-plus of novel glycosyltransferases. *The Journal of Biological Chemistry, 282*(51), 36777–36781. http://dx.doi.org/10.1074/jbc.R700036200.

Wu, J. L., Liu, C. G., Wang, X. L., & Huang, Z. H. (2012). Preparation and characterization of nanoparticles based on histidine-hyaluronic acid conjugates as doxorubicin carriers. *Journal of Materials Science: Materials in Medicine, 23*(8), 1921–1929. http://dx.doi.org/10.1007/s10856-012-4665-8.

Xu, P., Quick, G. K., & Yeo, Y. (2009). Gene delivery through the use of a hyaluronate-associated intracellularly degradable crosslinked polyethyleneimine. *Biomaterials, 30*(29), 5834–5843. http://dx.doi.org/10.1016/j.biomaterials.2009.07.012.

Yang, R., Yan, Z., Chen, F., Hansson, G. K., & Kiessling, R. (2002). Hyaluronic acid and chondroitin sulphate A rapidly promote differentiation of immature DC with upregulation of costimulatory and antigen-presenting molecules, and enhancement of NF-kappaB and protein kinase activity. *Scandinavian Journal of Immunology, 55*(1), 2–13.

Yerushalmi, N., Arad, A., & Margalit, R. (1994). Molecular and cellular studies of hyaluronic acid-modified liposomes as bioadhesive carriers for topical drug delivery in wound healing. *Archives of Biochemistry and Biophysics, 313*(2), 267–273. http://dx.doi.org/10.1006/abbi.1994.1387.

Yuan, Z. X., Zhang, Z. R., Zhu, D., Sun, X., Gong, T., Liu, J., et al. (2009). Specific renal uptake of randomly 50% N-acetylated low molecular weight chitosan. *Molecular Pharmaceutics, 6*(1), 305–314. http://dx.doi.org/10.1021/mp800078a.

Zahalka, M. A., Okon, E., Gosslar, U., Holzmann, B., & Naor, D. (1995). Lymph node (but not spleen) invasion by murine lymphoma is both CD44- and hyaluronate-dependent. *The Journal of Immunology, 154*(10), 5345–5355.

Zardawi, I. M. (2003). Granulomatous inflammation after Hylan G-F 20 viscosupplementation of the knee. *The Journal of Bone and Joint Surgery, 85-A*(12), 2484, author reply 2484–2485.

Zardawi, I. M., & Chan, I. (2001). Synvisc perisynovitis. *Pathology, 33*(4), 519–520.

Zeng, C., Toole, B. P., Kinney, S. D., Kuo, J. W., & Stamenkovic, I. (1998). Inhibition of tumor growth in vivo by hyaluronan oligomers. *International Journal of Cancer, 77*(3), 396–401.

Zhang, L., Gu, F. X., Chan, J. M., Wang, A. Z., Langer, R. S., & Farokhzad, O. C. (2008). Nanoparticles in medicine: Therapeutic applications and developments. *Clinical Pharmacology and Therapeutics, 83*(5), 761–769. http://dx.doi.org/10.1038/sj.clpt.6100400.

CHAPTER TWELVE

Hyaluronan/RHAMM Interactions in Mesenchymal Tumor Pathogenesis: Role of Growth Factors

Katerina Kouvidi, Dragana Nikitovic, Aikaterini Berdiaki, George N. Tzanakakis[1]
Laboratory of Anatomy-Histology-Embryology, School of Medicine, University of Crete, Heraklion, Greece
[1]Corresponding author: e-mail address: tzanakak@med.uoc.gr

Contents

1. Introduction 320
2. The Role of Hyaluronan and Its Receptors in Fibrosarcoma 323
   2.1 HA production in mesenchymal tumors 324
   2.2 RHAMM in fibrosarcoma development and progression 327
3. GF Signaling and ECM Organization in Fibrosarcoma Pathogenesis 332
   3.1 Transforming growth factor-β 333
   3.2 Fibroblast growth factor 334
   3.3 Platelet-derived growth factor 335
   3.4 Insulin-like growth factor and epidermal growth factor 336
4. Concluding Remarks 337
References 338

Abstract

Fibrosarcoma belongs to the sarcoma cancer group, which are spindle cell malignancies of mesenchymal origin, and owe their name to the predominant cell line that is present within the tumor. The extracellular matrix (ECM) is a complicated structure that surrounds and supports cells within tissues. Its main components are proteoglycans, collagens, glycoproteins, hyaluronan (HA), and several matrix-degrading enzymes. During cancer progression, significant changes can be observed in the structural and mechanical properties of ECM components. The ECM provides a physical scaffold to which tumor cells attach and migrate. Thus, it is required for key cellular events such as cell motility, adhesion, proliferation, invasion, and metastasis. Importantly, fibrosarcomas were shown to have a high content and turnover of ECM components including HA, proteoglycans, collagens, fibronectin, and laminin. In this review, we will focus on the HA component of fibrosarcoma ECM and critically discuss its role and involved mechanisms during fibrosarcoma pathogenesis.

Advances in Cancer Research, Volume 123
ISSN 0065-230X
http://dx.doi.org/10.1016/B978-0-12-800092-2.00012-5

1. INTRODUCTION

Cancer is a pathological condition characterized by uncontrolled cell growth, tumor formation, and loss of tissue organization. There are many different forms of cancer, and each one is classified by the type of cell that is initially affected. As a cell transforms from normal to cancerous, the biological necessity for survival and perpetuation leads to fundamental changes in the cell's behavior. The formation of a primary tumor can be either caused by genetic alterations or by environmental factors. Such alterations are caused by genetic damage in oncogenes and tumor suppressor genes, affecting the regulation of basic cell functions, such as proliferation, differentiation, and apoptosis (Yokota, 2000). In continuation, malignant cells invade the surrounding tissues, leading to the formation of primary tumor through specific modulation of various tumor cell matrix interactions (Timar et al., 2002). Cancer cells often gain the ability to migrate from the primary site via lymphatic routes and through the blood stream, escaping from the normal mechanisms of control, to establish metastases at distant sites (Pantel & Brakenhoff, 2004).

Fibrosarcoma is a rare malignant mesenchymal tumor derived from fibroblasts. The main characteristic of this type of tumor is that it contains a large number of immature proliferating fibroblasts or undifferentiated anaplastic spindle cells in a storiform pattern. Fibrosarcoma belongs to the sarcoma cancer group, which are spindle cell malignancies of mesenchymal origin, and owe their name to the predominant cell line that is present within the tumor. Primary fibrosarcoma occurs in soft tissues such as muscles, connective tissues, blood vessels, and lipid storage tissues and may present diverse degrees of differentiation (low, intermediate, or high) (Kransdorf, 1995). As far as the etiology of sarcoma development is concerned, it has not yet been fully determined. In certain types of sarcomas, such as rhabdomyosarcoma and fibrosarcoma, variations between ethnic groups in combination with their occurrence in a number of heritable syndromes, suggest that genetic background plays an important role (Stiller & Parkin, 1994). The association of genetic factors with sarcoma tumorigenesis was partly based on comparative genomic hybridization that showed gains in DNA sequence copy number within tumors and high-level amplifications of 12q14–q22 in adult-type fibrosarcomas (Schmidt et al., 2002). Fibrosarcoma occurs in both men and women with equal frequency, and most commonly affects patients between the third and sixth decade of life. In these age groups, fibrosarcoma often appears as a secondary tumor due to transformation of a preexisting benign lesion, such as enchondroma, chronic osteomyelitis, or fibrous dysplasia, and its prognosis is poor (Hoshi et al., 2006; Ruggieri, Sim, Bond, & Unni, 1994; Yabut, Kenan, Sissons, & Lewis, 1988). In the case of congenital fibrosarcoma, spindle cell tumor of the soft tissue, which usually arises before the age of 2 years, the prognosis is relatively good. It is thus important to differentiate congenital fibrosarcoma from more aggressive spindle cell sarcomas that occur during childhood and may have an identical morphology but poor prognosis, similar to the adult-type fibrosarcoma (Knezevich et al., 1998). In the past, fibrosarcoma represented approximately two-thirds or more of all sarcoma diagnoses. During the last decades, improved techniques and methodology in tissue study, such as immunohistochemistry (testing of specific proteins within tumors) and cytogenetics (analysis of chromosomes), have resulted in a major reduction of the number of fibrosarcoma cases being diagnosed annually (Wingo, Tong, & Bolden, 1995). In spite of these methodological advances, fibrosarcoma is still in some cases being diagnosed by exclusion of other types of soft tissue tumors (Wibmer, Leithner, Zielonke, Sperl, & Windhager, 2010). Therefore, the identification of new positive diagnostic markers and further study of molecular targets is crucial for the development of a "targeted" therapy against fibrosarcoma (Bahrami & Folpe, 2010).

The extracellular matrix (ECM) is a complicated structure that surrounds and supports cells within tissues. Its main components are proteoglycans, collagens, glycoproteins, hyaluronan (HA), and several matrix-degrading enzymes (Schaefer & Schaefer, 2010; Theocharis, Skandalis, Tzanakakis, & Karamanos, 2010). In some tissues, mainly of epithelial origin, these molecules are located at the basement membranes creating a suprastructure that facilitates the organization of cell entities (Rathinam & Alahari, 2010). Cells interact with ECM molecules through specific receptors, mainly integrins, but also through syndecans, CD44, and receptor for hyaluronan-mediated motility (RHAMM; Maxwell, McCarthy, & Turley, 2008; Sanderson, Yang, Suva, & Kelly, 2004). The attachment of these receptors to the ECM induces signal transduction pathways that lead to a variety of functional responses directing cellular organization of both the cytoskeleton and chromatin structures (Xu, Boudreau, & Bissell, 2009). The ECM regulates almost all cellular behavior and is thus indispensable for crucial developmental processes (Rebustini et al., 2009; Wiseman et al., 2003). Importantly, the ECM establishes bidirectional interactions with growth factors (GFs), which on one hand trigger various signaling pathways but on the other hand modulate the organization of the very ECM (Rajeswari, Kesavan, & Jayakar, 2011).

During cancer progression, significant changes can be observed in the structural and mechanical properties of ECM components. The ECM provides a physical scaffold to which tumor cells attach and migrate and is required for key cellular events such as cell motility, adhesion, proliferation, invasion, and metastasis. It has been reported that alteration of cell shape and changes in the interactions with the ECM are considered as important hallmarks of cancer cells. The architecture and other physical properties of tumor-associated ECM differ from that of the normal tissue stroma. An important example in tumors of epithelial origin is that linear collagen I fibers are localized adjacent to the tumor tissue rather than being relaxed and nonoriented (Levental et al., 2009; Provenzano et al., 2006). Malfunction of the ECM regulates cancer progression by promoting cellular transformation and metastasis. Moreover, altered expression of ECM molecules also deregulates the behavior of stromal cells, promotes tumor-associated angiogenesis, and inflammation and leads to generation of a tumorigenic microenvironment (Lu, Weaver, & Werb, 2012).

Importantly, fibrosarcoma cells were shown to have a high content and turnover of ECM components including HA, proteoglycans, collagens, fibronectin, and laminin (Berdiaki et al., 2009, 2008; Piepkorn, Hovingh, & Linker, 1988). It is noteworthy that during tumorigenesis both the signaling pathways activated by *GFs* that regulate the synthesis and degradation of ECM components, as well as the reciprocal regulation of GF function by ECM components, are suppressed, facilitating the induction of oncogenesis (Berdiaki et al., 2009; Nikitovic et al., 2008, 2011).

GFs are biologically active polypeptides capable of controlling cell growth and cellular differentiation. Their function is similar to that of hormones, which are capable of emitting regulatory signals. The distinction between GFs and hormones is frequently arbitrary. Growth factor receptors (GFRs) are usually transmembrane proteins that consist of an extracellular ligand binding domain with a high specificity to natural ligands, a transmembrane part, and an intracellular catalytic domain. The latter domain has the ability to undergo autophosphorylation, usually at multiple tyrosine or serine/threonine residues through which downstream signaling is activated (Horbelt, Denkis, & Knaus, 2012; Shi & Massague, 2003). The amino acid sequence of these molecules indicates their evolution from a single ancestral protein and in this way they can be clustered into families (Mercola & Stiles, 1988).

The GF molecules act in an autocrine, paracrine, juxtacrine, or retrocrine manner and they usually show high specificity toward the ligands they bind (Kim, Turnbull, & Guimond, 2011; Mythreye & Blobe, 2009). GFR isoforms can form homodimers or heterodimers. Once stabilized by the extracellular presence of corresponding ligands, GFR dimers undergo intracellular domain activation (Heldin & Westermark, 1999; Moustakas & Heldin, 2009). The phosphorylation of GFR intracellular domains triggers a cascade of successive phosphorylation of intracellular signaling substrates. This results in the transmission of the extracellular signal to the cell nucleus, where the regulation of gene expression takes place, affecting DNA synthesis and cell growth (Hill, 2009).

2. THE ROLE OF HYALURONAN AND ITS RECEPTORS IN FIBROSARCOMA

HA is an anionic, nonsulfated glycosaminoglycan (GAG) and its molecular weight *in vivo* is in the 5000–20,000,000 Da range. It is a polymer of disaccharides composed of alternating *N*-acetylglucosamine (GlcNAc) and glucuronic acid (GlcA) units (Weigel, Hascall, & Tammi, 1997). HA differs from the other members of the GAG family because it neither contains sulfate groups nor is it covalently linked with a core protein (Tammi et al., 2008). This GAG is synthesized by three types of integral membrane proteins denominated HA synthases: HAS1, HAS2, and HAS3. These enzymes synthesize different HA sizes by repeatedly adding glucuronic acid and *N*-acetylglucosamine to the nascent polysaccharide while it is extruded through the cell membrane into the extracellular space (Schulz, Schumacher, & Prehm, 2007). Specifically, HAS1 and HAS2 produce very high molecular weight HA (2000 kDa) (Itano & Kimata, 2002). The degradation of HA within tissues, on the other hand, is performed by enzymes known as hyaluronidases (HYAL). In humans, there are at least seven types of hyaluronidase-like enzymes. HYAL hydrolyzes the β(1–4) glycoside bond between *N*-acetyl-D-glucosamine and D-glucuronic acid, which results in the production of fragments of different sizes (Jiang, Liang, & Noble, 2011). Thus, HA is susceptible to degradation, in humans, with a half-life of about 2–5 min within the bloodstream, less than 1 day in skin, and 2–3 weeks in cartilage (Fraser, Appelgren, & Laurent, 1983; Idobe, Murawaki, Ikuta, Koda, & Kawasaki, 1998).

HA plays a major role in tissue homeostasis and biomechanical integrity as it has the capacity to bind large amounts of water and form viscous gels. Furthermore, it interacts with proteoglycans and other extracellular macromolecules forming a template that is important in the assembly of extracellular and pericellular matrices (Toole, 1997). These physicochemical properties of HA are responsible for its frequent use in tissue engineering (Balazs & Denlinger, 1989).

HA also plays an important role in various pathological conditions. Thus, high levels of HA in different cancer tissues have been suggested to be connected with cancer progression (Zhang, Underhill, & Chen, 1995). Indeed, the crucial role of HA in tumor progression is well established and has been shown to affect the process of blood vessel formation, among others (Slevin et al., 2007; Slevin, Kumar, & Gaffney, 2002). Moreover, HA has the ability to modulate tissue repair *in vivo*, by regulating the specific responses of epithelial cells and fibroblasts to injury (Itano et al., 2002; Jiang et al., 2005; Zoltan-Jones, Huang, Ghatak, & Toole, 2003). Furthermore, HA can activate inflammatory cells in order to induce the immune response (McKee et al., 1996; Teriete et al., 2004).

HA interacts with a large number of HA-binding proteins, called hyaladherins, and as a result, gains new biological identities (Day & Prestwich, 2002; Toole, 2004). In more detail, HA binds to its specific cell-surface receptors, including CD44, RHAMM, and ICAM-1, activating the transduction of a wide range of intracellular signals (Turley, Noble, & Bourguignon, 2002). These HA receptor interactions are implicated in pathological conditions, as they regulate cellular processes such as morphogenesis, wound healing, and inflammation (Segura et al., 2005; Underhill, 1992).

It is noteworthy that the size of HA chains affects its biological functions. HA chains of various sizes interact differentially with specific HA receptors, affecting cellular behavior in a distinct manner. Indeed, oligosaccharides that result from HA degradation, and low-molecular-weight HA (LMWHA) are able to induce the processes of inflammation and angiogenesis. HMWHA (1000 kDa), on the other hand, is present in intact tissues and is antiangiogenic as well as immunosuppressive (Klagas et al., 2009; McKee et al., 1997; West, Hampson, Arnold, & Kumar, 1985).

2.1. HA production in mesenchymal tumors

A significant number of studies show that HA deposition is elevated in various types of cancer tissues including colon, breast, lung, and prostate cancer (Lokeshwar et al., 2001; Pirinen et al., 2001; Toole, Wight, & Tammi, 2002). High amounts of HA have been linked with tumor progression since HA has been found to be overexpressed in malignant tumors as compared to benign tissues (Toole et al., 2002). This is also the case in soft tissue tumors as illustrated by a significant difference between the benign dermatofibroma, where specimens were poorly stained for HA within the tumor stroma, and the malignant dermatofibrosarcoma protuberans, in which there was strong HA staining (Calikoglu, Augsburger, Chavaz, Saurat, & Kaya, 2003). Importantly, activated stromal cells adjacent to the cancer cells are able to produce a tissue structure rich in HA in order to promote tumor growth, as well as to secrete factors that enhance cancer cell migration into the new matrix (Edward, Gillan, Micha, & Tammi, 2005; Knudson & Toole, 1988).

Thorough investigation of HA metabolism has revealed a central role for HA in cancer cell functions. It is widely accepted that HAS mRNA levels determine the synthesis of HA (Karvinen et al., 2003; Pasonen-Seppanen et al., 2003). In the case of fibrosarcoma (Berdiaki et al., 2009, 2008) and in other mesenchymal type tumors as well (Nikitovic et al., 2006), HAS mRNA expression is directly correlated to content of HA. As far as the degradation of HA is concerned, elevated extracellular levels of partially catabolized HA oligomers are found in certain malignancies (Stern, 2008). Indeed, evidence suggests that in many disease processes including cancer, the HA-degrading action of HYALs is linked with an increase in the permeability of connective tissues and with the decrease in the viscosity of body fluids (Girish & Kemparaju, 2007; Stair-Nawy, Csoka, & Stern, 1999).

Interestingly, in a human HT1080 fibrosarcoma cell line, the overexpression of HAS2 promoted anchorage-independent growth and tumorigenicity of the cells (Kosaki, Watanabe, & Yamaguchi, 1999). In addition, the migration ability of these cells was increased in response to high HAS1 and HAS2 expression levels. A decrease in HYAL-2 expression, on the other hand, was accompanied by a significant reduction in the migratory capacity of HT1080 cells (Berdiaki et al., 2009) In the case of dermatofibrosarcoma, protuberans an injection of HYAL within the lesion area was linked with decreased margin width and the wound after the operation was smaller than initially estimated (Menon, Smith, Crittenden, & Skelton, 1999; Smith et al., 1998). Furthermore, HYAL, in the presence of actinomycin D, is suggested to increase the resistance of the L929 fibrosarcoma cell line to tumor necrosis factor and Fas cytotoxicity (Chang, 1995).

HA also supports tumorigenesis by promoting tumor angiogenesis. Stromal HA accumulation promotes the recruitment of monocytes and macrophages, which are important for angiogenesis. These inflammatory cells bind within the tissues by adhesion to specific HA structures, allowing tumor-associated macrophages to release angiogenic and lymphangiogenic GFs (de la Motte, Hascall, Drazba, Bandyopadhyay, & Strong, 2003; Lin et al., 2006). Certain disease states, including inflammation and cancer, through increased reactive oxygen species (ROS) generation, stimulate HYAL expression on one hand whereas generated ROS can chemically degrade HMWHA to low-molecular-weight fragments (reviewed in Corsini, Galbiati, Nikitovic, & Tsatsakis, 2013; Nikitovic, Corsini, Kouretas, Tsatsakis, & Tzanakakis, 2013). These HA fragments can activate specific HA-binding proteins, including CD44 and Toll-like receptor 4 (TLR4), whose expression is upregulated in vascular disease, to enhance actin cytoskeletal reorganization and inhibition of endothelial cell–cell contacts (Lennon & Singleton, 2011).

Furthermore, it has recently been demonstrated that the progression and vascularization of carcinomas may be dependent on the HA production by epithelial or stromal cells. There is striking evidence that stromal accumulation of HA is correlated with the formation of an increased blood and lymphatic vasculature. Indeed, without the stromal cells, the tumors progressed more slowly because of their reduced blood and lymphatic vessel content (Koyama et al., 2007, 2008). The major role of stroma-derived HA in tumor vascularization was demonstrated when HAS2 null fibroblasts were co-implanted into nude mice with epithelial tumor cells, and the resultant tumors presented attenuated tumor angiogenesis and lymphangiogenesis with impaired macrophage activation (Kobayashi et al., 2010). These data support the hypothesis that the quantity as well as the size of HA chains in tumors play a fundamental role in the regulation of cancer cell processes during the different stages of tumor pathogenesis.

The majority of studies to date suggest that LMWHA fragments are pro-proliferative, -angiogenic, and increase motility. In contrast, results obtained in the fibrosarcoma cell model suggest that HMWHA may be pro-proliferative and enhance motility. This may be an unusual property of fibrosarcoma tumors that is opposite effects observed in tumors that originate within the epithelial compartment. Also, these results do not address the possibility that an accelerated rate of HA synthesis suggested to be pro-tumorigenic in fibrosarcoma may depend upon subsequent breakdown or turnover of the HA polymers to lower molecular weight oligosaccharides. This could be due to (1) HYALs activity within the fibrosarcoma cells, (2) other stromal cells, or (3) macrophages and other inflammatory cells that are recruited to the site and release ROS or HYAL upon arrival. These oligosaccharides could, in turn, be the signal that promotes angiogenesis, fibrosarcoma motility, and/or proliferation. The role of HA and its receptors in fibrosarcoma cell function is presented schematically in Fig. 12.1.

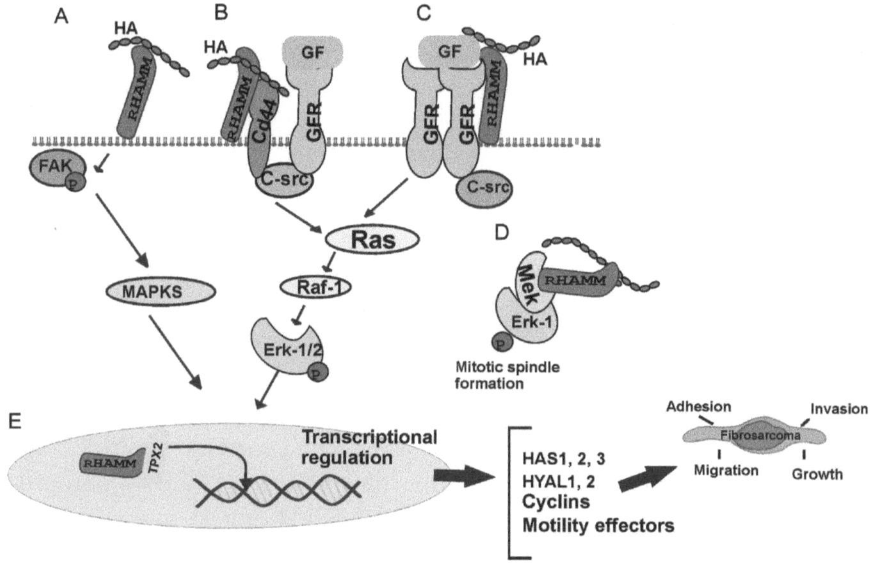

Figure 12.1 The role of HA and its receptors in fibrosarcoma cell functions. (A) HA-RHAMM interactions in a FAK/MAPK-dependent manner modulate transcription of cell cycle modulators, (B) RHAMM/CD44/HA complex formation facilitates GF/GFR binding and downstream signaling, (C) RHAMM facilitates GFR dimerization and respective signaling cascade, (D) cytoplasmic RHAMM/MEK/ERK complex regulates mitotic spindle formation, and (E) nuclear RHAMM/TPX2 complex formation directly regulates target gene transcription to affect key fibrosarcoma cell functions. (See the color plate.)

2.2. RHAMM in fibrosarcoma development and progression

RHAMM was first identified as a soluble protein in the cell culture of subconfluent fibroblasts (Turley, 1982) and was subsequently cloned from mesenchymal cells (Hardwick et al., 1992). RHAMM is distributed on the cell surface, within the cytoplasm, in the nucleus, or secreted to the ECM (Assmann, Marshall, Fieber, Hofmann, & Hart, 1998; Entwistle, Hall, & Turley, 1996). Various RHAMM isoforms are produced by alternative splicing and these transcript variants are suggested to be expressed in a cell type specific manner (Cheung, Cruz, & Turley, 1999). RHAMM protein is glycosylphosphatidylinositol anchored to the cell membrane as it lacks a typical transmembrane domain (Hardwick et al., 1992; Turley, 1992). Cell-surface RHAMM is not highly expressed in normal tissues but is overexpressed in many advanced cancers (Adamia, Pilarski, Belch, & Pilarski, 2013; Assmann et al., 1998; Turley et al., 2002). The overexpression of RHAMM during tumor development indicates that it may act as a factor of prognostic significance (Assmann et al., 2001; Wang et al., 1998).

At the cell surface, RHAMM affects HA- and GF-induced MAPK (ERK1, 2) signaling via associations with transmembrane HA receptors such as CD44 as well as PTK receptors including PDGFR and Ron-receptor tyrosine kinase (Hamilton et al., 2007; Turley et al., 2002). This signaling function of cell membrane RHAMM is required for tumor cell motility and invasion (Hamilton et al., 2007; Tolg et al., 2006). On the contrary, intracellular RHAMM binds to actin filaments, podosomes, the centrosome, microtubules, and the mitotic spindle (Assmann, Jenkinson, Marshall, & Hart, 1999; Maxwell, Keats, Belch, Pilarski, & Reiman, 2005; Turley et al., 2002), where it regulates crucial cellular processes in tumorigenesis (Maxwell et al., 2005; Mohapatra, Yang, Wright, Turley, & Greenberg, 1996). Indeed, Telmer, Tolg, McCarthy, and Turley, (2011) have recently suggested that intracellular RHAMM can form complexes with ERK2, MEK1, and ERK1/2 substrates due to its direct binding to ERK1 and proposed a model in which RHAMM acts as a scaffold protein to specific substrates, controlling activation and targeting of ERK1/2 (Telmer et al., 2011). This evidence supports the idea that the function of RHAMM strictly depends on its expression and cellular localization.

RHAMM was found to regulate the functions of fibroblasts, especially those correlated to the malignant transformation of these mesenchymal cells. Thus, restricting cell-surface RHAMM to the extracellular compartment by linking recombinant protein to beads, combined with expression of mutant active mitogen-activated kinase 1 (Mek1), rescued aberrant signaling through CD44–ERK1/2 complexes in scratch wound assays of RHAMM (−/−) fibroblasts. The processes of fibroblast migration and differentiation, and ERK1/2 activation, are also found to be defective during repair of RHAMM(−/−) excisional skin wounds, causing aberrant granulation tissue *in vivo*. Thus, in such wound repair functions, RHAMM may be an important regulator of CD44–ERK1/2 fibroblast motogenic signaling. Furthermore, in another study on fibroblasts, it was shown that native HA-activated NF-κB and activating protein 1, during wound repair. With the use of siRNA against CD44 it was demonstrated that this HA receptor is partly implicated in the effects, without ruling out the involvement of other receptors such as RHAMM (David-Raoudi et al., 2008). These two HA receptors interact and are involved in many common signaling pathways including those of VEGF, HGF, Src, ERK1/2, and Fos that regulate cell migration. Other CD44/RHAMM networks that are associated with proliferation and cancer include MAFG, DYNLL1, MAFK, and FAM83D, which are molecules implicated in the regulation of mitotic spindle formation (Turley & Naor, 2012). It has been proposed that RHAMM and CD44 receptors cooperate to induce cell growth within cementifying fibromas. More specifically, RHAMM interacts with ERK, increasing the proliferative ability of these cancer cells through a mechanism that involves the interaction of CD44 with the epidermal growth factor receptor (EGFR; Itano et al., 2002). Moreover, cell-surface RHAMM can facilitate invasive functions similar to CD44 and can even substitute for CD44 functions (Nedvetzki et al., 2004).

The major role that RHAMM seems to play in the regulation of fibroblastoid cell functions is further illustrated by the following study. It has been suggested that RHAMM targets and anchors MEK1/ERK1/2 to tubulin and that these MAPKs phosphorylate the tubulin-associated proteins that regulate microtubule dynamics in fibroblasts (Tolg et al., 2010). Microtubule formation has been correlated with functions that affect cell cycle progression and motility/invasion leading to cancer progression. These results support the idea that microtubules are a potential oncogenic target of transforming RHAMM protein forms, such as RHAMMΔ163, and are relevant targets of investigation in fibrosarcoma pathogenesis. The significant role of RHAMM in fibroblast cell motility is illustrated by a report which demonstrates that RHAMM (−/−) fibroblasts fail to resurface scratch wounds >3 mm or invade HA-supplemented collagen gels in culture (Tolg et al., 2006).

Furthermore, the interaction of HA with its RHAMM receptor is considered to be fundamental in the process of fibrosarcoma tumorigenesis. Early studies have shown that the overexpression of RHAMM in fibroblasts is transforming, required for H-ras transformation (Hall et al., 1995), thus implying that RHAMM has a unique role in orchestrating events that are essential for the induction of transformation. In an early report, RHAMM/HA signaling was found to be necessary for the stimulation of fibrosarcoma cell motility induced by transforming growth factor-beta 1 (TGFβ1). Indeed, signaling is performed through the formation of the RHAMM–HA complex as antibodies that inhibit RHAMM–HA-binding inhibit TGFβ1-induced increase in fibrosarcoma cell migration (Samuel et al., 1993). TGFβ1 is also capable of regulating RHAMM expression. Indeed, it has been demonstrated that TGFβ1 can induce RHAMM mRNA expression by stimulating multiple protein interactions at a unique cis-element in the 3′-untranslated region of the receptor (Amara, Entwistle, Kuschak, Turley, & Wright, 1996). Moreover, the expression of a dominant suppressor mutant of RHAMM resulted in a reverse in the transformation induced by ras, stabilizing focal adhesions.

Thus, it is suggested that RHAMM acts on focal adhesions downstream of ras or via a parallel pathway that converges at the level of ras (Hall & Turley, 1995). RHAMM has also been proposed to be involved in the development of fibromatosis, which is an aggressive mesenchymal tumor (Tolg, Poon, Fodde, Turley, & Alman, 2003).

Importantly, key signaling pathways implicated in fibrosarcoma pathogenesis appear to be under RHAMM regulation. Tyrosine kinase pp60c-src is associated with RHAMM and is essential for RHAMM-mediated cell motility whereas a strong correlation between specific ras mutations and the fibrosarcoma cell behavior has been reported (Guerrero et al., 2002). It is important to note that in this cancer type, the presentation of K-ras mutations is quite frequent. Moreover, an association between specific sarcoma types and mutations in codon 13 (Hong et al., 2000; Yoo & Robinson, 1999) and in codon 12 has been established (Gill et al., 1991; Hill et al., 1997). In addition, K-ras 12-derived tumors were shown to resemble fibrosarcomas, whereas K-ras 13-derived tumors were shown to resemble malignant fibrous histiocytomas (MFH) (Gill et al., 1991). The K12 tumors have a different expression profile or different activation pattern of other Ras downstream pathways, JNK, MAPK, AKT, Bcl-2, FAK, and cyclin B1, which could be correlated to their functional differences (Guerrero et al., 2002). The above investigations underline the significance of the ras-signaling pathways in mesenchymal tumor progression.

Regarding the effects of HA on fibrosarcoma cell adhesion and the mechanism of its action, it was shown that HA/RHAMM interaction regulates fibrosarcoma cell adhesion through the activation of FAK and ERK1/2 signaling pathways (Fig. 12.1; Kouvidi et al., 2011). The addition of HA preparations of various molecular weight to fibrosarcoma (HT1080) cells, led to an HA size-dependent modulation of their migration capacity (Berdiaki et al., 2009). These results correspond well to an earlier study finding that in ras-transformed fibroblasts, HA alters focal adhesion organization through RHAMM by signaling transient protein–tyrosine phosphorylation, regulating the motility of the cells (Hall, Wang, Lange, & Turley, 1994). The major signaling pathway that determines the tumorigenic and metastatic phenotype of these cells involves FAK phosphorylation, followed by net dephosphorylation and focal adhesion turnover, inducing cell motility (Hall et al., 1994).

RHAMM has additional effects on the regulation of expression of cell cycle mediators, which is important because studies demonstrate a correlation between cell cycle mediator expression and fibrosarcoma progression. In the case of H-ras-transformed fibroblasts, soluble RHAMM provokes mitotic arrest by suppressing Cdc2 and cyclin B1 expression (Mohapatra et al., 1996). Soluble RHAMM arrests ras-transformed fibroblasts at G2/M without affecting their progression through S-phase (Oda et al., 2003) *via* RHAMM–HA binding. Variations of cell cycle regulators could be used as potential prognostic factors in related myxofibrosarcoma (Oda et al., 2003). Low-grade myxofibrosarcoma (MFS) differs from low-grade fibromyxoid sarcoma (LGFMS) in that they have diverse metastatic properties. These two types could be distinguished through the comparison of conventional clinicopathological and immunohistochemical features and the assessment of the immunohistochemical expression of MIB-1, cyclin E, p21, and p27 (Oda et al., 2003). Importantly, Skp2 and cyclin-dependent kinase subunit 1 (Cks1) are implicated in posttranscriptional degradation of the tumor suppressor factor p27 (Kip1). The latter could be of high prognostic significance as it interacts with cell cycle regulators including Skp2, which is overexpressed in myxofibrosarcomas and is highly representative of the biological aggressiveness in this type of tumor (Huang et al., 2006). A novel pool of AP-1 target genes was identified by microarray analysis, including the tumor suppressor TSCL-1, and regulators of actin cytoskeletal dynamics such as the gelsolin-like actin capping protein CapG. The fact that AP-1 regulates the expression of genes involved in tumor cell motility and cytoskeletal morphology in a human fibrosarcoma cell line, HT1080, advances the study of tumorigenic cell locomotion (Bahassi el et al., 2004). In addition, RHAMM plays a significant role in the direct modulation of cell growth. It has been demonstrated that RHAMM regulates proliferation of cells with sparse cell–cell contacts, and thus it could be speculated that RHAMM blockade could function as a potential therapeutic target for this neoplasm, whose treatment continues to be a challenge (Tolg et al., 2003). Moreover, in osteoblastic MC3T3-E1 cells, RHAMM overexpression induces proliferation and suppresses differentiation through phosphorylation of ERK1/2. Thus, it can be proposed that the imbalance in the processes of differentiation to proliferation induced by RHAMM overexpression may be associated with the pathogenesis of human cementifying fibroma or other types of bone neoplasms (Hatano et al., 2012).

Importantly RHAMM can be used as a putative therapeutic target in cancer. For example, TCR-transgenic lymphocytes specific for RHAMM decrease the size of the tumor *in vivo* in various solid and leukemia tumor models (Spranger et al., 2012). Furthermore, in acute myeloid leukemia patients, immunotherapies like peptide vaccination or adoptive transfer of RHAMM-specific T cells could improve their immune response and their survival (Casalegno-Garduno et al., 2012). Moreover, the antitumor activity of HA synthesis inhibitor 4-methylumbelliferone was shown to be performed through the downregulation of proliferation, invasion, and motility in *in vitro* prostate cancer models (Lokeshwar et al., 2010). Finally, it has been demonstrated *in vivo* that sulfated HA decreased tumor growth by acting as a physical obstacle in the formation of a complex between phosphoinositide 3-kinase (PI3K) and HA receptors. HA not only inhibited PI3K signaling but also downregulated the transcription of HA receptors, CD44, and RHAMM (Benitez et al., 2011).

3. GF SIGNALING AND ECM ORGANIZATION IN FIBROSARCOMA PATHOGENESIS

There is a complex cell cycle network of signaling pathways that interact with each other and further control cell biological functions. GFs play a central role in the activation of downstream signaling cascades by acting early in G1 phase and regulating cell growth (Berridge, 2012). The GFRs regulate the activation of molecules that are members of the three main signaling pathways: the phosphoinositide 3-kinase (PI-3K) family, the mitogen-activated protein kinase (MAPK) family and the Smad protein family. PI 3-kinases are enzymes with the ability to phosphorylate the position 3 hydroxyl group of the inositol ring upon activation by GFRs (Cantley, 2002). The MAPK pathway includes many proteins called extracellular signal-regulated kinases (ERK), which add phosphate groups to downstream substrates mediating their activation (Chang & Karin, 2001; Heldin & Westermark, 1999). SMADs are intracellular proteins that mainly transmit extracellular signals, originating from TGFβ superfamily members, to the nucleus, thereby activating downstream gene transcription (Heldin, Miyazono, & ten Dijke, 1997; Hill, 2009).

The GF signaling pathways have been shown to interact with other ECM receptors including integrins, RHAMM, or CD44 forming thus a GF–ECM synergy. GFs have the ability to alter HA metabolism and thereby regulate cancer cell behavior. FGF-2 stimulates, in a cell-specific manner the migration capacity of fibrosarcoma cells by decreasing HYAL-2 expression in HT1080 cells and by increasing HAS1 and HAS2 expression (Berdiaki et al., 2009), whereas in B6FS fibrosarcoma cells, the addition of TGFB2 and PDGF-BB increased HA production (Berdiaki et al., 2008). Therefore, we will focus on the cooperation of the different GFs with HA and its receptors and their role in the pathogenesis of fibrosarcoma.

3.1. Transforming growth factor-β

TGFβ is a polypeptide that plays an important role in development, ECM organization, cellular differentiation, homeostasis, immune regulation, and tissue repair. TGFβ is secreted by many cell types and has the ability to regulate their behavior (Border & Noble, 1994; Kubo, Ihn, Yamane, & Tamaki, 2006). This GF has three different protein isoforms TGFβ1, TGFβ2, and TGFβ3 that bind to the TGFβ receptors I, II, and III (TGFβRI, TGFβRII, TGFβRIII, or betaglycan) (Lutz & Knaus, 2002; Massague, 2000). Type I (RI) and type II (RII) receptors have a highly conserved intracellular serine/threonine binding domain and they are both required for the regulation of the numerous effects of TGFβ (Lopez-Casillas, Wrana, & Massague, 1993; Massague, 1998, 2000). TGFβRIII, which is structurally different as it lacks a catalytic domain, acts as a coreceptor increasing ligand binding and enhancing TGFRI and II downstream signaling through the Smad pathway (Lopez-Casillas et al., 1993). Notably, TGFβ1 has a dual role in the process of tumorigenesis. Specifically, it is able to suppress tumor growth in the early stages but it can also stimulate cancer cellular functions such as invasion, metastasis, and angiogenesis by its autocrine and paracrine actions (Muraoka-Cook, Dumont, & Arteaga, 2005; Pardali & Moustakas, 2007; Saunier & Akhurst, 2006). TGFβ signaling is involved in the pathogenesis of fibrosis and the protein expression levels of TGFβRI and TGFβRII are diagnostic and prognostic for dermatofibromas (Border & Noble, 1994; Kubo et al., 2006; Massague, 1998, 2000).

TGFβ isoforms are implicated in the regulation of HA metabolism of fibrosarcoma cells in various ways. Overproduction of HA through the expression of a mammalian HAS2 promotes cell proliferation leading to the enhancement of tumorigenicity of HT1080 cells (Kosaki et al., 1999). Furthermore, it has been established that TGFβ1 stimulates multiple protein interactions at a unique cis-element in the 3′-untranslated region of RHAMM mRNA thus regulating its expression (Amara et al., 1996). Evidence shows that TGFβ induces locomotory response of fibrosarcoma cells by triggering the transcription, synthesis, and membrane expression of RHAMM and the secretion of HA. Thus, TGFβ1 is a necessary stimulator of cell motility through the RHAMM/HA pathway (Samuel et al., 1993). This is noteworthy in view of the fact that cells overexpressing RHAMM show elevated cell locomotion and focal adhesion loss, resemble Ras-transformed fibroblasts, and promote the tumorigenic phenotype leading to metastatic fibrosarcoma (Hall et al., 1995). In addition, it was shown that HA regulates fibrosarcoma cell adhesion via its interaction with RHAMM,

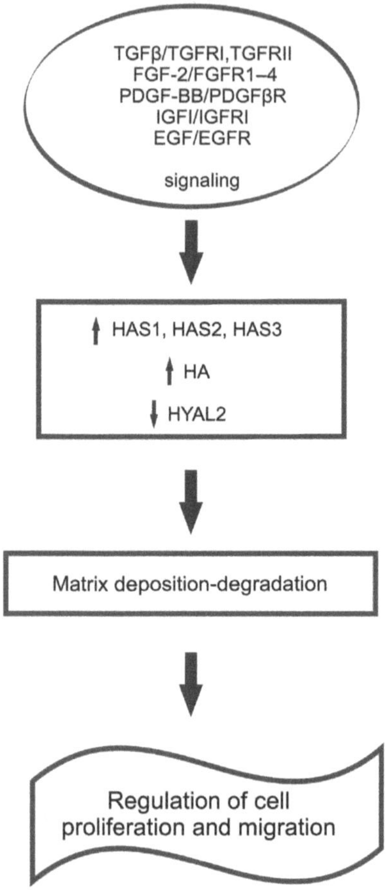

Figure 12.2 GF/GFR signaling through the modulation of HA metabolism and HA receptor expression affect fibrosarcoma cell function.

activating the FAK and ERK1/2 signaling pathways (Kouvidi et al., 2011). All the above data suggest that the RHAMM/HA complex is one of the main perpetrators of the TGFβ1 signaling axis in fibrosarcoma. The regulation of HA metabolism in fibrosarcoma by GF signaling is depicted in Fig. 12.2.

3.2. Fibroblast growth factor

Fibroblast growth factors or FGFs are heparin-binding proteins that signal through their respective FGF receptors (FGFR1–4) mediating different cellular processes (Eswarakumar, Lax, & Schlessinger, 2005; Jaye, Schlessinger, & Dionne, 1992). Interactions of the FGF–FGFRs with cell-surface-associated heparin sulfate proteoglycans modulate their downstream signaling. Experiments conducted on the fibrosarcoma cell model have shown that endogenous expression of FGF isoforms and their receptors promotes fibrosarcoma angiogenesis and metastasis (Berdiaki et al., 2009; Ding, Liu, Sun, Fenton, & Okunieff, 2002). Moreover, studies on the HT1080 human fibrosarcoma cell line (Hill, 2009) have demonstrated the key role of FGF-2 in the regulation of these cells' ECM component expression. Specific modulations in the ECM content were found to be correlated to changes in important biological functions such as HT1080 cell migration and angiogenesis (Berdiaki et al., 2009; Lee et al., 2010).

Specifically, it has been demonstrated that the HA content was increased in B6FS fibrosarcoma cells after the addition of FGF-2, probably due to the enhanced expression of both HAS1 and HAS2 (Berdiaki et al., 2009). Moreover, the treatment of HT1080 fibrosarcoma cells with FGF-2 increased the expression of HAS1 and HAS2 isoforms, while HYAL-2 expression was decreased, enhancing HMWHA deposition to the pericellular matrix. This induction of HA production was correlated with a significant decrease of HT1080 cell migration capacity. The motility of the cells was found to be regulated mainly through HAS1, as concluded by siRNA knockdown of HAS1 and HAS2. Furthermore, HA treatment of HT1080 cells increased their proliferation ability, demonstrating the existence of an FGF2-dependent mechanism in fibrosarcoma cell function (Berdiaki et al., 2009).

3.3. Platelet-derived growth factor

Platelet-derived growth factor (PDGF) is a dimeric glycoprotein composed of two A (PDGF-AA) or two B (PDGF-BB) chains or a combination of those two (PDGF-AB). Recently two new members of homodimers PDGF-CC and PDGF-DD have been added to the PDGF family (Heldin, Eriksson, & Ostman, 2002; Li & Eriksson, 2003). These PDGF ligands bind two structurally related tyrosine kinase receptors, α and β, which in turn mediate their biological action (Cao, 2005; Heldin & Westermark, 1999).

PDGF signaling regulates cell growth and metastasis, and PDGF isoforms are potent angiogenic factors (Mathiak, Yenisey, Grant, Sharma, & Iozzo, 1997). PDGFs are found to be overexpressed in numerous malignant tissues (Cao, 2005; Pietras, Sjoblom, Rubin, Heldin, & Ostman, 2003). A characteristic example is the overexpression of PDGF-BB in a murine fibrosarcoma model that is unresponsive to exogenous PDGFs. In this model, PDGF-BB overexpression resulted in the induction of tumor angiogenesis and thus increased tumor growth (Cao, 2005; Cao et al., 2004; Vincent & Rafii, 2004).

PDGF-BB was found to regulate the deposition of HA in a distinct manner among fibrosarcoma cell lines of different differentiation status (Berdiaki et al., 2008). PDGF has been shown to physically interact with GAGs, which are abundant in the fibrosarcoma cell microenvironment and regulate the growth of fibrosarcoma cell lines (Fager et al., 1995; Fthenou et al., 2006). In more detail, Fthenou et al. have demonstrated significant co-stimulatory effects of chondroitin sulfate A and PDGF-BB in the modulation of the growth of HT1080 and B6FS fibrosarcoma cell lines (Fthenou et al., 2006). This co-stimulatory effect was perpetrated through more efficient signaling of tyrosine kinase receptors (Fthenou et al., 2006). On the other hand, it has been observed that HA suppresses PDGF beta-receptor activation, regulating in this manner the migration capacity of primary human dermal fibroblasts. It has been suggested that HA acts through CD44, which has been shown to colocalize and to interact with PDGF beta-receptor (Li, Heldin, & Heldin, 2006). Interestingly, the inhibitory effect of HA on PDGF beta-receptor activation was not seen in the presence of the tyrosine phosphatase inhibitor pervanadate. Therefore Li et al. (2006) suggest that HA suppresses PDGF beta-receptor activation by recruiting a CD44-associated tyrosine phosphatase to the receptor. Moreover, PDGF-BB also increased the amount and activity of HAS2 protein, but not of HYAL1 and HYAL2 proteins through MEK1/2 (MAPK ERK 1/2, PI3K (phosphoinositide 3-kinase), and NF-kappa B signaling in fibroblasts) (Li, Asteriou, Bernert, Heldin, & Heldin, 2007). These studies raise the interesting possibility that extracellular GAGs, likely in an expression dependent manner, regulate PDGFR activation and downstream signaling in fibroblastoid type cells.

3.4. Insulin-like growth factor and epidermal growth factor

The insulin-like growth factors (IGF1 and IGF2) are proteins with strong similarity to insulin. They bind to their two specific receptors: insulin-like growth factor receptor I (IGFRI) and insulin-like growth factor receptor II (IGFRII), affecting tumor growth (Werner & LeRoith, 1996). Moreover, they activate various signaling cascades mainly through their interaction with IGFRI to modulate chemotaxis, mitosis, apoptosis, and differentiation (Baserga, 2000; Jones & Clemmons, 1995; Morrison, Tognon, Garnett,

Deal, & Sorensen, 2002). It has been shown that IGF-I is able to induce the growth of tumors directly by stimulating their growth rates. This resulted from a study conducted in athymic nude mice in which stimulation of tumor growth by recombinant human IGF-I was dependent on the dose and the level of IGF-I receptor expression (Butler et al., 1998). Importantly, IGF-1 was found to affect the metabolism of HA in fibroblastoid cells where it increased HAS1, HAS2, and HAS3 mRNA levels, as well as HA production (Kuroda, Utani, Hamasaki, & Shinkai, 2001).

Epidermal growth factor (EGF) is a low-molecular-weight polypeptide that specifically binds to its cell membrane EGFRs. EGFR is a member of a group of transmembrane glycoprotein receptors, including EGFR (ErbB-1), ErbB-2 (HER2), ErbB-3 (HER3), and ErbB-4 (HER 4) (Bargmann, Hung, & Weinberg, 1986; Kraus, Issing, Miki, Popescu, & Aaronson, 1989; Lin et al., 1984). Binding of EGF triggers the formation of EGFR homo- or heterodimers, allowing in turn the intracellular tyrosine kinase domains to autophosphorylate (Ferguson et al., 2003; Garrett et al., 2002; Zhang, Gureasko, Shen, Cole, & Kuriyan, 2006). It has been established that EGF upon binding to the EGFRs stimulates various cellular processes such as proliferation, differentiation, motility, and adhesion (Carpenter, 1984; Gialeli, Theocharis, Kletsas, Tzanakakis, & Karamanos, 2013; Li, Huang, Jiang, & Frank, 2008; Slovak et al., 1991). EGF signaling plays an important role in fibrosarcoma pathogenesis. Evidence shows that EGFR expression and activation in soft tissue sarcoma cell lines is linked with malignant progression (Fiorelli et al., 2011; Kwak et al., 2006; Sato et al., 2005). Studies correlating EGFR expression with patient survival suggest that EGFR could be a candidate prognostic factor in soft tissue sarcomas (Sato et al., 2005). Interestingly, HA-mediated CD44-EGFR colocalization and signaling through ERK1/2 is obligatory in fibroblast functionality and found to be impaired with aging (Midgley, Bowen, Phillips, & Steadman, 2014). Little is known about the regulation of HA metabolism by IGF and EGF. Therefore, further studies are necessary in order to reveal the potential role of these GFs in the pathogenesis and progression of fibrosarcoma.

4. CONCLUDING REMARKS

In conclusion, this review has highlighted the roles that RHAMM/HA have in fibrosarcoma development and progression. These molecules contribute to fibrosarcoma tumorigenesis and in various facets of fibrosarcoma and stromal cell biological functions. GFR, GF, and interactions have crucial roles in the regulation of HA deposition, organization, and remodeling as well as RHAMM expression during fibrosarcoma development. On the other hand the various HA fragments themselves are essential cellular effectors that integrate and respond to the tumor environmental stimuli through the regulation of GF signaling pathways. Therefore, integration of GF–HA signaling regulates key fibrosarcoma functions and is implicated in fibrosarcoma tumorigenesis. The intrinsic characteristics of RHAMM/HA, GFs and their receptors, and the intracellular signaling cascades triggered by their interactions as described in this review, provide a rationale for targeted pharmacologic interventions in conditions and diseases related to abnormalities in ECM such as fibrosarcoma.

REFERENCES

Adamia, S., Pilarski, P. M., Belch, A. R., & Pilarski, L. M. (2013). Aberrant splicing, hyaluronan synthases and intracellular hyaluronan as drivers of oncogenesis and potential drug targets. *Current Cancer Drug Targets, 13*(4), 347–361.

Amara, F. M., Entwistle, J., Kuschak, T. I., Turley, E. A., & Wright, J. A. (1996). Transforming growth factor-beta1 stimulates multiple protein interactions at a unique cis-element in the 3'-untranslated region of the hyaluronan receptor RHAMM mRNA. *The Journal of Biological Chemistry, 271*(25), 15279–15284.

Assmann, V., Gillett, C. E., Poulsom, R., Ryder, K., Hart, I. R., & Hanby, A. M. (2001). The pattern of expression of the microtubule-binding protein RHAMM/IHABP in mammary carcinoma suggests a role in the invasive behaviour of tumour cells. *The Journal of Pathology, 195*(2), 191–196. http://dx.doi.org/10.1002/path.941.

Assmann, V., Jenkinson, D., Marshall, J. F., & Hart, I. R. (1999). The intracellular hyaluronan receptor RHAMM/IHABP interacts with microtubules and actin filaments. *Journal of Cell Science, 112*(Pt 22), 3943–3954.

Assmann, V., Marshall, J. F., Fieber, C., Hofmann, M., & Hart, I. R. (1998). The human hyaluronan receptor RHAMM is expressed as an intracellular protein in breast cancer cells. *Journal of Cell Science, 111*(Pt 12), 1685–1694.

Bahassi el, M., Karyala, S., Tomlinson, C. R., Sartor, M. A., Medvedovic, M., & Hennigan, R. F. (2004). Critical regulation of genes for tumor cell migration by AP-1. *Clinical & Experimental Metastasis, 21*(4), 293–304.

Bahrami, A., & Folpe, A. L. (2010). Adult-type fibrosarcoma: A reevaluation of 163 putative cases diagnosed at a single institution over a 48-year period. *The American Journal of Surgical Pathology, 34*(10), 1504–1513. http://dx.doi.org/10.1097/PAS.0b013e3181ef70b6.

Balazs, E. A., & Denlinger, J. L. (1989). Clinical uses of hyaluronan. *Ciba Foundation Symposium, 143*, 265–275, discussion 275-280, 281-265.

Bargmann, C. I., Hung, M. C., & Weinberg, R. A. (1986). The neu oncogene encodes an epidermal growth factor receptor-related protein. *Nature, 319*(6050), 226–230. http://dx.doi.org/10.1038/319226a0.

Baserga, R. (2000). The contradictions of the insulin-like growth factor 1 receptor. *Oncogene, 19*(49), 5574–5581. http://dx.doi.org/10.1038/sj.onc.1203854.

Benitez, A., Yates, T. J., Lopez, L. E., Cerwinka, W. H., Bakkar, A., & Lokeshwar, V. B. (2011). Targeting hyaluronidase for cancer therapy: Antitumor activity of sulfated hyaluronic acid in prostate cancer cells. *Cancer Research, 71*(12), 4085–4095. http://dx.doi.org/10.1158/0008-5472.CAN-10-4610.

Berdiaki, A., Nikitovic, D., Tsatsakis, A., Katonis, P., Karamanos, N. K., & Tzanakakis, G. N. (2009). bFGF induces changes in hyaluronan synthase and hyaluronidase isoform expression and modulates the migration capacity of fibrosarcoma cells. *Biochimica et Biophysica Acta, 1790*(10), 1258–1265. http://dx.doi.org/10.1016/j.bbagen.2009.06.013.

Berdiaki, A., Zafiropoulos, A., Fthenou, E., Katonis, P., Tsatsakis, A., Karamanos, N. K., et al. (2008). Regulation of hyaluronan and versican deposition by growth factors in fibrosarcoma cell lines. *Biochimica et Biophysica Acta, 1780*(2), 194–202. http://dx.doi.org/10.1016/j.bbagen.2007.10.005.

Berridge, M. J. (2012). Calcium signalling remodelling and disease. *Biochemical Society Transactions, 40*(2), 297–309. http://dx.doi.org/10.1042/BST20110766.

Border, W. A., & Noble, N. A. (1994). Transforming growth factor beta in tissue fibrosis. *New England Journal of Medicine, 331*(19), 1286–1292. http://dx.doi.org/10.1056/NEJM199411103311907.

Butler, A. A., Blakesley, V. A., Poulaki, V., Tsokos, M., Wood, T. L., & LeRoith, D. (1998). Stimulation of tumor growth by recombinant human insulin-like growth factor-I (IGF-I) is dependent on the dose and the level of IGF-I receptor expression. *Cancer Research, 58*(14), 3021–3027.

Calikoglu, E., Augsburger, E., Chavaz, P., Saurat, J. H., & Kaya, G. (2003). CD44 and hyaluronate in the differential diagnosis of dermatofibroma and dermatofibrosarcoma protuberans. *Journal of Cutaneous Pathology, 30*(3), 185–189.

Cantley, L. C. (2002). The phosphoinositide 3-kinase pathway. *Science, 296*(5573), 1655–1657. http://dx.doi.org/10.1126/science.296.5573.1655.

Cao, Y. (2005). Direct role of PDGF-BB in lymphangiogenesis and lymphatic metastasis. *Cell Cycle, 4*(2), 228–230.

Cao, R., Bjorndahl, M. A., Religa, P., Clasper, S., Garvin, S., Galter, D., et al. (2004). PDGF-BB induces intratumoral lymphangiogenesis and promotes lymphatic metastasis. *Cancer Cell, 6*(4), 333–345. http://dx.doi.org/10.1016/j.ccr.2004.08.034.

Carpenter, G. (1984). Properties of the receptor for epidermal growth factor. *Cell, 37*(2), 357–358.

Casalegno-Garduno, R., Meier, C., Schmitt, A., Spitschak, A., Hilgendorf, I., Rohde, S., et al. (2012). Immune responses to RHAMM in patients with acute myeloid leukemia after chemotherapy and allogeneic stem cell transplantation. *Clinical & Developmental Immunology, 2012*, 146463. http://dx.doi.org/10.1155/2012/146463.

Chang, N. S. (1995). Hyaluronidase induces murine L929 fibrosarcoma cells resistant to tumor necrosis factor and Fas cytotoxicity in the presence of actinomycin D. *Cell Biochemistry and Biophysics, 27*(2), 109–132.

Chang, L., & Karin, M. (2001). Mammalian MAP kinase signalling cascades. *Nature, 410*(6824), 37–40. http://dx.doi.org/10.1038/35065000.

Cheung, W. F., Cruz, T. F., & Turley, E. A. (1999). Receptor for hyaluronan-mediated motility (RHAMM), a hyaladherin that regulates cell responses to growth factors. *Biochemical Society Transactions, 27*(2), 135–142.

Corsini, E., Galbiati, V., Nikitovic, D., & Tsatsakis, A. M. (2013). Role of oxidative stress in chemical allergens induced skin cells activation. *Food and Chemical Toxicology, 61*, 74–81. http://dx.doi.org/10.1016/j.fct.2013.02.038.

David-Raoudi, M., Tranchepain, F., Deschrevel, B., Vincent, J. C., Bogdanowicz, P., Boumediene, K., et al. (2008). Differential effects of hyaluronan and its fragments on fibroblasts: Relation to wound healing. *Wound Repair and Regeneration, 16*(2), 274–287. http://dx.doi.org/10.1111/j.1524-475X.2007.00342.x.

Day, A. J., & Prestwich, G. D. (2002). Hyaluronan-binding proteins: Tying up the giant. *The Journal of Biological Chemistry, 277*(7), 4585–4588. http://dx.doi.org/10.1074/jbc.R100036200.

de la Motte, C. A., Hascall, V. C., Drazba, J., Bandyopadhyay, S. K., & Strong, S. A. (2003). Mononuclear leukocytes bind to specific hyaluronan structures on colon mucosal smooth muscle cells treated with polyinosinic acid:polycytidylic acid: Inter-alpha-trypsin inhibitor is crucial to structure and function. *The American Journal of Pathology*, *163*(1), 121–133.

Ding, I., Liu, W., Sun, J., Fenton, B., & Okunieff, P. (2002). Comparison and modulation of angiogenic responses by FGFs, VEGF and SCF in murine and human fibrosarcomas. *Comparative Biochemistry and Physiology. Part A, Molecular & Integrative Physiology*, *132*(1), 17–25.

Edward, M., Gillan, C., Micha, D., & Tammi, R. H. (2005). Tumour regulation of fibroblast hyaluronan expression: A mechanism to facilitate tumour growth and invasion. *Carcinogenesis*, *26*(7), 1215–1223. http://dx.doi.org/10.1093/carcin/bgi064.

Entwistle, J., Hall, C. L., & Turley, E. A. (1996). HA receptors: Regulators of signalling to the cytoskeleton. *Journal of Cellular Biochemistry*, *61*(4), 569–577. http://dx.doi.org/10.1002/(SICI)1097-4644(19960616)61:4<569::AID-JCB10>3.0.CO;2-B.

Eswarakumar, V. P., Lax, I., & Schlessinger, J. (2005). Cellular signaling by fibroblast growth factor receptors. *Cytokine & Growth Factor Reviews*, *16*(2), 139–149. http://dx.doi.org/10.1016/j.cytogfr.2005.01.001.

Fager, G., Camejo, G., Olsson, U., Ostergren-Lunden, G., Lustig, F., & Bondjers, G. (1995). Binding of platelet-derived growth factor and low density lipoproteins to glycosaminoglycan species produced by human arterial smooth muscle cells. *Journal of Cellular Physiology*, *163*(2), 380–392. http://dx.doi.org/10.1002/jcp.1041630218.

Ferguson, K. M., Berger, M. B., Mendrola, J. M., Cho, H. S., Leahy, D. J., & Lemmon, M. A. (2003). EGF activates its receptor by removing interactions that autoinhibit ectodomain dimerization. *Molecular Cell*, *11*(2), 507–517.

Fiorelli, A., Ricciardi, C., Pannone, G., Santoro, A., Bufo, P., Santini, M., et al. (2011). Interplay between steroid receptors and neoplastic progression in sarcoma tumors. *Journal of Cellular Physiology*, *226*(11), 2997–3003. http://dx.doi.org/10.1002/jcp.22645.

Fraser, J. R., Appelgren, L. E., & Laurent, T. C. (1983). Tissue uptake of circulating hyaluronic acid. A whole body autoradiographic study. *Cell and Tissue Research*, *233*(2), 285–293.

Fthenou, E., Zafiropoulos, A., Tsatsakis, A., Stathopoulos, A., Karamanos, N. K., & Tzanakakis, G. N. (2006). Chondroitin sulfate A chains enhance platelet derived growth factor-mediated signalling in fibrosarcoma cells. *The International Journal of Biochemistry & Cell Biology*, *38*(12), 2141–2150. http://dx.doi.org/10.1016/j.biocel.2006.06.006.

Garrett, T. P., McKern, N. M., Lou, M., Elleman, T. C., Adams, T. E., Lovrecz, G. O., et al. (2002). Crystal structure of a truncated epidermal growth factor receptor extracellular domain bound to transforming growth factor alpha. *Cell*, *110*(6), 763–773.

Gialeli, Ch., Theocharis, A. D., Kletsas, D., Tzanakakis, G. N., & Karamanos, N. K. (2013). Expression of matrix macromolecules and functional properties of EGF-responsive colon cancer cells are inhibited by panitumumab. *Investigational New Drugs*, *31*(3), 516–524. http://dx.doi.org/10.1007/s10637-012-9875-x.

Gill, S., Stratton, M. R., Patterson, H., Spurr, N. K., Fisher, C., Gusterson, B. A., et al. (1991). Detection of transforming genes by transfection of DNA from primary soft-tissue tumours. *Oncogene*, *6*(9), 1651–1656.

Girish, K. S., & Kemparaju, K. (2007). The magic glue hyaluronan and its eraser hyaluronidase: A biological overview. *Life Sciences*, *80*(21), 1921–1943. http://dx.doi.org/10.1016/j.lfs.2007.02.037.

Guerrero, S., Figueras, A., Casanova, I., Farre, L., Lloveras, B., Capella, G., et al. (2002). Codon 12 and codon 13 mutations at the K-ras gene induce different soft tissue sarcoma types in nude mice. *FASEB Journal*, *16*(12), 1642–1644. http://dx.doi.org/10.1096/fj.02-0050fje.

Hall, C. L., & Turley, E. A. (1995). Hyaluronan: RHAMM mediated cell locomotion and signaling in tumorigenesis. *Journal of Neuro-Oncology*, *26*(3), 221–229.

Hall, C. L., Wang, C., Lange, L. A., & Turley, E. A. (1994). Hyaluronan and the hyaluronan receptor RHAMM promote focal adhesion turnover and transient tyrosine kinase activity. *The Journal of Cell Biology*, *126*(2), 575–588.

Hall, C. L., Yang, B., Yang, X., Zhang, S., Turley, M., Samuel, S., et al. (1995). Overexpression of the hyaluronan receptor RHAMM is transforming and is also required for H-ras transformation. *Cell*, *82*(1), 19–26.

Hamilton, S. R., Fard, S. F., Paiwand, F. F., Tolg, C., Veiseh, M., Wang, C., et al. (2007). The hyaluronan receptors CD44 and Rhamm (CD168) form complexes with ERK1,2 that sustain high basal motility in breast cancer cells. *The Journal of Biological Chemistry*, *282*(22), 16667–16680. http://dx.doi.org/10.1074/jbc.M702078200.

Hardwick, C., Hoare, K., Owens, R., Hohn, H. P., Hook, M., Moore, D., et al. (1992). Molecular cloning of a novel hyaluronan receptor that mediates tumor cell motility. *The Journal of Cell Biology*, *117*(6), 1343–1350.

Hatano, H., Shigeishi, H., Kudo, Y., Higashikawa, K., Tobiume, K., Takata, T., et al. (2012). Overexpression of receptor for hyaluronan-mediated motility (RHAMM) in MC3T3-E1 cells induces proliferation and differentiation through phosphorylation of ERK1/2. *Journal of Bone and Mineral Metabolism*, *30*(3), 293–303. http://dx.doi.org/10.1007/s00774-011-0318-0.

Heldin, C. H., Eriksson, U., & Ostman, A. (2002). New members of the platelet-derived growth factor family of mitogens. *Archives of Biochemistry and Biophysics*, *398*(2), 284–290. http://dx.doi.org/10.1006/abbi.2001.2707.

Heldin, C. H., Miyazono, K., & ten Dijke, P. (1997). TGF-beta signalling from cell membrane to nucleus through SMAD proteins. *Nature*, *390*(6659), 465–471. http://dx.doi.org/10.1038/37284.

Heldin, C. H., & Westermark, B. (1999). Mechanism of action and in vivo role of platelet-derived growth factor. *Physiological Reviews*, *79*(4), 1283–1316.

Hill, C. S. (2009). Nucleocytoplasmic shuttling of Smad proteins. *Cell Research*, *19*(1), 36–46. http://dx.doi.org/10.1038/cr.2008.325.

Hill, M. A., Gong, C., Casey, T. J., Menon, A. G., Mera, R., Gillespie, A. T., et al. (1997). Detection of K-ras mutations in resected primary leiomyosarcoma. *Cancer Epidemiology, Biomarkers & Prevention*, *6*(12), 1095–1100.

Hong, H. H., Devereux, T. R., Melnick, R. L., Moomaw, C. R., Boorman, G. A., & Sills, R. C. (2000). Mutations of ras protooncogenes and p53 tumor suppressor gene in cardiac hemangiosarcomas from B6C3F1 mice exposed to 1,3-butadiene for 2 years. *Toxicologic Pathology*, *28*(4), 529–534.

Horbelt, D., Denkis, A., & Knaus, P. (2012). A portrait of Transforming Growth Factor beta superfamily signalling: Background matters. *The International Journal of Biochemistry & Cell Biology*, *44*(3), 469–474. http://dx.doi.org/10.1016/j.biocel.2011.12.013.

Hoshi, M., Matsumoto, S., Manabe, J., Tanizawa, T., Shigemitsu, T., Izawa, N., et al. (2006). Malignant change secondary to fibrous dysplasia. *International Journal of Clinical Oncology*, *11*(3), 229–235. http://dx.doi.org/10.1007/s10147-006-0559-4.

Huang, H. Y., Kang, H. Y., Li, C. F., Eng, H. L., Chou, S. C., Lin, C. N., et al. (2006). Skp2 overexpression is highly representative of intrinsic biological aggressiveness and independently associated with poor prognosis in primary localized myxofibrosarcomas. *Clinical Cancer Research*, *12*(2), 487–498. http://dx.doi.org/10.1158/1078-0432.CCR-05-1497.

Idobe, Y., Murawaki, Y., Ikuta, Y., Koda, M., & Kawasaki, H. (1998). Post-prandial serum hyaluronan concentration in patients with chronic liver disease. *Internal Medicine*, *37*(7), 568–575.

Itano, N., Atsumi, F., Sawai, T., Yamada, Y., Miyaishi, O., Senga, T., et al. (2002). Abnormal accumulation of hyaluronan matrix diminishes contact inhibition of cell growth and promotes cell migration. *Proceedings of the National Academy of Sciences of the United States of America, 99*(6), 3609–3614. http://dx.doi.org/10.1073/pnas.052026799.

Itano, N., & Kimata, K. (2002). Mammalian hyaluronan synthases. *IUBMB Life, 54*(4), 195–199. http://dx.doi.org/10.1080/15216540214929.

Jaye, M., Schlessinger, J., & Dionne, C. A. (1992). Fibroblast growth factor receptor tyrosine kinases: Molecular analysis and signal transduction. *Biochimica et Biophysica Acta, 1135*(2), 185–199.

Jiang, D., Liang, J., Fan, J., Yu, S., Chen, S., Luo, Y., et al. (2005). Regulation of lung injury and repair by Toll-like receptors and hyaluronan. *Nature Medicine, 11*(11), 1173–1179. http://dx.doi.org/10.1038/nm1315.

Jiang, D., Liang, J., & Noble, P. W. (2011). Hyaluronan as an immune regulator in human diseases. *Physiological Reviews, 91*(1), 221–264. http://dx.doi.org/10.1152/physrev.00052.2009.

Jones, J. I., & Clemmons, D. R. (1995). Insulin-like growth factors and their binding proteins: Biological actions. *Endocrine Reviews, 16*(1), 3–34. http://dx.doi.org/10.1210/edrv-16-1-3.

Karvinen, S., Pasonen-Seppanen, S., Hyttinen, J. M., Pienimaki, J. P., Torronen, K., Jokela, T. A., et al. (2003). Keratinocyte growth factor stimulates migration and hyaluronan synthesis in the epidermis by activation of keratinocyte hyaluronan synthases 2 and 3. *The Journal of Biological Chemistry, 278*(49), 49495–49504. http://dx.doi.org/10.1074/jbc.M310445200.

Kim, S. H., Turnbull, J., & Guimond, S. (2011). Extracellular matrix and cell signalling: The dynamic cooperation of integrin, proteoglycan and growth factor receptor. *The Journal of Endocrinology, 209*(2), 139–151. http://dx.doi.org/10.1530/JOE-10-0377.

Klagas, I., Goulet, S., Karakiulakis, G., Zhong, J., Baraket, M., Black, J. L., et al. (2009). Decreased hyaluronan in airway smooth muscle cells from patients with asthma and COPD. *The European Respiratory Journal, 34*(3), 616–628. http://dx.doi.org/10.1183/09031936.00070808.

Knezevich, S. R., Garnett, M. J., Pysher, T. J., Beckwith, J. B., Grundy, P. E., & Sorensen, P. H. (1998). ETV6-NTRK3 gene fusions and trisomy 11 establish a histogenetic link between mesoblastic nephroma and congenital fibrosarcoma. *Cancer Research, 58*(22), 5046–5048.

Knudson, W., & Toole, B. P. (1988). Membrane association of the hyaluronate stimulatory factor from LX-1 human lung carcinoma cells. *Journal of Cellular Biochemistry, 38*(3), 165–177. http://dx.doi.org/10.1002/jcb.240380304.

Kobayashi, N., Miyoshi, S., Mikami, T., Koyama, H., Kitazawa, M., Takeoka, M., et al. (2010). Hyaluronan deficiency in tumor stroma impairs macrophage trafficking and tumor neovascularization. *Cancer Research, 70*(18), 7073–7083. http://dx.doi.org/10.1158/0008-5472.CAN-09-4687.

Kosaki, R., Watanabe, K., & Yamaguchi, Y. (1999). Overproduction of hyaluronan by expression of the hyaluronan synthase Has2 enhances anchorage-independent growth and tumorigenicity. *Cancer Research, 59*(5), 1141–1145.

Kouvidi, K., Berdiaki, A., Nikitovic, D., Katonis, P., Afratis, N., Hascall, V. C., et al. (2011). Role of receptor for hyaluronic acid-mediated motility (RHAMM) in low molecular weight hyaluronan (LMWHA)-mediated fibrosarcoma cell adhesion. *The Journal of Biological Chemistry, 286*(44), 38509–38520. http://dx.doi.org/10.1074/jbc.M111.275875.

Koyama, H., Hibi, T., Isogai, Z., Yoneda, M., Fujimori, M., Amano, J., et al. (2007). Hyperproduction of hyaluronan in neu-induced mammary tumor accelerates angiogenesis through stromal cell recruitment: Possible involvement of versican/PG-M. *The American Journal of Pathology, 170*(3), 1086–1099. http://dx.doi.org/10.2353/ajpath.2007.060793.

Koyama, H., Kobayashi, N., Harada, M., Takeoka, M., Kawai, Y., Sano, K., et al. (2008). Significance of tumor-associated stroma in promotion of intratumoral lymphangiogenesis: Pivotal role of a hyaluronan-rich tumor microenvironment. *The American Journal of Pathology*, *172*(1), 179–193. http://dx.doi.org/10.2353/ajpath.2008.070360.

Kransdorf, M. J. (1995). Malignant soft-tissue tumors in a large referral population: Distribution of diagnoses by age, sex, and location. *AJR. American Journal of Roentgenology*, *164*(1), 129–134. http://dx.doi.org/10.2214/ajr.164.1.7998525.

Kraus, M. H., Issing, W., Miki, T., Popescu, N. C., & Aaronson, S. A. (1989). Isolation and characterization of ERBB3, a third member of the ERBB/epidermal growth factor receptor family: Evidence for overexpression in a subset of human mammary tumors. *Proceedings of the National Academy of Sciences of the United States of America*, *86*(23), 9193–9197.

Kubo, M., Ihn, H., Yamane, K., & Tamaki, K. (2006). The expression levels and the differential expression of transforming growth factor-beta receptors in dermatofibroma and dermatofibrosarcoma protuberans. *The British Journal of Dermatology*, *154*(5), 919–925. http://dx.doi.org/10.1111/j.1365-2133.2005.06904.x.

Kuroda, K., Utani, A., Hamasaki, Y., & Shinkai, H. (2001). Up-regulation of putative hyaluronan synthase mRNA by basic fibroblast growth factor and insulin-like growth factor-1 in human skin fibroblasts. *Journal of Dermatological Science*, *26*(2), 156–160.

Kwak, H. J., Park, M. J., Cho, H., Park, C. M., Moon, S. I., Lee, H. C., et al. (2006). Transforming growth factor-beta1 induces tissue inhibitor of metalloproteinase-1 expression via activation of extracellular signal-regulated kinase and Sp1 in human fibrosarcoma cells. *Molecular Cancer Research*, *4*(3), 209–220. http://dx.doi.org/10.1158/1541-7786.MCR-05-0140.

Lee, H. J., Kim, S. A., Lee, H. J., Jeong, S. J., Han, I., Jung, J. H., et al. (2010). Paeonol oxime inhibits bFGF-induced angiogenesis and reduces VEGF levels in fibrosarcoma cells. *PLoS One*, *5*(8), e12358. http://dx.doi.org/10.1371/journal.pone.0012358.

Lennon, F. E., & Singleton, P. A. (2011). Hyaluronan regulation of vascular integrity. *American Journal of Cardiovascular Disease*, *1*(3), 200–213.

Levental, K. R., Yu, H., Kass, L., Lakins, J. N., Egeblad, M., Erler, J. T., et al. (2009). Matrix crosslinking forces tumor progression by enhancing integrin signaling. *Cell*, *139*(5), 891–906. http://dx.doi.org/10.1016/j.cell.2009.10.027.

Li, L., Asteriou, T., Bernert, B., Heldin, C. H., & Heldin, P. (2007). Growth factor regulation of hyaluronan synthesis and degradation in human dermal fibroblasts: Importance of hyaluronan for the mitogenic response of PDGF-BB. *The Biochemical Journal*, *404*(2), 327–336. http://dx.doi.org/10.1042/BJ20061757.

Li, X., & Eriksson, U. (2003). Novel PDGF family members: PDGF-C and PDGF-D. *Cytokine & Growth Factor Reviews*, *14*(2), 91–98.

Li, L., Heldin, C. H., & Heldin, P. (2006). Inhibition of platelet-derived growth factor-BB-induced receptor activation and fibroblast migration by hyaluronan activation of CD44. *The Journal of Biological Chemistry*, *281*(36), 26512–26519. http://dx.doi.org/10.1074/jbc.M605607200.

Li, X., Huang, Y., Jiang, J., & Frank, S. J. (2008). ERK-dependent threonine phosphorylation of EGF receptor modulates receptor downregulation and signaling. *Cellular Signalling*, *20*(11), 2145–2155. http://dx.doi.org/10.1016/j.cellsig.2008.08.006.

Lin, C. R., Chen, W. S., Kruiger, W., Stolarsky, L. S., Weber, W., Evans, R. M., et al. (1984). Expression cloning of human EGF receptor complementary DNA: Gene amplification and three related messenger RNA products in A431 cells. *Science*, *224*(4651), 843–848.

Lin, E. Y., Li, J. F., Gnatovskiy, L., Deng, Y., Zhu, L., Grzesik, D. A., et al. (2006). Macrophages regulate the angiogenic switch in a mouse model of breast cancer. *Cancer Research*, *66*(23), 11238–11246. http://dx.doi.org/10.1158/0008-5472.CAN-06-1278.

Lokeshwar, V. B., Lopez, L. E., Munoz, D., Chi, A., Shirodkar, S. P., Lokeshwar, S. D., et al. (2010). Antitumor activity of hyaluronic acid synthesis inhibitor 4-methylumbelliferone in prostate cancer cells. *Cancer Research, 70*(7), 2613–2623. http://dx.doi.org/10.1158/0008-5472.CAN-09-3185.

Lokeshwar, V. B., Rubinowicz, D., Schroeder, G. L., Forgacs, E., Minna, J. D., Block, N. L., et al. (2001). Stromal and epithelial expression of tumor markers hyaluronic acid and HYAL1 hyaluronidase in prostate cancer. *The Journal of Biological Chemistry, 276*(15), 11922–11932. http://dx.doi.org/10.1074/jbc.M008432200.

Lopez-Casillas, F., Wrana, J. L., & Massague, J. (1993). Betaglycan presents ligand to the TGF beta signaling receptor. *Cell, 73*(7), 1435–1444.

Lu, P., Weaver, V. M., & Werb, Z. (2012). The extracellular matrix: A dynamic niche in cancer progression. *The Journal of Cell Biology, 196*(4), 395–406. http://dx.doi.org/10.1083/jcb.201102147.

Lutz, M., & Knaus, P. (2002). Integration of the TGF-beta pathway into the cellular signalling network. *Cellular Signalling, 14*(12), 977–988.

Massague, J. (1998). TGF-beta signal transduction. *Annual Review of Biochemistry, 67*, 753–791. http://dx.doi.org/10.1146/annurev.biochem.67.1.753.

Massague, J. (2000). How cells read TGF-beta signals. *Nature Reviews. Molecular Cell Biology, 1*(3), 169–178. http://dx.doi.org/10.1038/35043051.

Mathiak, M., Yenisey, C., Grant, D. S., Sharma, B., & Iozzo, R. V. (1997). A role for perlecan in the suppression of growth and invasion in fibrosarcoma cells. *Cancer Research, 57*(11), 2130–2136.

Maxwell, C. A., Keats, J. J., Belch, A. R., Pilarski, L. M., & Reiman, T. (2005). Receptor for hyaluronan-mediated motility correlates with centrosome abnormalities in multiple myeloma and maintains mitotic integrity. *Cancer Research, 65*(3), 850–860.

Maxwell, C. A., McCarthy, J., & Turley, E. (2008). Cell-surface and mitotic-spindle RHAMM: Moonlighting or dual oncogenic functions? *Journal of Cell Science, 121*(Pt 7), 925–932. http://dx.doi.org/10.1242/jcs.022038.

McKee, C. M., Lowenstein, C. J., Horton, M. R., Wu, J., Bao, C., Chin, B. Y., et al. (1997). Hyaluronan fragments induce nitric-oxide synthase in murine macrophages through a nuclear factor kappaB-dependent mechanism. *The Journal of Biological Chemistry, 272*(12), 8013–8018.

McKee, C. M., Penno, M. B., Cowman, M., Burdick, M. D., Strieter, R. M., Bao, C., et al. (1996). Hyaluronan (HA) fragments induce chemokine gene expression in alveolar macrophages. The role of HA size and CD44. *The Journal of Clinical Investigation, 98*(10), 2403–2413. http://dx.doi.org/10.1172/JCI119054.

Menon, P., Smith, K. J., Crittenden, J., & Skelton, H. (1999). Adjuvant therapy with hyaluronidase prior to excision of dermatofibrosarcoma protuberans. *Dermatologic Surgery, 25*(3), 205–209.

Mercola, M., & Stiles, C. D. (1988). Growth factor superfamilies and mammalian embryogenesis. *Development, 102*(3), 451–460.

Midgley, A. C., Bowen, T., Phillips, A. O., & Steadman, R. (2014). MicroRNA-7 inhibition rescues age-associated loss of epidermal growth factor receptor and hyaluronan-dependent differentiation in fibroblasts. *Aging Cell, 13*(2), 235–244. http://dx.doi.org/10.1111/acel.12167.

Mohapatra, S., Yang, X., Wright, J. A., Turley, E. A., & Greenberg, A. H. (1996). Soluble hyaluronan receptor RHAMM induces mitotic arrest by suppressing Cdc2 and cyclin B1 expression. *The Journal of Experimental Medicine, 183*(4), 1663–1668.

Morrison, K. B., Tognon, C. E., Garnett, M. J., Deal, C., & Sorensen, P. H. (2002). ETV6-NTRK3 transformation requires insulin-like growth factor 1 receptor signaling and is associated with constitutive IRS-1 tyrosine phosphorylation. *Oncogene, 21*(37), 5684–5695. http://dx.doi.org/10.1038/sj.onc.1205669.

Moustakas, A., & Heldin, C. H. (2009). The regulation of TGFbeta signal transduction. *Development, 136*(22), 3699–3714. http://dx.doi.org/10.1242/dev.030338.

Muraoka-Cook, R. S., Dumont, N., & Arteaga, C. L. (2005). Dual role of transforming growth factor beta in mammary tumorigenesis and metastatic progression. *Clinical Cancer Research, 11*(2 Pt 2), 937s–943s.

Mythreye, K., & Blobe, G. C. (2009). Proteoglycan signaling co-receptors: Roles in cell adhesion, migration and invasion. *Cellular Signalling, 21*(11), 1548–1558. http://dx.doi.org/10.1016/j.cellsig.2009.05.001.

Nedvetzki, S., Gonen, E., Assayag, N., Reich, R., Williams, R. O., Thurmond, R. L., et al. (2004). RHAMM, a receptor for hyaluronan-mediated motility, compensates for CD44 in inflamed CD44-knockout mice: A different interpretation of redundancy. *Proceedings of the National Academy of Sciences of the United States of America, 101*(52), 18081–18086. http://dx.doi.org/10.1073/pnas.0407378102.

Nikitovic, D., Assouti, M., Sifaki, M., Katonis, P., Krasagakis, K., Karamanos, N. K., et al. (2008). Chondroitin sulfate and heparan sulfate-containing proteoglycans are both partners and targets of basic fibroblast growth factor-mediated proliferation in human metastatic melanoma cell lines. *The International Journal of Biochemistry & Cell Biology, 40*(1), 72–83. http://dx.doi.org/10.1016/j.biocel.2007.06.019.

Nikitovic, D., Chalkiadaki, G., Berdiaki, A., Aggelidakis, J., Katonis, P., Karamanos, N. K., et al. (2011). Lumican regulates osteosarcoma cell adhesion by modulating TGFbeta2 activity. *The International Journal of Biochemistry & Cell Biology, 43*(6), 928–935. http://dx.doi.org/10.1016/j.biocel.2011.03.008.

Nikitovic, D., Corsini, E., Kouretas, D., Tsatsakis, A., & Tzanakakis, G. (2013). ROS-major mediators of extracellular matrix remodeling during tumor progression. *Food and Chemical Toxicology, 61*, 178–186. http://dx.doi.org/10.1016/j.fct.2013.06.013.

Nikitovic, D., Zafiropoulos, A., Katonis, P., Tsatsakis, A., Theocharis, A. D., Karamanos, N. K., et al. (2006). Transforming growth factor-beta as a key molecule triggering the expression of versican isoforms v0 and v1, hyaluronan synthase-2 and synthesis of hyaluronan in malignant osteosarcoma cells. *IUBMB Life, 58*(1), 47–53. http://dx.doi.org/10.1080/15216540500531713.

Oda, Y., Takahira, T., Kawaguchi, K., Yamamoto, H., Tamiya, S., Matsuda, S., et al. (2003). Altered expression of cell cycle regulators in myxofibrosarcoma, with special emphasis on their prognostic implications. *Human Pathology, 34*(10), 1035–1042.

Pantel, K., & Brakenhoff, R. H. (2004). Dissecting the metastatic cascade. *Nature Reviews. Cancer, 4*(6), 448–456. http://dx.doi.org/10.1038/nrc1370.

Pardali, K., & Moustakas, A. (2007). Actions of TGF-beta as tumor suppressor and prometastatic factor in human cancer. *Biochimica et Biophysica Acta, 1775*(1), 21–62. http://dx.doi.org/10.1016/j.bbcan.2006.06.004.

Pasonen-Seppanen, S., Karvinen, S., Torronen, K., Hyttinen, J. M., Jokela, T., Lammi, M. J., et al. (2003). EGF upregulates, whereas TGF-beta downregulates, the hyaluronan synthases Has2 and Has3 in organotypic keratinocyte cultures: Correlations with epidermal proliferation and differentiation. *The Journal of Investigative Dermatology, 120*(6), 1038–1044. http://dx.doi.org/10.1046/j.1523-1747.2003.12249.x.

Piepkorn, M., Hovingh, P., & Linker, A. (1988). Evidence for independent metabolism and cell surface localization of cell surface localization of cellular proteoglycans and glycosaminoglycan free chains. *Journal of Cellular Physiology, 135*(2), 189–199. http://dx.doi.org/10.1002/jcp.1041350206.

Pietras, K., Sjoblom, T., Rubin, K., Heldin, C. H., & Ostman, A. (2003). PDGF receptors as cancer drug targets. *Cancer Cell, 3*(5), 439–443.

Pirinen, R., Tammi, R., Tammi, M., Hirvikoski, P., Parkkinen, J. J., Johansson, R., et al. (2001). Prognostic value of hyaluronan expression in non-small-cell lung cancer: Increased stromal expression indicates unfavorable outcome in patients with adenocarcinoma. *International Journal of Cancer, 95*(1), 12–17.

Provenzano, P. P., Eliceiri, K. W., Campbell, J. M., Inman, D. R., White, J. G., & Keely, P. J. (2006). Collagen reorganization at the tumor-stromal interface facilitates local invasion. *BMC Medicine, 4*(1), 38. http://dx.doi.org/10.1186/1741-7015-4-38.

Rajeswari, J., Kesavan, K., & Jayakar, B. (2011). Phytochemical and pharmacological evaluation of prop roots of Pandanus fascicularis Lam. *Asian Pacific Journal of Tropical Medicine, 4*(8), 649–653. http://dx.doi.org/10.1016/S1995-7645(11)60165-X.

Rathinam, R., & Alahari, S. K. (2010). Important role of integrins in the cancer biology. *Cancer Metastasis Reviews, 29*(1), 223–237. http://dx.doi.org/10.1007/s10555-010-9211-x.

Rebustini, I. T., Myers, C., Lassiter, K. S., Surmak, A., Szabova, L., Holmbeck, K., et al. (2009). MT2-MMP-dependent release of collagen IV NC1 domains regulates submandibular gland branching morphogenesis. *Developmental Cell, 17*(4), 482–493. http://dx.doi.org/10.1016/j.devcel.2009.07.016.

Ruggieri, P., Sim, F. H., Bond, J. R., & Unni, K. K. (1994). Malignancies in fibrous dysplasia. *Cancer, 73*(5), 1411–1424.

Samuel, S. K., Hurta, R. A., Spearman, M. A., Wright, J. A., Turley, E. A., & Greenberg, A. H. (1993). TGF-beta 1 stimulation of cell locomotion utilizes the hyaluronan receptor RHAMM and hyaluronan. *The Journal of Cell Biology, 123*(3), 749–758.

Sanderson, R. D., Yang, Y., Suva, L. J., & Kelly, T. (2004). Heparan sulfate proteoglycans and heparanase—Partners in osteolytic tumor growth and metastasis. *Matrix Biology, 23*(6), 341–352. http://dx.doi.org/10.1016/j.matbio.2004.08.004.

Sato, O., Wada, T., Kawai, A., Yamaguchi, U., Makimoto, A., Kokai, Y., et al. (2005). Expression of epidermal growth factor receptor, ERBB2 and KIT in adult soft tissue sarcomas: A clinicopathologic study of 281 cases. *Cancer, 103*(9), 1881–1890. http://dx.doi.org/10.1002/cncr.20986.

Saunier, E. F., & Akhurst, R. J. (2006). TGF beta inhibition for cancer therapy. *Current Cancer Drug Targets, 6*(7), 565–578.

Schaefer, L., & Schaefer, R. M. (2010). Proteoglycans: From structural compounds to signaling molecules. *Cell and Tissue Research, 339*(1), 237–246. http://dx.doi.org/10.1007/s00441-009-0821-y.

Schmidt, H., Taubert, H., Wurl, P., Kappler, M., Lange, H., Bartel, F., et al. (2002). Gains of 12q are the most frequent genomic imbalances in adult fibrosarcoma and are correlated with a poor outcome. *Genes, Chromosomes & Cancer, 34*(1), 69–77.

Schulz, T., Schumacher, U., & Prehm, P. (2007). Hyaluronan export by the ABC transporter MRP5 and its modulation by intracellular cGMP. *The Journal of Biological Chemistry, 282*(29), 20999–21004. http://dx.doi.org/10.1074/jbc.M700915200.

Segura, T., Anderson, B. C., Chung, P. H., Webber, R. E., Shull, K. R., & Shea, L. D. (2005). Crosslinked hyaluronic acid hydrogels: A strategy to functionalize and pattern. *Biomaterials, 26*(4), 359–371. http://dx.doi.org/10.1016/j.biomaterials.2004.02.067.

Shi, Y., & Massague, J. (2003). Mechanisms of TGF-beta signaling from cell membrane to the nucleus. *Cell, 113*(6), 685–700.

Slevin, M., Krupinski, J., Gaffney, J., Matou, S., West, D., Delisser, H., et al. (2007). Hyaluronan-mediated angiogenesis in vascular disease: Uncovering RHAMM and CD44 receptor signaling pathways. *Matrix Biology, 26*(1), 58–68. http://dx.doi.org/10.1016/j.matbio.2006.08.261.

Slevin, M., Kumar, S., & Gaffney, J. (2002). Angiogenic oligosaccharides of hyaluronan induce multiple signaling pathways affecting vascular endothelial cell mitogenic and wound healing responses. *The Journal of Biological Chemistry*, *277*(43), 41046–41059. http://dx.doi.org/10.1074/jbc.M109443200.

Slovak, M. L., Mirski, S. E., Cole, S. P., Gerlach, J. H., Yohem, K. H., & Trent, J. M. (1991). Tumourigenic multidrug-resistant HT1080 cells do not overexpress receptors for epidermal growth factor. *British Journal of Cancer*, *64*(2), 296–298.

Smith, K., Menon, P., Rolfe, A., Welch, M., Barrett, T., Crittenden, M. J., et al. (1998). Pretreatment with intralesional hyaluronidase prior to excision of dermatofibrosarcoma protuberans. *Cancer Letters*, *131*(1), 117–126.

Spranger, S., Jeremias, I., Wilde, S., Leisegang, M., Starck, L., Mosetter, B., et al. (2012). TCR-transgenic lymphocytes specific for HMMR/Rhamm limit tumor outgrowth in vivo. *Blood*, *119*(15), 3440–3449. http://dx.doi.org/10.1182/blood-2011-06-357939.

Stair-Nawy, S., Csoka, A. B., & Stern, R. (1999). Hyaluronidase expression in human skin fibroblasts. *Biochemical and Biophysical Research Communications*, *266*(1), 268–273. http://dx.doi.org/10.1006/bbrc.1999.1802.

Stern, R. (2008). Hyaluronidases in cancer biology. *Seminars in Cancer Biology*, *18*(4), 275–280. http://dx.doi.org/10.1016/j.semcancer.2008.03.017.

Stiller, C. A., & Parkin, D. M. (1994). International variations in the incidence of childhood soft-tissue sarcomas. *Paediatric and Perinatal Epidemiology*, *8*(1), 107–119.

Tammi, R. H., Kultti, A., Kosma, V. M., Pirinen, R., Auvinen, P., & Tammi, M. I. (2008). Hyaluronan in human tumors: Pathobiological and prognostic messages from cell-associated and stromal hyaluronan. *Seminars in Cancer Biology*, *18*(4), 288–295. http://dx.doi.org/10.1016/j.semcancer.2008.03.005.

Telmer, P. G., Tolg, C., McCarthy, J. B., & Turley, E. A. (2011). How does a protein with dual mitotic spindle and extracellular matrix receptor functions affect tumor susceptibility and progression? *Communicative & Integrative Biology*, *4*(2), 182–185. http://dx.doi.org/10.4161/cib.4.2.14270.

Teriete, P., Banerji, S., Noble, M., Blundell, C. D., Wright, A. J., Pickford, A. R., et al. (2004). Structure of the regulatory hyaluronan binding domain in the inflammatory leukocyte homing receptor CD44. *Molecular Cell*, *13*(4), 483–496.

Theocharis, A. D., Skandalis, S. S., Tzanakakis, G. N., & Karamanos, N. K. (2010). Proteoglycans in health and disease: Novel roles for proteoglycans in malignancy and their pharmacological targeting. *The FEBS Journal*, *277*(19), 3904–3923. http://dx.doi.org/10.1111/j.1742-4658.2010.07800.x.

Timar, J., Lapis, K., Dudas, J., Sebestyen, A., Kopper, L., & Kovalszky, I. (2002). Proteoglycans and tumor progression: Janus-faced molecules with contradictory functions in cancer. *Seminars in Cancer Biology*, *12*(3), 173–186. http://dx.doi.org/10.1016/S1044-579X(02)00021-4.

Tolg, C., Hamilton, S. R., Morningstar, L., Zhang, J., Zhang, S., Esguerra, K. V., et al. (2010). RHAMM promotes interphase microtubule instability and mitotic spindle integrity through MEK1/ERK1/2 activity. *The Journal of Biological Chemistry*, *285*(34), 26461–26474. http://dx.doi.org/10.1074/jbc.M110.121491.

Tolg, C., Hamilton, S. R., Nakrieko, K. A., Kooshesh, F., Walton, P., McCarthy, J. B., et al. (2006). Rhamm-/- fibroblasts are defective in CD44-mediated ERK1,2 motogenic signaling, leading to defective skin wound repair. *The Journal of Cell Biology*, *175*(6), 1017–1028. http://dx.doi.org/10.1083/jcb.200511027.

Tolg, C., Poon, R., Fodde, R., Turley, E. A., & Alman, B. A. (2003). Genetic deletion of receptor for hyaluronan-mediated motility (Rhamm) attenuates the formation of aggressive fibromatosis (desmoid tumor). *Oncogene*, *22*(44), 6873–6882. http://dx.doi.org/10.1038/sj.onc.1206811.

Toole, B. P. (1997). Hyaluronan in morphogenesis. *Journal of Internal Medicine*, *242*(1), 35–40.

Toole, B. P. (2004). Hyaluronan: From extracellular glue to pericellular cue. *Nature Reviews. Cancer*, *4*(7), 528–539. http://dx.doi.org/10.1038/nrc1391.

Toole, B. P., Wight, T. N., & Tammi, M. I. (2002). Hyaluronan-cell interactions in cancer and vascular disease. *The Journal of Biological Chemistry*, *277*(7), 4593–4596. http://dx.doi.org/10.1074/jbc.R100039200.

Turley, E. A. (1982). Purification of a hyaluronate-binding protein fraction that modifies cell social behavior. *Biochemical and Biophysical Research Communications*, *108*(3), 1016–1024.

Turley, E. A. (1992). Hyaluronan and cell locomotion. *Cancer Metastasis Reviews*, *11*(1), 21–30.

Turley, E. A., & Naor, D. (2012). RHAMM and CD44 peptides-analytic tools and potential drugs. *Frontiers in Bioscience (Landmark Edition)*, *17*, 1775–1794.

Turley, E. A., Noble, P. W., & Bourguignon, L. Y. (2002). Signaling properties of hyaluronan receptors. *The Journal of Biological Chemistry*, *277*(7), 4589–4592. http://dx.doi.org/10.1074/jbc.R100038200.

Underhill, C. (1992). CD44: The hyaluronan receptor. *Journal of Cell Science*, *103*(Pt 2), 293–298.

Vincent, L., & Rafii, S. (2004). Vascular frontiers without borders: Multifaceted roles of platelet-derived growth factor (PDGF) in supporting postnatal angiogenesis and lymphangiogenesis. *Cancer Cell*, *6*(4), 307–309. http://dx.doi.org/10.1016/j.ccr.2004.09.024.

Wang, C., Thor, A. D., Moore, D. H., 2nd., Zhao, Y., Kerschmann, R., Stern, R., et al. (1998). The overexpression of RHAMM, a hyaluronan-binding protein that regulates ras signaling, correlates with overexpression of mitogen-activated protein kinase and is a significant parameter in breast cancer progression. *Clinical Cancer Research*, *4*(3), 567–576.

Weigel, P. H., Hascall, V. C., & Tammi, M. (1997). Hyaluronan synthases. *The Journal of Biological Chemistry*, *272*(22), 13997–14000.

Werner, H., & LeRoith, D. (1996). The role of the insulin-like growth factor system in human cancer. *Advances in Cancer Research*, *68*, 183–223.

West, D. C., Hampson, I. N., Arnold, F., & Kumar, S. (1985). Angiogenesis induced by degradation products of hyaluronic acid. *Science*, *228*(4705), 1324–1326.

Wibmer, C., Leithner, A., Zielonke, N., Sperl, M., & Windhager, R. (2010). Increasing incidence rates of soft tissue sarcomas? A population-based epidemiologic study and literature review. *Annals of Oncology*, *21*(5), 1106–1111. http://dx.doi.org/10.1093/annonc/mdp415.

Wingo, P. A., Tong, T., & Bolden, S. (1995). Cancer statistics, 1995. *CA: A Cancer Journal for Clinicians*, *45*(1), 8–30.

Wiseman, B. S., Sternlicht, M. D., Lund, L. R., Alexander, C. M., Mott, J., Bissell, M. J., et al. (2003). Site-specific inductive and inhibitory activities of MMP-2 and MMP-3 orchestrate mammary gland branching morphogenesis. *The Journal of Cell Biology*, *162*(6), 1123–1133. http://dx.doi.org/10.1083/jcb.200302090.

Xu, R., Boudreau, A., & Bissell, M. J. (2009). Tissue architecture and function: Dynamic reciprocity via extra- and intra-cellular matrices. *Cancer Metastasis Reviews*, *28*(1–2), 167–176. http://dx.doi.org/10.1007/s10555-008-9178-z.

Yabut, S. M., Jr., Kenan, S., Sissons, H. A., & Lewis, M. M. (1988). Malignant transformation of fibrous dysplasia. A case report and review of the literature. *Clinical Orthopaedics and Related Research*, *228*, 281–289.

Yokota, J. (2000). Tumor progression and metastasis. *Carcinogenesis*, *21*(3), 497–503.

Yoo, J., & Robinson, R. A. (1999). H-ras and K-ras mutations in soft tissue sarcoma: Comparative studies of sarcomas from Korean and American patients. *Cancer*, *86*(1), 58–63.

Zhang, X., Gureasko, J., Shen, K., Cole, P. A., & Kuriyan, J. (2006). An allosteric mechanism for activation of the kinase domain of epidermal growth factor receptor. *Cell*, *125*(6), 1137–1149. http://dx.doi.org/10.1016/j.cell.2006.05.013.
Zhang, L., Underhill, C. B., & Chen, L. (1995). Hyaluronan on the surface of tumor cells is correlated with metastatic behavior. *Cancer Research*, *55*(2), 428–433.
Zoltan-Jones, A., Huang, L., Ghatak, S., & Toole, B. P. (2003). Elevated hyaluronan production induces mesenchymal and transformed properties in epithelial cells. *The Journal of Biological Chemistry*, *278*(46), 45801–45810. http://dx.doi.org/10.1074/jbc.M308168200.

CHAPTER THIRTEEN

CD147: Regulator of Hyaluronan Signaling in Invasiveness and Chemoresistance

G. Daniel Grass[*,1], **Lu Dai**[†], **Zhiqiang Qin**[‡], **Chris Parsons**[†,‡], **Bryan P. Toole**[*,§,1]

*Department of Regenerative Medicine & Cell Biology, Medical University of South Carolina, Charleston, South Carolina, USA
†Department of Medicine, Stanley S. Scott Cancer Center, Louisiana State University Health Sciences Center, New Orleans, Louisiana, USA
‡Department of Microbiology, Immunology & Parasitology, Stanley S. Scott Cancer Center, Louisiana State University Health Sciences Center, New Orleans, Louisiana, USA
§Hollings Cancer Center, Medical University of South Carolina, Charleston, South Carolina, USA
[1]Corresponding authors: e-mail address: grass@musc.edu; toolebp@musc.edu

Contents

1. Introduction 352
2. Structure and Pleiotropic Functions of CD147 353
3. CD147-Induced HA Synthesis and Signaling 356
4. CD147–HA Axis in Cellular Invasion 356
   4.1 Epithelial–mesenchymal transitions 357
   4.2 Regulation of invadopodia formation and activity by CD147 358
   4.3 Regulation of invadopodia activity and invasiveness by CD147-induced HA–CD44 and EGFR signaling 359
5. CD147–HA Axis in Chemoresistance 361
6. Induction of the CD147–HA Axis by Kaposi's Sarcoma-Associated Herpesvirus 363
7. Conclusions 364
References 366

Abstract

Major determinants that influence negative outcome in cancer patients are the abilities of cancer cells to resist current therapies and to invade surrounding host tissue, consequently leading to local and metastatic dissemination. Hyaluronan (HA), a prominent constituent of the tumor microenvironment, not only provides structural support but also interacts with cell surface receptors, especially CD44, that influence cooperative signaling pathways leading to chemoresistance and invasiveness. CD147 (emmprin; basigin) is a member of the Ig superfamily that has also been strongly implicated in chemoresistance and invasiveness. CD147 both regulates HA synthesis and interacts with the HA receptors, CD44, and LYVE-1. Increased CD147 expression induces formation of multiprotein complexes containing CD44 (or LYVE-1) as well as members of the

Advances in Cancer Research, Volume 123
ISSN 0065-230X
http://dx.doi.org/10.1016/B978-0-12-800092-2.00013-7

membrane-type matrix metalloproteinase, receptor tyrosine kinase, ABC drug transporter, or monocarboxylate transporter families, which become assembled in specialized lipid raft domains along with CD147 itself. In each case, multivalent HA-receptor interactions are essential for formation or stabilization of the lipid raft complexes and for downstream signaling pathways or transporter activities that are driven by these complexes. We conclude that cooperativity between HA, HA receptors, and CD147 may be a major driver of the interconnected pathways of invasiveness and chemoresistance widely critical to malignancy.

1. INTRODUCTION

Cancer is driven by various genetic aberrations and microenvironmental cues. An often overlooked component of the local microenvironment is the extracellular matrix (ECM), which is composed of a complex network of proteins and carbohydrates that form a favorable niche to facilitate tumor progression (Lu, Weaver, & Werb, 2012). A major constituent of the ECM is the linear nonsulfated glycosaminoglycan, hyaluronan (HA). In addition to participating as a vital structural molecule in several tissues, HA plays an instructional role in cell signaling by interacting with cell surface receptors (Toole, 2004, 2009). Accumulating evidence, reviewed extensively in this collection of chapters, supports a role for HA in various stages of cancer progression. CD44, the most studied HA receptor in cancer cells, is often associated with several other cell surface proteins including signaling proteins, transporters, and proteases (Bourguignon, 2008; Heldin, Basua, Kozlovaa, & Porsch, 2014; Toole, 2009; Zoller, 2011). HA–CD44 interactions are multivalent, leading to clustering of CD44 in the plasma membrane, often within lipid microdomains commonly termed lipid rafts (Murai et al., 2011; Oliferenko et al., 1999). Thus, HA–CD44 interactions facilitate formation of a pericellular scaffold that stabilizes CD44-containing complexes on the cell surface (see Fig. 13.1).

CD147 (emmprin; basigin) is a cell surface glycoprotein with various roles in cancer biology (Kennedy & Dewhirst, 2010; Nabeshima et al., 2006; Weidle, Scheuer, Eggle, Klostermann, & Stockinger, 2010). Among the several functions of CD147 is stimulation of HA synthesis (Marieb et al., 2004), thus promoting HA-receptor signaling and stabilization of cell surface complexes containing HA receptors. Moreover, CD147 has been shown to interact with the HA receptors, CD44 (Grass, Tolliver, Bratoeva, & Toole, 2013; Slomiany, Grass, et al., 2009), and LYVE-1 (Qin, Dai, Bratoeva, et al., 2011).

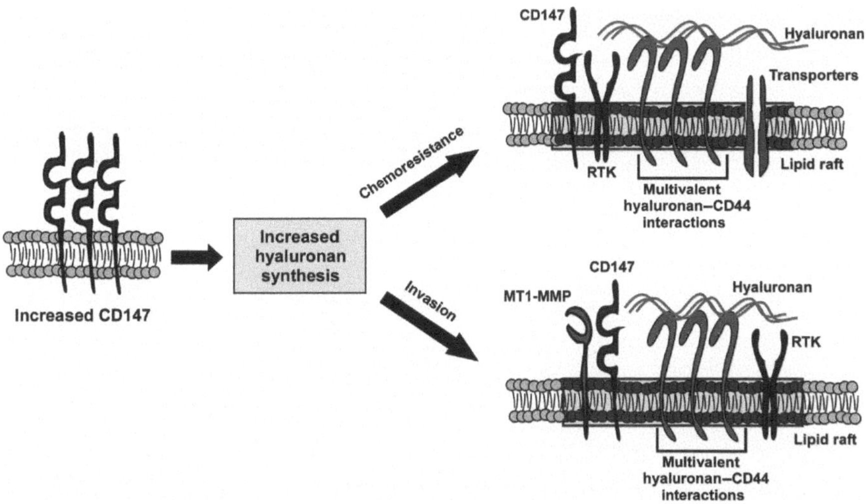

Figure 13.1 *Regulation of chemoresistance and invasiveness by CD147-induced HA–CD44 interactions.* Increased CD147 expression stimulates production of HA, which interacts multivalently with CD44, leading to assembly or stabilization of signaling and transporter complexes that are localized within specialized lipid raft domains in the plasma membrane. These complexes can contain MMPs, e.g., MT1-MMP, receptor tyrosine kinases, e.g., ErbB2 and EGFR, and transporters, e.g., ABC-family drug transporters and MCTs. These complexes drive many activities associated with malignant cells such as anti-apoptotic signaling, cell proliferation, altered metabolic phenotype, invasiveness, and chemoresistance. (See the color plate.)

In this chapter, we will focus on the role of CD147-induced activation of HA-receptor signaling in invasiveness and chemotherapy resistance.

2. STRUCTURE AND PLEIOTROPIC FUNCTIONS OF CD147

CD147 is a ubiquitously expressed glycoprotein to which many roles have been attributed in various physiologic and pathologic contexts. CD147 was initially identified and characterized independently under various names across different species and tissues, using diverse experimental methods (Muramatsu, 2012; Seulberger, Unger, & Risau, 1992). It is now clear that all of these proteins are in fact the same and are now commonly termed CD147, emmprin or basigin. In this review we will employ CD147, since it is now established that CD147 has pleiotropic functions.

The human gene locus for CD147 is on chromosome 19p13.3 (Kaname et al., 1993) and contains 10 exons (Belton, Chen, Mesquita, & Nowak, 2008; Liao et al., 2011). CD147 is flanked on the 5′-end with a CpG island-rich sequence in the promoter region, which contains a TATA box (Liang, Major, & Bocan, 2002). The mouse gene is located on chromosome 10 and is smaller in size compared to the human gene due to shorter intronic sequences interspersed between size-conserved exons (Cheng, Li, Kamholz, & Burns, 1994; Guo et al., 1998; Miyauchi et al., 1995). CD147 gene expression is regulated by various factors, but Sp1 appears to be the main transcriptional regulator (Kong, Liao, Chen, et al., 2011; Kong, Liao, Fei, et al., 2011).

Analyses by several groups have demonstrated that CD147 is a single-chain type I transmembrane glycoprotein and member of the Ig superfamily (Biswas et al., 1995; Miyauchi et al., 1990; Yu et al., 2008). Based on data in the Entrez Gene Database, the human CD147 gene (*BSG*) is transcribed as four splice variants, designated CD147-1, -2, -3, and -4. The most abundantly expressed and widely distributed variant of CD147 is CD147-2, and this form will be designated as CD147 hereafter. The mRNA transcript encodes a 269-amino acid protein that is composed of a 21-residue signal sequence, a 186-residue extracellular portion consisting of 2 Ig-like domains at the N-terminus, a 21-residue transmembrane domain, and a 41-residue cytoplasmic domain at the C-terminus (Weidle et al., 2010). The protein has three conserved N-glycosylation sites in the extracellular Ig domains, which provide attachment sites for highly branched sugars (Huang et al., 2013; Tang, Chang, & Hemler, 2004). The other isoforms are CD147-1, a retina-specific variant with three Ig-like domains (Hanna et al., 2003; Ochrietor et al., 2003), and two variants, CD147-3 and CD147-4, which each lack the N-terminal Ig-like domain (Belton et al., 2008; Liao et al., 2011).

CD147, then termed basigin, was initially cloned from F9 embryonal carcinoma cells (Miyauchi et al., 1990). It was first characterized functionally by Biswas and colleagues (Biswas, 1984; Ellis, Nabeshima, & Biswas, 1989) using a co-culture system of human lung carcinoma cells and fibroblasts, in which they identified a tumor cell-derived collagenase-stimulatory factor termed TCSF, which was present on the cell surface and in conditioned media of cancer cells, and which induced expression of collagenase (matrix metalloproteinase-1 [MMP-1)]). In subsequent years, TCSF was also found to stimulate message and protein levels of MMP-2 and MMP-3 and was renamed *E*xtracellular *M*atrix *M*etallo*pr*oteinase *In*ducer (emmprin) (Biswas et al., 1995; Kataoka, DeCastro, Zucker, & Biswas, 1993). Early studies also characterized CD147 as a cell recognition molecule (5A11) involved in neuronal–glial patterning and aggregation (Fadool & Linser,

1993). Studies in knock-out mice have demonstrated a role for CD147 in fertility, spatial learning, memory and sensory perception, and retinal function (Muramatsu & Miyauchi, 2003; Ochrietor & Linser, 2004). CD147 is also involved in immune, inflammatory and wound healing responses, as well as pathological conditions such as atherosclerosis and arthritis (Agrawal & Yong, 2011; Gabison, Huet, Baudouin, & Menashi, 2009; Joghetaei et al., 2013; Landskron & Tasken, 2013; Yurchenko, Constant, Eisenmesser, & Bukrinsky, 2010). In several cell types, CD147 is necessary for the trafficking and functional activity of monocarboxylate transporters (MCTs), specifically MCT-1, MCT-3, and MCT-4, and it is an accessory subunit of these MCTs (Halestrap, 2011). MCT-2 utilizes a related glycoprotein, embigin, in a similar manner (Wilson et al., 2005). CD147 also facilitates viral infection and the invasion of some microorganisms into host cells (Dai, Bai, et al., 2013; Muramatsu, 2012).

Although it is clear that CD147 participates in various aspects of biology and pathology, the role of CD147 in cancer remains the most studied. CD147 is highly enriched on the surfaces of most malignant cancer cell types, whereas expression in normal tissue is usually low and compartmentalized (Li et al., 2009; Riethdorf et al., 2006). As stated above, CD147 is a dominant inducer of both soluble and membrane-type MMPs (MT-MMPs), and has been shown to regulate expression of MMP-1, MMP-2, MMP-3, MMP-9, MT1-MMP, and MT2-MMP in various systems (Grass, Bratoeva, & Toole, 2012; Yan, Zucker, & Toole, 2005). CD147 stimulates MMP production in fibroblasts, endothelial cells, macrophages, epithelial cells and tumor cells, via heterophilic or homophilic cell–cell interactions (Weidle et al., 2010). In addition to MMPs, CD147 also aids in regulating the urokinase-type plasminogen activator system (Lescaille et al., 2012; Quemener et al., 2007). An *in vivo* study, in which CD147 was overexpressed in weakly malignant breast cancer cells, demonstrated that increased CD147 promotes tumor growth, local invasion, and metastases (Zucker et al., 2001). Additionally, CD147 is a predominant marker (90%) in bone metastases isolated from breast cancer patients (Klein et al., 2002; Reimers et al., 2004), is enriched in malignant pleural effusions (Davidson et al., 2004), and is expressed on circulating tumor cells (Pituch-Noworolska et al., 2007). Our laboratory has shown that CD147 is also highly enriched on $CD133^+$ primary ovarian cancer stem cells obtained from patient-derived ascites (Slomiany, Dai, Tolliver, et al., 2009), and that subpopulations of cancer cells with high-CD147 expression have cancer stem cell properties, including high levels of chemoresistance, invasiveness, and tumorigenicity (Dai, Guinea, et al., 2013). It is clear that CD147 is an important mediator of several cancer cell properties, especially invasion, and chemoresistance, which facilitate progression of numerous cancer types.

3. CD147-INDUCED HA SYNTHESIS AND SIGNALING

We found that experimental up-regulation of CD147 increases HA synthesis in breast cancer cells (Marieb et al., 2004). This was due to increased message levels for all three HA synthases (HASs), with HAS2 being the major isoform induced in these cells. Subsequent studies in other cell types, using a variety of techniques, have also demonstrated that CD147 aids in regulating HA synthesis (Edward, Quinn, & Sands, 2011; Ghatak, Misra, & Toole, 2005; Grass et al., 2013; Kato et al., 2011; Qin, Dai, Bratoeva, et al., 2011). Currently, the mechanism whereby CD147 regulates HAS transcription and subsequent HA production is not clear. However, one possibility stems from the above-mentioned observation that CD147 is essential for trafficking of MCTs to the cell surface and consequently for efflux of lactate from the cell. Lactate stimulates HA synthesis (Stern, Shuster, Neudecker, & Formby, 2002), and several genes related to HA metabolism have lactate-response elements (Formby & Stern, 2003). Thus, CD147-mediated efflux of lactate via MCT transporters may stimulate HA synthesis. Alternatively, CD147-induced proteolytic activity may free ECM-sequestered growth factors that modulate HA synthesis (Chow, Tauler, & Mulshine, 2010; Li, Asteriou, Bernert, Heldin, & Heldin, 2007; Pienimaki et al., 2001). Interestingly, subpopulations of cancer cells with high levels of cell surface CD147 (Dai, Guinea, et al., 2013) express high levels of hyaluronidase-2, as well as HAS2 and HA (unpublished results). These observations are consistent with the importance of HA turnover in cancer progression (Lokeshwar & Selzer, 2008; McAtee, Barycki, & Simpson, 2014).

4. CD147–HA AXIS IN CELLULAR INVASION

Cell invasion, defined as migration through or within a tissue barrier, constitutes an essential mechanism employed by different cell types in various biological and pathological contexts. In carcinomas, cells depart from the primary tumor by utilizing diverse migratory and invasive mechanisms (Chaffer & Weinberg, 2011). Many of the initial steps in the metastatic cascade involve transitions in cellular programs that support increased cellular motility and invasion, such as mesenchymal, amoeboid, and collective migration processes (Sahai, 2005). Numerous early studies focusing on induction of MMPs by CD147 also demonstrated the potential importance of CD147 in promoting invasiveness. In this section, we discuss various ways that cooperativity between CD147 and HA–CD44 signaling may impact cell invasiveness.

4.1. Epithelial–mesenchymal transitions

The epithelial–mesenchymal transition (EMT) is involved in regulation of developmental processes, wound repair/tissue regeneration, cancer invasion, and metastasis. EMT is characterized by dissociation of cell–cell contacts, alteration of the cytoskeletal network, and increased proteolytic activity, consequently leading to cell invasiveness, antiapoptotic signaling, and anchorage-independent growth (anoikis) (Kalluri & Weinberg, 2009; Teng, Zeisberg, & Kalluri, 2007). In hepatocellular carcinoma, upregulation of CD147 initiates a TGF-β signaling cascade resulting in increased Slug expression, cadherin switching, and changes in hepatocyte morphology indicative of an EMT (Wu et al., 2011). CD147 may also prime cells to undergo an EMT by modulating the E-cadherin–β-catenin complex through stimulation of nuclear translocation of β-catenin (Sidhu et al., 2010). A recent study showed that tumor-derived CD147 promotes a cancer-associated fibroblast phenotype, which subsequently initiates a feedback loop promoting an EMT in the tumor cells that provided the initial CD147 stimulus (Xu, Lu, Qiu, Chen, & Fan, 2013).

CD147 may also promote the mesenchymal phenotype by increasing HA synthesis. HA is an important regulator of EMT, as initially demonstrated in studies of the HAS2 knock-out mice where defective EMT events were shown to be the major cause of embryonic lethality (Camenisch et al., 2000). It has been shown that HAS2 overexpression in breast epithelial cells induces mesenchymal properties that are dependent on HA–CD44 interactions (Zoltan-Jones, Huang, Ghatak, & Toole, 2003), and that experimentally increased HA synthesis in a spontaneous breast cancer mouse model results in carcinoma cells that appear to have acquired mesenchymal properties (Koyama et al., 2007). Anchorage-independent growth is necessary for a cell to undergo a complete EMT (Guadamillas, Cerezo, & del Pozo, 2011). Experimental overexpression of CD147 induces anchorage-independent growth in breast epithelial cells by a process that is dependent on HA-receptor interactions (Marieb et al., 2004). The specific mechanisms by which CD147 promotes EMT-like events are still elusive and most likely are cell type-specific.

4.2. Regulation of invadopodia formation and activity by CD147

Invasive cell phenotypes exhibit a continuum of different forms of membrane protrusions and contraction states, which are regulated by diverse signaling and physical matrix cues (Friedl & Wolf, 2009). When invasive cells are cultured on an ECM substratum, they form highly dynamic protrusive structures called invadopodia, first identified in Rous sarcoma virus-transformed cells cultured on fibronectin (Chen, Chen, Parsons, & Parsons, 1985). Growing evidence now indicates that invadopodia are critical for cancer cell invasiveness *in vitro* and *in vivo* (Gligorijevic et al., 2012; Paz, Pathak, & Yang, 2013). Similar cell protrusions, usually termed podosomes, have been identified in osteoclasts, macrophages, dendritic cells, endothelial cells, and vascular smooth muscle cells. Thus, the term "invadopodia" is usually reserved for cells with transformed properties and the term "podosome" for normal cells (Murphy & Courtneidge, 2011).

Invadopodia contain an actin-rich core, modulators of actin polymerization such as cortactin, cell adhesion molecules, signaling phosphoproteins, and proteases (Murphy & Courtneidge, 2011; Paz et al., 2013). Following cell attachment to the ECM, invadopodia are assembled in regions of plasma membrane enriched in specialized lipid microdomains termed lipid rafts (Caldieri et al., 2009; Yamaguchi et al., 2009), and as invadopodia mature various types of proteases become enriched at their tips (Schoumacher, Goldman, Louvard, & Vignjevic, 2010). These proteases include both soluble and membrane-tethered proteases, but MT1-MMP (MMP-14) has emerged as an essential protein regulating the ECM-degrading activity of invadopodia (Frittoli, Palamidessi, Disanza, & Scita, 2011; Poincloux, Lizarraga, & Chavrier, 2009).

Multiple mechanisms have been proposed to explain how cancer cells deliver active MT1-MMP to invadopodia at the cell surface. Such trafficking routes include complex interactions with various cytoskeletal, vesicular, and signaling components (Frittoli et al., 2011; Murphy & Courtneidge, 2011; Poincloux et al., 2009). Our laboratory showed that overexpression of CD147 in noninvasive, nontransformed breast epithelial cells induces formation of invadopodia that incorporate lipid raft-associated CD147–MT1-MMP complexes and degrade underlying ECM; concomitantly, the epithelial cells become invasive (Grass et al., 2012). Additionally, we found that down-regulation of CD147 expression in invasive breast cancer cells attenuates invadopodia activity and invasiveness, thus providing strong evidence that CD147 is a critical regulator of invadopodia activity and cell invasiveness.

Interestingly, we noted that overexpression of CD147 also induced formation of filopodia, enriched with lipid rafts, which project from the apical aspects of the breast epithelial cells (Grass et al., 2013). These filopodia appear to be similar to projections that have been observed following HAS overexpression; these are also enriched in lipid rafts and CD44 (Kultti et al., 2006). Further investigation is required to determine if CD147-induced filopodial projections are dependent on HA synthesis.

4.3. Regulation of invadopodia activity and invasiveness by CD147-induced HA–CD44 and EGFR signaling

We have recently investigated the signaling pathways underlying the induction of invadopodia activity and invasiveness by CD147. We found that CD147 regulates these phenomena by activation of an EGFR–Ras–ERK pathway, which is at least in part dependent on CD147-induced increases in HA synthesis and interactions with CD44 (Grass et al., 2013). Previous studies have shown that CD44 is associated with and involved in formation of invadopodia (Vikesaa et al., 2006), and that CD44 is often incorporated into lipid rafts (Hryniewicz-Jankowska, Augoff, Biernatowska, Podkalicka, & Sikorski, 2014; Murai et al., 2011; Oliferenko et al., 1999). We showed that upregulation of CD147 leads to formation of lipid raft-associated complexes of CD147 with EGFR and CD44, without apparent changes in overall cell surface expression of CD44 or EGFR. Thus, we propose that increased CD147 expression causes a redistribution of endogenous EGFR, CD44, and CD147 into lipid raft domain complexes that are stabilized by increased HA–CD44 interaction and that these complexes, along with the above-described CD147–MT1-MMP complex (Grass et al., 2012), are important components of active invadopodia. In support of this, we found that disruption of HA-receptor interactions with small HA oligosaccharides or knockdown of CD44 attenuated CD147-induced EGFR–ERK activation. The effect of the oligosaccharides was limited in duration, since activation of the EGFR and ERK pathway returned towards baseline after treatment for 12 h (Grass et al., 2013). This may be due to continued constitutive synthesis of HA or initiation of adaptive mechanisms to reactivate downregulated signaling pathways, e.g., via formation of a

CD44–Rhamm complex previously shown to regulate ERK activity and sustain high-basal motility in breast cancer cells (Hamilton et al., 2007).

Although partial knockdown of CD44 causes a decrease in CD147-induced invasiveness, it does not significantly alter CD147-induced MT1-MMP levels (Grass et al., 2013). This suggests that there are diverging pathways elicited by CD147: those mediated by HA–CD44 interactions and those that may utilize other CD147-binding partners. To date the upstream mechanisms whereby CD147 stimulates MT1-MMP expression, HA production and EGFR signaling are not known. However, it is known that CD147 is critical for trafficking and activity of MCTs, both in normal cells (Philp, Ochrietor, Rudoy, Muramatsu, & Linser, 2003) and cancer cells (Gallagher, Castorino, Wang, & Philp, 2007). MCTs are important in lactate secretion and acidification of the microenvironment, which in turn promote invasion and metastasis (Damaghi, Wojtkowiak, & Gillies, 2013; Parks, Chiche, & Pouyssegur, 2013). Also, MCTs have been found at the leading edge of migrating cells (Gallagher, Castorino, & Philp, 2009) and are associated with aggressive tumors (Parks et al., 2013). We have found that CD147, CD44, and MCT1/4 form complexes on the cell surface, and that the functional activity and cell surface stability of these MCTs are dependent on HA-receptor interactions as well as CD147 (Slomiany, Grass, et al., 2009). Thus, it is plausible that CD147, CD44, and MCTs cooperate at the leading edge of cancer cells to facilitate acidification of the pericellular milieu and promote invasiveness. Evidence has been published supporting the possibility that a major pro-invasive, protumorigenic activity of CD147 is regulation of lactate secretion via interaction with MCTs (Izumi et al., 2011; Le Floch et al., 2011; Schneiderhan et al., 2009).

We also showed that oncogenic Ras activity increases HA synthesis [as previously shown (Hamilton et al., 2007)], CD147 protein expression, assembly of lipid raft-associated EGFR–CD44–CD147 complexes, and formation of active invadopodia. Ras-induced HA synthesis, ERK activation, and invadopodia activity were at least in part dependent on CD147 (Grass et al., 2013), suggesting that Ras cooperates with CD147 to increase HA-receptor interactions and downstream signaling events. Previous observations indicate that HAS2 activity is necessary for EGFR activation and cooperates with Ras in cellular invasion *in vivo* (Camenisch, Schroeder, Bradley, Klewer, & McDonald, 2002; Camenisch et al., 2000). It has also been shown that HA increases Ras activity (Fitzgerald, Bowie, Skeffington, & O'Neill, 2000; Serbulea et al., 1999) and that Ras influences CD44 alternative splicing, which results in additional Ras activation

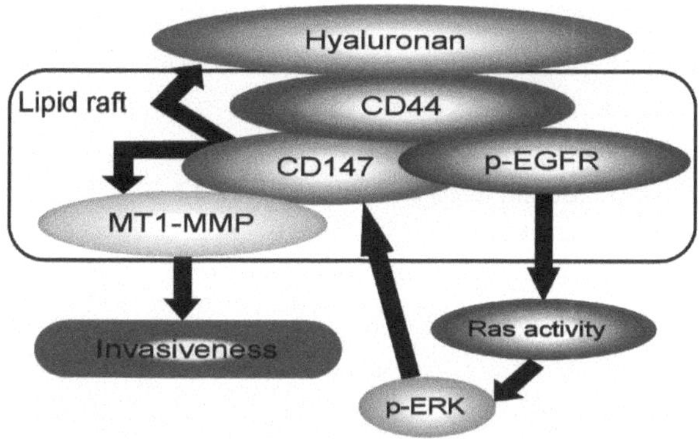

Figure 13.2 *Regulation of invadopodia formation and invasiveness by CD147.* Increased CD147 expression stimulates MT1-MMP expression, formation of CD147–MT1-MMP complexes associated within lipid raft domains, and HA production. Consequent to the latter, increased HA–CD44 interaction stimulates assembly of lipid raft-associated complexes containing CD44, EGFR, and CD147. These complexes drive the EGFR–Ras–ERK signaling pathway, which in turn participates in a positive feedback loop that results in further increases in CD147 expression, MT1-MMP activity, and HA–CD44 interaction. These events lead to increase assembly and activity of invadopodia, thus promoting invasiveness. (See the color plate.)

(Cheng, Yaffe, & Sharp, 2006). Our findings that CD147 increases HA synthesis and EGFR–Ras–ERK activation, and that Ras increases CD147 expression, imply that CD147 integrates into a HA–CD44–EGFR–Ras signaling module by propagating a positive feedback loop (see Fig. 13.2). In further support of this, we have found that subpopulations of cancer cells with constitutively high cell surface expression of CD147 exhibit elevated HA production, cell surface expression of EGFR and CD44, and ERK activity compared to subpopulations with low cell surface CD147 expression (Dai, Guinea, et al., 2013; Grass et al., 2013).

5. CD147–HA AXIS IN CHEMORESISTANCE

Several studies have shown that CD147 expression in cancer cells correlates with and regulates drug resistance and resistance to apoptosis (Dai, Guinea, et al., 2013; Hao et al., 2012; Kang et al., 2013; Misra, Ghatak, Zoltan-Jones, & Toole, 2003; Qin, Dai, Bratoeva, et al., 2011; Yang et al., 2003; Zhou et al., 2013). Using human carcinoma cell lines, we showed that experimental upregulation of CD147 leads to anchorage-independent growth, i.e., resistance to apoptosis due to substrate detachment, and causes increased resistance to treatment with chemotherapeutic drugs, concomitant with increased HA production. These properties were reversed by treatment with a HA-receptor antagonist, i.e., small HA oligosaccharides, indicating that CD147-induced anchorage-independent growth and chemoresistance are dependent on HA-receptor interaction (Marieb et al., 2004; Misra et al., 2003). These studies also showed that increased CD147 stimulates phosphorylation of Akt, ERK, and BAD through stimulation of HA–CD44 interaction. In another study we showed that HA–CD44 interaction regulates drug resistance via stimulation of ErbB2 and PI3-kinase activity, which in turn regulates expression of the ABC-family drug transporter, P-glycoprotein (ABCB1) (Misra, Ghatak, & Toole, 2005). In addition, CD147 increases ErbB2 and PI3-kinase activity via stimulation of HA–CD44 interaction (Ghatak et al., 2005), thus leading to increased drug transporter expression and drug resistance. We have also compared the properties of cells with high versus low surface levels of CD147 expression after separation by fluorescent activated cell sorting (FACS) from several cancer cell types. In accordance with the studies described above, the cells with high cell surface CD147 expression produce more HA, exhibit greater transporter activity, and are much more drug-resistant than cells with low cell surface CD147 expression (Dai, Guinea, et al., 2013).

It is very likely that one of the mechanisms whereby CD147 stimulates drug resistance is via downstream antiapoptotic signaling pathways that are promoted by increased interaction of HA with its receptors, especially CD44. However, as discussed above, one of the established roles of CD147 is its involvement in trafficking and activity of MCTs (Halestrap, 2011). Increased MCT activity usually leads to increased lactate secretion, which has been linked to various aspects of malignancy, including chemoresistance (Kennedy & Dewhirst, 2010). We have shown that CD147 and CD44 form complexes with MCT1 and MCT4, and that antagonists of either CD147 or HA-receptor interaction inhibit MCT activity (Slomiany, Grass, et al., 2009). Evidence has also been published indicating that CD147 (Wang et al., 2008; Zhou et al., 2013) and CD44 (Miletti-Gonzalez et al., 2005; Slomiany, Dai, Bomar, et al., 2009) form complexes with drug transporters. Thus, as also described in the section on cell invasion, cooperative interactions between CD147 and CD44 may be important in the formation or stabilization of plasma membrane complexes that regulate numerous downstream activities, including drug resistance.

An important development over the past decade has been the study of highly malignant and chemoresistant subpopulations of many types of cancers and established tumor cell lines, commonly termed cancer stem cells. It has been proposed that such subpopulations may be responsible for resistance of recurrent tumors to treatment with chemotherapeutic agents and may be highly metastatic (Chaffer & Weinberg, 2011; Clevers, 2011; O'Connor et al., 2014). CD44 is one of the most common markers for cancer stem cells (Zoller, 2011), and we have found that primary ovarian cancer stem cells isolated from patient ascites produce more HA than the non stem cell subpopulation, and that CD44 is present in close association with CD147, receptor tyrosine kinases, drug transporters, and MCTs in the plasma membrane of the cancer stem cells; accordingly, treatment of the cancer stem cells with a HA–CD44 antagonist, i.e., small HA oligosaccharides, inhibits the association of these proteins, reduces drug transporter activity, and prevents their growth *in vivo* (Slomiany, Dai, Tolliver, et al., 2009). In light of these studies, we compared the properties of cells with high versus low surface levels of CD147 expression after FACS from several cancer cell types. The cells with high cell surface CD147 expression exhibit characteristic cancer stem cell properties, i.e., anchorage-independent growth, spheroid formation, drug resistance, and invasiveness *in vitro*, as well as higher tumorigenicity *in vivo* than cells expressing low levels of cell surface CD147 (Dai, Guinea, et al., 2013). In accordance with these results, a recent proteomic study showed that CD147 is critical for chemoresistance in breast cancer stem cells (Kang et al., 2013).

6. INDUCTION OF THE CD147–HA AXIS BY KAPOSI'S SARCOMA-ASSOCIATED HERPESVIRUS

CD147 facilitates infection of host cells by several viruses and other microorganisms (Dai, Bai, et al., 2013; Muramatsu, 2012). Kaposi's sarcoma-associated herpesvirus (KSHV) is the etiologic agent for unique malignancies arising preferentially in patients infected with the human immunodeficiency virus or in those receiving organ transplants, including Kaposi's sarcoma (KS), and primary effusion lymphoma (PEL) (Cesarman, Chang, Moore, Said, & Knowles, 1995; Chang et al., 1994). Cytotoxic chemotherapy remains the standard approach for these tumors, but morbidity and mortality remain high, due largely to drug resistance and off-target toxicities. KSHV-infected "spindle cells" of endothelial cell origin represent the characteristic feature of KS lesions. We have reported that KSHV induces CD147 expression in primary human endothelial cells, which in turn promotes synthesis of pro-angiogenic and pro-invasive factors, e.g., VEGF-A and several MMPs, and acquisition of cell invasiveness (Dai, Bratoeva, Toole, Qin, & Parsons, 2012; Qin, Dai, Slomiany, Toole, & Parsons, 2010). Fibroblasts are also found in KS tumors (Akula et al., 2003), and we previously reported that KSHV-infected, skin-derived fibroblasts promote endothelial cell invasion of ECM through paracrine mechanisms (Qin, Dai, Toole, Robertson, & Parsons, 2011) suggesting a potential role for KSHV-infected fibroblasts in promoting KS pathogenesis. More recently, we reported that KSHV induces a "tumor-associated" phenotype for primary human fibroblasts, including their production of pro-angiogenic cytokines and increased intrinsic invasiveness, through upregulation of CD147 (Dai, Qin, et al., 2012). KSHV also stimulates HA synthesis in endothelial cells and fibroblasts (unpublished data), and we are currently investigating potential linkages between HA-receptor signaling and the pro-invasive activity of CD147 in these cells.

We have also found that CD147 and HA are produced by KSHV-infected PEL cells, and that the levels of CD147, HA, and the HA receptor, LYVE-1, correlate directly with their resistance to cytotoxic agents currently employed for PEL therapy (Qin, Dai, Bratoeva, et al., 2011). Moreover, we identified protein complexes containing LYVE-1, CD147, and the drug transporter, BCRP/ABCG2 on the surface of PEL cells and found that CD147 mediates KSHV-induced chemoresistance in a LYVE-1- and BCRP-dependent manner. Interestingly, BCRP expression in PEL cells is dependent on both CD147 and LYVE-1, and knockdown of LYVE-1 or CD147 in PEL cells results in increased apoptosis in the presence of cytotoxic agents (Qin, Dai, Bratoeva, et al., 2011). This work represented the initial identification of a putative mechanism for cancer pathogenesis associated with LYVE-1, and the first to identify a functional interdependence for CD147 and LYVE-1. Importantly, we found that incubation of PEL cells with small HA oligosaccharides that interfere with endogenous HA-receptor interactions leads to loss of BCRP expression and chemosensitization. Furthermore, ectopic expression of CD147 in chemosensitive PEL cell lines increases their chemoresistance, and this effect is reversed by cotreatment with HA oligosaccharides (Qin, Dai, Bratoeva, et al., 2011). We have previously reviewed future directions for understanding KSHV regulation of CD147 expression and downstream pathogenesis, including invasiveness, and chemoresistance, for KSHV-infected cells (Dai, Bai, et al., 2013).

7. CONCLUSIONS

A striking aspect of the research published on CD147 and HA-receptor interactions in cancer is their role in assembly or stabilization of plasma membrane complexes, often present in specialized lipid microdomains termed lipid rafts (see Fig. 13.1). Among these are complexes containing receptor tyrosine kinases that regulate several downstream pathways, e.g., PI3 kinase/Akt and MAPK pathways, leading to malignant cellular phenotypes such as increased proliferation and invasion. Also among these are complexes containing drug transporters and MCTs that regulate chemoresistance and altered metabolic pathways characteristic of cancer cells (Toole, 2009). The exact manner in which these complexes are assembled and regulated is not known. However, it is tempting to postulate that the multivalent nature of HA may be critical. By crosslinking numerous HA receptor molecules, which in turn interact with critical members of the cytoskeleton and intracellular signaling pathways, HA may facilitate stable assembly of multicomponent complexes in an active configuration, e.g., in lipid rafts, which are known to be critical in protumorigenic signaling pathways (Head, Patel, & Insel, 2014; Hryniewicz-Jankowska et al., 2014; Staubach & Hanisch, 2011). This would explain the sensitivity of numerous malignant cellular properties to perturbations of HA-receptor interactions.

Also difficult to explain are the pleiotropic functions of CD147, although some of these functions appear to be explained by its role in stimulation of HA synthesis and, consequently, HA-receptor interactions and their downstream outcomes. A major function of CD147 that stands out as a potentially unifying mechanism with respect to its role in tumor progression is as an accessory subunit of MCT1, MCT3, and MCT4 (Halestrap, 2011; Ochrietor & Linser, 2004). It has been shown that the cell surface expression of these MCTs and CD147 are interdependent and that the function of CD147 in tumor growth may be dependent on its interaction with MCTs (Gallagher et al., 2007; Le Floch et al., 2011). MCT4 is of particular interest since both MCT4 and CD147 are upregulated under hypoxic conditions and participate in creating an acidic pericellular environment (Ke et al., 2012; Ullah, Davies, & Halestrap, 2006). Hypoxic and acidic microenvironments are interrelated conditions in malignant cancers that promote invadopodia activity (Brisson, Reshkin, Gore, & Roger, 2012) and tumor progression (Damaghi et al., 2013; Parks et al., 2013). In addition, the mechanism whereby CD147 promotes HA-receptor interactions and downstream signaling may be mediated by its role in secretion of lactate by MCTs, since lactate stimulates HA production and possibly HA receptor complex formation (Formby & Stern, 2003; Slomiany, Grass, et al., 2009; Stern et al., 2002).

REFERENCES

Agrawal, S. M., & Yong, V. W. (2011). The many faces of EMMPRIN—Roles in neuroinflammation. *Biochimica et Biophysica Acta-Molecular Basis of Disease, 1812*(2), 213–219.

Akula, S. M., Naranatt, P. P., Walia, N.-S., Wang, F.-Z., Fegley, B., & Chandran, B. (2003). Kaposi's sarcoma-associated herpesvirus (human herpesvirus 8) infection of human fibroblast cells occurs through endocytosis. *Journal of Virology, 77*(14), 7978–7990.

Belton, R. J., Chen, L., Mesquita, F. S., & Nowak, R. A. (2008). Basigin-2 is a cell surface receptor for soluble basigin ligand. *Journal of Biological Chemistry, 283*(26), 17805–17814.

Biswas, C. (1984). Collagenase stimulation in cocultures of human fibroblasts and human tumor cells. *Cancer Letters, 24*, 201–207.

Biswas, C., Zhang, Y., DeCastro, R., Guo, H., Nakamura, T., Kataoka, H., et al. (1995). The human tumor cell-derived collagenase stimulatory factor (renamed EMMPRIN) is a member of the immunoglobulin superfamily. *Cancer Research, 55*, 434–439.

Bourguignon, L. Y. (2008). Hyaluronan-mediated CD44 activation of RhoGTPase signaling and cytoskeleton function promotes tumor progression. *Seminars in Cancer Biology, 18*(4), 251–259.

Brisson, L., Reshkin, S. J., Gore, J., & Roger, S. (2012). pH regulators in invadosomal functioning: Proton delivery for matrix tasting. *European Journal of Cell Biology, 91*(11–12), 847–860.

Caldieri, G., Giacchetti, G., Beznoussenko, G., Attanasio, F., Ayala, I., & Buccione, R. (2009). Invadopodia biogenesis is regulated by caveolin-mediated modulation of membrane cholesterol levels. *Journal of Cellular and Molecular Medicine, 13*(8B), 1728–1740.

Camenisch, T. D., Schroeder, J. A., Bradley, J., Klewer, S. E., & McDonald, J. A. (2002). Heart-valve mesenchyme formation is dependent on hyaluronan-augmented activation of ErbB2–ErbB3 receptors. *Nature Medicine, 8*(8), 850–855.

Camenisch, T. D., Spicer, A. P., Brehm-Gibson, T., Biesterfeldt, J., Augustine, M. L., Anthony Calabro, J., et al. (2000). Disruption of hyaluronan synthase-2 abrogates normal cardiac morphogenesis and hyaluronan-mediated transformation of epithelium to mesenchyme. *Journal of Clinical Investigation, 106*(3), 349–360.

Cesarman, E., Chang, Y., Moore, P. S., Said, J. W., & Knowles, D. M. (1995). Kaposi's sarcoma-associated herpesvirus-like DNA sequences in AIDS-related body-cavity-based lymphomas. *New England Journal of Medicine, 332*(18), 1186–1191.

Chaffer, C. L., & Weinberg, R. A. (2011). A perspective on cancer cell metastasis. *Science, 331*, 1559–1564.

Chang, Y., Cesarman, E., Pessin, M. S., Lee, F., Culpepper, J., Knowles, D. M., et al. (1994). Identification of herpesvirus-like DNA sequences in AIDS-associated Kaposi's sarcoma. *Science, 266*(5192), 1865–1869.

Chen, W. T., Chen, J. M., Parsons, S. J., & Parsons, J. T. (1985). Local degradation of fibronectin at sites of expression of the transforming gene product of pp60src. *Nature Reviews. Molecular Cell Biology, 316*(6024), 156–158.

Cheng, Y., Li, X., Kamholz, J., & Burns, F. R. (1994). Organization of the mouse GP42/Basigin gene: A member of the Ig superfamily. *Biochimica et Biophysica Acta, 1217*, 307–311.

Cheng, C., Yaffe, M. B., & Sharp, P. A. (2006). A positive feedback loop couples Ras activation and CD44 alternative splicing. *Genes & Development, 20*, 1715–1720.

Chow, G., Tauler, J., & Mulshine, J. L. (2010). Cytokines and growth factors stimulate hyaluronan production: Role of hyaluronan in epithelial to mesenchymal-like transition in non-small cell lung cancer. *Journal of Biomedicine and Biotechnology, 2010*, 485468.

Clevers, H. (2011). The cancer stem cell: Premises, promises and challenges. *Nature Medicine, 17*(3), 313–319.

Dai, L., Bai, L., Xu, Z., Reiss, K., Del Valle, L., Kaleeba, J., et al. (2013). Emmprin and KSHV: New partners in viral cancer pathogenesis. *Cancer Letters, 337*, 161–166.

Dai, L., Bratoeva, M., Toole, B. P., Qin, Z., & Parsons, C. (2012). KSHV activation of VEGF secretion and invasion for endothelial cells is mediated through viral upregulation of emmprin-induced signal transduction. *International Journal of Cancer, 131*(4), 834–843.

Dai, L., Guinea, M. C., Slomiany, M. G., Bratoeva, M., Grass, G. D., Tolliver, L. B., et al. (2013). CD147-dependent heterogeneity in malignant and chemoresistant properties of cancer cells. *American Journal of Pathology, 182*(2), 577–585.

Dai, L., Qin, Z., Defee, M., Toole, B. P., Kirkwood, K. L., & Parsons, C. (2012). Kaposi sarcoma-associated herpesvirus (KSHV) induces a functional tumor-associated phenotype for oral fibroblasts. *Cancer Letters, 318*, 214–220.

Damaghi, M., Wojtkowiak, J. W., & Gillies, R. J. (2013). pH sensing and regulation in cancer. *Frontiers in Physiology, 4*, 370.

Davidson, B., Konstantinovsky, S., Nielsen, S., Dong, H. P., Berner, A., Vyberg, M., et al. (2004). Altered expression of metastasis-associated and regulatory molecules in effusions from breast cancer patients: A novel model for tumor progression. *Clinical Cancer Research, 10*, 7335–7346.

Edward, M., Quinn, J. A., & Sands, W. (2011). Keratinocytes stimulate fibroblast hyaluronan synthesis through the release of stratifin: A possible role in the suppression of scar tissue formation. *Wound Repair and Regeneration, 19*(3), 379–386.

Ellis, S. M., Nabeshima, K., & Biswas, C. (1989). Monoclonal antibody preparation and purification of a tumor cell collagenase-stimulatory factor. *Cancer Research, 49*, 3385–3391.

Fadool, J. M., & Linser, P. J. (1993). 5A11 antigen is a cell recognition molecule which is involved in neuronal–glial interactions in avian neural retina. *Developmental Dynamics, 196*, 252–262.

Fitzgerald, K. A., Bowie, A. G., Skeffington, B. S., & O'Neill, L. A. J. (2000). Ras, protein kinase Cζ, and IkB kinases 1 and 2 are downstream effectors of CD44 during the activation of NF-kB by hyaluronic acid fragments in T-24 carcinoma cells. *Journal of Immunology, 164*(4), 2053–2063.

Formby, B., & Stern, R. (2003). Lactate-sensitive response elements in genes involved in hyaluronan catabolism. *Biochemical and Biophysical Research Communications, 305*, 203–208.

Friedl, P., & Wolf, K. (2009). Plasticity of cell migration: A multiscale tuning model. *Journal of Cell Biology, 188*(1), 11–19.

Frittoli, E., Palamidessi, A., Disanza, A., & Scita, G. (2011). Secretory and endo/exocytic trafficking in invadopodia formation: The MT1-MMP paradigm. *European Journal of Cell Biology, 90*(2–3), 108–114.

Gabison, E. E., Huet, E., Baudouin, C., & Menashi, S. (2009). Direct epithelial–stromal interaction in corneal wound healing: Role of EMMPRIN/CD147 in MMPs induction and beyond. *Progress in Retinal and Eye Research, 28*, 19–33.

Gallagher, S. M., Castorino, J. J., & Philp, N. J. (2009). Interaction of monocarboxylate transporter 4 with β_1-integrin and its role in cell migration. *American Journal of Physiology. Cell Physiology, 296*, C414–C421.

Gallagher, S. M., Castorino, J. J., Wang, D., & Philp, N. J. (2007). Monocarboxylate transporter 4 regulates maturation and trafficking of CD147 to the plasma membrane in the metastatic breast cancer cell line MDA-MB-231. *Cancer Research, 67*(9), 4182–4189.

Ghatak, S., Misra, S., & Toole, B. P. (2005). Hyaluronan constitutively regulates ErbB2 phosphorylation and signaling complex formation in carcinoma cells. *Journal of Biological Chemistry, 280*(10), 8875–8883.

Gligorijevic, B., Wyckoff, J., Yamaguchi, H., Wang, Y., Roussos, E. T., & Condeelis, J. (2012). N-WASP-mediated invadopodium formation is involved in intravasation and lung metastasis of mammary tumors. *Journal of Cell Science, 125*(3), 724–734.

Grass, G. D., Bratoeva, M., & Toole, B. P. (2012). Regulation of invadopodia formation and activity by CD147. *Journal of Cell Science, 125*(3), 777–788.

Grass, G. D., Tolliver, L., Bratoeva, M., & Toole, B. P. (2013). CD147, CD44, and the epidermal growth factor receptor (EGFR) signaling pathway cooperate to regulate breast epithelial cell invasiveness. *Journal of Biological Chemistry, 288*(36), 26089–26104.

Guadamillas, M. C., Cerezo, A., & del Pozo, M. A. (2011). Overcoming anoikis—Pathways to anchorage-independent growth in cancer. *Journal of Cell Science, 124*, 3189–3197.

Guo, H., Majmudar, G., Jensen, T. C., Biswas, C., Toole, B. P., & Gordon, M. K. (1998). Characterization of the gene for human EMMPRIN, a tumor cell surface inducer of matrix metalloproteinases. *Gene, 220*, 99–108.

Halestrap, A. P. (2011). The monocarboxylate transporter family—Structure and functional characterization. *IUBMB Life, 64*(1), 1–9.

Hamilton, S. R., Fard, S. F., Paiwand, F. F., Tolg, C., Veiseh, M., Wang, C., et al. (2007). The hyaluronan receptors CD44 and Rhamm (CD168) form complexes with ERK1,2 that sustain high basal motility in breast cancer cells. *Journal of Biological Chemistry, 282*(22), 16667–16680.

Hanna, S. M., Kirk, P., Holt, O. J., Puklavec, M. J., Brown, M. H., & Barclay, A. N. (2003). A novel form of the membrane protein CD147 that contains an extra Ig-like domain and interacts homophilically. *BMC Biochemistry, 4*, 17.

Hao, J., Madigan, M. C., Khatri, A., Power, C. A., Hung, T.-T., Beretov, J., et al. (2012). In vitro and in vivo prostate cancer metastasis and chemoresistance can be modulated by expression of either CD44 or CD147. *PLoS One*, 7(8), e40716.

Head, B. P., Patel, H. H., & Insel, P. A. (2014). Interaction of membrane/lipid rafts with the cytoskeleton: Impact on signaling and function: Membrane/lipid rafts, mediators of cytoskeletal arrangement and cell signaling. *Biochimica et Biophysica Acta, Biomembranes, 1838*(2), 532–545.

Heldin, P., Basua, K., Kozlovaa, I., & Porsch, H. (2014). HAS2 and CD44 signaling in breast tumorigenesis. *Advances in Cancer Research, 123*, 211–229.

Hryniewicz-Jankowska, A., Augoff, K., Biernatowska, A., Podkalicka, J., & Sikorski, A. F. (2014). Membrane rafts as a novel target in cancer therapy. *Biochimica et Biophysica Acta-Reviews on Cancer, 1845*(2), 155–165.

Huang, W., Luo, W.-J., Zhu, P., Tang, J., Yu, X.-L., Cui, H.-Y., et al. (2013). Modulation of CD147-induced matrix metalloproteinase activity: Role of CD147 N-glycosylation. *Biochemical Journal, 449*, 437–448.

Izumi, H., Takahashi, M., Uramato, H., Nakayama, Y., Oyama, T., Wang, K.-Y., et al. (2011). Monocarboxylate transporters 1 and 4 are involved in the invasion activity of human lung cancer cells. *Cancer Science, 102*(5), 1007–1013.

Joghetaei, N., Stein, A., Byrne, R. A., Schulz, C., King, L., May, A. E., et al. (2013). The extracellular matrix metalloproteinase inducer (EMMPRIN, CD147)—A potential novel target in atherothrombosis prevention? *Thrombosis Research, 131*(6), 474–480.

Kalluri, R., & Weinberg, R. A. (2009). The basics of epithelial–mesenchymal transition. *Journal of Clinical Investigation, 119*(6), 1420–1428.

Kaname, T., Miyauchi, T., Kuwano, A., Matsuda, Y., Muramatsu, T., & Kajiii, T. (1993). Mapping basigin (BSG), a member of the immunoglobulin superfamily, to 19p13.3. *Cytogenetics and Cell Genetics, 64*, 195–197.

Kang, M. J., Kim, H.-P., Lee, K.-s., Yoo, Y.-D., Kwon, Y.-T., Kim, K. M., et al. (2013). Proteomic analysis reveals that CD147/EMMPRIN confers chemoresistance in cancer stem cell-like cells. *Proteomics, 13*(10–11), 1714–1725.

Kataoka, H., DeCastro, R., Zucker, S., & Biswas, C. (1993). Tumor cell-derived collagenase-stimulatory factor increases expression of interstitial collagenase, stromelysin, and 72-kDa gelatinase. *Cancer Research, 53*, 3154–3158.

Kato, N., Kosugi, T., Sato, W., Ishimoto, T., Kojima, H., Sato, Y., et al. (2011). Basigin/CD147 promotes renal fibrosis after unilateral ureteral obstruction. *American Journal of Pathology, 178*(2), 572–579.

Ke, X., Fei, F., Chen, Y., Xu, L., Zhang, Z., Huang, Q., et al. (2012). Hypoxia upregulates CD147 through a combined effect of HIF-1a and Sp1 to promote glycolysis and tumor progression in epithelial solid tumors. *Carcinogenesis*, *33*(8), 1598–1607.

Kennedy, K. M., & Dewhirst, M. W. (2010). Tumor metabolism of lactate: The influence and therapeutic potential for MCT and CD147 regulation. *Future Oncology*, *6*(1), 127–148.

Klein, C. A., Seidl, S., Petat-Dutter, K., Offner, S., Geigl, J., Schmidt-Kittler, O., et al. (2002). Combined transcriptome and genome analysis of single micrometastatic cells. *Nature Biotechnology*, *20*, 387–392.

Kong, L.-M., Liao, C.-G., Chen, L., Yang, H.-S., Zhang, S.-H., Zhang, Z., et al. (2011). Promoter hypomethylation up-regulates CD147 expression through increasing Sp1 binding and associates with poor prognosis in human hepatocellular carcinoma. *Journal of Cellular and Molecular Medicine*, *15*(6), 1415–1428.

Kong, L.-M., Liao, C.-G., Fei, F., Guo, X., Xing, J.-L., & Chen, Z.-N. (2011). Transcription factor Sp1 regulates expression of cancer-associated molecule CD147 in human lung cancer. *Cancer Science*, *101*(6), 1463–1470.

Koyama, H., Hibi, T., Isogai, Z., Yoneda, M., Fujimori, M., Amano, J., et al. (2007). Hyperproduction of hyaluronan in neu-induced mammary tumor accelerates angiogenesis through stromal cell recruitment: Possible involvement of versican/PG-M. *American Journal of Pathology*, *170*(3), 1086–1099.

Kultti, A., Rilla, K., Tiihonen, R., Spicer, A. P., Tammi, R. H., & Tammi, M. I. (2006). Hyaluronan synthesis induces microvillus-like cell surface protrusions. *Journal of Biological Chemistry*, *281*(23), 15821–15828.

Landskron, J., & Tasken, K. (2013). CD147 in regulatory T cells. *Cellular Immunology*, *282*(1), 17–20.

Le Floch, R., Chiche, J., Marchiq, I., Naiken, T., Ilk, K., Murray, C. M., et al. (2011). CD147 subunit of lactate/H^+ symporters MCT1 and hypoxia-inducible MCT4 is critical for energetics and growth of glycolytic tumors. *PNAS*, *108*(40), 16663–16668.

Lescaille, G., Menashi, S., Cavelier-Balloy, B., Khayati, F., Quemener, C., Podgorniak, M. P., et al. (2012). EMMPRIN/CD147 up-regulates urokinase-type plasminogen activator: Implications in oral tumor progression. *BMC Cancer*, *12*(1), 115.

Li, L., Asteriou, T., Bernert, B., Heldin, C.-H., & Heldin, P. (2007). Growth factor regulation of hyaluronan synthesis and degradation in human dermal fibroblasts: Importance of hyaluronan for the mitogenic response of PDGF-BB. *Biochemical Journal*, *404*, 327–336.

Li, Y., Xu, J., Chen, L., Zhong, W.-D., Zhang, Z., Mi, L., et al. (2009). HAb18G (CD147), a cancer-associated biomarker and its role in cancer detection. *Histopathology*, *54*, 677–687.

Liang, L., Major, T., & Bocan, T. (2002). Characterization of the promoter of human extracellular matrix metalloproteinase inducer (EMMPRIN). *Gene*, *282*, 75–86.

Liao, C.-G., Kong, L.-M., Song, F., Xing, J.-L., Wang, L.-X., Sun, Z.-J., et al. (2011). Characterization of basigin isoforms and the inhibitory function of basigin-3 in human hepatocellular carcinoma proliferation and invasion. *Molecular and Cellular Biology*, *31*(13), 2591–2604.

Lokeshwar, V. B., & Selzer, M. G. (2008). Hyaluronidase: Both a tumor promoter and suppressor. *Seminars in Cancer Biology*, *18*(4), 281–287.

Lu, P., Weaver, V. M., & Werb, Z. (2012). The extracellular matrix: A dynamic niche in cancer progression. *Journal of Cell Biology*, *196*(4), 395–406.

Marieb, E. A., Zoltan-Jones, A., Li, R., Misra, S., Ghatak, S., Cao, J., et al. (2004). Emmprin promotes anchorage-independent growth in human mammary carcinoma cells by stimulating hyaluronan production. *Cancer Research*, *64*(4), 1229–1232.

McAtee, C. O., Barycki, J. J., & Simpson, M. A. Emerging roles for hyaluronidase in cancer metastasis and therapy. *Advances in Cancer Research, 123*, 1–34.
Miletti-Gonzalez, K. E., Chen, S., Muthukumaran, N., Saglimbeni, G. N., Wu, X., Yang, J., et al. (2005). The CD44 receptor interacts with P-glycoprotein to promote cell migration and invasion in cancer. *Cancer Research, 65*(15), 6660–6667.
Misra, S., Ghatak, S., & Toole, B. P. (2005). Regulation of MDR1 expression and drug resistance by a positive feeback loop involving hyaluronan, phosphoinositide 3-kinase, and ErbB2. *Journal of Biological Chemistry, 280*(21), 20310–20315.
Misra, S., Ghatak, S., Zoltan-Jones, A., & Toole, B. P. (2003). Regulation of multidrug resistance in cancer cells by hyaluronan. *Journal of Biological Chemistry, 278*(28), 25285–25288.
Miyauchi, T., Jimma, F., Igakura, T., Yu, S., Ozawa, M., & Muramatsu, T. (1995). Structure of the mouse basigin gene, a unique member of the immunoglobulin superfamily. *Journal of Biochemistry, 118*(4), 717–724.
Miyauchi, T., Kanekura, T., Yamaoka, A., Ozawa, M., Miyazawa, S., & Muramatsu, T. (1990). Basigin, a new broadly distributed member of the immunoglobulin superfamily, has strong homology with both the immunoglobulin V domain and the beta-chain of major histocompatability complex class II antigen. *Journal of Biochemistry, 107*, 16–323.
Murai, T., Maruyama, Y., Mio, K., NIshiyama, H., Suga, M., & Sato, C. (2011). Low cholesterol triggers membrane microdomain-dependent CD44 shedding and suppresses tumor cell migration. *Journal of Biological Chemistry, 286*(3), 1999–2007.
Muramatsu, T. (2012). Basigin: A multifunctional membrane protein with an emerging role in infections by malaria parasites. *Expert Opinion on Therapeutic Targets, 16*(10), 999–1011.
Muramatsu, T., & Miyauchi, T. (2003). Basigin (CD147): A multifunctional transmembrane protein involved in reproduction, neural function, inflammation and tumor invasion. *Histology and Histopathology, 18*, 981–987.
Murphy, D. A., & Courtneidge, S. A. (2011). The 'ins' and 'outs' of podosomes and invadopodia: Characteristics, formation and function. *Nature Reviews. Molecular Cell Biology, 12*, 413–426.
Nabeshima, K., Iwasaki, H., Koga, K., Hojo, H., Suzumiya, J., & Kikuchi, M. (2006). Emmprin (basigin/CD147): Matrix metalloproteinase modulator and multifunctional cell recognition molecule that plays a critical role in cancer progression. *Pathology International, 56*, 359–367.
Ochrietor, J. D., & Linser, P. J. (2004). 5A11/basigin gene products are necessary for proper maturation and function of the retina. *Developmental Neuroscience, 26*, 380–387.
Ochrietor, J. D., Moroz, T. P., Ekeris, L. v, Clamp, M. F., Jefferson, S. C., deCarvalho, A. C., et al. (2003). Retina-specific expression of 5A11/basigin-2, a member of the immunoglobulin gene superfamily. *Investigative Ophthalmology and Visual Science, 44*(9), 4086–4096.
O'Connor, M. L., Xiang, D., Shiqdar, S., Macdonald, J., Li, Y., Wang, T., et al. (2014). Cancer stem cells: A contentious hypothesis now moving forward. *Cancer Letters, 344*(2), 180–187.
Oliferenko, S., Paiha, K., Harder, T., Gerke, V., Schwarzler, C., Schwarz, H., et al. (1999). Analysis of CD44-containing lipid rafts: Recruitment of annexin II and stabilization by the actin cytoskelton. *Journal of Cell Biology, 146*(4), 843–854.
Parks, S. K., Chiche, J., & Pouyssegur, J. (2013). Disrupting proton dynamics and energy metabolism for cancer therapy. *Nature Reviews. Cancer, 13*(9), 611–623.
Paz, H., Pathak, N., & Yang, J. (2013). Invading one step at a time: The role of invadopodia in tumor metastasis. *Oncogene*, 1–10. [epub ahead of print].
Philp, N. J., Ochrietor, J. D., Rudoy, C., Muramatsu, T., & Linser, P. J. (2003). Loss of MCT1, MCT3, and MCT4 expression in the retinal pigment epithelium and neural retina of the 5A11/basigin-null mouse. *Investigative Ophthalmology and Visual Science, 44*(3), 1305–1311.

Pienimaki, J.-P., Rilla, K., Fulop, C., Sironen, R. K., Karvinen, S., Pasonen, S., et al. (2001). Epidermal growth factor activates hyaluronan synthase 2 in epidermal keratinocytes and increases pericellular and intracellular hyaluronan. *Journal of Biological Chemistry, 276*, 20428–20435.

Pituch-Noworolska, A., Drabik, G., Szantanek, R., Biatas, M., Kotodziejczyk, P., Szczepanik, A., et al. (2007). Immunophenotype of isolated tumor cells in blood, bone marrow and lymph nodes of patients with gastric cancer. *Polish Journal of Pathology, 58*, 93–97.

Poincloux, R., Lizarraga, F., & Chavrier, P. (2009). Matrix invasion by tumour cells: A focus on MT1-MMP trafficking to invadopodia. *Journal of Cell Science, 122*, 3015–3024.

Qin, Z., Dai, L., Bratoeva, M., Slomiany, M. G., Toole, B. P., & Parsons, C. (2011). Cooperative roles for emmprin and LYVE-1 in the regulation of chemoresistance for primary effusion lymphoma. *Leukemia, 25*(10), 1598–1609.

Qin, Z., Dai, L., Slomiany, M. G., Toole, B. P., & Parsons, C. (2010). Direct activation of emmprin and associated pathogenesis by an oncogenic herpesvirus. *Cancer Research, 70*(10), 3884–3889.

Qin, Z., Dai, L., Toole, B. P., Robertson, E., & Parsons, C. (2011). Regulation of Nm23-H1 and cell invasiveness by Kaposi's sarcoma-associated herpesvirus. *Journal of Virology, 85*(7), 3596–3606.

Quemener, C., Gabison, E. E., Naimi, B., Lescaille, G., Bougatef, F., Podgorniak, M. P., et al. (2007). Extracellular matrix metalloproteinase inducer up-regulates the urokinase-type plasminogen activator system promoting tumor cell invasion. *Cancer Research, 67*(1), 9–15.

Reimers, N., Zafrakas, K., Assmann, V., Egen, C., Riethdorf, L., Riethdorf, S., et al. (2004). Expression of extracellular matrix metalloproteinase inducer on micrometastatic and primary mammary carcinoma cells. *Clinical Cancer Research, 10*, 3422–3428.

Riethdorf, S., Reimers, N., Assmann, V., Kornfeld, J.-W., Terracciano, L., Sauter, G., et al. (2006). High incidence of EMMPRIN expression in human tumors. *International Journal of Cancer, 119*, 1800–1810.

Sahai, E. (2005). Mechanisms of cancer invasion. *Current Opinion in Genetics and Development, 15*, 87–96.

Schneiderhan, W., Scheler, M., Holzmann, K. H., Marx, M., Gschwend, J. E., Bucholz, M., et al. (2009). CD147 Silencing inhibits lactate transport and reduces malignant potential of pancreatic cancer cells in in vivo and in vitro models. *Gut, 58*, 1391–1398.

Schoumacher, M., Goldman, R. D., Louvard, D., & Vignjevic, D. M. (2010). Actin, microtubules, and vimentin intermediate filaments cooperate for elongation of invadopodia. *Journal of Cell Biology, 189*(3), 541–556.

Serbulea, M., Kakumu, S., Thant, A. A., Miyazaki, K., Machida, K., Senga, T., et al. (1999). Hyaluronan activates mitogen-activated protein kinase via Ras-signaling pathway. *International Journal of Oncology, 14*(4), 733–738.

Seulberger, H., Unger, C. M., & Risau, W. (1992). HT7, neurothelin, basigin, gp42 and OX-47-many names for one developmentally regulated immuno-globulin-like surface glycoprotein on blood–brain barrier endothelium, epithelial tissue barriers and neurons. *Neuroscience Letters, 140*(1), 93–97.

Sidhu, S. S., Nawroth, R., Retz, M., Lemjabbar-Alaoui, H., Dasari, V., & Basbaum, C. (2010). EMMPRIN regulates the canonical Wnt/β-catenin signaling pathway, a potential role in accelerating lung tumorigenesis. *Oncogene, 29*(29), 4145–4156.

Slomiany, M. G., Dai, L., Bomar, P. A., Knackstedt, T. J., Kranc, D. A., Tolliver, L. B., et al. (2009). Abrogating drug resistance in malignant peripheral nerve sheath tumors by disrupting hyaluronan–CD44 interactions with small hyaluronan oligosaccharides. *Cancer Research, 69*(12), 4992–4998.

Slomiany, M. G., Dai, L., Tolliver, L. B., Grass, G. D., Zeng, Y., & Toole, B. P. (2009). Inhibition of functional hyaluronan–CD44 interactions in CD133-positive primary human ovarian carcinoma cells by hyaluronan oligosaccharides. *Clinical Cancer Research, 15*(24), 7593–7601.

Slomiany, M. G., Grass, G. D., Robertson, A. D., Yang, X. Y., Maria, B. L., Beeson, C., et al. (2009). Hyaluronan, CD44, and emmprin regulate lactate efflux and membrane localization of monocarboxylate transporters in human breast carcinoma cells. *Cancer Research, 69*(4), 1293–1301.

Staubach, S., & Hanisch, F.-G. (2011). Lipid rafts: Signaling and sorting platforms of cells and their roles in cancer. *Expert Review of Proteomics, 8*(2), 263–277.

Stern, R., Shuster, S., Neudecker, B. A., & Formby, B. (2002). Lactate stimulates fibroblast expression of hyaluronan and CD44: The Warburg effect revisited. *Experimental Cell Research, 276*, 24–31.

Tang, W., Chang, S. B., & Hemler, M. E. (2004). Links between CD147 function, glycosylation, and caveolin-1. *Molecular Biology of the Cell, 15*, 4043–4050.

Teng, Y., Zeisberg, M., & Kalluri, R. (2007). Transcriptional regulation of epithelial–mesenchymal transition. *Journal of Clinical Investigation, 117*, 304–306.

Toole, B. P. (2004). Hyaluronan: From extracellular glue to pericellular cue. *Nature Reviews. Cancer, 4*(7), 528–539.

Toole, B. P. (2009). Hyaluronan–CD44 interactions in cancer: Paradoxes and possibilities. *Clinical Cancer Research, 15*, 7462–7468.

Ullah, M. S., Davies, A. J., & Halestrap, A. P. (2006). The plasma membrane lactate transporter MCT4, but not MCT1, is up-regulated by hypoxia through a HIF-1a-dependent mechanism. *Journal of Biological Chemistry, 281*(14), 9030–9037.

Vikesaa, J., Hansen, T. V. O., Jonson, L., Borup, R., Wewer, U. M., Christiansen, J., et al. (2006). RNA-binding IMPs promote cell adhesion and invadopodia formation. *EMBO Journal, 25*, 1456–1468.

Wang, W.-J., Li, Q.-Q., Xu, J.-D., Cao, X.-X., Li, H.-X., Tang, F., et al. (2008). Interaction between CD147 and P-glycoprotein and their regulation by ubiquitination in breast cancer cells. *Chemotherapy, 54*, 291–301.

Weidle, U. H., Scheuer, W., Eggle, D., Klostermann, S., & Stockinger, H. (2010). Cancer-related issues of CD147. *Cancer Genomics & Proteomics, 7*, 157–170.

Wilson, M. C., Meredith, D., Fox, J. E. M., Manoharan, C., Davies, A. J., & Halestrap, A. P. (2005). Basigin (CD147) is the target for organomercurial inhibition of monocarboxylate transporter isoforms 1 and 4: The ancillary protein for the insensitive MCT2 is embigin (gp70). *Journal of Biological Chemistry, 280*, 27213–27221.

Wu, J., Ru, N.-Y., Zhang, Y., Li, Y., Wei, D., Ren, Z., et al. (2011). HAb18G/CD147 promotes epithelial–mesenchymal transition through TGF-beta signaling and is transcriptionally regulated by Slug. *Oncogene, 30*(43), 4410–4427.

Xu, J., Lu, Y., Qiu, S., Chen, Z. N., & Fan, Z. (2013). A novel role of EMMPRIN/CD147 in transformation of quiescent fibroblasts to cancer-associated fibroblasts by breast cancer cells. *Cancer Letters, 335*, 380–386.

Yamaguchi, H., Takeo, Y., Yoshida, S., Kouchi, Z., Nakamura, Y., & Fukami, K. (2009). Lipid rafts and caveolin-1 are required for invadopodia formation and extracellular matrix degradation by human breast cancer cells. *Cancer Research, 69*(22), 8594–8602.

Yan, L., Zucker, S., & Toole, B. P. (2005). Roles of the multifunctional glycoprotein, emmprin (basigin;CD147), in tumour progression. *Thrombosis & Haemostasis, 93*, 199–204.

Yang, J.-M., Xu, Z., Wu, H., Zhu, H., Wu, X., & Hait, W. N. (2003). Overexpression of extracellular matrix metalloproteinase inducer in multidrug resistant cancer cells. *Molecular Cancer Research, 1*, 420–427.

Yu, X.-L., Hu, T., Du, J.-M., Ding, J.-P., Yang, X.-M., Zhang, J., et al. (2008). Crystal structure of HAb18G/CD147: Implications for immunoglobulin superfamily homophilic adhesion. *Journal of Biological Chemistry*, *283*(26), 18056–18065.

Yurchenko, V., Constant, S., Eisenmesser, E., & Bukrinsky, M. (2010). Cyclophilin–CD147 interactions: A new target for anti-inflammatory therapeutics. *Clinical and Experimental Immunology*, *160*, 305–317.

Zhou, S., Liao, L., Chen, C., Zeng, W., Liu, S., Su, J., et al. (2013). CD147 mediates chemoresistance in breast cancer via ABCG2 by affecting its cellular localization and dimerization. *Cancer Letters*, *337*(2), 285–292.

Zoller, M. (2011). CD44: Can a cancer-initiating cell profit from an abundantly expressed molecule? *Nature Reviews. Cancer*, *11*, 254–267.

Zoltan-Jones, A., Huang, L., Ghatak, S., & Toole, B. P. (2003). Elevated hyaluronan production induces mesenchymal and transformed properties in epithelial cells. *Journal of Biological Chemistry*, *278*(46), 45801–45810.

Zucker, S., Hymowitz, M., Rollo, E. E., Mann, R., Conner, C. E., Cao, J., et al. (2001). Tumorigenic potential of extracellular matrix metalloproteinase inducer. *American Journal of Pathology*, *158*, 1921–1928.

INDEX

Note: Page numbers followed by "*f*" indicate figures and "*t*" indicate tables.

A

Aberrant HAS1 splicing
  B lineage malignancies, 72–73
  clinical impact, 75–76
  splice variants, 73–75
Acute myeloid leukemia (AML), 51–52, 175–176
Adenosine monophosphate activated protein kinase (AMPK)
  α-catalytic subunit, 107
  drug metformin, 107
  HAS2, 103–105
  HAS genes, 103–105
  human cancers, 107–108
  in vitro experiments, 105
  LKB1, 107
  metabolism, 106, 106*f*
  mitochondrial glutaminolysis, 107–108
  NADPH, 107–108
  plasma membrane, 105
  posttranslational modifications, 105–106
AER. *See* Apical ectodermal ridge (AER)
Aldehyde dehydrogenenase (ALDH), 214–215, 264
Alternating *N*-acetylglucosamine (GlcNAc), 323
AML. *See* Acute myeloid leukemia (AML)
AMPK. *See* Adenosine monophosphate activated protein kinase (AMPK)
Angiogenesis
  BM neovascularization, 172
  CD44, 171–172
  endothelial barrier function, 192
  HAase, 53
  HA-binding proteins, 201–202
  HMW HA, 171
  HUVEC, 171
  Hyal1, 16
  lymphangiogenesis, 171
  monocytes and macrophages, 325–326
  prostate tumors, 16
Apical ectodermal ridge (AER), 242

B

Basigin. *See* CD147
B lineage malignancies
  MM, 72–73
  WM, 73
BM. *See* Bone marrow (BM)
Bone marrow (BM)
  endothelial cells, 166–173
  hematological malignancy, 174–177
  macrophages, 163–166
  monoclonal plasma cells, 72–73
  MSC (*see* Mesenchymal stem cells (MSCs))
  nerve cells, 173–174
  stem cells, 151–152
Breast tumorigenesis
  bioactive network, 211–212
  cancer, 213–215
  CD44, 213
  ECM, 211–212
  fibronectin, 212
  glycosaminoglycans, 212–213
  growth factors and hormones, 212–213
  HA-CD44 interactions, 220–222
  stromal HA, tumor progression, 215–218

C

Cancer
  ALDH, 214–215
  breast carcinoma, 136, 136*f*
  CD24, 213–214, 214*f*
  chemotherapic drugs, 134–135
  chronic inflammation, 97–98
  CSCs, 214–215
  EMT, 215
  ErbB2, 213–214
  GFP, 136–137, 137*f*

Cancer (*Continued*)
and HAS genes, 218–220
human, 107–108
human mammary gland, 213–214
immunostaining, HASs, 136
MDCK cells, 136–137, 137*f*
myoepithelial cells, 213–214
O-GlcNAcylation, 108–112
progression, 127–128, 200–201, 202*f*, 278, 322
trastuzumab therapy, 134–135
tumor vasculature, 134–135
Cancer metastasis
cellular chemokinesis, 16–17
endothelial barrier function, HA regulation, 198–200
HA and hyaluronidase, 3–5
Hyal expression (*see* Hyaluronidase (Hyal))
naked mole rats, 5–6
targeting, Hyal, 22–24
tumor cell hyaluronidases, 199–200
vesicle shedding, 14–15
Cancer-specific splicing patterns, 71–72
Cancer stem cells (CSCs)
CD44v3, 263–265
chemotherapeutic drugs, 261–262
cisplatin, 267
ErbB2, 268–269
IAPs, 266
miR-21, 267–268
miR-302, 263
multidrug resistance, 261–262
Nanog/Stat-3 signaling, 267–268
NFκB, 268–269
paclitaxel, 261–262
p53 tumor suppressor, 268
stemness functions, 265–266
TICs, 262–263
and tumor progression, 269, 270*f*
CD147
description, 352–353
HA synthesis and signaling, 356
invadopodia formation and activity, 358–359
membrane-type MMPs, 355–356
structure and pleiotropic functions, 353–356
CD147–HA axis
in cellular invasion, 356–361
in chemoresistance, 361–363
induction by Kaposi's sarcoma-associated herpesvirus, 363–364
CD147-induced HA synthesis and signaling, 356
CD44 receptor
ALDH1, 264
breast cancer metastasis, 221–222
CD147-induced, 359–361
cell-microenvironment, 220–221
cell-surface glycoproteins, 256–258
cell-surface receptors, 238–244
cellular environment, 234
CSCs, 256–258
delivering antitumor therapies, 50
description, 280
E-cadherin, 222
EC barrier, 195, 197–198
ECM, 234–238, 263
EMT, 221
epithelial cells, 256–258
epithelial isoform, 233–234
ESRP genes, 221
exons, 233
expression, 168–169, 175–176
HA–EVs, 136–137
head and neck cancer, 263–265
human cancers, 177
immunoperoxidase staining, 256–258, 258*f*
intestinal stem cells, 233
ligation, 175–176
LMW HA, 171–172
matrix metalloproteinase, 222
metastatic lymph nodes, 263–264
metastatic tumors, 256–258
microenvironmental signals, 232–233
MSC-mediated tissue regeneration, 157–158
multidomain signal integration, 244–248
oHA, 43–44
protein, 50–51
signal transduction, 232–233
siRNA delivery, 49
splice variants, 163–164
stem cells, 263, 263*f*

structure on normal cells, 281–283
TGFα signaling, 221–222
transmembrane glycoproteins, 233
tumor cells, 232
vaccines, 48–49
Cell recruitment, BM
CD44/HA interactions, 170–171
hyaluronidase treatment, 170–171
lymphocytes, 170–171
Cell-surface receptors, CD44
GPCRs, 242–243
HA, 233–234
HGF, 238–241
MIF, 244
RTKs, 241–242
serine-glycine, 238
TGF family, 242–243
Chemical inhibitors, HA synthesis
mannose, 43
4-MU, 39–43
UDP-*N*-acetylhexosamines, 43
Chemoresistance, CD147–HA axis, 361–363
Chronic lymphatic leukemias (CLL), 177
Chronic myeloid leukemia (CML), 51–52, 176–177
Chronic osteomyelitis, 320–321
CLL. *See* Chronic lymphatic leukemias (CLL)
CML. *See* Chronic myeloid leukemia (CML)
Coreceptor function, CD44
BMP-7, 245
chondrocytes, 245
CXCL12, 246
dermal fibroblasts, 245–246
PDGFRβ activation, 245–246
CSCs. *See* Cancer stem cells (CSCs)

D

Dermatofibrosarcoma, 324–325
Drug delivery, HA
amphilic nanoparticles, 45–46
anticancer drugs, 46
carbon nanotubes, 47
DOX and paclitaxel, 45–46, 47
HA-octadecyl(C18)-based nanosystem., 47
HA-TRAIL complexes, 48
nanomedicine, 39–43
nanoparticles, 45–47, 48
nontumor toxicity, 47
siRNAs, 48

E

ECM. *See* Extracellular matrix (ECM)
ECs. *See* Endothelial cells (ECs)
EGFR. *See* Epidermal growth factor receptor (EGFR)
EMT. *See* Epithelial-mesenchymal transition (EMT)
Enchondroma, 320–321
Endothelial barrier function
cancer metastasis, 198–200
cancer progression, 200–201
tumor angiogenesis, 197–198
Endothelial cells (ECs)
angiogenesis, 171–172
atherosclerosis, 167–168
blood vessels, 169–170
cell recruitment, 170–171
C-reactive protein, 169
extravascular space, 166–167, 168–169
glycocalyx, 167–168
HAS expression, 167
HUVEC, 167–168
hyaluronan-poly-L-lysine, 301–302
hydrocortisone, 169–170
interstitial flow model, 168–169
mechanotransduction, 168–169
RF24 cell line, 167–168
spindle cells, 363–364
vascular integrity, 172–173
Epidermal growth factor (EGF), 337
Epidermal growth factor receptor (EGFR)
soft tissue sarcoma cell, 337
transmembrane glycoprotein, 337
Epithelial-mesenchymal transition (EMT)
CD147, 357–358
HA, 97
HAS2, 103–105
Epithelial splicing regulatory protein (ESRP), 221
ErbB2, 280–281
ESRP. *See* Epithelial splicing regulatory protein (ESRP)

EVs. *See* Extracellular vesicles (EVs)
Extracellular matrix (ECM)
cancer progression, 322
CD44, 285
cells interaction, 321
colon cancer cells, 237–238
fibrosarcoma cells, 322
fibrosarcoma pathogenesis, 332–337
growth factors (GFs), 322–323
hyaluronic acid, 234–237
OPN, 237–238
physical properties, 322
properties, 278
role, 278–279
serglycin, 237
structure, 321
TGFβ, 333
tumor cells, 198–199
Extracellular vesicles (EVs)
atomic force microscopy, 125–126
body fluids, 125
CD44, 138
electron microscopy, 125–126
exosomes and microvesicles, 124–125
plasma membrane, 124
potential biomarkers, tumors, 137–138
shedding (*see* Shedding, EVs)
target cells, 126
tumor progression, 126–129
Ezrin–radixin-moesin (ERM), 9–10, 239–240

F

Fibroblast growth factor (FGFs)
FGF–FGFRs interactions, 334–335
HYAL-2 expression, 335
Fibroblasts, MSCs
chemotherapy drugs, 160–161
dermal fibroblasts, 159–160
Down syndrome, 161
fibrosarcoma, 160–161, 320–321
HAS expression, 160–161
hematopoietic microenvironment, 159
hormones, 160
4MU, 159–160
paracrine, 159–160
RHAMM, 327
smooth muscle cells, 160
stromal cells, 159
TGFβ1, 159–160
Fibrosarcoma
description, 320–321
GF signaling and ECM organization, 332–337
immunohistochemistry, 320–321
RHAMM in, 327–332
as secondary tumor, 320–321
Fibrous dysplasia, 320–321

G

Gab-1, 280–281
GFAT1. *See* Glutamine: fructose-6-phosphate amidotransferase 1 (GFAT1)
GF signaling and ECM organization, fibrosarcoma, 332–337
Glucuronic acid (GlcA), 323
Glutamine: fructose-6-phosphate amidotransferase 1 (GFAT1), 110–111
Glycosaminoglycan (GAG), 6–7, 11, 17–18, 96, 336
Glycosaminoglycans
HA, 96
vascular integrity, 192–193
Glycosylation, 281–283
Goldilocks hypothesis, 85–88
GPCRs. *See* G protein-coupled receptors (GPCRs)
G protein-coupled receptors (GPCRs), 242–243
Grb2, 280–281

H

HA. *See* Hyaluronan (HA)
HA-degrading enzyme, 101
HA oligosaccharides (oHA)
antitumor activity, 43–44
B16-F10 murine melanoma model, 44–45
CD44, 303–304
HA-receptor interactions, 359–360
intratibial growth model, 44–45
osteosarcoma models, 44–45
PI3-kinase/Akt signaling, 43–44
RHAMM expression, 45

HAS1/HAS2/HAS3, 279–280
HAS mRNA expression, 325
HAS2 O-GlcNAcylation, 112
HAS1 splice-variant transcripts (HAS1Vs)
B and hematopoietic stem cells, MM, 76–77, 78*f*
cellular cytoskeletal network, 81–82, 82*f*, 83*f*
de novo synthesis, 80–81
elongating HA, 80–81
extracellular/intracellular HA, 77
extracellular matrix, 82–83
and genetic instability, 80
and HAS3, 76–77
HAS1FL, 81–83
intermolecular interactions, 79–80
intracellular structures, 78*f*, 81–82, 82*f*
and mitotic catastrophe, 83–88
MM-plasma cells, 76–77
motility and malignant spread, 77–79
and oncogenic events, 79
HAS1Vs. *See* HAS1 splice-variant transcripts (HAS1Vs)
HA synthases (HAS) genes
Akt activation, 218–219
bone marrow stromal cells, 219–220
breast carcinomas, 220
HAS1 and HAS2, 218
HYALs, 218, 219*t*
in vivo and *in vitro*, 218
luminal cancer, 220
Stat3 activation, 218
TIMP1, 218–219
tumor-stromal interactions, 220
HA synthesis and signaling, CD147-induced, 356
Hematological malignancy, BM
AML, 175–176
butyric acid, 177
CD44, 177
CLL, 177
CML, 176–177
dexamethasone, 174–175
human/mouse xenograft model, 175–176
leukemia stem cells, 174
lymphoma cell lines, 177
MM cells, 174–175
retinoic acid, 177
RHAMM, 175–176
serum HA levels, 174
TGFα1, 176–177
Hematopoietic stem/progenitor cells (HSPC), 154–155
Hepatocyte growth factor (HGF)
amino acids, 238
Bsp73ASs6 cells, 239
CD44v3-v10 isoforms, 238
colorectal cancer, 241
ERMs, 239–240
hepatectomy, 240–241
metastatic process, 241
Met signaling, 238–239
mouse genetics, 240–241
RTKs, 240
signal transduction, 239–240
HGF. *See* Hepatocyte growth factor (HGF)
High-molecular weight (HMW), 171
High molecular weight HA (HMW-HA), 324, 326
actin "ring" structure, 194
cancer progression, 200–201
endothelial cell (EC) barrier, 194–195
treatment, 16–17
vascular integrity, 194
HMW. *See* High-molecular weight (HMW)
HMW-HA. *See* High molecular weight HA (HMW-HA)
HMW hyaluronan, 293
HSPC. *See* Hematopoietic stem/progenitor cells (HSPC)
Human genome splicing, 68–69
Human Hyals
Arg265, 18–19
chondroitin sulfate hydrolase, 20
crystal structure, 18–19
C-terminal EGF-like domain, 22
Cys207 and Cys221, 18–19
Glu131 protonates, 18–19
HA cleavage, 20, 21*f*
PH-20, 20
substrate-binding cleft bisects, 18–19
Tyr202 and Ser245, 18–19
virus entry receptor, 19–20
Human umbilical vein endothelial cells (HUVEC), 171

HUVEC. *See* Human umbilical vein endothelial cells (HUVEC)
Hyal. *See* Hyaluronidase (Hyal)
Hyaladherins, 324
Hyal1 expression
breast cancer, 7
colorectal cancer, 7
expression and functional importance, 6–7
genitourinary cancer, 7
ovarian cancer, 7–8
pancreatic and lung cancers, 8
splice variants, 8
Hyal2 expression
description, 8–9
methylation profile, 9
overexpression, 9–10
RON tyrosine kinase, 9–10
splice variant, 9
Hyal3 expression, 10–11
Hyal4 expression, 11
Hyaluronan (HA)
and AMPK (*see* Adenosine monophosphate activated protein kinase (AMPK))
biological activities, 96–97
BM (*see* Bone marrow (BM))
cancer-resist mole rats, 5–6
catabolism, 101–102, 104*f*
cell signaling, 279–280
cellular signaling, 122
CSCS, 261–269
description, 2, 96, 279–280
ECM, 153
endocytic uptake/hyaluronidase-catalyzed hydrolysis, 2
epithelial cells, 123–124
EVs (*see* Extracellular vesicles (EVs))
glycosaminoglycan, 153
growth factors, 153
HAS1, HAS2 and HAS3, 2
high-molecular-weight, 96–97
hyaluronan-based hydrogels, 283–284
and hyaluronidase accelerate human cancers, 3–5
hyaluronidases, 153
immunostaining, 122–123
malignant growth, 122–123
polymers and oligosaccharides, 3
production in mesenchymal tumors, 324–326
receptors, 97, 98
receptors in fibrosarcoma, 323–332
smaller oligosaccharides/monodisperse, 96–97
stromal fibroblasts, 122–123, 123*f*
synthases, 279–280
tumor microenvironment, 97, 98
tumor progression (*see* Tumor progression)
Hyaluronan-based hydrogelsn, 283–284
Hyaluronan–CD44 interactions
advantages of nanomedicine, 295–296
application, 300–304
expression, 281–283
glycosylation, 281–283
nanoparticle properties, 296–297
targeted drug delivery, 295
targeting, 282*f*
chemotherapeutics, 299–300
in tumors, 279–284
therapeutic interventions/strategies, 284–295
tissue-specific deletion, 297–299
Hyaluronan-mediated motility receptor (RHAMM), 175–176
Hyaluronan/RHAMM interactions. *See* RHAMM in fibrosarcoma development and progression
Hyaluronan synthase 1 (HAS1)
aberrant splice variants, 73–75
clinical impact, 75–76
genetic variations, 76
Hyaluronan (HA) synthesis
oxysterols, 99–101
posttranslational modifications, 101, 102*t*
tumorigenesis and cancer progression, 101
UDP-GlcUA, 99
UDP-sugar precursors, 98–99
UGDH overexpression, 99–101
Hyaluronic acid
actin cytoskeleton, 235–237
binding domains, 234–235
coreceptor function, 234–235, 236*f*
cytoplasmic domain, 235–237

glycosylation, 235
mutagenesis, 234–235
phosphorylation, ERM, 234–235, 236*f*
RTKs, 234–235, 236*f*
Hyaluronidase (Hyal)
chondroitinase activity, 18
chondroitin sulfate, 17
degraded GAG products, 17–18
enzyme activity, 17–18
fragments *vs.* oligos, 16–17
human cancers, 3–5
Hyal2, 8–10
Hyal1 expression (*see* Hyal1 expression)
Hyal3, Hyal4 and PH-20, 10–11
sulfated polysaccharides, 17
vesicle shedding, 14–15
vesicle trafficking and cell motility, 11–14, 13*f*
Hyaluronidases (HYAL), 279–280, 323, 325
Hyper-O-GlcNAcylation, 111

I

Inhibitors of the apoptosis family of proteins (IAPs), 266
Insulin-like growth factors (IGF1 and IGF2), 336–337
Interferon (IFN), 165
Invadopodia formation and activity
by CD147, 358–359
by CD147-induced HA-CD44 and EGFR signaling, 359–361

K

Kaposi's sarcoma-associated herpesvirus
CD147–HA axis, 363–364

L

Leukemia inhibitory factor (LIF), 161–162
LIF. *See* Leukemia inhibitory factor (LIF)
Lipid-based nanoparticles (LNPs), 302
Lipopolysaccharide (LPS), 163–164, 194
LMW-HA. *See* Low molecular weight fragments (LMW-HA)
Low-grade fibromyxoid sarcoma (LGFMS), 330–331
Low molecular weight fragments (LMW-HA)
bone marrow (BM), 175–176
EC barrier disruption, 197–199
HUVEC proliferation, 171–172
IFNβ production, 165
macrophages, 164
molecular weight, HA, 164
pathophysiologic vicious cycle, 164–165
stress fiber formation, EC, 194, 195*f*
LPS. *See* Lipopolysaccharide (LPS)

M

Macrophage Fas-induced apoptosis (MAFIA), 163
Macrophages, BM
adenosine, 164–165
CD44 splice variants, 163–164
cell-cell network, 163
HMW, 163–164
IFN, 165
LPS, 163–164
MAFIA, 163
osteoclasts, 166
ROS, 164
THP-1, 165
MAFIA. *See* Macrophage Fas-induced apoptosis (MAFIA)
Matrix metalloproteinases (MMPs)
CD147, 360–361, 361*f*
collagenase, 354–355
Mesenchymal stem cells (MSCs)
CD44, 157–158
cell production, 155–157
chaperone effects, 153–154
fibroblasts, 159–161
HAS expression, 154, 154*f*
hematopoiesis, 155–158, 156*f*
HSPC, 154–155
mouse and human, 154, 154*f*
4MU, 155–157, 156*f*
nonadherent cells, 154–157, 155*f*, 156*f*
osteoblasts, 161–162
smooth muscle cells, 155–157
TGFβ, 155–157
tHAS1/2/3 KO, 154–155, 155*f*, 157–158, 158*f*
tissue regeneration, 157–158
4-Methylumbelliferone (4-MU)
antitumor activity, 40–42, 41*f*
concentrations, 40–42
glucuronate scavenger, 40–42

4-Methylumbelliferone (4-MU) (*Continued*)
inhibit tumor growth and metastasis, 40–42
orally bioavailable dietary supplement, 43
Sorafenib, 42–43
UGA, 39–40
uses, 40
xenograft studies, 40–43
MIF. *See* Migration inhibitory factor (MIF)
Migration inhibitory factor (MIF), 244
Mitotic catastrophe, 83–88
MM. *See* Multiple myeloma (MM)
MMPs. *See* Matrix metalloproteinases (MMPs)
Monocarboxylate transporters (MCTs), 354–355
MSCs. *See* Mesenchymal stem cells (MSCs)
4-MU. *See* 4-methylumbelliferone (4-MU)
Multidomain signal integration, CD44
coreceptor function, 245–246
cytoplasm, 247–248
ligand-receptor, 244–245
RTKs, 246–247
Multiple myeloma (MM)
description, 72–73
HAS1Vs, 75, 76–77
malignant B cells, 73–74
patients serum samples, 174–175
plasma cells, 76–79
RHAMM, 83–84, 85
Myxofibrosarcoma (MFS), 330–331

N

Nanomedicine
advantages, 295–296
nanoparticle properties, 296–297
polysaccharides, 300–304
targeted drug delivery, 295
targeting chemotherapeutics to CD44, 299–300
tissue-specific deletion, CD44 signaling, 297–299
Nanoparticle (NP) properties, 296–297

O

OGA. *See* O-GlcNAcase (OGA)
O-GlcNAcase (OGA), 108, 110, 111
O-GlcNAc transferase (OGT), 108–109, 110, 111
O-GlcNAcylation
aerobic glycolysis, 109–110
cancer-associated proteins, 110
HAS2, 112
hyper-O-GlcNAcylation, 111
β-*N*-acetylglucosamine, 108
OGT and OGA expression, 111
OGT transcripts, 110
oncogenes, 110–111
protein machinery, 108–109
and UDP-GlcNAc, 108, 109*f*, 110–111
OGT. *See* O-GlcNAc transferase (OGT)
oHA. *See* HA oligosaccharides (oHA)
Oncogenic events and properties, 79, 83–88
OPN. *See* Osteopontin (OPN)
Osteoblasts, MSCs
bone homeostasis, 161
chondrocytes, 162
HMW, 161–162
LIF, 161–162
PTH, 162
Osteopontin (OPN), 237–238
O-sulfated HA (sHA), 53, 54*f*

P

Parathyroid hormone (PTH), 162
PDGFRβ, 280–281
Pericellular matrix, 216–217
PH-20 expression, 11, 20, 23–24
PI3 kinase/Akt pathways, 280–281
Platelet-derived growth factor (PDGF), 335–336
Pleiotropic functions, CD147, 353–356
Polysaccharides (hyaluronan)
dendritic cells (DC) and T-cells, 303–304
lipid-based nanoparticles (LNPs), 302
NPs, 301
PEG layer, 300–301
siRNA–hyaluronan–PEI complex, 301–302
Pre-mRNA splicing
aberrant splicing, 69–70, 70*f*
mutations, 70–71
"trans" and "cis" elements, 69
PTH. *See* Parathyroid hormone (PTH)

Q

12q14–q22 in adult-type fibrosarcomas, 320–321

R

Rac1, 280–281
Ras, 280–281
Reactive oxygen species (ROS), 164
Receptor for hyaluronan-mediated motility (RHAMM)
  description, 83–84
  expression, 45
  goldilocks hypothesis, 85–88
  HA fragment-induced angiogenesis, 197–198
  leukemia-associated antigen, 175–176
  overexpression and isoform imbalance, 84–85
  targeting, 51–52
Receptor tyrosine kinases (RTKs)
  AER, 242
  CD44v6, 241
  EGFR family, 242
  MMP-7, 242
  tumor angiogenesis, 241
Regulation, HA
  cellular glucose supply, 131–132
  disaccharide units, 192–193
  disruption/dysregulation, endothelial layer, 192–193
  endothelial barrier function, 197–200
  HMW-HA, 192–193, 200–201
  synthesis and secretion, 193–194
  vascular endothelium, 192
  vascular integrity, 194–196
RHAMM. *See* Receptor for hyaluronan-mediated motility (RHAMM)
RHAMM in fibrosarcoma development and progression
  excisional skin wounds, 328–329
  H-ras transformation, 329–330
  isoforms, 327
  LGFMS, 330–331
  protein forms, 329
  Ras downstream pathways, 330
RhoA, 280–281
ROS. *See* Reactive oxygen species (ROS)
RTKs. *See* Receptor tyrosine kinases (RTKs)

S

Shedding, EVs
  body fluids, 133
  cancer, 134–137
  cell body, 129–130
  cellular glucose, 131–132
  colorectal cancer cells, 129
  EGFP–HAS3, 129
  glycocalyx, 129–130
  HA metabolism, 129
  human mesothelial cells, 133, 134*f*
  intestinal enterocytes, 131
  MCF-7, 129–130, 130*f*
  microvilli, 129–130, 131*f*
  plasma membrane protrusions, 129–130, 130*f*
  platelet dust, 131
  tissue regeneration, 133
Splicing
  cancer-specific splicing patterns, 71–72
  HAS1 (*see* Hyaluronan synthase 1 (HAS1))
  human genome, 68–69
  pre-mRNA, 69–71
Stem cells, BM
  fetus, 151–152
  HSC, 151–152
  microenvironment, 152
  multipotency, 151–152

T

TAMs. *See* Tumor-associated macrophages (TAMs)
Targeting HA
  angiogenic fragments, 36–37, 38
  CD44, 48–51
  chemical inhibitors (*see* Chemical inhibitors, HA synthesis)
  drug delivery, 45–48
  genetic knockdown studies, 38–39
  oligosaccharides, 43–45
  RHAMM, 51–52
  tumor growth and progression, 36–37

Targeting of hyaluronidase (HAase)
cancer therapy, 22–24
controlling tumor growth and progression, 54
O-sulfated HA (sHA), 53, 54*f*
tumor imaging, 25–26
TGF-beta receptor type I, 280–281
Therapeutic interventions/strategies
anticancer effects, hyaluronidase, 284–285
lymphoma cell dissemination, 285–293
primary tumor growth, 285–293
targeting hyaluronan-CD44 system, 286*t*
TICs. *See* Tumor-initiating cells (TICs)
TIMP1. *See* Tissue metalloproteinase inhibitor 1 (TIMP1)
Tissue metalloproteinase inhibitor 1 (TIMP1), 218–219
Transforming growth factor-β (TGFβ)
fibrosarcoma cell motility, 329–330
GF/GFR signaling, 334*f*
locomotory, 333–334
polypeptide, 333
protein isoforms, 333
RHAMM, 329–330
RHAMM/HA pathway, 333–334
type I (RI) and type II (RII) receptors, 333
Transmembrane glycoprotein receptors, 337
Tumor angiogenesis
EC surface receptors, 197–198
LMW-HA, 197–198
resealing, endothelial barrier, 198
solid tumors, 197
Tumor-associated macrophages (TAMs), 217–218
Tumor-initiating cells (TICs), 262–263
Tumor progression
angiogenesis, 128
ankyrin function, 258–259
antitumor immunity, 128–129
autocrine/paracrine, 215–216, 216*f*
cancers, 126–127
carcinomas, 215–216
CD44, 256–258
cytoskeletal function, 258
epithelial cancer, 215–216
lung cancer cells, 127–128
malignant cells, 127–128
malignant tumors, 256
mammalian tissues, 256
mammary tumors, 217–218
MMPs, 260–261
multidrug resistance, 127–128
myxoma, 217
pericellular matrix, 216–217
RhoGTPases signaling, 259–260
signaling pathways, 215–216, 216*f*
stromal interactions, 216–217
TAMs, 217–218
thrombosis, 128
umbilical cord, 217
versican, 216–217

U

Ubiquitination, 279–280
UDP-GlcUA, 131–132
glucuronidation, 99
lowers HA deposition, 101
NADH, 99
UGDH, 99
UDP-glucose dehydrogenase (UGDH), 99–101
UDP-glucuronic acid (GlcUA), 96, 99, 100*f*, 104*f*
UDP-glucuronic acid (UGA), 39–40, 43
UDP-*N*-acetylglucosamine (GlcNAc), 96, 100*f*, 104*f*, 108, 109*f*
UGA. *See* UDP-glucuronic acid (UGA)
UGDH. *See* UDP-glucose dehydrogenase (UGDH)
UGT. *See* Uridine diphosphoglucuronosyltransferases (UGT)
Uridine diphosphoglucuronosyltransferases (UGT), 99

V

Vascular integrity
actin stress fiber formation, 195
acute lung injury, 194
bone marrow endothelial cells (BM EC), 172–173
CD44s, 194–195
EC barrier disruption, 195, 196*f*

ECs
CD44, 172–173
glycocalyx, 173
glycosaminoglycan, 192–193
pulmonary vascular hyperpermeability, 172–173
tHAS1/2/3 KO mice, 173
thromboembolic risk, 173
HA regulation, endothelial barrier function, 194, 195*f*
HMW-HA, 194, 195
Vav2, 280–281

W

Waldenstrom macroglobulinemia (WM), 73, 74–75, 80
Warburg effect, 107, 109–110
WM. *See* Waldenstrom macroglobulinemia (WM)

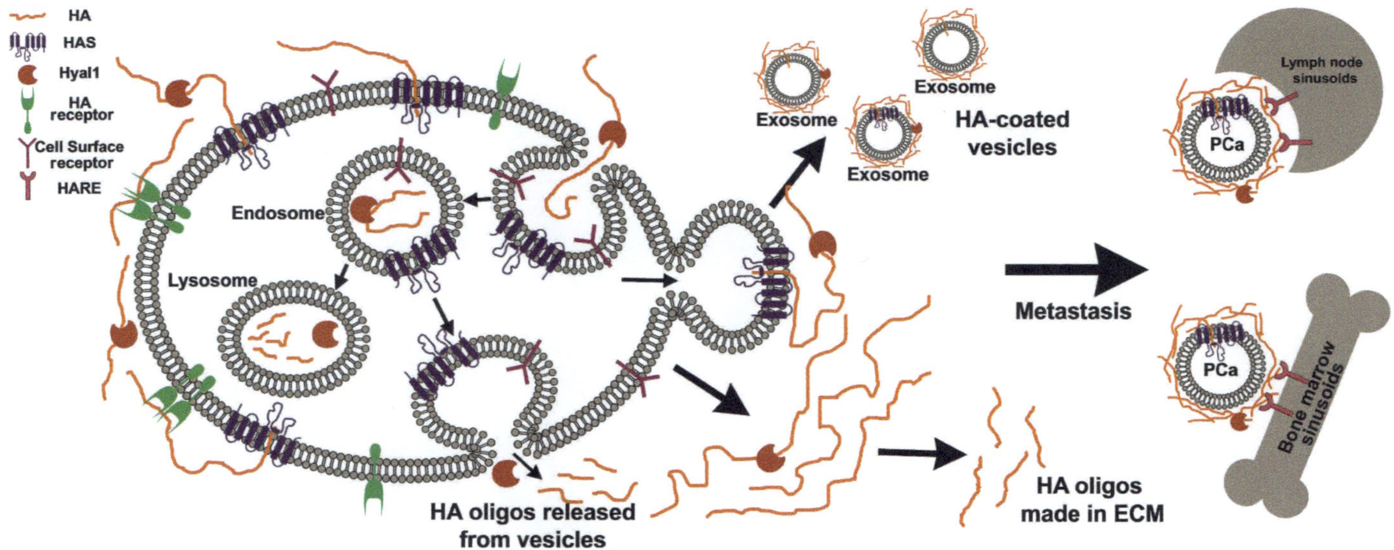

Figure 1.1, Caitlin O. McAtee *et al.* (See Page 13 of this volume.)

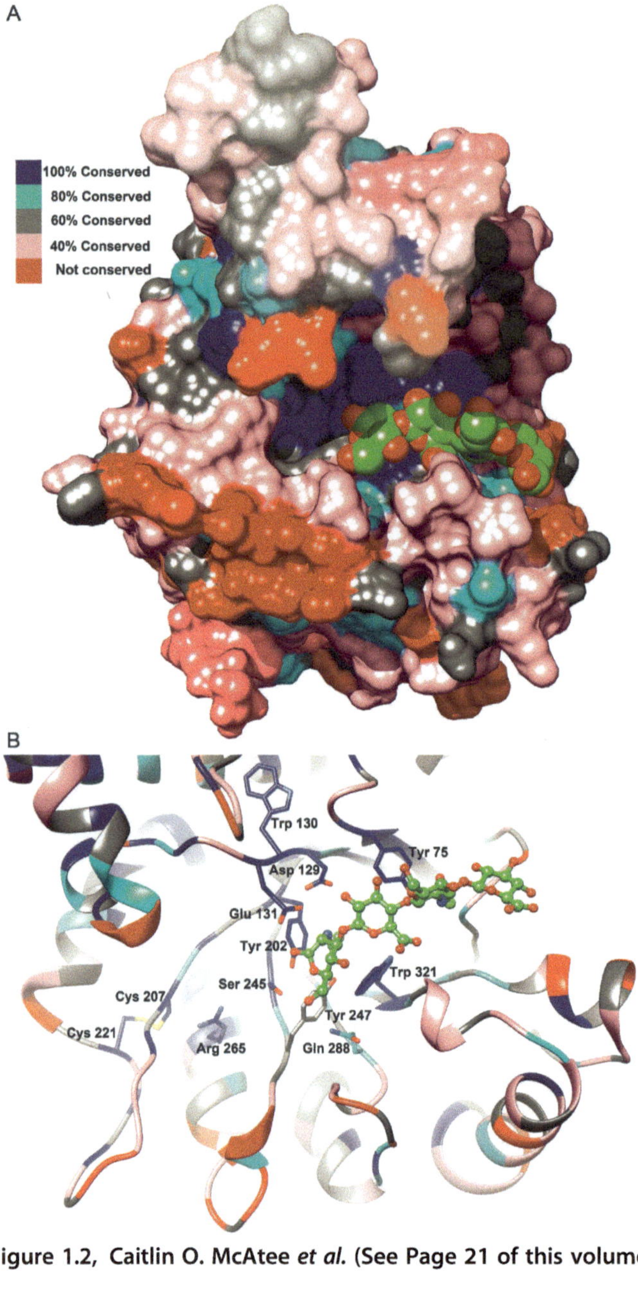

Figure 1.2, Caitlin O. McAtee *et al.* (See Page 21 of this volume.)

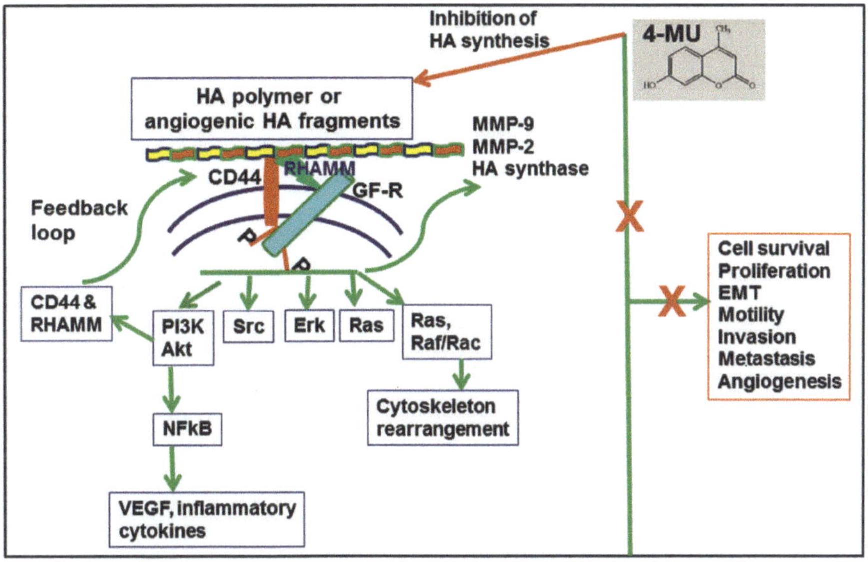

Figure 2.1, Vinata B. Lokeshwar *et al.* (See Page 41 of this volume.)

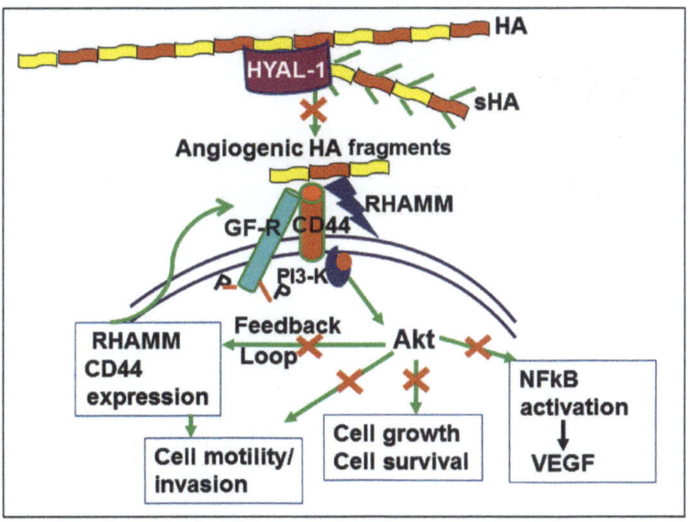

Figure 2.2, Vinata B. Lokeshwar *et al.* (See Page 54 of this volume.)

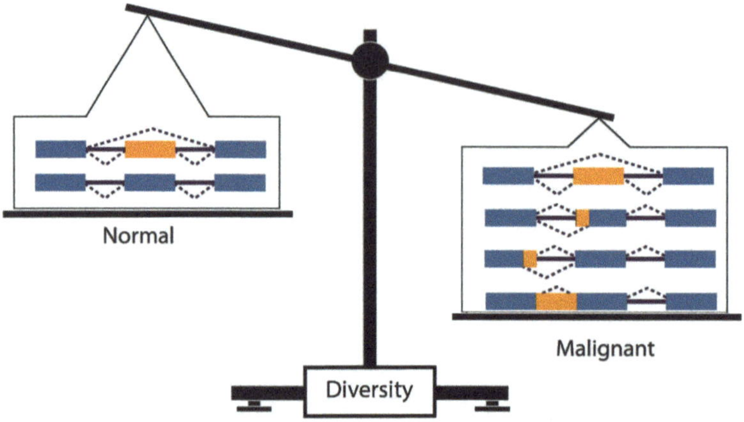

Figure 3.1, Sophia Adamia *et al.* (See Page 70 of this volume.)

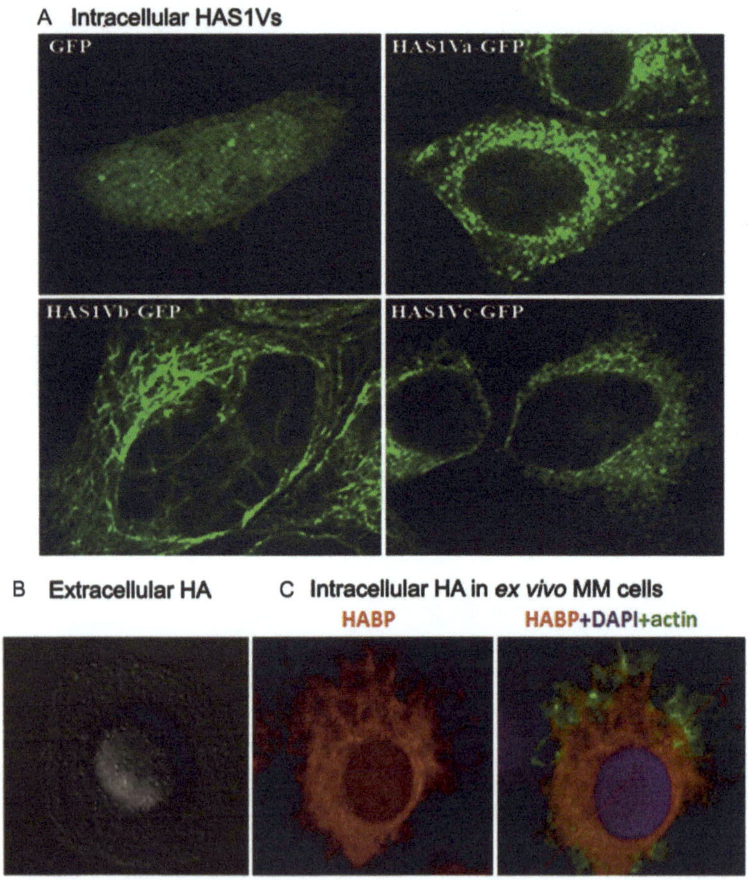

Figure 3.3, Sophia Adamia *et al.* (See Page 78 of this volume.)

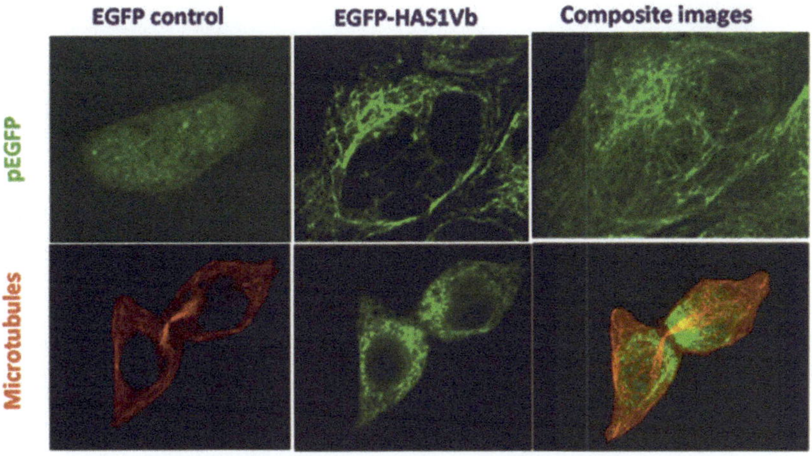

Figure 3.4, Sophia Adamia *et al.* (See Page 82 of this volume.)

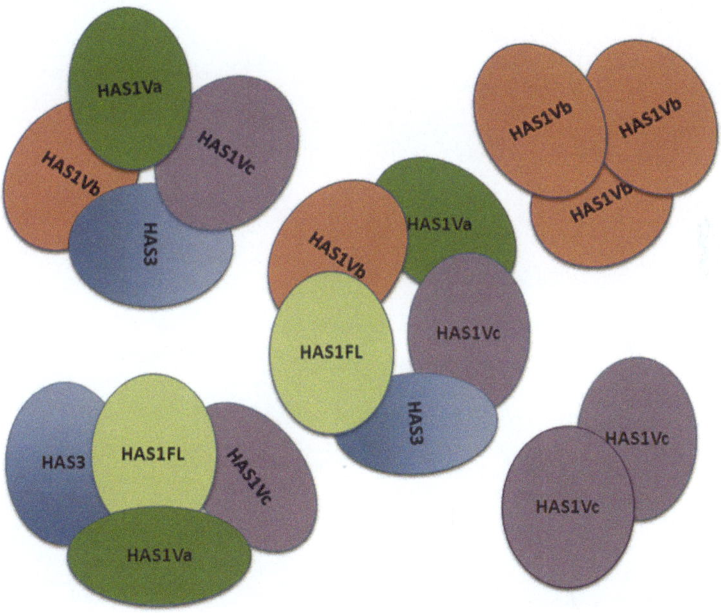

Figure 3.5, Sophia Adamia *et al.* (See Page 83 of this volume.)

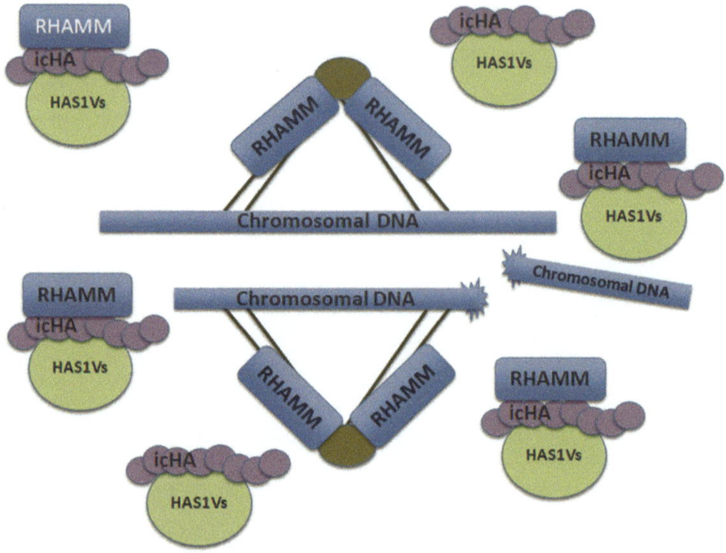

Figure 3.6, Sophia Adamia *et al.* (See Page 87 of this volume.)

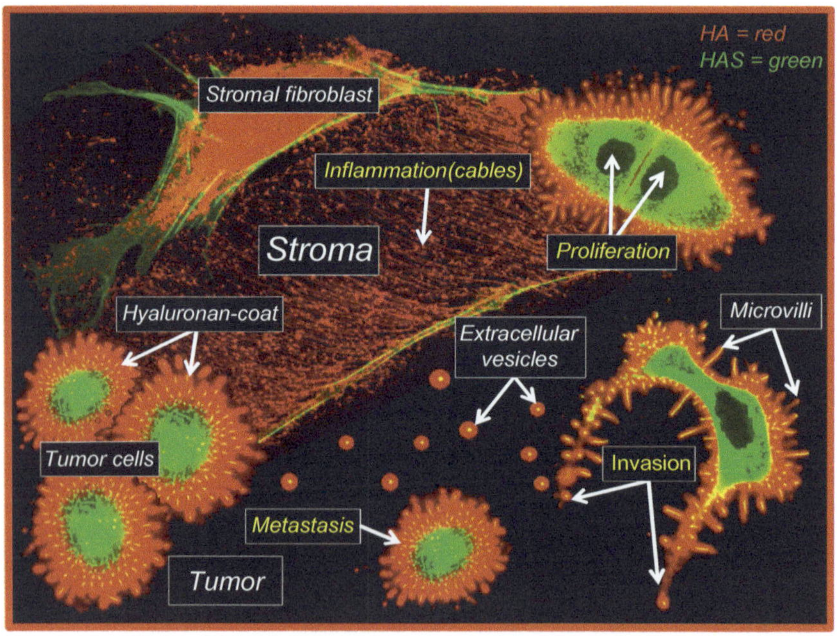

Figure 5.1, Kirsi Rilla *et al.* (See Page 123 of this volume.)

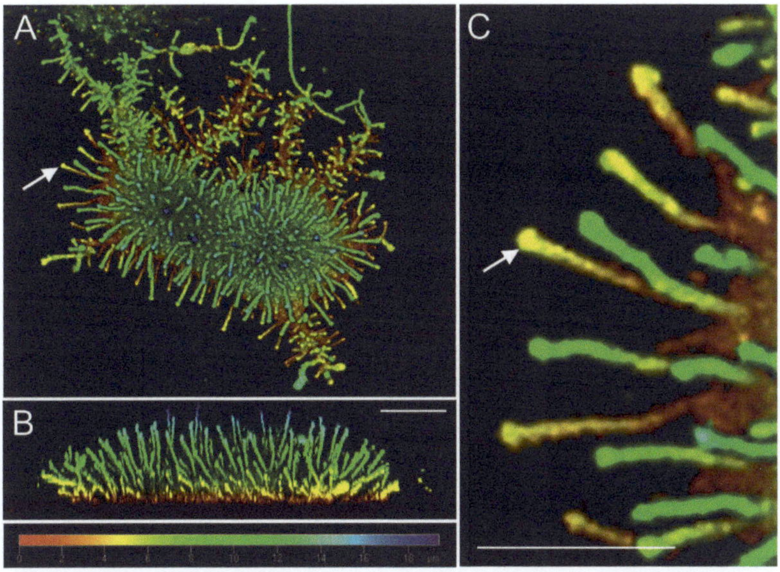

Figure 5.2, Kirsi Rilla *et al.* (See Page 130 of this volume.)

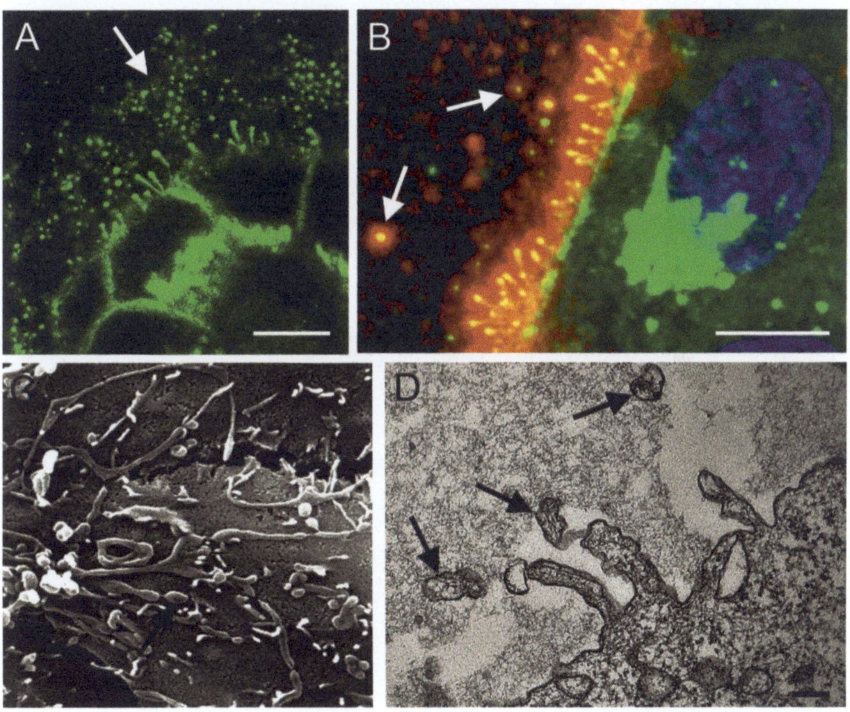

Figure 5.3, Kirsi Rilla *et al.* (See Page 131 of this volume.)

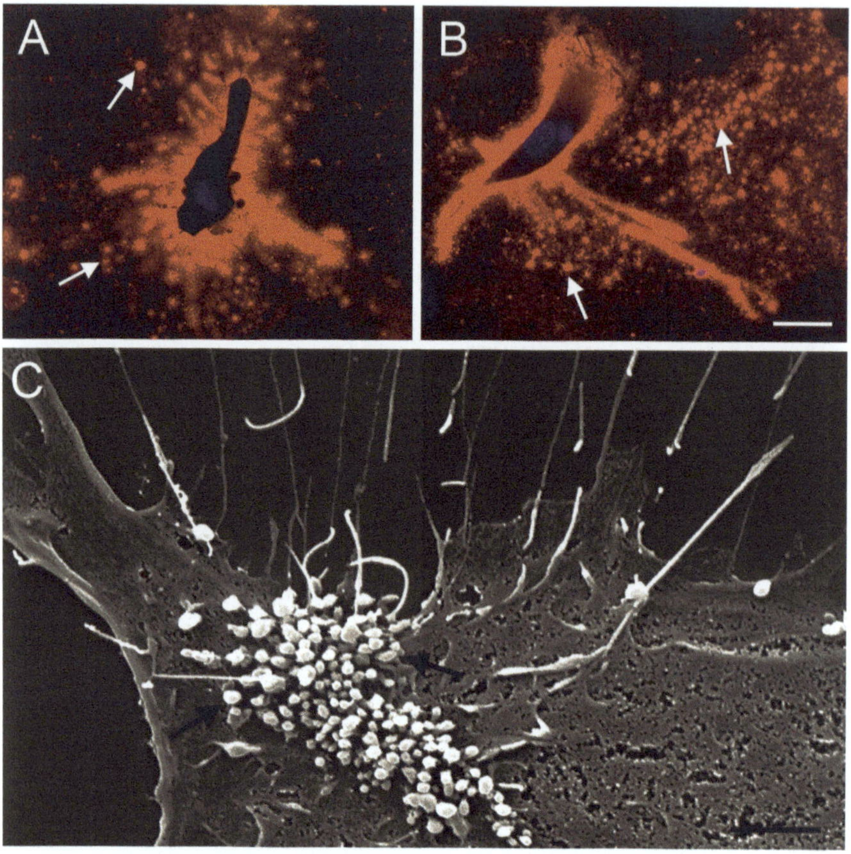

Figure 5.4, Kirsi Rilla *et al.* (See Page 134 of this volume.)

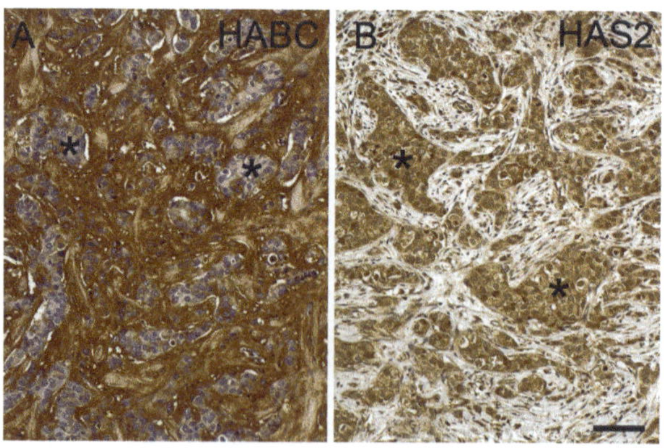

Figure 5.5, Kirsi Rilla *et al.* (See Page 136 of this volume.)

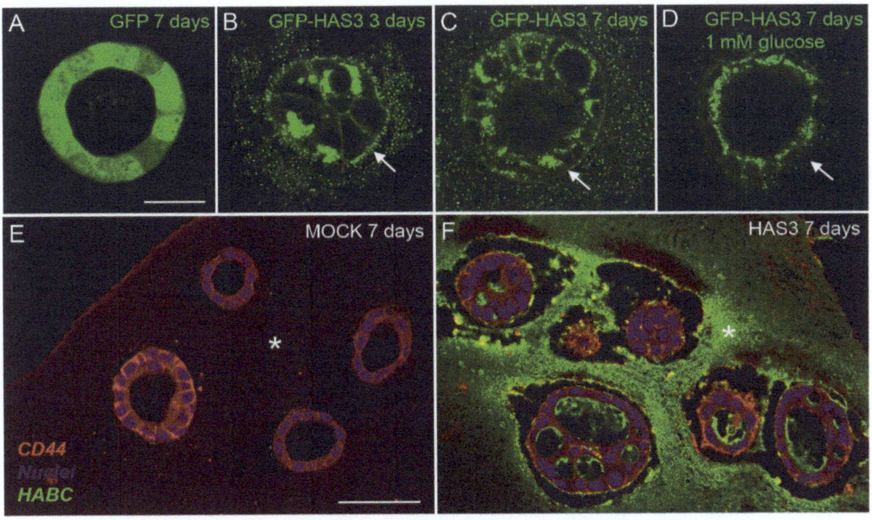

Figure 5.6, Kirsi Rilla *et al.* (See Page 137 of this volume.)

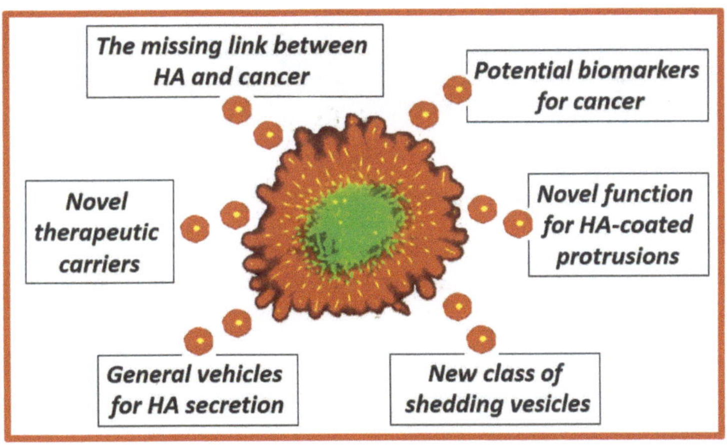

Figure 5.7, Kirsi Rilla *et al.* (See Page 139 of this volume.)

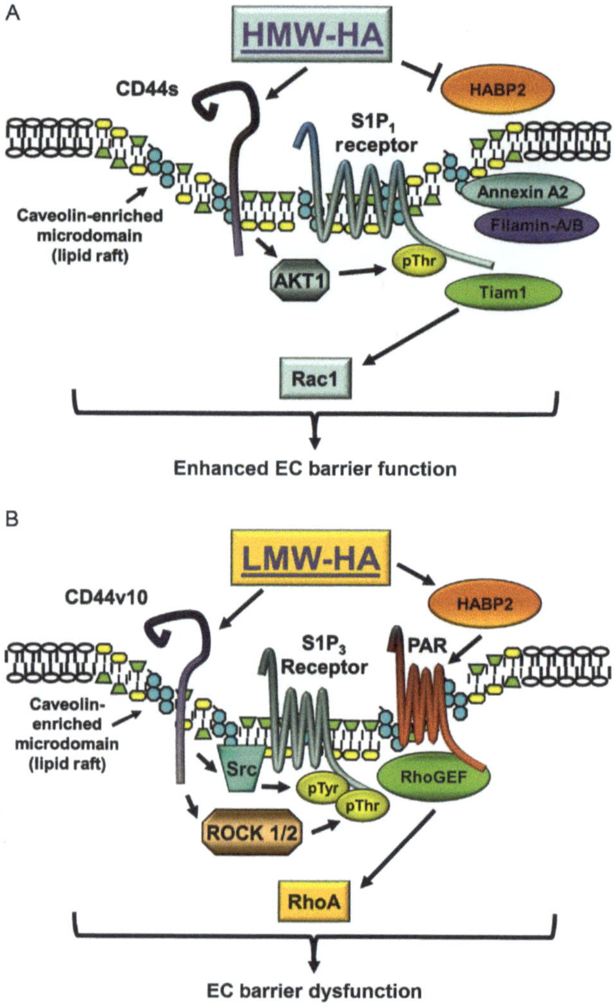

Figure 7.2, Patrick A. Singleton (See Page 196 of this volume.)

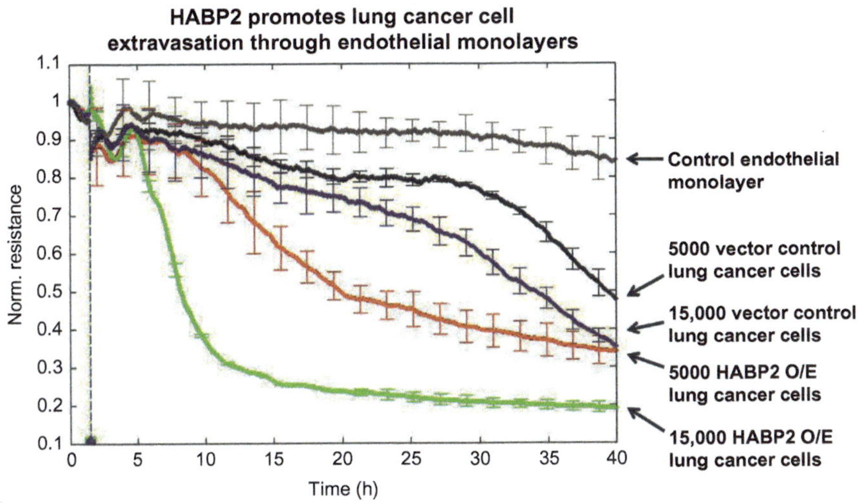

Figure 7.3, Patrick A. Singleton (See Page 200 of this volume.)

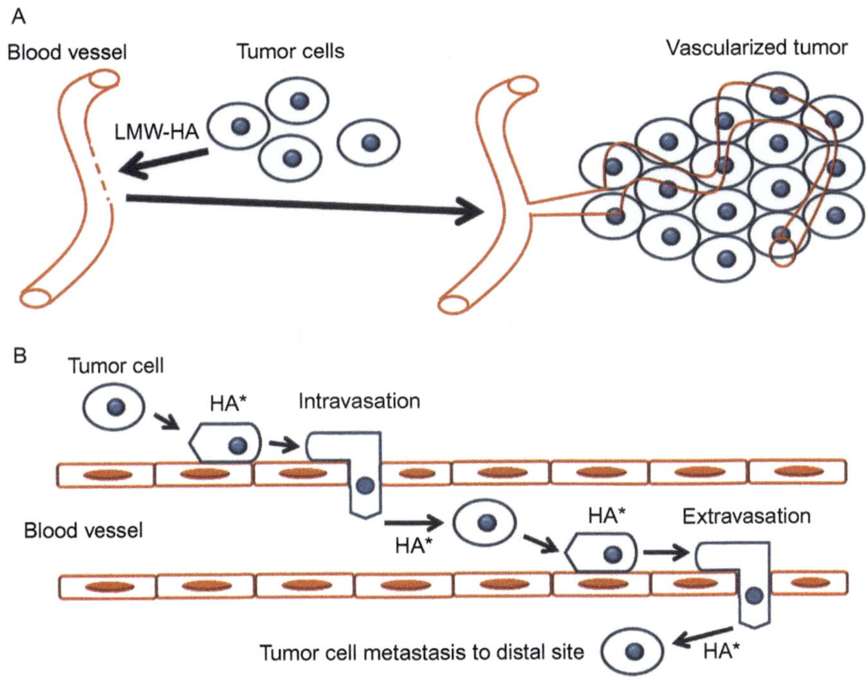

Figure 7.4, Patrick A. Singleton (See Page 202 of this volume.)

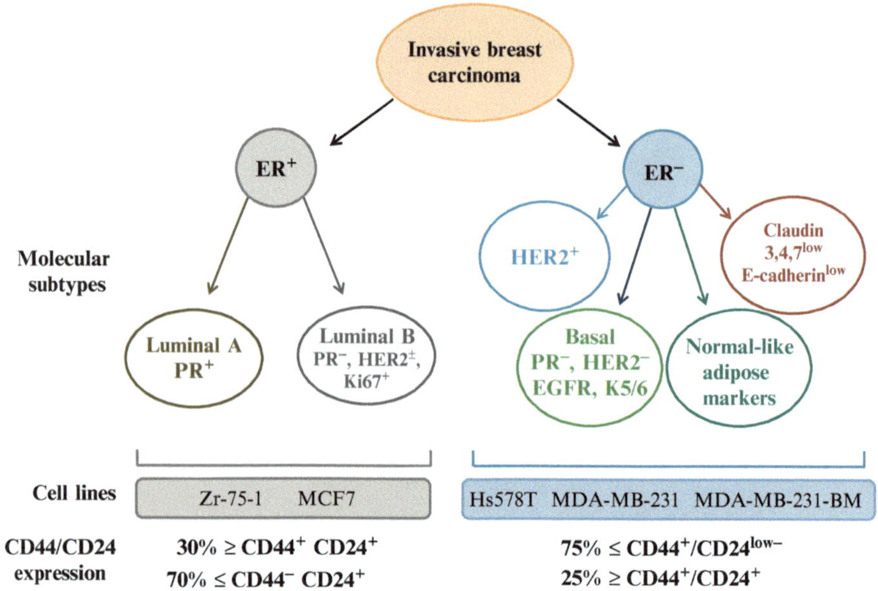

Figure 8.1, Paraskevi Heldin *et al.* (See Page 214 of this volume.)

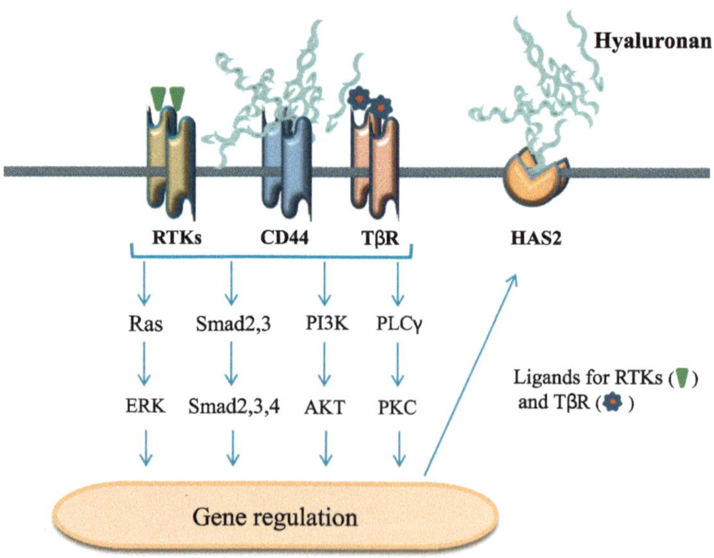

Figure 8.2, Paraskevi Heldin *et al.* (See Page 216 of this volume.)

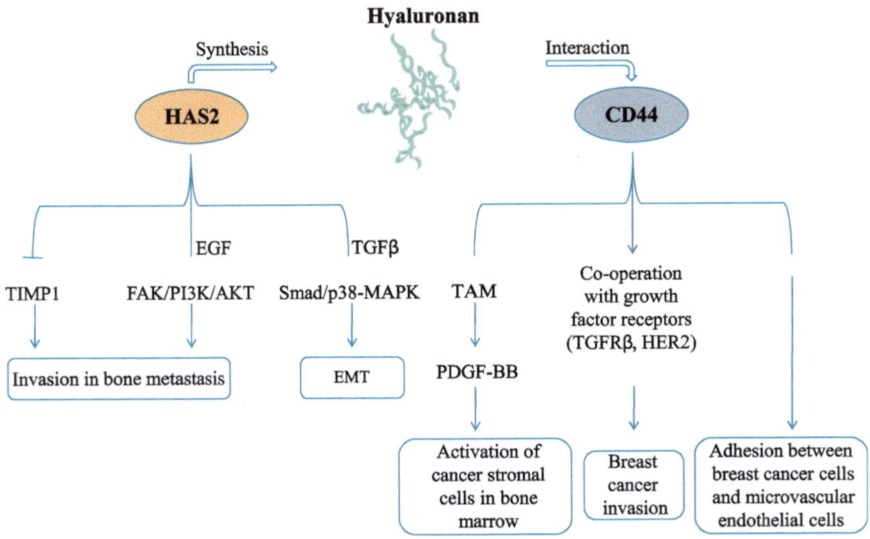

Figure 8.3, Paraskevi Heldin *et al.* (See Page 223 of this volume.)

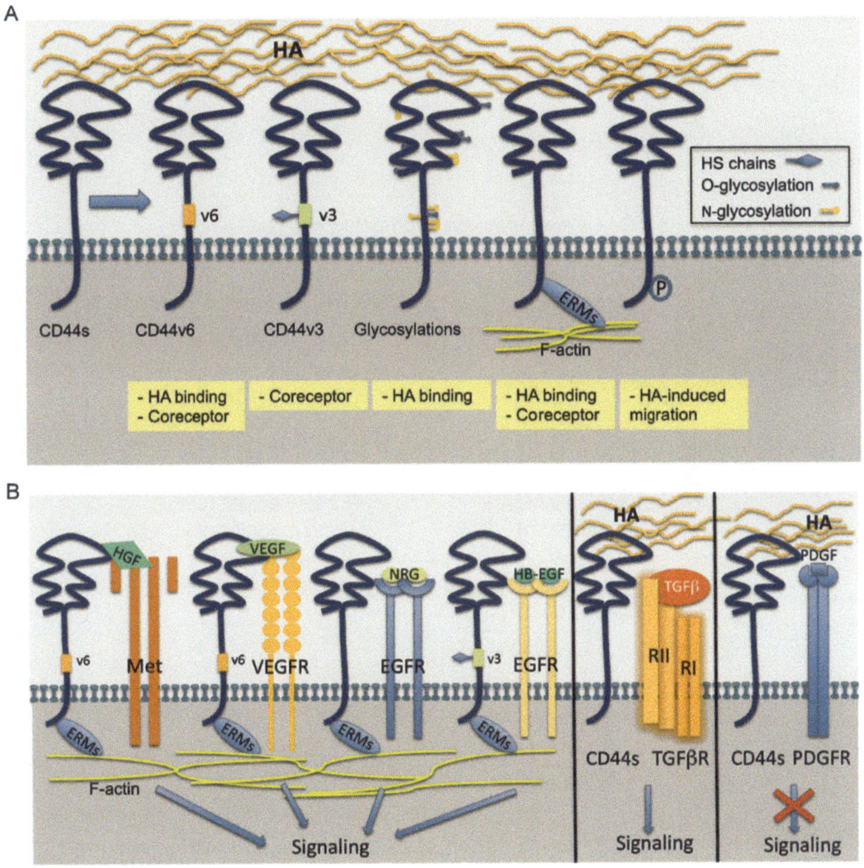

Figure 9.1, Véronique Orian-Rousseau and Jonathan Sleeman (See Page 236 of this volume.)

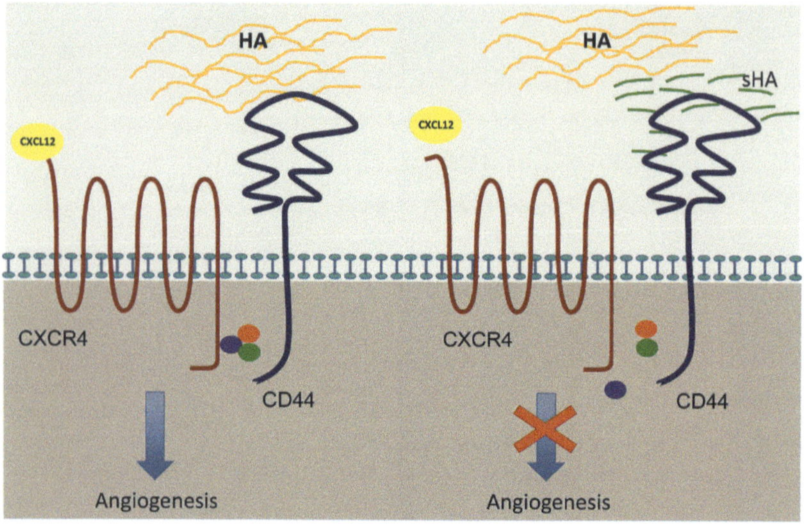

Figure 9.2, Véronique Orian-Rousseau and Jonathan Sleeman (See Page 244 of this volume.)

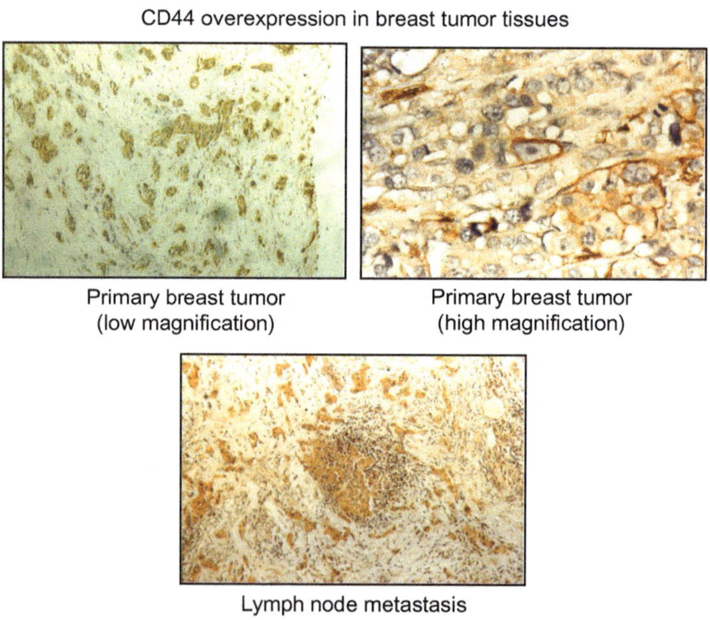

Figure 10.1, Lilly Y.W. Bourguignon *et al.* (See Page 258 of this volume.)

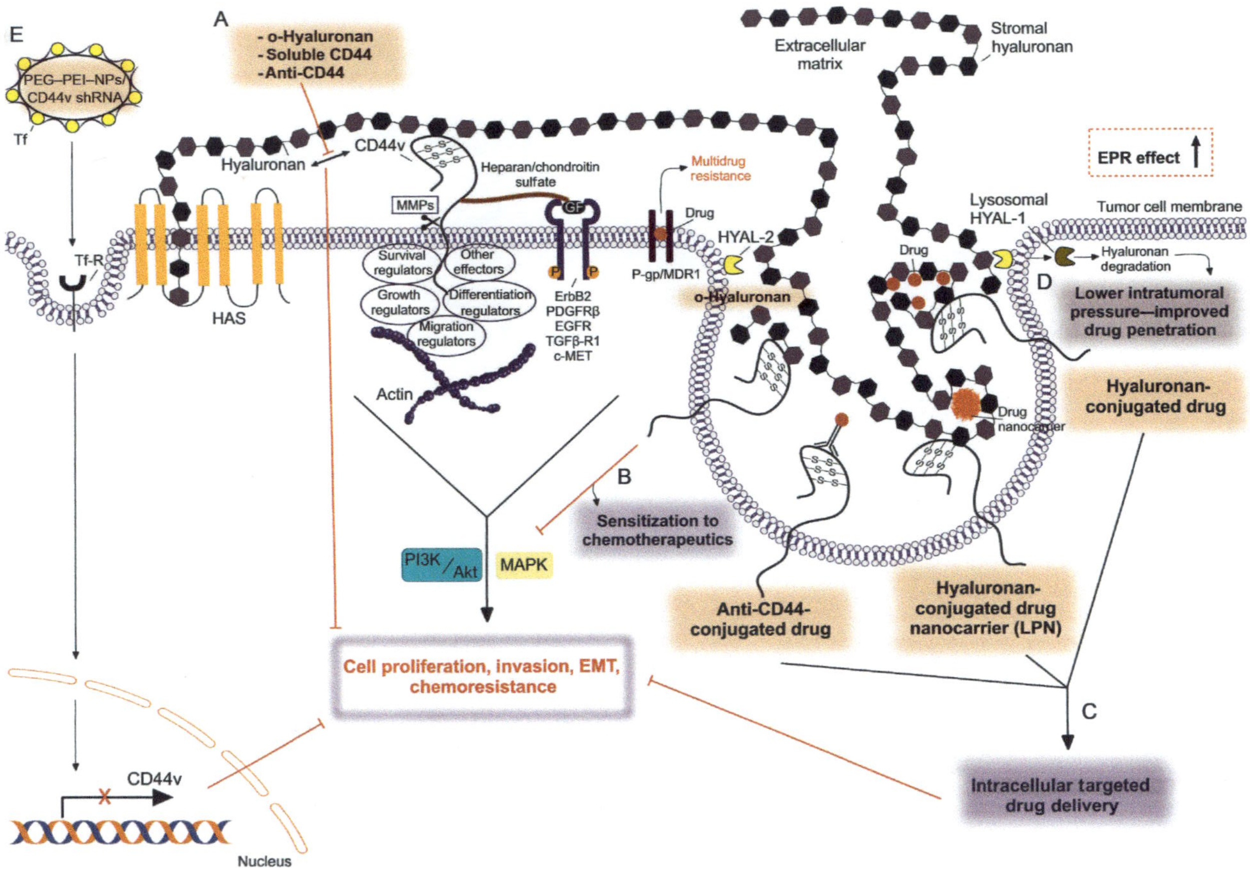

Figure 11.1, Spyros S. Skandalis *et al.* (See Page 282 of this volume.)

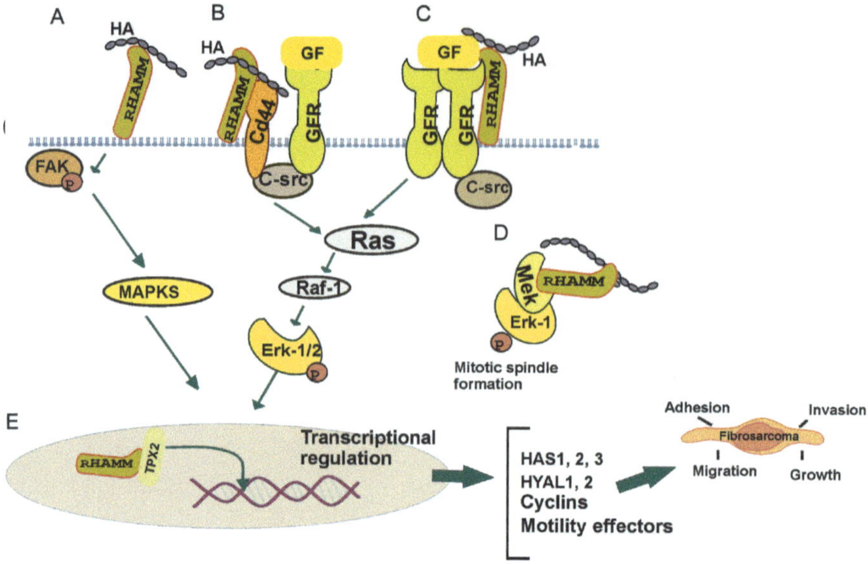

Figure 12.1, Katerina Kouvidi *et al.* (See Page 327 of this volume.)

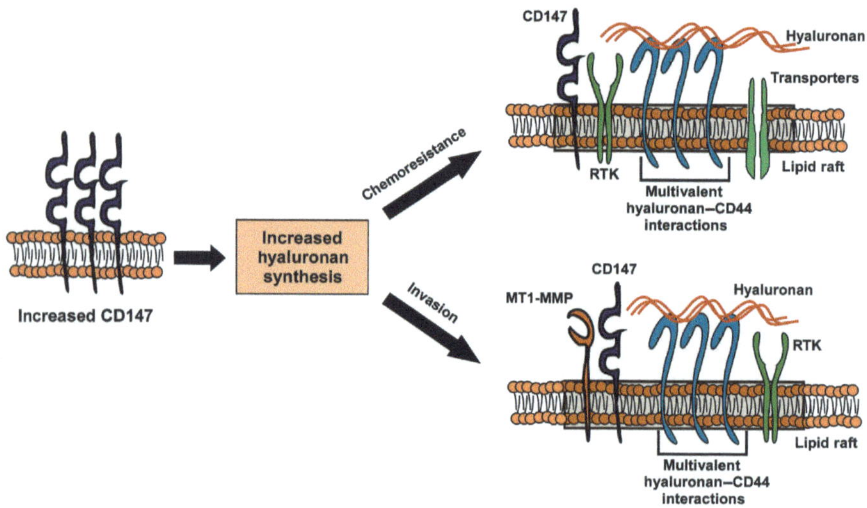

Figure 13.1, G. Daniel Grass *et al.* (See Page 353 of this volume.)

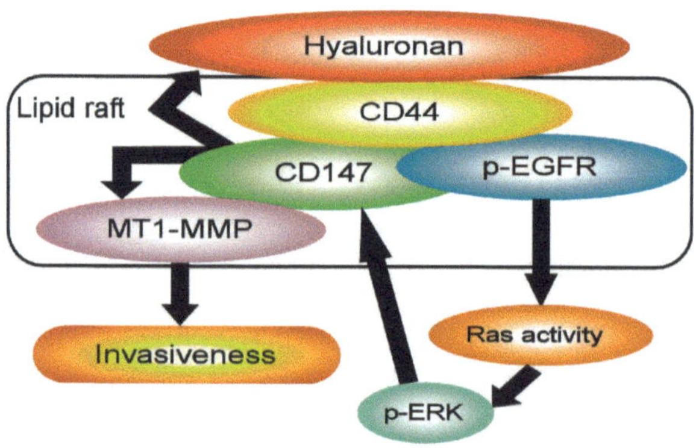

Figure 13.2, G. Daniel Grass *et al.* (See Page 361 of this volume.)

Kolibri 360 GmbH
Ferdinand-Jühlke-Straße 7
99095 Erfurt, Deutschland
produktsicherheit@kolibri360.de